T0302090

APERIODIC ORDER
Volume I: A Mathematical Invitation

Quasicrystals are non-periodic solids that were discovered in 1982 by Dan Shechtman, Nobel Prize Laureate in Chemistry 2011. The underlying mathematics, known as the theory of Aperiodic Order, is the subject of this comprehensive multi-volume series.

This first volume provides a graduate-level introduction to the many facets of this relatively new area of mathematics. Special attention is given to methods from algebra, discrete geometry and harmonic analysis, while the main focus is on topics motivated by physics and crystallography. In particular, the authors provide a systematic exposition of the mathematical theory of kinematic diffraction.

Numerous illustrations and worked examples help the reader to bridge the gap between theory and application. The authors also point to more advanced topics to show how the theory interacts with other areas of pure and applied mathematics.

Encyclopedia of Mathematics and Its Applications

This series is devoted to significant topics or themes that have wide application in mathematics or mathematical science and for which a detailed development of the abstract theory is less important than a thorough and concrete exploration of the implications and applications.

Books in the **Encyclopedia of Mathematics and Its Applications** cover their subjects comprehensively. Less important results may be summarised as exercises at the ends of chapters. For technicalities, readers are referred to the reference list, which is expected to be comprehensive. As a result, volumes are encyclopaedic references or manageable guides to major subjects.

All the titles listed below can be obtained from good booksellers or from Cambridge University Press. For a complete series listing visit www.cambridge.org/mathematics

ENCYCLOPEDIA OF MATHEMATICS AND ITS APPLICATIONS

Aperiodic Order

Volume 1: A Mathematical Invitation

MICHAEL BAAKE

Universität Bielefeld, Germany

UWE GRIMM

The Open University, Milton Keynes

CAMBRIDGE
UNIVERSITY PRESS

CAMBRIDGE
UNIVERSITY PRESS

University Printing House, Cambridge CB2 8BS, United Kingdom

One Liberty Plaza, 20th Floor, New York, NY 10006, USA

477 Williamstown Road, Port Melbourne, VIC 3207, Australia

314-321, 3rd Floor, Plot 3, Splendor Forum, Jasola District Centre, New Delhi - 110025, India

79 Anson Road, #06-04/06, Singapore 079906

Cambridge University Press is part of the University of Cambridge.

It furthers the University's mission by disseminating knowledge in the pursuit of
education, learning and research at the highest international levels of excellence.

www.cambridge.org
Information on this title: www.cambridge.org/9780521869911

© Michael Baake and Uwe Grimm 2013

First published 2013
Reprinted with corrections 2014

A catalogue record for this publication is available from the British Library

ISBN 978-0-521-86991-1 Hardback

Contents

Foreword by Roger Penrose

In a famous address to the 1900 International Congress of Mathematicians, held in Paris, the great mathematician David Hilbert announced a list of 23 unsolved mathematical problems, many of which shaped the subsequent course of mathematics for the 20th century. It would have been clear at the time that several of the problems concerned issues of profound mainstream mathematical interest. Some of the others may have seemed, then, more like curious mathematical side-issues; yet Hilbert showed a remarkable sensitivity in realising that within such problems were matters of genuine potential, mathematical subtlety and importance.

In this latter category was Problem 18, which raises issues of the filling of space with congruent shapes. Among other matters (such as the *Kepler conjecture* concerning the close-packing of spheres[1]) was the question of whether there exists a polyhedron which tiles Euclidean 3-space, but only in a way that it is not the fundamental domain of any space group — that is to say, must every tiling by that polyhedron be necessarily isohedral, which would mean that every instance of the polyhedron is obtainable from every other, through a Euclidean motion of the entire tiling pattern into itself (i.e., all polyhedra in the tiling would thereby be on an 'equal footing' with respect to the pattern as a whole). Such shapes which tile space, but only in ways that are *not* isohedral, are now known as *anisohedral* prototiles (where the word 'prototile' simply means a *tile shape*, in current terminology). In asking this question in three (and more) dimensions, Hilbert was probably assuming — incorrectly, as it turned out — that no anisohedral prototile could exist in two dimensions. No doubt it was felt at that time that little could be mathematically interesting or difficult concerning tiling with plane shapes. How greatly our views on this matter have now changed!

The first 3-dimensional anisohedral tile was found by Karl Reinhardt in 1928, and then in 1935 Heinrich Heesch found a 2-dimensions example. My own first serious interest into the area of plane tilings was exhibited in an article I wrote with my father, in 1958, which presented a small collection of

---

[1]Solved, in 1998, by a computer-aided proof by Thomas Hales, following an approach suggested by László Fejes Tóth in 1953.

somewhat unusual 'Puzzles for Christmas' [PP58b]. Puzzle 6 asked for the (unique) tiling patterns provided by each of seven different prototiles, the last five of these being anisohedral — although I was not at that time aware either of the terminology or of the inherent interest in such features.

I had played with such things, on and off, for several years before that, and it had been in 1954 that I had first come across the graphic work of Maurits C. Escher, with his many representations of tiling patterns using birds, fish and many other curious and fascinating designs (all isohedral, in fact). This was while attending the 12th International Congress of Mathematicians held in Amsterdam (as a second-year graduate student at Cambridge). I was amazed by much that I saw there, and as a result, I decided to try my hand at creating 'Escher-like' pictures of one kind or another, having written another article with my father on one aspect of these [PP58a]. We cited Escher's work as depicted in the Amsterdam exhibition and, as a result of this, my father entered into a correspondence with Escher.

In around 1962, I happened to be driving through the Netherlands, and I telephoned Escher on the off-chance that I might be able to visit him. He most obligingly invited me (and my then wife Joan) to tea, showing me a great number of his amazing creations, one of which I was offered as a gift. In exchange, I left a number of identically-shaped wooden pieces with him, set as a puzzle for him to see how to use these prototiles to tile the entire Euclidean plane. The task was not an immediately obvious one, the shape requiring 12 different orientations (six being inverted) before repeating, the tiling being necessarily non-isohedral. Escher used a modified version of this tiling pattern in his final (and only non-isohedral) print.[2]

The exploration of anisohedral prototiles has moved on enormously since the early work of Reinhardt and Heesch. All the anisohedral prototiles that I have referred to so far are what are called *2-isohedral*, where a k-isohedral prototile would be one which can tile the entire space in a way that the pattern involves k transitivity classes, but not in a way where the pattern involves fewer than k transitivity classes (so that the tile necessarily takes up k distinct relationships to the pattern as a whole). In 2003, Joseph Myers exhibited a remarkable 10-isohedral tile — apparently the current record.[3]

Subsequent to my early concern with such things as anisohedral prototiles, I became interested in the possibility of prototiles that tile the plane in hierarchical ways (such as those that later became known as *rep-tiles*), and this route led me to the aperiodic sets of prototiles now referred to as 'Penrose tiles' where the term *aperiodic* refers to the fact that whereas they do tile

---

[2]'Ghosts'; see, for example, top left of [CEPT86, p. 395], and my own article 'Escher and the visual representation of mathematical ideas' in the same volume.

[3]See the excellent article [Goo11] by Chaim Goodman-Strauss.

the entire plane, they do so only without translational symmetry (so that, in a sense, they constitute an '∞-isohedral' pair). My own approach to these issues was very direct and geometrical, and not the result of any deep mathematical underpinning, such as was later provided by Nicolaas de Bruijn's [dBr81] very fruitful and revealing subsequent procedure for obtaining such tiling patterns by slicing and projecting higher-dimensional cubic lattices. Although there was no such sophisticated mathematical input into the initial discovery of these particular prototiles, my vague awareness of the profound logical considerations of Hao Wang, as developed by Robert Berger, and leading to Raphael Robinson's aperiodic 6-prototile set [Rob71] could well have influenced my anticipation that the original non-periodic pentagon pattern that I had found [Pen74, Fig. 4] could be *forced* as the tiling arrangement of an aperiodic 6-prototile set, and subsequently that this set could be reduced to an aperiodic 2-prototile set.

Of more direct (although largely unconscious) influence was my much earlier acquaintance with Johannes Kepler's very remarkable 1619 picture exhibiting various non-crystallographic tiling patterns [Kep19]. I did not have these in mind when I first found my own aperiodic sets early in 1974, but I had seen Kepler's designs many years before, and I believe that they strongly influenced my attitude to the fruitfulness of pentagonal tilings. It was only some years after 1974 that I realised the extraordinarily close relationship between Kepler's configuration 'Aa' and my own pentagonal tiling, where the configuration of line segments constituting Kepler's entire configuration, *without any exceptions*, can be found within those of my own pentagonal tiling. I have always been intrigued as to how Kepler intended his pattern 'Aa' to be continued, and I believe it is quite within plausible possibility that he had in mind some sort of hierarchical continuation similar to that of my own pattern.

Kepler was clearly interested in the different kinds of symmetry that could co-exist with the packing of shapes together in a systematic way, possibly in relation to some kind of atomic underpinning that might be relevant to biological as well as crystalline structures. The non-crystalline symmetry underlying my own tilings is indeed intimately related to their lack of periodicity, a feature distinguishing them from the earlier aperiodic sets of Berger and Robinson. Many Islamic patterns also contain elements with the non-crystallographic 5-fold and 10-fold symmetry, as well as 8-fold and 12-fold, and I had found these fascinating, but I was not aware of any such patterns containing extended regions with such symmetry, and I do not think that these things influenced me significantly in my own quasi-symmetric designs.

I suspect that Kepler's influence on me might have been similar to that which my own tiling patterns could have had on Dan Shechtman. He once told me that when he first came across his puzzling 10-fold-symmetric diffraction patterns (those that led him to his revolutionary insights that won him the 2011 Nobel Prize in Chemistry), he did not have my tiling patterns in mind. Yet, he said he had been aware of them, and I like to think that they may have unconsciously influenced him to be favourably disposed towards the presence of a genuinely novel type of 'crystallographic' atomic arrangement with an underlying 10-fold or 5-fold symmetry, as turned out to be the case.

My own tilings, and those found by Robert Ammann very shortly afterwards, were fortunately discovered in time to be incorporated and beautifully described by Branko Grünbaum and Geoffrey Shephard in their classic and, at that time, comprehensive 1987 work *Tilings and Patterns*. Before this, there had been little organisation in the subject or in its terminology. The current work, by Michael Baake and Uwe Grimm, provides a most worthy continuation of Grünbaum and Shephard's achievement, providing, as it does, an excellent overview of the subject of aperiodic order that has grown up since that time. The mathematical underpinnings of the aperiodic order underlying the non-crystallographic patterns exhibited by these early aperiodic tile sets has been vastly extended. In addition to the 5-fold (or 10-fold) and Ammann (and Beenker) 8-fold quasi-symmetric patterns, which arrived in time for inclusion in *Tilings and Patterns*, we now have, illustrated here, several 12-fold examples found by Schlottmann, Gähler, Socolar and others.

Moreover, we find here a study not only of these quasi-symmetries, which appear to underlie the actual atomic arrangements of quasi-crystals, but also of many other types of quasi-symmetric patterns. For example, there is the 7-fold case, as exhibited by the 3-prototile set constructed by Ludwig Danzer, which is aperiodic if we restrict the tilings to conform to a certain atlas of vertex stars, or else if we require that we construct the tiling by following specific *inflation* rules. Most of the early aperiodic tilings (including the reptiles, the Robinson 6-prototile set, Amman's examples, and my own), were originally based upon inflation rules where, in effect, the prototiles can be collected into larger versions of themselves, and the entire pattern is built up in this hierarchical way. More recent tilings that can be obtained by inflation are also described here, such as the remarkable hexagonal prototile of Joan Taylor and a somewhat similar earlier one of my own — both being aperiodic via second-order matching rules. The inflation scaling factor is referred to as the *inflation number*, and for the 5-, 8-, 10-, and 12-fold quasi-symmetries, this inflation number is an example of what is referred to as a *Pisot* (or Pisot–Vijayaraghavan) number (a real algebraic integer larger than 1 whose conjugates all lie within the unit circle in the complex plane). More unsettled-looking tilings, such as Danzer's, arise with *non*-Pisot inflation numbers.

This illustrates a connection between tilings and *number theory* and *complex analysis*, and many connections with other areas of mathematics are also provided here, such as that arising from the work of Nicolaas de Bruijn referred to above. Intriguing relationships to various other areas of mathematics are also demonstrated here. Group theory has an obvious importance, and Peter Kramer's group theoretic approach to the projection method was particularly noteworthy in relation to 3-dimensional quasicrystals. But we also find roles for Fourier analysis, a subject which is highly relevant to the subject of crystallography, where diffraction patterns play key roles. Thus, we find here many windows into powerful areas of mathematics which are often as unexpected as they are fascinating. Most importantly, there is the additional attraction that so much of it can be illustrated in actual visual images that are beautiful to behold and for which much of their mathematical quality can be discerned by simply looking at them. We have surely moved a long way from the simple question of whether an anisohedral prototile can exist! One can but wonder what David Hilbert would have made of it all.

Roger Penrose

Preface

The theory of aperiodic order is still a relatively young field of mathematics. It has grown rapidly over the past three decades following the experimental discovery of aperiodically ordered materials. This is the introductory volume of a book series that attempts to provide a comprehensive account of the field, which is still developing. We entitled this volume 'a mathematical invitation' because we hope that it will inspire readers to enter the fascinating (and largely still to be explored) world of aperiodic order and provide the background that enables them to follow the current developments. Subsequent volumes will address particular as well as complementary topics in more depth, in the form of selected survey articles contributed by expert authors. While the scope and the details of the later volumes are still evolving, the second volume will focus on crystallography and almost periodicity, and the third is planned to expand on model sets and dynamical systems.

It was our aim to keep this introductory volume at a relatively elementary level, and to make the subject accessible to both mathematicians and physicists, which requires a certain compromise for the exposition. Consequently, we ask readers with a mathematical background for their patience for our largely constructive and often example-motivated approach, which sometimes also requires substantial explicit calculations. In particular, we do not strive at a formal exposition at maximal generality, but rather at emphasising the meaning of the formalism at each step. Similarly, people with a background in the physical sciences are kindly asked to bear with us on our journey through aperiodic order, which is intentionally more rigorous than the standard physics literature. While we aim at a certain level of completeness (and at substantiating some of the 'folklore'), we will not spell out all proofs, but sometimes refer to the original literature.

For a number of years, there have been parallel developments in mathematics and physics, which took limited notice of each other. It is one of our aims to bring these developments together, which in particular means a more physics oriented selection of topics with a stronger mathematically oriented exposition. Within physics, we were influenced mostly by the work of Peter Kramer, who pioneered the symmetry-oriented construction of non-periodic tilings by the projection method. Within mathematics, we would like to highlight the early contributions by the late Peter Pleasants, in particular his number-theoretic approach to the projection method. With hindsight, it is the most natural continuation of the pioneering (but only later appreciated) work of Yves Meyer, and has influenced our way of thinking and our presentation significantly. Roger Penrose's famous fivefold tiling provided a paradigm that not only shows the aesthetic appeal of aperiodically ordered structures, but has become an influential guiding example for many later and current investigations, also with regard to applications. We also build on the geometric insight of Robert Ammann and Ludwig Danzer, who constructed many important tilings in two and three dimensions. More recently, the advance of the field, in particular in its mathematical flavour, was perhaps most stimulated by the contributions of Jeffrey Lagarias and Robert Moody, which will also be reflected by our exposition.

A number of people read and commented on drafts of various parts of this book, which resulted in numerous corrections and improvements. In particular, we would like to thank Shelomo Ben-Abraham, Paolo Bugarin, David Damanik, Aernout van Enter, Dirk Frettlöh, Franz Gähler, Svenja Glied, Manuela Heuer, Christian Huck, Tobias Jakobi, Daniel Lenz, Reinhard Lück, Claudia Lütkehölter, Markus Moll, Robert Moody, Natascha Neumärker, Johan Nilsson, Arthur (Robbie) Robinson, Johannes Roth, Boris Solomyak, Joan Taylor, Venta Terauds and Peter Zeiner for their valuable input. Special thanks go to Dirk Frettlöh for contributing to the parts on Danzer's icosahedral ABCK tiling, to Franz Gähler for his help with the literature on local rules, and to Egon Schulte for his encouragement during the writing process. All remaining errors in the manuscript are ours.

Finally, it is our pleasure to thank the Department of Mathematics and Physics at the University of Tasmania, Hobart, for its hospitality and for providing a stimulating working environment to get this project off the ground. During the period of writing, we received support from various sources, including the German Research Foundation (DFG), the Erwin Schrödinger International Institute for Mathematical Physics (ESI) in Vienna, the Engineering and Physical Sciences Research Council (EPSRC), the Leverhulme Trust and the Royal Society.

The revised second printing gave us the opportunity to include a number of minor corrections and additions. We are grateful to Robert Moody and Nicolae Strungaru for their suggestions. Moreover, a list of definitions has been added, and the references have been updated.

Michael Baake and Uwe Grimm

CHAPTER 1

Introduction

In April 1982, while on sabbatical at the National Bureau of Standards
in Washington, DC, Dan Shechtman from the Technion at Haifa made a profound discovery, for which he was awarded the Wolf Prize in Physics in 1999
and the Nobel Prize in Chemistry in 2011. When inspecting various samples
of a rapidly solidified AlMn alloy with an electron microscope in diffraction
mode, he noticed a phase that showed clear and sharp Bragg reflexes together with a rather perfect icosahedral symmetry, similar to the one shown
in Figure 1.1. While a Bragg spectrum is a typical fingerprint of a crystal,
fivefold or icosahedral symmetry is incompatible with the latter. He concluded that this phase must possess long-range order (to explain the Bragg
reflexes) without being a perfect crystal (to be able to accommodate the unusual symmetry). It took Shechtman two years to convince colleagues until
the result was finally published in [SBGC84]. Even after the paper appeared
in print, prominent scientists (including Nobel Laureate Linus Pauling) expressed their scepticism, though they found themselves in a rapidly shrinking
minority as other phases with similar properties were discovered.

In fact, Ishimasa, Nissen and Fukano [INF85] at the ETH Zürich found
twelvefold (or dodecagonal) symmetry in a sample of a NiCr alloy before[1] they
became aware of Shechtman's discovery, while Bendersky [Ben85] demonstrated the existence of another AlMn phase with tenfold (or decagonal)
symmetry soon after (at this stage, we do not distinguish between fivefold
and tenfold symmetry). A little later, Kuo and his coworkers [WCK87] completed the list of presently known non-crystallographic symmetries with the
discovery of VNiSi and CrNiSi phases displaying eightfold (or octagonal)
symmetry. Structures of this type are nowadays referred to as *quasicrystals*,
a term that was coined by Levine and Steinhardt in [LS84]; see [SO87] for
a compilation of early publications and [Jar88, JG89, Jan94, SSH02, Tre03]
for initial accounts of the theory of quasicrystals.

While no further non-crystallographic symmetries have been observed so
far in alloys, many different intermetallic phases with non-crystallographic (in

---

[1]They announced their result on a poster at the 'Workshop on Physics of Small
Particles' in Gwatt (Switzerland) in October 1984.

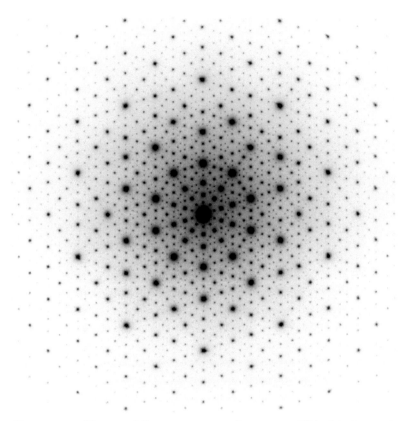

FIGURE 1.1. Electron diffraction pattern of a ternary Al Mn Pd alloy with icosahedral symmetry. Image © Conradin Beeli.

particular, icosahedral or decagonal) symmetry have been identified; for recent surveys, we refer to [Ste04, SD09]. Surprisingly, quasicrystalline phases are not restricted to solids, but can also occur in soft matter; see [LD07] and references therein. By now, various non-crystallographic symmetries (also beyond those mentioned above) have been exploited in engineered structures, such as phononic and photonic quasicrystals [SSW07]. Dodecagonal quasicrystalline phases also form in dense random tetrahedral packings [H-AE+09], which is one aspect of the fascinating question of how to pack tetrahedra [LZ12].

Let us return to the quasicrystals found in solids. The natural question of the atomic structure of these new alloys was disputed from various angles, displaying an entire spectrum of divided opinions. As early as 1982, and independently of the experimental discovery, Peter Kramer from Tübingen constructed a cell model with icosahedral symmetry [Kra82] based on ideas from group theory and projections from cells in higher dimensions. This

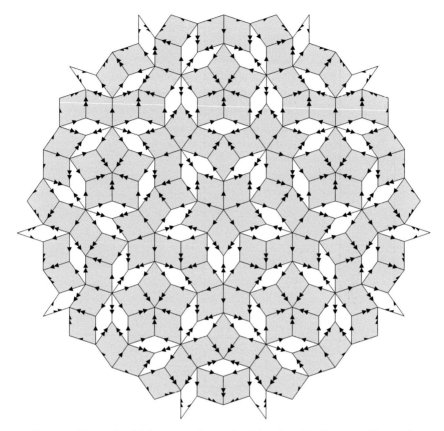

FIGURE 1.2. A fivefold symmetric patch of the rhombic Penrose tiling with edge matching rule decorations by two types of arrows.

was followed by a projection model from 6-space in 1984, developed jointly with his student Roberto Neri [KN84]. Other groups favoured quasiperiodic density waves [Bak86], thus building on previous work on incommensurate phases [dWo74, JJ77], or random tilings [Els85] (see also [Hen99] and references therein) that explain the notion of order and symmetry in a probabilistic fashion. This situation led to interesting and long-lasting debates, in particular about specific models and tilings, and about the question how to find out what is 'typical'.

In mathematics, various predecessors existed that now found rather unexpected applications. Perhaps best known is the fivefold symmetric planar tiling due to Roger Penrose from Oxford [Pen74], which became popular and widely known through an article by Martin Gardner [Gar77]. Figure 1.2 shows its rhombic version and Figure 1.3 a realisation at Texas A&M University. The tiling is highly ordered, although it has no non-zero period at all.

It was later shown to be compatible with a projection from 5-space and to have a pure Bragg spectrum by de Bruijn [dBr81], thus substantiating an experimental observation made by Alan Mackay [Mac82]. Fivefold symmetry in the plane and icosahedral symmetry in 3-space are of particular importance, both theoretically and in practice. Mathematically, this can be seen as a continuation of the tradition founded by Felix Klein's famous monograph [Kle84]. Icosahedral structures of various kinds continue to attract scientific interest. Recent examples include quasicrystals, fullerenes (or 'buckyballs') and tensegrities; see [CT08] for a catalogue of the latter.

As so often, the Penrose tiling itself had various predecessors, such as the 'Aa' tiling of Johannes Kepler (which indirectly inspired Penrose to find his construction[2]) or old ornaments, particularly in Islamic art [Mak92]. The many attempts show that fivefold symmetry puzzled generations of scientists, yet there is no evidence that any of Penrose's predecessors (including Dürer and Kepler) had a proof for the extensibility of their constructions to infinite tilings of the plane with fivefold symmetry and a certain degree of homogeneity. For a brief historical summary, we refer to [Lüc00], while [Jar89, Sen95, AG95, Moo97, Pat98, BM00b] cover a decent part of the mathematical development.

Let us say a bit more about the rhombic Penrose tiling. A large patch is shown in Figure 6.44 on page 238, which highlights the hierarchy of meandering 'paths' built from thick rhombuses. Each of the two prototiles

occurs in ten distinct orientations. This version includes edge markers which, upon putting tiles together, are supposed to form complete single or double arrows. This condition constitutes a set of local rules with three remarkable properties. The first is that the rules are compatible with at least one (in fact, more than one) gapless, face to face tiling that covers the entire plane. Secondly, they guarantee that none of these tilings has any (non-zero) period. Finally, they also enforce that any two of these tilings are locally indistinguishable. They provide an example of what we will later call a set of perfect aperiodic local rules. There are many other fascinating aspects of the Penrose tiling (such as its self-similarity, which underlies the proof of the claims, and its description as a projection essentially from 4-space), some of which will be explained in this book.

---

[2]This was mentioned by Roger Penrose during his Kepler lecture at the University of Tübingen, Faculty of Physics, on June 11, 1997; compare the Foreword to this book.

FIGURE 1.3. Roger Penrose and his famous rhombus tiling in the foyer of the Mitchell Institute for Fundamental Physics and Astronomy at Texas A&M University. Photograph by user Solarflare100, available at http://en.wikipedia.org/wiki/Penrose_tiling under a Creative Commons Attribution 3.0 Unported License.

To appreciate the role of the symmetry in this matter, let us sketch a simple intuitive argument why a periodic point set in the plane, with minimal distance between distinct points, cannot have fivefold rotational symmetry. To see this, assume to the contrary that there is such a periodic point set with fivefold symmetry, where it is also assumed (for simplicity) that some points are rotation centres. Considering two rotation centres of minimal distance, one can rotate them around each other, as sketched in Figure 1.4. Unlike two-, three-, four- or sixfold symmetry, a fivefold rotation produces point

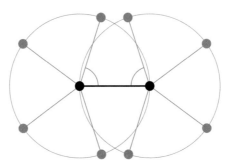

FIGURE 1.4. Sketch of the incompatibility between points of fivefold rotational symmetry at minimal distance (black) and lattice periodicity in the Euclidean plane.

pairs at a smaller distance, thus contradicting the starting assumption (this will be made more precise in Chapters 3 and 5). The same problem occurs for any rotation of order $n \geqslant 7$ in the plane.

Related tilings with other symmetries have also been constructed and studied intensively. Many early examples are due to Robert Ammann [Sen04], in particular examples with aperiodic local rules. A well-known tiling with eightfold symmetry is the octagonal or Ammann–Beenker tiling illustrated in Figure 1.5.

Another predecessor, which went largely unnoticed for a long time, is Yves Meyer's book [Mey72] on the connection between algebraic number theory and harmonic analysis. In fact, it essentially contains the abstract theory of the projection method, in the full generality of locally compact Abelian groups, though this was only systematically revealed by Jeffrey Lagarias [Lag96] and Robert Moody [Moo97a] much later. This can also be considered as an extension of the theory of almost periodic functions, which is largely due to Harald Bohr [Boh47], to the setting of point sets. Nevertheless, the original focus was a different one, and it is a remarkable fact that these abstractly investigated systems turn out to be precisely what is needed to describe the structure of perfect quasicrystals.

By the nature of the subject, a large variety of notions and concepts has been used during its development. Some of them appeared to be mutually incompatible, which resulted in long-lasting disputes between different schools. Thirty years after Shechtman's discovery, it has become apparent that one needs a unified framework to accommodate and reconcile these different viewpoints, and the mathematical theory of *measures* is able to provide just that. This is our main motivation to present an exposition that develops the subject in this direction and ultimately works with measures in a systematic manner.

FIGURE 1.5. A 'real world' tessellation (in Greifswald, Germany) based on the eightfold Ammann–Beenker tiling. Photograph © Stan Sherer.

In order to keep the mathematical machinery adequately elementary, the book effectively consists of two parts. The first comprises Chapters 1–7 and covers the ground up to the cut and project method. We start with a general chapter on mathematical tools, with a focus on discrete geometry and algebraic methods. Chapter 3 covers some standard material from lattice theory, selected for our purpose and presented in a way to facilitate the entrance into the non-periodic world.

Here and below, we face the difficult task of selecting from the vast amount of material. It is thus inevitable that we have to refer to other sources for some of the proofs and further details, or that we sometimes only give a sketch of a proof. For this reason, we use the symbol $\square$ to indicate 'end of proof' as well as 'absence of proof'. However, we have tried to cover some less common or unfamiliar results in a more complete fashion, in order to provide a reference for future work. While the core of the book is designed to be accessible, we sketch more advanced material in additional remarks, which usually include hints on the underlying arguments and further references.

Since one-dimensional structures are significantly easier to grasp, the chapter on lattices is followed by a detailed exposition of symbolic substitution sequences and the corresponding geometric inflation rules. Nevertheless, the reader should be aware that this only offers a glimpse at what is known collectively about such systems, to which we will refer via several standard monographs. Before we extend this to planar and higher-dimensional inflation tilings in Chapter 6, by way of important guiding examples, we develop appropriate and sufficiently general notions of discrete geometry in Chapter 5, which also contains a brief discussion of local rules (more on this topic will follow in the second volume of the book series). The treatment of systems with icosahedral symmetry focuses on the discussion of several important examples. In addition, we provide some useful material on the icosahedral group as an appendix. The geometric point of view is completed with the cut and project method in Chapter 7, which we develop step by step from a one-dimensional example. It is constructed in such a way that the generalisation to higher dimensions (and to the more algebraic setting of Meyer) is largely self-explanatory.

The second part of the book is mainly concerned with spectral properties of periodic and non-periodic systems, as accessible via mathematical diffraction theory. It starts with a selected review of methods from Fourier and harmonic analysis in Chapter 8, which includes a brief summary of quasiperiodic and almost periodic functions. The following substantial chapter on diffraction theory, which expands on two previous reviews [BG11, BG12], covers the general concepts, the treatment of crystallographic systems and that of cut and project sets. Once again, we develop the latter from a paradigmatic example, which reveals a natural path to the generalisations, where we restrict to pure point diffractive systems. Here, we also discuss the concept of homometry, which is important for the corresponding inverse problem.

In Chapter 10, we extend our analysis to (deterministic) systems with mixed spectra. In particular, we include complete treatments of the Thue–Morse and the Rudin–Shapiro system, which are (deterministic) paradigms for purely singular continuous and purely absolutely continuous diffraction. The theory of lattice subsets is then also developed, with a discussion of the set of visible lattice points. All this has an extension to the general setting of Meyer sets, which we briefly outline. As we go along, we keep mentioning the connections to dynamical systems and their spectra, for which we also provide an informal appendix.

Finally, Chapter 11 starts the corresponding development for systems with structural disorder, which means that we enter the territory of probability theory and stochastic processes. In comparison with the previous topics, the understanding of stochastic systems is much less developed, but has steadily been on the increase in recent years. At this stage, we selected systems and examples that mainly require classical tools of probability theory, such as the renewal process, Bernoulli and Markov systems, and employ the strong law of large numbers and some elementary results on Gibbs measures. This is perhaps the area with the largest gap between the physical intuition and results with mathematical rigour. In particular, some of the most obvious questions about random tiling ensembles are still open.

A quintessential summary of our tour through aperiodic order could be as follows. There is a good understanding of systems with perfect crystalline or quasicrystalline order, and a decent one of systems with sufficiently strong disorder. In between, however, we have only reached a limited understanding, although various examples with mixed spectrum have been worked out explicitly. In particular, it is unlikely that the known examples that are covered in this book suffice as a system of stepping stones to bridge this gap. Evidence for this claim is provided by systems such as the pinwheel tiling (which is deterministic, but still rather enigmatic; see Figure 6.58 on page 248 for a 'real world' example) or the non-crystallographic planar random tilings (which are understood heuristically but not rigorously). Further support for the underlying difficulties is given by dynamical systems of algebraic origin (such as Ledrappier's example or the ($\times 2, \times 3$) system), or the huge class of Meyer sets with entropy, which is largely unexplored.

It is fair to say that a classification of a hierarchy of (aperiodic) order has not only not been achieved yet, but is actually not even in sight. We would be delighted if our mathematical invitation inspires some progress in this direction.

CHAPTER 2

Preliminaries

In this chapter, we summarise the basic notions and elementary results that we will be using throughout the book. This will cover various aspects of discrete geometry, but also quite a few algebraic connections. We assume the reader to be mildly familiar with some fundamental concepts of group theory and a little algebra, in particular elementary algebraic number theory.

2.1. Point sets

The symbols $\mathbb{C}$, $\mathbb{R}$, $\mathbb{Q}$ and $\mathbb{Z}$ denote the complex, real, rational and integer numbers. The natural numbers are $\mathbb{N} = \{1, 2, 3, \ldots\}$, while $\mathbb{N}_0 = \{0\} \cup \mathbb{N}$. The symbol $\varnothing$ denotes the empty set, and the subset symbol $\subset$ is understood to include equality of sets. The symbol $\dot\cup$ denotes the disjoint union, so that $A \dot\cup B = A \cup B$ with $A \cap B = \varnothing$. Euclidean d-space is written as $\mathbb{R}^d$, and is equipped with its usual metric topology. We prefer this to the (otherwise common) notation $\mathbb{E}^d$ because the group structure of $\mathbb{R}^d$, which gives the origin $0 \in \mathbb{R}^d$ a special meaning, is often important for our discussion.

A set consisting of one point is called a *singleton set*, and countable unions of singleton sets are called *point sets*. In particular, every finite set is a point set. A point set $\Lambda \subset \mathbb{R}^d$ is *discrete* if each element $x \in \Lambda$ has an open neighbourhood $U = U(x) \subset \mathbb{R}^d$ that does not contain any other point of Λ. For each $x \in \Lambda$, there is then an $r > 0$ such that $B_r(x)$, the open ball of radius r around x, satisfies $B_r(x) \cap \Lambda = \{x\}$. Note that the radius r may depend on x. In contrast, Λ is *uniformly discrete* if there is an open neighbourhood U of $0 \in \mathbb{R}^d$ such that $(x + U) \cap (y + U) = \varnothing$ holds for all distinct $x, y \in \Lambda$. Here, $x + U := \{x + u \mid u \in U\}$ and, more generally,

$$U \pm V := \{u \pm v \mid u \in U,\ v \in V\}$$

for arbitrary $U, V \subset \mathbb{R}^d$. This is also known as the *Minkowski sum* and *difference*. We use the convention that $\varnothing \pm U = \varnothing$.

In $\mathbb{R}^d$, uniform discreteness implies the existence of a minimum distance between distinct points. In particular, the set of integers $\mathbb{Z} \subset \mathbb{R}$ is uniformly discrete, while the set $A = \{\frac{1}{n} \mid n \in \mathbb{N}\}$ is discrete, but not uniformly discrete. Note that the closure is $\overline{A} = \{0\} \cup A$, which fails to be discrete. A set Λ is called *locally finite* if, for all compact $K \subset \mathbb{R}^d$, the intersection $K \cap \Lambda$ is a finite set (or empty). This is only possible when Λ is a point set. Clearly, Λ is locally finite if and only if it is discrete and closed. Beyond finite point sets, this condition is met by $\mathbb{Z}$ for instance, but violated by the set A from above. Next, Λ is *relatively dense* if a compact $K \subset \mathbb{R}^d$ exists such that $\Lambda + K = \mathbb{R}^d$.

Definition 2.1. A point set $\Lambda \subset \mathbb{R}^d$ is a *Delone set*, or *Delone* for short, if it is both uniformly discrete and relatively dense.

Note that Delaunay is a common alternative spelling for Delone. For any given Delone set $\Lambda \subset \mathbb{R}^d$, one can choose two radii r and R so that $U = B_r(0)$ and $K = \overline{B_R(0)}$ are appropriate 'confining' neighbourhoods (of uniform discreteness and relative denseness) for Λ. For this reason, Λ is also called an (r, R)-set. Given $\mathbb{R}^d$, we use $\mathcal{D}$ to denote the set of all Delone sets, and $\mathcal{D}_{(r,R)}$ for all Delone sets that are compatible with $U = B_r(0)$ and $K = \overline{B_R(0)}$. Note that $\mathcal{D}_{(r,R)} = \varnothing$ whenever $r > R$. One can now define the *packing radius* r_{p} and the *covering radius* R_{c} of $\Lambda \in \mathcal{D}$ as

$$(2.1) \qquad \begin{aligned} r_{\mathsf{p}} &= \sup\{r > 0 \mid \Lambda \in \bigcup_{R>0} \mathcal{D}_{(r,R)}\}, \\ R_{\mathsf{c}} &= \inf\{R > 0 \mid \Lambda \in \bigcup_{r>0} \mathcal{D}_{(r,R)}\}. \end{aligned}$$

The geometric interpretation of these quantities should be clear.

Definition 2.2. A point set $\Lambda \subset \mathbb{R}^d$ is a *Meyer set*, or *Meyer* for short, if Λ is relatively dense and $\Lambda - \Lambda$ is uniformly discrete.

Every Meyer set is a Delone set, because uniform discreteness of $\Lambda - \Lambda$ implies that of Λ, while the converse is generally not true. The importance of a classification scheme for such sets was emphasised in [Lag96]. Here, we can only present a selection of properties and results, and refer to [Lag96, Moo97a, Lag99] for further material.

Example 2.1 (*Delone versus Meyer sets*). Consider the set

$$\Lambda = \{n + \tfrac{1}{n} \mid n \in \mathbb{Z} \setminus \{0\}\},$$

which is uniformly discrete (with packing radius $r = \frac{1}{4}$) and relatively dense (with covering radius $R = 2$), hence Delone. However, 1 (which is neither in Λ nor in $\Lambda - \Lambda$) is an accumulation point of $\Lambda - \Lambda$, so that $\Lambda - \Lambda$ is not uniformly discrete, and Λ is not Meyer. Note that $\mathbb{Z} \dot\cup \Lambda$ is still locally finite, but not uniformly discrete, hence neither Meyer nor Delone. $\Diamond$

Lemma 2.1. *Let $\Lambda \subset \mathbb{R}^d$ be a Delone set, such that $\Lambda - \Lambda \subset \Lambda + F$ for some finite set $F \subset \mathbb{R}^d$. Then, Λ is a Meyer set.*

PROOF. As a Delone set, Λ is locally finite and relatively dense. To establish the uniform discreteness of $\Delta := \Lambda - \Lambda$, it is sufficient to show that 0 is an isolated point of $\Delta - \Delta$. Using $\Delta \subset \Lambda + F$ twice, one has

$$0 \in \Delta - \Delta \subset (\Lambda + F) - (\Lambda + F) = \Delta + (F - F) \subset \Lambda + F'$$

where $F' = F + F - F$ is still a finite set. Consequently, $\Lambda + F'$ is locally finite, wherefore 0 is an isolated point of this set, hence also of $\Delta - \Delta$. □

Remark 2.1 (*General Meyer sets*). The converse of Lemma 2.1 is true in $\mathbb{R}^d$ [Lag99, Thm. 3.1], and in the more general context of an arbitrary *locally compact Abelian group* (LCAG) that is also compactly generated, see Section 2.3.3 and [BLM07, Appendix], but the argument is considerably more involved. The original definition of a Meyer set, compare [Moo97a], employs the property $\Lambda - \Lambda \subset \Lambda + F$. This provides an appropriate general setting for arbitrary LCAGs, as originally used in [Mey72, Thm. II.X]. ◊

A *cluster* of a point set Λ is the intersection $K \cap \Lambda$ for some compact $K \subset \mathbb{R}^d$. If K is convex, the cluster is also called a *patch*. Note that the same cluster (or patch) can be obtained from the intersection with different sets K. In this sense, the shape of K is immaterial. For many purposes, clusters are only considered up to translations, thus identifying clusters that are translates of one another. In this case, one point of the interior of K is marked (such as the centre of a ball) and supposed to be part of the cluster, except for the empty cluster (the latter often being excluded).

Definition 2.3. A discrete point set $\Lambda \subset \mathbb{R}^d$ has *finite local complexity* (FLC) with respect to translations when the collection $\{(t + K) \cap \Lambda \mid t \in \mathbb{R}^d\}$, for any compact $K \subset \mathbb{R}^d$, contains only finitely many clusters up to translations.

Later, we will also consider FLC up to Euclidean motions, which is useful for systems such as the pinwheel tiling of Figure 6.33. For now, we restrict the notion to translations only, and simply refer to it as FLC. It is sufficient to verify the finiteness condition for K being restricted to arbitrary closed balls. Moreover, one has the following characterisation.

Proposition 2.1 ([Lag99, Schl00]). *A discrete point set $\Lambda \subset \mathbb{R}^d$ has finite local complexity if and only if $\Lambda - \Lambda$ is locally finite.*

PROOF. Let $r > 0$ be fixed. Λ being FLC means that $\overline{B_r(t)} \cap \Lambda$ with $t \in \Lambda$ produces only finitely many patches up to translations. Any pair $x, y \in \Lambda$ with $|x - y| \leqslant r$ must occur in one of them, wherefore $(\Lambda - \Lambda) \cap \overline{B_r(0)}$ is finite. Since $r > 0$ was arbitrary, and any compact set $K \subset \mathbb{R}^d$ is contained in $B_r(0)$ for some r, the Minkowski difference $\Lambda - \Lambda$ is locally finite.

Conversely, $\Lambda - \Lambda$ locally finite implies $(\Lambda - \Lambda) \cap \overline{B_r(0)}$ to be a finite set, for any $0 < r < \infty$. We can thus write $\Lambda - \Lambda = \{z_i \mid i \in \mathbb{N}_0\}$ with $z_0 = 0$ and $|z_i| \leqslant |z_{i+1}|$ for all $i \geqslant 0$. Any non-empty patch in Λ of a fixed diameter $2r$ must then be of the form $\{x + z_i \mid i \in I\}$ with $x \in \Lambda$ and $I \subset \{i \mid |z_i| \leqslant r\}$, where the latter set is finite. Clearly, there are only finitely many possibilities for the choice of I. Since $r > 0$ was arbitrary, Λ must be FLC. $\qquad\square$

Example 2.2 (*FLC versus Meyer property*). It is clear by Proposition 2.1 that every Meyer set is FLC. To see that the converse implication is generally false, consider the set $\Lambda = -\mathbb{N} \,\dot\cup\, \{0\} \,\dot\cup\, \sqrt{2}\,\mathbb{N}$. This set is obviously FLC and hence also uniformly discrete (with packing radius $\frac{1}{2}$), and one finds

$$\Lambda - \Lambda = \{m + n\sqrt{2} \mid m, n \in \mathbb{Z} \text{ and } mn \geqslant 0\}.$$

Since m and n cannot have opposite signs, one can verify that $\Lambda - \Lambda$ is locally finite. It fails to be uniformly discrete because $\sqrt{2}$ is irrational, so that

$$\inf\big\{|u - v| \,\big|\, u, v \in \Lambda - \Lambda,\, u \neq v\big\} = \inf\big\{|k - \ell\sqrt{2}| \,\big|\, k, \ell \in \mathbb{N}\big\} = 0.$$

Consequently, Λ cannot be a Meyer set. $\qquad\diamond$

A Delone set of finite local complexity is called a Delone set of *finite type* in [Lag99]. Moreover, a Delone set (or, more generally, any point set) is called *finitely generated*, when the Abelian group $\langle \Lambda - \Lambda \rangle_{\mathbb{Z}}$ is finitely generated. Here, $\langle A \rangle_{\mathbb{Z}}$ denotes the smallest Abelian group that contains all finite integer linear combinations of elements of the set A. The set Λ from Example 2.1 is neither FLC (as 1 is an accumulation point of $\Lambda - \Lambda$) nor finitely generated (as $\langle \Lambda - \Lambda \rangle_{\mathbb{Z}}$ contains 1 and hence also $1/p$ for any prime p).

The concepts introduced so far are related via

$$\Lambda \text{ Meyer } \Longrightarrow \; \Lambda \text{ FLC and Delone}$$
(2.2) $$\Longrightarrow \; \Lambda \text{ finitely generated Delone}$$
$$\Longrightarrow \; \Lambda \text{ Delone}$$

where no arrow is reversible. The second implication (which is not obvious) was shown by Lagarias [Lag99, Thm. 2.1], together with a systematic classification scheme. Let us add a counterexample to demonstrate that the FLC property alone does not imply a set to be Delone.

Example 2.3 (*FLC versus Delone property*). Consider a sequence $(a_n)_{n \in \mathbb{N}_0}$ of non-negative numbers with $a_0 = 0$, $a_i < a_{i+1}$ for all $i \geqslant 0$ and $a_n \longrightarrow \infty$ as $n \to \infty$. Define

$$S_n = \sum_{i=0}^{n} a_i$$

and consider the set $\Lambda = \{0\} \,\dot\cup\, \{\pm S_n \mid n \in \mathbb{N}\}$, which is FLC by construction (due to the growth condition on the a_i).

Now, let all a_i be rationally independent, for instance by choosing $a_i = \alpha^i$ with $\alpha > 1$ transcendental. This implies

$$\langle a_i \mid i \in \mathbb{N}_0 \rangle_{\mathbb{Z}} \subset \langle \Lambda - \Lambda \rangle_{\mathbb{Z}},$$

so that $\langle \Lambda - \Lambda \rangle_{\mathbb{Z}}$ is not finitely generated. Consequently, the FLC set Λ can neither be a Delone set (as Λ is not relatively dense) nor a subset of a finitely generated Delone set Λ', because embedding Λ into Λ' would imply $\langle \Lambda - \Lambda \rangle_{\mathbb{Z}} \subset \langle \Lambda' - \Lambda' \rangle_{\mathbb{Z}}$ and hence a contradiction.   ◇

Lemma 2.2. *If $\Lambda \subset \mathbb{R}^d$ is a relatively dense subset of a Meyer set, it is a Meyer set itself.*

PROOF. Let $\Lambda \subset \Lambda'$ with Λ' being a Meyer set. We know that Λ is relatively dense. Moreover, it is clear that $(\Lambda - \Lambda) \subset (\Lambda' - \Lambda')$, which is thus uniformly discrete.   □

Example 2.4 (*Meyer sets with entropy*). Consider the set $\Lambda = 2\mathbb{Z} \cup S$, where S is an arbitrary subset of $2\mathbb{Z} + 1$. Since $\Lambda \subset \mathbb{Z}$ is relatively dense, it is a Meyer set by Lemma 2.2. This does not depend on any detail of S. In particular, one can select points from $2\mathbb{Z} + 1$ by repeated coin tossing, which produces a stochastic Meyer set with positive entropy; compare Example 10.5. We shall say more about entropy in Section 4.8.2 and in Chapter 11.   ◇

Let us close this section by recalling the definitions of various important point sets that will reappear frequently; see [Cas71] for background material.

Definition 2.4. A point set $\Gamma \subset \mathbb{R}^d$ is called a *point lattice* or simply a *lattice* in $\mathbb{R}^d$ if there exist d vectors $b_1, \ldots, b_d$ such that

$$\Gamma = \mathbb{Z}b_1 \oplus \cdots \oplus \mathbb{Z}b_d := \left\{ \sum_{i=1}^{d} m_i b_i \mid \text{all } m_i \in \mathbb{Z} \right\},$$

together with the requirement that its $\mathbb{R}$-span is $\mathbb{R}^d$, meaning that $\langle \Gamma \rangle_{\mathbb{R}} = \mathbb{R}^d$. The set $\{b_1, \ldots, b_d\}$ is then called a *basis* of the lattice Γ.

It is clear from this definition that a lattice $\Gamma \subset \mathbb{R}^d$ is a discrete subgroup of $\mathbb{R}^d$, both viewed as Abelian groups. Moreover, our definition is chosen so that the factor group $\mathbb{R}^d / \Gamma$ is a compact Abelian group, namely a d-dimensional torus. Alternatively, one can thus characterise a lattice Γ as a discrete, co-compact subgroup of $\mathbb{R}^d$. This more abstract version also defines lattices in general LCAGs (see Section 2.3.3 below).

Lemma 2.3. *Any lattice $\Gamma \subset \mathbb{R}^d$ is a Meyer set. Consequently, it is also a Delone set of finite local complexity.*

PROOF. Since Γ is an Abelian group, we have $\Gamma - \Gamma = \Gamma$. It is thus sufficient to show that Γ is uniformly discrete and relatively dense. For any given basis of Γ, let K be the closed parallelotope spanned by the d basis vectors, which is a compact set. Any closed ball that contains K is suitable to show relative denseness of Γ (since $\Gamma + K = \mathbb{R}^d$), while any open ball in the interior of K can be used to derive uniform discreteness by shifting its centre to a lattice

point (since the interior of K contains no lattice point); see [Cas71, Ch. III]
for background.   □

Since a translate of a lattice need not be a lattice again (due to losing the
group property), one often needs the following generalisation (for instance to
describe the idealised positions of atoms in a crystal).

Definition 2.5. A non-empty point set $\Lambda \subset \mathbb{R}^d$ is called a *crystallographic
point packing* in $\mathbb{R}^d$ if there is a lattice Γ in $\mathbb{R}^d$ and a finite point set F such
that $\Lambda = \Gamma + F$.

Proposition 2.2. *Any crystallographic point packing $\Lambda \subset \mathbb{R}^d$ is Meyer.*

PROOF. Since $\Lambda = \Gamma + F$ with a lattice Γ and a finite set F, Λ is still
locally finite, wherefore uniform discreteness follows from lattice periodicity.
Relative denseness of Λ is inherited from that of Γ, so Λ is Delone. Observing

$$\Lambda - \Lambda = (\Gamma + F) - (\Gamma + F) = \Gamma + F - F = \Lambda + F',$$

where $F' = -F$ is again a finite set, establishes the Meyer property by an
application of Lemma 2.1.   □

A characteristic property of a point set $\Lambda \subset \mathbb{R}^d$ is the average number of
points per unit volume, provided this quantity exists. With $\Lambda_r := \Lambda \cap \overline{B_r(0)}$,
we call the numbers

$$\overline{\mathrm{dens}}(\Lambda) := \limsup_{r\to\infty} \frac{\mathrm{card}(\Lambda_r)}{\mathrm{vol}(B_r)} \quad \text{and} \quad \underline{\mathrm{dens}}(\Lambda) := \liminf_{r\to\infty} \frac{\mathrm{card}(\Lambda_r)}{\mathrm{vol}(B_r)}$$

the *upper* and the *lower natural density* of Λ. The term 'natural' refers to
the use of centred balls for the averaging, and will often be omitted. Note
that $0 \leqslant \underline{\mathrm{dens}}(\Lambda) \leqslant \overline{\mathrm{dens}}(\Lambda) \leqslant \infty$, which is a well-defined inequality.

Definition 2.6. If the upper and the lower natural density of a point set
$\Lambda \subset \mathbb{R}^d$ coincide, the corresponding value is called the *natural density* of Λ
and denoted by $\mathrm{dens}(\Lambda)$.

In general, a replacement of the centred balls by other averaging sets may
change the upper and lower densities. When

$$(2.3) \qquad \rho_a(r) := \frac{\mathrm{card}\big(\Lambda \cap \overline{B_r(a)}\big)}{\mathrm{vol}(B_r)},$$

as $r \to \infty$, converges uniformly in $a \in \mathbb{R}^d$, $\mathrm{dens}(\Lambda)$ is referred to as the
uniform density of Λ.

Example 2.5 (*Tied versus uniform densities*). Let us consider the point set
$S = -\mathbb{N} \cup 2\mathbb{N} \subset \mathbb{R}$. Clearly, for any fixed $a \in \mathbb{R}$, one has $\lim_{r\to\infty} \rho_a(r) = \frac{3}{4}$,
which is sometimes called the *tied* density [BMP00, Appendix] of S. On
the other hand, one has $\lim_{r\to\infty} \rho_r(r) = \frac{1}{2}$ and $\lim_{r\to\infty} \rho_{-r}(r) = 1$, and

similarly for other r-dependent choices of the centre a. In fact, via suitable choices of sequences $(r_i)_{i\in\mathbb{N}}$ and $(a_i)_{i\in\mathbb{N}}$, one can arrange to obtain any limit between $\frac{1}{2}$ and 1. The absence of such phenomena is one advantage of uniform convergence. ◇

In higher dimensions, an averaging over cubes rather than balls in the case of uniform convergence has no influence on the value of the density. Both are examples of averages along van Hove sequences, which form a large class of suitable averaging sequences. Their common feature is that the surface to volume ratio of the sets tends to 0 as their diameter diverges. They will be introduced more precisely later, in Definition 2.9 and Remark 2.5.

Let $K \subset \mathbb{R}^d$ be a compact set and fix the (non-empty) cluster (or patch) $P = \Lambda \cap K$ of a (locally finite) point set Λ. Consider the quotient

$$\frac{\operatorname{card}\{t \in B_r(a) \mid (-t + \Lambda) \cap K = P\}}{\operatorname{vol}(B_r)},$$

which is the number of clusters of type P per volume, with reference point in $B_r(a)$. If, as $r \to \infty$ with $a \in \mathbb{R}^d$ fixed, the quotient converges, the limit is called the *absolute frequency* of the cluster P in Λ. The density of Λ can also be understood as the absolute frequency of the singleton cluster. Upper and lower frequencies can be defined in analogy to upper and lower densities, and always exist. If the above convergence is uniform in a, the limit is called the *uniform absolute cluster frequency* of P.

Definition 2.7. A point set $\Lambda \subset \mathbb{R}^d$ is said to have *uniform cluster frequencies* (UCF) if all clusters of Λ possess uniform absolute frequencies.

Example 2.6 (*Frequencies in lattices*). Consider a lattice $\Gamma \subset \mathbb{R}^d$ according to Definition 2.4, with basis $\{b_1, \ldots, b_d\}$. By standard (although somewhat technical) arguments, compare [Cas71], Γ has the uniform density

$$\operatorname{dens}(\Gamma) = \left|\det(b_1, \ldots, b_d)\right|^{-1},$$

which reflects the fact that a lattice has one point per fundamental domain.

Moreover, if we consider the cluster $\Gamma \cap K \neq \varnothing$ for a compact $K \subset \mathbb{R}^d$, its absolute frequency is uniform, too, and always equals $\operatorname{dens}(\Gamma)$. This follows from the group property of Γ, whence the lattice translates of any cluster provide a complete collection of all clusters of the same type in Γ. ◇

Remark 2.2 (*Absolute versus relative frequencies*). When working with point sets, it is often useful to define frequencies per point rather than per unit volume. This is particularly obvious in the previous example. We call the frequency defined per point the *relative frequency*. When the (natural) density of a given point set exists, it equals the quotient of absolute and relative frequencies. ◇

Before we continue our general discussion of point sets, let us briefly introduce the notion of a Cantor set. It appears in various disguises throughout the theory of aperiodic order. Recall that a non-empty set is *totally disconnected* if all connected components are singleton sets. Moreover, a set is called *perfect* when it is closed and has no isolated points (recall that a point $x \in X$ is isolated if and only if $\{x\}$ is an open set in the relative topology of X). A non-empty subset of $\mathbb{R}$ is a (classical) *Cantor set* if it is compact, perfect, and totally disconnected.

The best-known example is the set $\mathcal{C} \subset [0,1]$ that emerges from the middle-thirds removal procedure; see the inlay of Figure 8.5 on page 321 for an illustration and [Car00] for general background and further references. As a subset of $\mathbb{R}$, $\mathcal{C}$ with its induced (or relative) topology is also a metrisable space. More generally, any topological space homeomorphic to $\mathcal{C}$ is also called a Cantor set (or a Cantor space).

Definition 2.8. A topological space $X \neq \varnothing$ is called a *Cantor set* (or space) when it is compact, perfect, totally disconnected and metrisable.

In general, X is metrisable if it is regular (meaning that, for any set $A \subset X$ and point $x \in X \setminus A$, there are disjoint neighbourhoods of A and x in X) and has a countable base for its topology. Both properties are clear for $\mathbb{R}^d$, where the countable base is provided by open balls with rational radii and centres with rational coordinates. The latter property is also referred to as second countability; see [Bou66, vQu79] for more information. We shall meet further examples of Cantor sets later, such as the p-adic integers in Example 2.10 or the hulls of aperiodic substitutions in Chapter 4.

2.2. Voronoi and Delone cells

Let $\varnothing \neq \Lambda \subset \mathbb{R}^d$ be a locally finite point set. The *Voronoi domain* or *Voronoi cell* of a point $a \in \Lambda$ is defined as

$$(2.4) \qquad V(a) := \big\{ x \in \mathbb{R}^d \mid \|x - a\| \leqslant \|x - b\| \text{ for all } b \in \Lambda \big\},$$

where $\|.\|$ denotes Euclidean distance. Due to the local finiteness of Λ, the cell $V(a)$ is a closed set with non-empty interior, though it need not be compact. In fact, $V(a)$ is the intersection of at most countably many closed half-spaces. A simple example is shown in Figure 2.1, where the Voronoi cell is a polygon. If $\Lambda \subset \mathbb{R}^d$ is a locally finite and relatively dense point set, all its Voronoi cells are (bounded) polytopes, while the polytope property in general can be rather subtle.

Each point of Λ has its own well-defined Voronoi cell. The collection of all Voronoi cells constitutes a face to face tiling of $\mathbb{R}^d$ (possibly with non-compact tiles). In particular, two distinct Voronoi cells can only intersect in

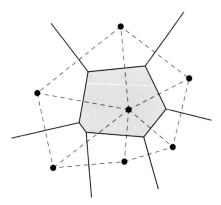

FIGURE 2.1. Voronoi cell (shaded) of a point in the plane as imposed by six surrounding points. The dashed lines represent the six triangular Delone cells that are dual to the vertices of the Voronoi cell.

boundary facets. As a consequence, the collection of Voronoi cells forms a Euclidean cell complex, called the *Voronoi complex*; see [KS89, Schl93a] and references therein. This complex has a well-defined geometric dual, which is another Euclidean cell complex known as a *Delone complex*. It can be made unique by the requirement that the original points of Λ become the vertex points (or 0-boundaries) of the Delone complex. To each facet of the Voronoi complex of dimension ℓ, with $0 \leqslant \ell \leqslant d$, there is then a unique dual facet of dimension $d - \ell$ in the Delone complex, and vice versa. The two facets of a dual pair are mutually orthogonal (with obvious meaning for 0-boundaries), but need not have a point in common. In Figure 2.1, the Delone complex is indicated by the dashed lines. It contains one instance where a 1-boundary does not intersect its dual 1-boundary, and one vertex of the Voronoi cell that lies outside the Delone cell dual to this vertex; compare Figure 2.3 for a more generic situation.

Clearly, these concepts are also applicable to crystallographic point packings. Figure 2.2 illustrates the mutually dual pair of the triangular lattice and the hexagonal packing, interpreted with their roles interchanged.

Example 2.7 (*Primitive cubic lattices*). For the integer lattice $\mathbb{Z}^d$, which is also known as the primitive (hyper)cubic lattice in Euclidean d-space, the Voronoi cells are $V(0) = \left[-\frac{1}{2}, \frac{1}{2}\right]^d$ and its lattice translates. Likewise, $[0, 1]^d$ and its lattice translates form the dual Delone complex. This is an exceptional case where both complexes are built from congruent cells. ◇

Example 2.8 (*Point sets with equal Voronoi complexes*). Distinct point sets, even infinite ones, can possess the same Voronoi complex. As an example,

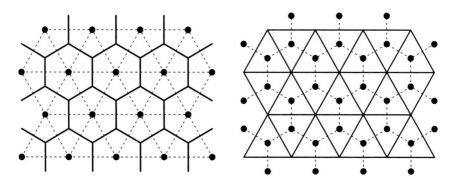

FIGURE 2.2. Voronoi complex for the points of the triangular lattice (left) and for the points of the hexagonal point packing (right). In both cases, the corresponding Delone complexes are also shown (dashed).

consider the integer lattice $\mathbb{Z}$, whose Voronoi complex (by Example 2.7) is built from the closed interval $V(0) = \left[-\frac{1}{2}, \frac{1}{2}\right]$ and its $\mathbb{Z}$-translates. Now choose any ε in the interior of $V(0)$, and consider the deformed point set

$$\Lambda_\varepsilon := \left\{ m + (-1)^m \varepsilon \mid m \in \mathbb{Z} \right\},$$

so that $\Lambda_0 = \mathbb{Z}$. As a consequence, Λ_ε with $0 < |\varepsilon| < \frac{1}{2}$ is periodic, with lattice of periods $2\mathbb{Z}$. Nevertheless, all these sets produce the same $\mathbb{Z}$-periodic Voronoi complex, namely the one of $\mathbb{Z}$ itself. A typical illustration is

$$\cdots \quad \bullet \mid \bullet \quad \mid \quad \bullet \mid \bullet \quad \mid \quad \bullet \mid \bullet \quad \mid \quad \bullet \mid \bullet \quad \mid \quad \bullet \mid \bullet \quad \cdots$$

where the vertical lines separate neighbouring Voronoi cells. Higher dimensional examples can be constructed similarly.

This example shows that, in general, one cannot reconstruct the original point set from its Voronoi complex alone. However, the Delone complex differs for all these cases, because the original points are its 0-boundaries by construction. This is the reason why one should always consider the dual pair of the Voronoi and the Delone complex of a point set in conjunction. ◊

In general, Voronoi and Delone complexes look less regular. This is particularly so for point sets that model an ideal gas, which mathematically means point sets from a homogeneous Poisson process [KT75]; see Figure 2.3 for a snapshot. Recall that a *Poisson process* in $\mathbb{R}^d$ of point density (also called intensity) ρ is a point process with the property that the number of points in any measurable set A is Poisson-$(\rho \operatorname{vol}(A))$-distributed and that the number of points in any disjoint sets $A_1, \ldots, A_m \subset \mathbb{R}^d$ are independent random variables. In a certain sense, this process produces some of the most disordered point sets. We shall say more about this process in Chapter 11.

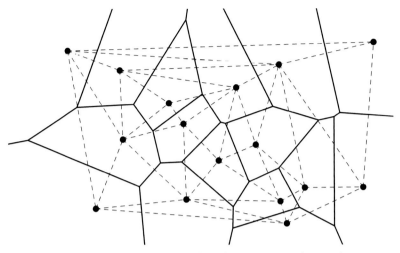

FIGURE 2.3. Dual pair of Voronoi (solid) and Delone (dashed) complex for a typical finite point set that was created by a Poisson process.

The theory of aperiodic order aims at exploring and understanding the universe of structures between crystallographic sets (at the totally ordered end of the spectrum) and Poisson point sets (at the totally stochastic end).

Remark 2.3 (*Laguerre domains*). The definition of the Voronoi cell in Eq. (2.4) uses the Euclidean distance between points. Clearly, this can be generalised by employing other distance concepts. One specific choice leads to the so-called Laguerre domains (or cells) together with their duals, which are well-defined. This proved to be a useful generalisation in the context of tiling theory, and in the theory of aperiodic order in particular [Schl93a]. An illustrative discussion with many examples can be found in [GS93]. ◇

Remark 2.4 (*Average coordination numbers for planar cell complexes*). Consider the Voronoi or Delone complex of a Delone set in $\mathbb{R}^2$, which is a tiling of the plane by polygons, where neighbouring ones meet edge to edge (and hence vertex to vertex). By taking a large circular patch and wrapping it on the unit sphere in 3-space, one obtains (possibly after some adjustments at the boundary, which is shrunk to a point in this process) a polyhedron that satisfies Euler's relation $V - E + F = 2$ for the total number of vertices (V), edges (E) and facets (F). Let e be the average number of edges per polygon facet, and c the average coordination number at the vertices, which means the average number of edges at a single vertex. Then, $E = eF/2$ and $E = cV/2$, so that $\left(\frac{2}{c} - 1 + \frac{2}{e} \right) = \frac{2}{E}$ by Euler's relation.

Taking larger and larger patches (so that $E \to \infty$ by our assumptions), and assuming that the averaged quantities exist in the limit, one finds the consistency condition

$$(2.5) \qquad\qquad c = \frac{2e}{e - 2}$$

between the average coordination and edge numbers. Standard examples are $c = e = 4$ for the square lattice, $c = 6$ with $e = 3$ for the triangular lattice, and $c = 3$ with $e = 6$ for the hexagonal packing; compare Figure 2.2.   ◊

Let us postpone further conceptual and structural notions for discrete point sets and tilings until Chapter 5 and turn our attention to some of the more algebraic aspects.

2.3. Groups

In a crystallographic setting, matrix groups are important. The *general linear group* in $\mathbb{R}^d$ consists of all invertible $d \times d$-matrices with real entries. It is denoted by $\mathrm{GL}(d, \mathbb{R})$, or $\mathrm{GL}(d)$ for short, while $\mathrm{O}(d)$ and $\mathrm{SO}(d)$ stand for the *orthogonal* and the *special orthogonal group* in d-space, the latter being the index 2 subgroup of orientation-preserving (linear) isometries (or rotations). For this reason, $\mathrm{SO}(d)$ is also called the *rotation group* of Euclidean d-space. When fields or rings other than $\mathbb{R}$ are needed, they will always be specified as an explicit argument. For an elementary general reference on matrix groups, we refer to [Cur84].

2.3.1. Finite groups

Several finite groups will show up repeatedly. We need the *cyclic group* C_n of order n, which is generated by an element g of order n, hence an element that only satisfies the relation $g^n = e$ (and thus $g^{\ell n} = e$ for all $\ell \in \mathbb{Z}$, but no other ℓ), where e denotes the neutral element. Although C_n is an Abelian group, it is written multiplicatively rather than additively; see [CM80] for a general reference. Another common and self-explaining notation is

$$C_n = \langle g \mid g^n = e \rangle.$$

In the same spirit, we write

$$D_n = \langle g, h \mid g^n = h^2 = (gh)^2 = e \rangle$$

for the *dihedral group* of order $2n$, which is non-commutative for $n > 2$, while $D_2 \simeq C_2 \times C_2$ is Klein's 4-group. The *symmetric* or *permutation group* of n symbols, which is of order $n!$, is written as S_n. For $n > 1$, its index 2 subgroup of even permutations is the *alternating group*, denoted by A_n.

Due to the history of the field, the symmetry groups of regular polytopes are important, in particular those of the five Platonic solids shown in Figure 2.4. Here, cube and octahedron, as well as dodecahedron and icosahedron, are mutually dual (in the sense that joining the facial midpoints

FIGURE 2.4. The five Platonic solids — tetrahedron, cube (also called hexahedron), octahedron, dodecahedron and icosahedron (from left to right).

gives the dual partner, up to similarity), while the tetrahedron is self-dual in this sense. Since mutually dual polytopes share the same symmetries, there are only three distinct symmetry groups, T_d, O_h and Y_h, where T_d does not contain the space inversion; for general background on regular polytopes, we refer to [Cox73, MS02].

The full *tetrahedral group* T_d contains 24 elements, which comprise all rotation and reflection symmetries of the regular tetrahedron. These symmetries can be identified with the permutations of the 4 vertices of the tetrahedron, so that $T_d \simeq S_4$. Clearly, T_d has a subgroup T of order 12 which contains the orientation-preserving elements of T_d. One finds $T \simeq A_4$, and a possible presentation in terms of generators and relations reads

$$(2.6) \qquad T = \langle r, s \mid r^3 = s^2 = (rs)^3 = e \rangle.$$

Similarly, the orientation-preserving symmetries of the cube (and also of the octahedron) form the group

$$(2.7) \qquad O = \langle r, t \mid r^3 = t^2 = (rt)^4 = e \rangle,$$

called the *octahedral group*, where the element r can be shared by T and O, due to their mutual geometric relation. One finds $O \simeq S_4$ as abstract groups, while the full octahedral group is the direct product $O_h = O \times C_2$; see [CM80, Sec. 1.5] for details. Choosing $s = (rt)^2$, one has $s^2 = e$, and (using $rtrtr = trrt$) also

$$(rs)^3 = rrtrtrrtrtrrtrt = rtrrttrrtrtrt = (rt)^4 = e,$$

which makes T a subgroup of O this way, and T_d one of O_h, in line with the geometric action of the groups in 3-space. Note, however, that T_d is *not* a direct product, and hence not a holohedry in 3-space; compare [BBN+78] for details.

Perhaps the most important finite group in this context is the symmetry group of the dodecahedron and the icosahedron, called the *icosahedral group* Y_h, together with its rotation subgroup Y. They are related by $Y_h = Y \times C_2$. The direct product structure of O_h and Y_h is due to the inversion symmetry of the cube and the icosahedron, which does not apply to the tetrahedron.

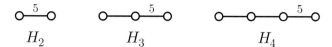

$$H_2 \qquad\qquad H_3 \qquad\qquad H_4$$

FIGURE 2.5. Coxeter diagrams for the non-crystallographic root systems of type H_2, H_3 and H_4. Each node corresponds to a generating reflection. Products of two generators which correspond to directly connected nodes have order 3 or 5 (indicated); all other (binary) products have order 2.

The group Y can be presented as

$$(2.8) \qquad\qquad Y = \langle r, u \mid r^3 = u^2 = (ru)^5 = e \rangle$$

and is isomorphic to the alternating group A_5. This is also an important group in Galois theory, as it is the first simple group in the series of alternating groups. The element r in Eqs. (2.6), (2.7) and (2.8) may represent the same threefold rotation, which highlights the geometric relations between the three groups; see [Cox73] for more. In particular, T (which can be considered as half of O) is a subgroup of Y, while O itself is not. An extensive discussion of the groups T, O and Y, tailored to the context of quasicrystals, is given in [KH89], while a summary of the icosahedral group is provided in Appendix A.

The groups considered here will appear below as the finite subgroups of $\mathrm{SO}(3)$ (up to isomorphism); see Theorem 3.4 on page 58.

2.3.2. ROOT SYSTEMS AND COXETER GROUPS

For a non-zero vector $a \in \mathbb{R}^d$, we define the reflection r_a in the (central) hyperplane orthogonal to a as the mapping $x \mapsto x - 2\frac{\langle x|a \rangle}{\langle a|a \rangle}a$, where $\langle x|y \rangle$ denotes the standard Euclidean scalar product.

A finite set $\Delta \subset \mathbb{R}^d \setminus \{0\}$ is called a finite *root system* if, for all $a \in \Delta$, $\mathbb{R}a \cap \Delta = \{\pm a\}$ and $r_a(\Delta) = \Delta$. The *rank* of Δ is the dimension of the subspace of $\mathbb{R}^d$ spanned by it. A root system is *decomposable* if it can be split into two mutually orthogonal subsystems; otherwise, it is *indecomposable*.

Example 2.9 (*Hypercubic root systems*). The set $\{\pm e_i \mid 1 \leqslant i \leqslant d\} \subset \mathbb{R}^d$ of $2d$ unit vectors forms a decomposable root system of rank d. It is related to the hypercubic lattice $\mathbb{Z}^d$ of Example 2.7, which is the integer span of the root system. The indecomposable components are the rank one subsystems $\{\pm e_i\}$. Since $\mathbb{Z}^d$ is a lattice, this root system is called crystallographic. $\diamond$

More generally, a finite root system Δ is called *crystallographic* if one has $2\frac{\langle a|b \rangle}{\langle a|a \rangle} \in \mathbb{Z}$ for all $a, b \in \Delta$; otherwise, Δ is called *non-crystallographic*. The finite reflection group generated by the r_a, with $a \in \Delta$, is called the *Weyl group* of Δ. It is thus a finite *Coxeter group*. All finite Coxeter groups arise in this way; see [Pat97, Sec. 2]. The Coxeter groups are classified and well

FIGURE 2.6. The icosidodecahedron (left, a semi-regular polyhedron) and the triacontahedron (right, also known as Kepler's body) are two mutually dual convex polyhedra with full icosahedral symmetry.

understood; see [Hum92, Sec. I.2] for a complete exposition. Here, we focus our attention on three related non-crystallographic cases, namely H_2 (often referred to as $I_2(5)$), H_3 and H_4, which are all linked to fivefold symmetry. Their Coxeter diagrams are shown in Figure 2.5. Explicit choices for the generating elements can be found in [Pat97].

The root system Δ_{H_2} can most easily be realised as the set of all 10th roots of unity, the latter being discussed in more detail in Section 2.5.2. The corresponding Weyl group is thus the symmetry group of the regular decagon. It contains 20 elements and is isomorphic with the dihedral group $D_{10} \simeq H_2$.

The root system Δ_{H_3} comprises 30 roots, which can be chosen as

$$(2.9) \qquad\qquad (\pm1, 0, 0), \ \tfrac{1}{2}(\pm1, \pm\tau', \pm\tau),$$

where $\tau = \tfrac{1}{2}(1 + \sqrt{5})$ is the golden ratio (with $\tau' = 1 - \tau$), together with all *even* permutations of the coordinates. The corresponding Weyl group has 120 elements and is isomorphic with the icosahedral group $Y_{\mathsf{h}} \simeq H_3$. The convex hull of the 30 roots is the *icosidodecahedron*, which is an Archimedean (convex and semi-regular) polyhedron with icosahedral symmetry; see Figure 2.6 or [Cox73, Plate I]. We shall meet Δ_{H_3} again in Example 2.20 in the context of icosahedral modules in 3-space.

A concrete representation of the non-crystallographic root system Δ_{H_4} is given by the vectors

$$(2.10) \qquad (\pm1, 0, 0, 0), \ \tfrac{1}{2}(\pm1, \pm1, \pm1, \pm1), \ \tfrac{1}{2}(0, \pm1, \pm\tau', \pm\tau),$$

together with their *even* coordinate permutations. These form the 120 vectors of Δ_{H_4}, all of unit length. The convex hull is known as the 600-cell, a regular polytope in 4-space whose 3-boundary is formed by 600 regular tetrahedra; see [Cox73, Sec. 8.5] for details. Its dual, another regular polytope, is the

120-cell. Both have the Weyl group H_4 of Δ_{H_4} of order $120^2 = 14400$ as their symmetry group.

2.3.3. LOCALLY COMPACT ABELIAN GROUPS

The content of this paragraph is a brief reminder of some properties of topological Abelian groups; see [HR97] and [Loo11, Ch. VII] for background material. Since much of our discussion proceeds along $\mathbb{R}^d$, the material is mainly needed for various generalisations that will appear later.

Let G be a group, and assume that it is equipped with a topology. If each point of G possesses a compact neighbourhood, the group G is called *locally compact*. If G is also Abelian, it is called a *locally compact Abelian group*, or LCAG for short. Clearly, $\mathbb{R}^d$ with its usual topology is an LCAG, as is any finite Abelian group (equipped with the discrete topology) or any compact Abelian group, such as the unit circle $\mathbb{S}^1$. Further examples are discrete groups such as $\mathbb{Z}^n$, again with the discrete topology.

An LCAG G is called *compactly generated* when a compact neighbourhood of $0 \in G$ exists that generates the entire group. Again, $\mathbb{R}^d$, $\mathbb{Z}^n$ and compact Abelian groups $\mathbb{K}$ are examples. Since compactly generated LCAGs are particularly important, it is worthwhile to recall their classification; see [HR97, Thm. 9.8] for details and a proof of the following result.

Theorem 2.1. *If G is a compactly generated LCAG, it is isomorphic, as a topological group, with $\mathbb{R}^d \times \mathbb{Z}^n \times \mathbb{K}$, for some $d, n \in \mathbb{N}_0$ and some compact Abelian group $\mathbb{K}$.* □

Example 2.10 (*p-adic integers*). Let p be a fixed (rational) prime. Any non-negative integer n then has a unique expansion as

$$n = \sum_{i=0}^{k} \ell_i \, p^i,$$

for some $k \in \mathbb{N}_0$, with $\ell_i \in \{0, 1, \dots, p-1\}$ for all i. Introducing a norm (see below) in which increasing powers of p become smaller, one can extend this to infinite series that converge. In particular, $\ell_i = p - 1$ for all $i \in \mathbb{N}_0$ then gives a representation of -1, because $\sum_{i=0}^{\infty} p^i = 1/(1-p)$. This way, all negative integers are covered by products of finite sums with -1. Taking the completion in the corresponding topology gives the set $\mathbb{Z}_p$ of p-adic integers, which is an example of a compact Abelian group. It is uncountable, and has the structure of a Cantor set; compare Definition 2.8 and the discussion around it. For general background, we refer to [Kob96, Gou97].

More generally, any $0 \neq x \in \mathbb{Q}$ can uniquely be written as $x = \frac{r}{s} p^a$ with (coprime) integers r, s that are not divisible by p and with a thereby unique exponent $a \in \mathbb{Z}$. Defining the p-adic norm of x as

$$|x|_p = p^{-a},$$

together with $|0|_p := 0$, the p-adic distance $\mathrm{d}_p(x, y) := |x - y|_p$ is a metric, known as the p-adic metric on $\mathbb{Q}$. In fact, d_p is an ultrametric, which means that we have the stronger version of the triangle inequality in the form $\mathrm{d}_p(x, y) \leqslant \max(\mathrm{d}_p(x, z), \mathrm{d}_p(y, z))$. The set of p-adic numbers $\mathbb{Q}_p$ is obtained as the completion of $\mathbb{Q}$ in the topology defined by this metric. $\mathbb{Q}_p$ is an LCAG (but not compactly generated), and a complete metric space. Its elements can be identified with the (d_p-converging) power series of the form

$$\pm \sum_{i=m}^{\infty} \ell_i \, p^i$$

where $m \in \mathbb{Z}$ and $0 \leqslant \ell_i \leqslant p - 1$ for all i. The p-adic integers form a compact subset that is also a complete metric space. They correspond to the power series with $m \geqslant 0$. $\mathbb{Z}_p$ can graphically be represented as the set of infinite paths starting at the root of a p-ary tree; see [Sin06, Ch. 3c] for details and alternative representations.   ◇

For some of our later examples, it is helpful to distinguish the unit circle $\mathbb{S}^1 = \{z \in \mathbb{C} \,|\, |z| = 1\}$, which is written multiplicatively, and the 1-torus $\mathbb{T}$, which is represented by $[0, 1)$ with addition modulo 1, even though they are isomorphic as compact Abelian groups.

Example 2.11 (*Simple solenoids*). Consider the 1-torus $\mathbb{T}$ and define

$$\Sigma_2 := \big\{ (x_0, x_1, x_2, \ldots) \in \mathbb{T}^{\mathbb{N}_0} \mid x_i = 2x_{i+1} \bmod 1 \text{ for all } i \in \mathbb{N}_0 \big\},$$

which is a closed subgroup of $\mathbb{T}^{\mathbb{N}_0}$ and hence an example of a compact Abelian group. The topology is induced by the Fréchet-type metric

$$\mathrm{d}_{\Sigma_2}(x, y) := \sum_{i \geqslant 0} \frac{|x_i - y_i|_{\mathbb{T}}}{2^i},$$

where $|.|_{\mathbb{T}}$ is the proper distance on $\mathbb{T}$. The group Σ_2 is known as the *dyadic solenoid* [vDan30, Sad08]. It can be considered as the simplest extension of $\mathbb{T}$ that makes the linear (odometer) mapping $x \mapsto 2x \bmod 1$ invertible. Indeed, if $x = (x_0, x_1, x_2, \ldots) \in \Sigma_2$, one has $2x = (2x_0 \bmod 1, x_0, x_1, x_2, \ldots)$ and $\frac{1}{2}x = (x_1, x_2, x_3, \ldots)$.

There is an interesting connection with the 2-adic integers from Example 2.10, which will become important in the model set construction of (deterministic) Toeplitz-type systems in Chapter 7. To this end, consider the LCAG $\mathbb{R} \times \mathbb{Z}_2$, which contains the lattice

$$\Gamma = \big\{ (n, \iota(n)) \mid n \in \mathbb{Z} \big\}.$$

This is the diagonal extension of $\mathbb{Z}$ to $\mathbb{R} \times \mathbb{Z}_2$, where $\iota\colon \mathbb{Z} \hookrightarrow \mathbb{Z}_2$ denotes the canonical embedding. As $\mathbb{Z}_2$ is compact, $(\mathbb{R} \times \mathbb{Z}_2)/\Gamma$ is a compact Abelian group, which turns out to be isomorphic with Σ_2. To see this, observe that $(x, \alpha) \equiv (x', \alpha') \bmod \Gamma$ if and only if $x' = x + n$ and $\alpha' = \alpha + \iota(n)$ for some $n \in \mathbb{Z}$. A short calculation reveals that we can choose $\mathbb{T} \times \mathbb{Z}_2$, with suitable addition, as a set of representatives for $(\mathbb{R} \times \mathbb{Z}_2)/\Gamma$. An element $(x, \alpha) \in \mathbb{T} \times \mathbb{Z}_2$ is mapped to an element of $z = (z_0, z_1, z_2, \ldots) \in \Sigma_2$ via

$$z_n = \frac{1}{2^n}\Big(x + \sum_{i=0}^{n-1} \alpha_i 2^i\Big).$$

This mapping is a continuous homomorphism and invertible.

The corresponding construction works for $\mathbb{Z}_p$ with p prime. This gives $\Sigma_p \simeq (\mathbb{R} \times \mathbb{Z}_p)/\Gamma$, with $\Gamma \simeq \mathbb{Z}$ as above. The solenoid can also be defined for an arbitrary integer [vDan30]. Viewing it as an inverse limit of topological spaces, it can be generalised considerably by using different mappings in each step; see [Wil74] for details in the context of dynamical systems.   ◊

2.3.4. LCAGs AND AVERAGING SEQUENCES

An LCAG G is called σ-*compact* if it is the union of countably many compact subsets. An equivalent characterisation [Bou66, Sec. I.9.9] is the existence of a nested sequence $\mathcal{A} = (A_i)_{i \in \mathbb{N}}$ of relatively compact, open subsets A_i of G such that $\overline{A_i} \subset A_{i+1}$ for all $i \in \mathbb{N}$ and $\bigcup_{i \in \mathbb{N}} A_i = G$. In view of Chapter 9, we call such a G-exhausting sequence $\mathcal{A}$ an *averaging sequence*.

Example 2.12 (*Averaging sequences*). The LCAG $\mathbb{R}^d$ is both σ-compact and compactly generated. A natural choice of an averaging sequence $\mathcal{A}$ is a sequence of centred open balls of increasing radii, such as $A_i = B_{r_i}(0)$ with $r_{i+1} > r_i$ and $r_i \xrightarrow{i \to \infty} \infty$.

Also $\mathbb{Z}^n$ is σ-compact, with one averaging sequence emerging from the intersection of $\mathbb{Z}^n$ with the balls of increasing radii (with obvious meaning in this case).

Any compact (and hence every finite) Abelian group G is σ-compact because G is both open and closed, so that one can simply take the constant sequence with $A_i = G$ in this case.   ◊

Any LCAG G carries an essentially unique G-invariant positive measure μ_G, called the *Haar measure* of G (see our later Sections 4.3 and 8.5 for more on measures). Multiples of invariant measures are invariant again, but this is the only remaining degree of freedom, so uniqueness can be achieved by a suitable normalisation. The standard convention for finite and compact groups is $\mu_G(G) = 1$, thus turning μ_G into a probability measure. For $\mathbb{Z}^n$, one usually takes the counting measure, which gives weight 1 to each single point. In $\mathbb{R}^d$, one uses Lebesgue measure, which is the unique translation invariant measure that gives volume 1 to the unit cube $[0,1]^d$, and analogous choices are made for other groups.

Given a σ-compact LCAG G with an averaging sequence $\mathcal{A}$ and a continuous, complex-valued, bounded function g on G, one can consider averages

$$M(g) := \lim_{i \to \infty} \frac{1}{\mu_G(A_i)} \int_{A_i} g \, \mathrm{d}\mu_G,$$

provided that the limit exists (if not, one restricts to suitable subsequences of $\mathcal{A}$). If G is compact, one simply has $M(g) = \mu_G(g)$, in linear functional notation, due to the above-mentioned conventions for the Haar measure. The existence of such averages will be an important ingredient to the general diffraction theory of Chapter 9.

When C and K are compact subsets of G, we consider

$$(2.11) \qquad \partial^K C := \left((C+K) \setminus C^\circ \right) \cup \left((\overline{G \setminus C} - K) \cap C \right),$$

where C° denotes the *interior* of C. This defines some form of a 'K-boundary' of C. Note, however, that this is generally not a boundary in a topological sense (the latter would always be empty when G is discrete).

Definition 2.9. A sequence $(D_m)_{m \in \mathbb{N}}$ of compact subsets of G is called a generalised *van Hove sequence* if, for every compact $K \subset G$, one has

$$\lim_{m \to \infty} \frac{\mu_G(\partial^K D_m)}{\mu_G(D_m)} = 0.$$

This definition emerges from Euclidean space where it means that the 'surface to bulk ratio' becomes sufficiently negligible in the limit $m \to \infty$. The existence of van Hove sequences in compactly generated LCAGs is easily seen via Theorem 2.1. One obvious choice for the group $\mathbb{R}^d \times \mathbb{Z}^n \times \mathbb{K}$ is

$$D_m = \{ (x,y,z) \in \mathbb{R}^d \times \mathbb{Z}^n \times \mathbb{K} \mid \max(|x|,|y|) \leqslant m \}$$

with $m \in \mathbb{N}$. For the more general case of σ-compact LCAGs, we refer to [Schl00, p. 145]. The importance of van Hove sequences emerges in various translational averages and their calculation via Birkhoff's ergodic theorem.

Remark 2.5 (*Følner sequences*). Let G be a general LCAG. A sequence $(F_i)_{i \in \mathbb{N}}$ of subsets of G that exhaust G is called a *Følner sequence* when $\mu_G\left((K+F_i) \triangle F_i \right) / \mu_G(F_i) \xrightarrow{i \to \infty} 0$ holds for any compact $K \subset G$. Here,

$$A \triangle B = (A \setminus B) \cup (B \setminus A)$$

denotes the *symmetric difference* of two sets.

Every van Hove sequence is Følner [BL04, Sec. 5], but the converse need not be true [Tem92, Appendix, Thm. 3.K]. The existence of a Følner sequence is equivalent to G being *amenable*, while van Hove sequences are more natural (and useful) for metric and statistical properties of (subsets of) G. $\Diamond$

Later, we also need the *Pontryagin dual* $\widehat{G}$ of an LCAG G, where $\widehat{\widehat{G}} = G$ via canonical isomorphism. Moreover, one has $\widehat{\mathbb{R}^d} = \mathbb{R}^d$, which is the most relevant example for us. Another important property is that the dual of a discrete group is compact and vice versa, for instance $\widehat{\mathbb{Z}} = \mathbb{S}^1$ and $\widehat{\mathbb{S}^1} = \mathbb{Z}$; we refer to [Kat04, Loo11, HR97] for further details.

2.4. Perron–Frobenius theory

Positive operators on vector spaces with an order relation have many powerful properties. In our context, they usually show up in the form of positive, finite-dimensional matrices. Since the applications are manifold and rather useful, we give a brief summary here (without proofs), also formulated for matrices. We refer to [Gan59, Sen06, Kit98] as general references, and to [SW99] for the more general theory of positive operators.

A matrix $M \in \mathrm{Mat}(d, \mathbb{R})$ is called *non-negative* when all its entries are non-negative numbers. A non-negative matrix is called *positive* when at least one entry is > 0, and it is called *strictly positive* when all its entries are > 0. A common notation (which we adopt here) for the three cases is $M \geqslant 0$, $M > 0$ and $M \gg 0$, respectively.

Definition 2.10. A non-negative matrix $M = (M_{ij})_{1 \leqslant i,j \leqslant d}$ is called *irreducible* if, for each index pair (i, j), there is an integer $k \in \mathbb{N}$ with $(M^k)_{ij} > 0$.

To explain the meaning of irreducibility, we employ the directed pseudograph $\mathcal{G} = \mathcal{G}_M$ associated to a non-negative matrix $M \in \mathrm{Mat}(d, \mathbb{R})$. Here, $\mathcal{G}$ consists of d vertices (one for each dimension of the matrix, labelled from 1 to d) and directed edges from vertex i to vertex j for all index pairs (i, j) with $M_{ij} > 0$. We call it a pseudograph to indicate that it may contain loops. The following result is standard.

Lemma 2.4. *A non-negative matrix $M \in \mathrm{Mat}(d, \mathbb{R})$ is irreducible if and only if its associated pseudograph $\mathcal{G}_M$ is strongly connected in the sense that, for each index pair (i, j) with $1 \leqslant i, j \leqslant d$, it contains a directed path from vertex i to vertex j.* $\square$

The associated graph $\mathcal{G}_M$ of an irreducible non-negative matrix M is useful for the determination of further important properties of M. In particular, it makes various algebraic properties of M immediately 'visible'.

Definition 2.11. For an irreducible non-negative matrix M with associated graph $\mathcal{G}_M$, let h denote the greatest common divisor (gcd) of the lengths of all (minimal) directed cycles in $\mathcal{G}_M$. The matrix M is then called *cyclic* with period h. When $h = 1$, the matrix M is called *cyclically primitive*[1].

If $M \geqslant 0$ is irreducible, one can show that $(M^q)_{i,j} > 0$ for some $q \leqslant m$ (when $i = j$) or $q \leqslant m - 1$ (when $i \neq j$), with m the degree of the minimal polynomial of M; see [Gan59, Ch. 13.2] for details. Testing a given matrix for irreducibility is thus a task of bounded effort. Also, the integer h from Definition 2.11 can be calculated directly from the entries of the matrix M.

The concept of irreducibility is useful, but sometimes too general. In line with the general practice, we thus define the following important subclass.

Definition 2.12. A non-negative matrix $M \in \mathrm{Mat}(d, \mathbb{R})$ is called *primitive* if there exists an integer $k \in \mathbb{N}$ such that $M^k \gg 0$.

Example 2.13 (*Irreducibility versus primitivity*). The matrix $M = \left(\begin{smallmatrix} 1 & 1 \\ 1 & 0 \end{smallmatrix}\right)$ is primitive, as $M^2 = M + \mathbb{1} \gg 0$, where $\mathbb{1}$ is the identity. In contrast, the matrix $M' = \left(\begin{smallmatrix} 0 & 1 \\ 1 & 0 \end{smallmatrix}\right)$ is merely irreducible, because all even powers of M' are $\mathbb{1}$, while all odd powers coincide with M' itself. The graph of M' is ①⇄②, which immediately reveals that it is cyclic with period 2. ◇

Primitivity, irreducibility and cyclic primitivity of non-negative matrices are related as follows; see [Sen06, Thm. 1.4] for a proof.

Lemma 2.5. *A non-negative matrix is primitive if and only if it is irreducible and cyclically primitive.* □

Remark 2.6 (*Wielandt's bound*). It is not immediately clear how difficult it is to test a non-negative matrix for primitivity. Fortunately, there is an upper bound for the matrix power to be inspected, which is due to Wielandt; see [Wie50] or [Pra94, Thm. 37.4.2]. It states that a primitive matrix $M \in \mathrm{Mat}(d, \mathbb{R})$ satisfies $M^{d^2 - 2d + 2} \gg 0$. So, if this power fails to be strictly positive, the matrix is not primitive. ◇

Though these properties look somewhat innocent, they have powerful consequences for the eigenvalues and eigenvectors of such matrices, and on the leading eigenvalue in particular. Recall that the *spectral radius* of a matrix M is the non-negative number

$$r(M) = \sup\{|\lambda| \,|\, \lambda \text{ is an eigenvalue of } M\}.$$

It satisfies $r(M) = \lim_{n\to\infty} \|M^n\|^{1/n}$, where $\|.\|$ is any matrix norm; see [RS80, Thm. VI.6]. In general, the spectral radius need not be an eigenvalue itself, as in the case of skew-symmetric matrices.

---

[1]Quite frequently, this property is called 'aperiodic' in the literature. Since the latter term has a different meaning in our context, we prefer the term 'cyclically primitive'.

Theorem 2.2 (Perron–Frobenius). *Let $M \in \mathrm{Mat}(d, \mathbb{R})$ be an irreducible non-negative matrix. Its spectral radius is then > 0 and a simple eigenvalue of M, called the Perron–Frobenius eigenvalue λ_{PF} of M. The corresponding eigenvector v can be chosen to have all entries > 0, written as $v \gg 0$. The set of eigenvalues λ of M with $|\lambda| = \lambda_{\mathrm{PF}}$ has the property that the ratios $\lambda/\lambda_{\mathrm{PF}}$ form a cyclic subgroup of $\mathbb{S}^1$. The latter is isomorphic with C_h, with h the period of M.* □

When M is cyclic with period h, the entire set of eigenvalues of M is invariant under a rotation through $2\pi/h$ around the origin of the plane. Consequently, if $\lambda \neq 0$ is an eigenvalue of M, then so is $\lambda \mathrm{e}^{2\pi \mathrm{i}/h}$; see [Sen06, Sec. 1.4]. The cyclic group C_h thus acts faithfully on the spectrum of M, which is quite a remarkable property.

From now on, we use the shorthand PF in the Perron–Frobenius context. Let us also point out that the PF eigenvector v of an irreducible non-negative matrix is the only one that can be chosen to satisfy $v \gg 0$ (or, in fact, $v > 0$). This follows from the structure of the eigenvalues of M together with the fact that the matrix has non-negative entries only and thus maps the positive cone into itself. Note that several linearly independent and strictly positive eigenvectors might exist when M fails to be irreducible (just think of a direct sum of two primitive matrices with the same PF eigenvalue).

Remark 2.7 (*Left and right eigenvectors*). Quite frequently, the real potential of PF theory only unfolds when one uses the left and right PF eigenvectors in conjunction. Properties such as irreducibility, cyclic primitivity and primitivity are inherited from a matrix to its transpose, with the new graph $\mathcal{G}_{M^T}$ being obtained from $\mathcal{G}_M$ by reversing all directed edges. The above results thus transfer to the left eigenvectors accordingly. When dealing with both eigenvectors (u and v say) at the same time, one convenient normalisation is $\langle u|v \rangle = 1$, with the scalar product $\langle u|v \rangle$ as before. ◇

Theorem 2.3 (Perron). *If $M \in \mathrm{Mat}(d, \mathbb{R})$ is a primitive non-negative matrix, Theorem 2.2 applies with the following additions. All eigenvalues λ of M with $\lambda \neq \lambda_{\mathrm{PF}}$ satisfy $|\lambda| < \lambda_{\mathrm{PF}}$, and the cyclic subgroup C_h of $\mathbb{S}^1$ is trivial. Moreover, the sequence $(\lambda_{\mathrm{PF}}^{-n} M^n)_{n \in \mathbb{N}}$ converges, as $n \to \infty$, to the matrix $vu = (v_i u_j)_{1 \leqslant i,j \leqslant d}$ with the right and left PF eigenvectors and their normalisation defined as in Remark 2.7.* □

The geometric intuition behind Perron's theorem is as follows. The closed positive cone is mapped into itself under the action of a primitive non-negative matrix M, with 0 being the only common boundary point. The image is thus a contracted cone with the same apex. This repeats with every iteration of the matrix M, thus leading to a sequence of cones that approximate a unique invariant direction. The latter is nothing but the PF eigenvector.

2.5. Number-theoretic tools

Since we need various concepts from elementary algebra and number theory throughout the text, we now recall some basic notions and augment our exposition by suitable references. As standard sources for background material, we refer to [HW08, BS66, AW99, Lan02], while some more specific results below are drawn from [Was97].

A real or complex number a is called *algebraic* if it is a root of a polynomial $P(x)$ with integer coefficients. When $P(x)$ can be chosen to have leading coefficient 1 (meaning the coefficient of $P(x)$ with the highest power in x), a is an *algebraic integer*. If, in addition, the constant coefficient of $P(x)$ is ± 1, a is called a *unit*. Numbers like e and π are not algebraic (they are examples of transcendental numbers). All rational numbers $\frac{p}{q}$ are algebraic (with $P(x) = qx - p$), and the algebraic integers within $\mathbb{Q}$ are the ordinary (or rational) integers $\mathbb{Z}$, with ± 1 being the only units.

More interesting are *quadratic irrationalities*, which are roots of quadratic polynomials with integer coefficients that are *irreducible* over $\mathbb{Z}$ (and hence also over $\mathbb{Q}$; see [Lan02, Thm. IV.2.3]). Irreducibility of a polynomial $P(x)$ of degree n over a ring R means that $P(x)$ cannot be written as a product of two polynomials over R of degrees smaller than n. Examples of quadratic irrationalities include $\sqrt{2}$ and $\sqrt{3}$ (both being algebraic integers, but not units) or the golden mean $\tau = \frac{1}{2}(1+\sqrt{5})$ (which is a unit, with polynomial $P(x) = x^2 - x - 1$). The last example highlights the polynomial as the defining entity for an algebraic integer, rather than the explicit expression of the number itself within the field $\mathbb{Q}(\sqrt{5})$.

An algebraic number a is the root of many different polynomials. If such a polynomial is irreducible over $\mathbb{Z}$ and of degree n, its n (complex) roots are simple. The polynomial can be chosen so that its coefficients have no non-trivial common divisor, which makes it essentially unique (up to an overall sign). The roots of this 'minimal' polynomial are then called the *algebraic conjugates* of a, and a is said to have *degree* n. Their relations are investigated in Galois theory [AW99]; see [Stew89] for an introductory account from a historical perspective.

Recall that an *algebraic number field* K is a subfield of $\mathbb{C}$ that is of finite degree $n = [K : \mathbb{Q}]$ over $\mathbb{Q}$, the latter referring to the dimension of K as a $\mathbb{Q}$-vector space. Each algebraic number field is of the form $K = \mathbb{Q}(\theta)$ for some algebraic number θ. Here, $\mathbb{Q}(\theta)$ denotes the smallest field extension of $\mathbb{Q}$ that contains θ; see [ST79, Thm. 2.2] for details. Clearly, θ must then be of degree n. Moreover, one can always choose θ to be an algebraic integer. The algebraic conjugates of θ, denoted by $\{\theta_1, \ldots, \theta_n\}$ with $\theta_1 = \theta$ say, define fields $\mathbb{Q}(\theta_i)$ and field isomorphisms $\sigma_i : K \longrightarrow \mathbb{Q}(\theta_i)$, with σ_1 being the identity. In many of our later examples, these isomorphisms will actually be field automorphisms of K itself, though this is not true in general. Given any $a \in K$, its field polynomial is defined as

$$f_a(x) = \prod_{i=1}^{n} (x - \sigma_i(a)),$$

which happens to have all coefficients in $\mathbb{Q}$; see [ST79, Thm. 2.4]. Note that this polynomial need not be irreducible (it is $(x-1)^n$ for $a = 1$ for instance). Of particular interest are the coefficients

$$(2.12) \qquad \mathrm{tr}_{K/\mathbb{Q}}(a) := \sum_{i=1}^{n} \sigma_i(a) \quad \text{and} \quad \mathrm{nr}_{K/\mathbb{Q}}(a) := \prod_{i=1}^{n} \sigma_i(a),$$

which are referred to as the *trace* and the *norm* of the algebraic number a. Whenever it is clear from the context, we will drop the subscript $K/\mathbb{Q}$.

2.5.1. QUADRATIC FIELDS

A particularly simple but relevant situation is that of a *quadratic number field* K, which is a degree-2 extension generated by a quadratic irrationality; see [Zag84] for a systematic exposition. In our context, we need both real and imaginary quadratic fields.

Before continuing, let us recall that a ring R is called an *integral domain* (InD) if it is a commutative ring with 1 that contains no zero divisor. It is called a *unique factorisation domain* (UFD) when, in addition, every non-zero element has an essentially unique factorisation into a product of a unit and powers of non-associated primes of R. If an integral domain R has the property that all ideals are principal ideals, it is a *principal ideal domain* (PID). Since this means that there is only one class of ideals in R, one also calls this the *class number one* case. Finally, R is a *Euclidean domain* (ED) when a (generalised) Euclidean algorithm exists within R that is based on a certain function $v : R \setminus \{0\} \longrightarrow \mathbb{N}_0$ which satisfies $v(x) \leqslant v(xy)$ for all non-zero $x, y \in R$. In particular, given $a, b \in R$ with $a \neq 0$, there are $q, r \in R$ with $b = aq + r$ such that $r = 0$ or $v(r) < v(a)$. These properties are connected by the implications

$$(2.13) \qquad \mathrm{ED} \implies \mathrm{PID} \implies \mathrm{UFD} \implies \mathrm{InD},$$

where none of the arrows is reversible; see [AW99, Ch. 2] for details.

Example 2.14 (*Real quadratic fields*)**.** Consider the quadratic field $\mathbb{Q}(\sqrt{2})$, which is the smallest field extension of $\mathbb{Q}$ that contains $\sqrt{2}$. It is of degree 2 over $\mathbb{Q}$, and can be represented as $\mathbb{Q}(\sqrt{2}) = \{\alpha + \beta\sqrt{2} \mid \alpha, \beta \in \mathbb{Q}\}$.

All elements of $\mathbb{Q}(\sqrt{2})$ are algebraic numbers. The unique non-trivial field automorphism is defined by $\sqrt{2} \mapsto -\sqrt{2}$, which leads to

$$\mathrm{tr}(\alpha + \beta\sqrt{2}) = 2\alpha \quad \text{and} \quad \mathrm{nr}(\alpha + \beta\sqrt{2}) = \alpha^2 - 2\beta^2.$$

The algebraic integers within $\mathbb{Q}(\sqrt{2})$ form the ring

$$\mathbb{Z}[\sqrt{2}] = \{m + n\sqrt{2} \mid m, n \in \mathbb{Z}\},$$

which is an easy exercise in polynomial arithmetic. $\mathbb{Z}[\sqrt{2}]$ is a PID [HW08], which by Eq. (2.13) also means that one has unique prime factorisation up to units in $\mathbb{Z}[\sqrt{2}]$. The units form the multiplicative group

$$\mathbb{Z}[\sqrt{2}]^{\times} = \{\pm(1+\sqrt{2})^m \mid m \in \mathbb{Z}\} \simeq C_2 \times C_{\infty},$$

where C_{∞} denotes the infinite cyclic group (which is isomorphic with $\mathbb{Z}$ as a group). It is generated by the *fundamental unit* of $\mathbb{Q}(\sqrt{2})$, which is $1 + \sqrt{2}$. Let us also mention that, using the norm, the unit group can be characterised as $\mathbb{Z}[\sqrt{2}]^{\times} = \{a \in \mathbb{Z}[\sqrt{2}] \mid \mathrm{nr}(a) = \pm 1\}$. The field $\mathbb{Q}(\sqrt{2})$ will frequently occur due to its connection with eightfold symmetry.

The analogous exercise (relevant for fivefold symmetry) for the quadratic field $\mathbb{Q}(\sqrt{5})$, where the non-trivial field automorphism (denoted by $'$) is given by $\sqrt{5} \mapsto -\sqrt{5}$, leads to the ring of integers

$$\mathbb{Z}[\tau] = \{m + n\tau \mid m, n \in \mathbb{Z}\},$$

which is again a PID. Here, with $\tau = \frac{1}{2}(1 + \sqrt{5})$ and $\tau' = 1 - \tau$, one has

$$\mathrm{tr}(m + n\tau) = 2m + n \quad \text{and} \quad \mathrm{nr}(m + n\tau) = m^2 + mn - n^2,$$

and analogously for elements of $\mathbb{Q}(\tau) = \mathbb{Q}(\sqrt{5})$. Since τ is the fundamental unit here, the unit group of $\mathbb{Z}[\tau]$ reads

$$\mathbb{Z}[\tau]^{\times} = \{a \in \mathbb{Z}[\tau] \mid \mathrm{nr}(a) = \pm 1\} = \{\pm\tau^m \mid m \in \mathbb{Z}\} \simeq C_2 \times C_{\infty};$$

see [HW08] for details, and [Dod83] for an extensive and explicit exposition of this important example.

Finally, we also need the quadratic field $\mathbb{Q}(\sqrt{3})$, where the field automorphism is defined by $\sqrt{3} \mapsto -\sqrt{3}$, so that

$$\mathrm{tr}(\alpha + \beta\sqrt{3}) = 2\alpha \quad \text{and} \quad \mathrm{nr}(\alpha + \beta\sqrt{3}) = \alpha^2 - 3\beta^2.$$

The ring of integers is the PID $\mathbb{Z}[\sqrt{3}]$, this time with unit group

$$\mathbb{Z}[\sqrt{3}]^{\times} = \{a \in \mathbb{Z}[\sqrt{3}] \mid \mathrm{nr}(a) = 1\} = \{\pm(2+\sqrt{3})^m \mid m \in \mathbb{Z}\} \simeq C_2 \times C_{\infty},$$

where $2 + \sqrt{3}$ is the fundamental unit of $\mathbb{Q}(\sqrt{3})$, with $\mathrm{nr}(2 + \sqrt{3}) = 1$. We shall meet this example in the context of twelvefold symmetry. $\diamond$

Example 2.15 (*Imaginary quadratic fields*). An interesting and widely used quadratic number field emerges from $\mathbb{Q}$ by adjoining the imaginary unit i. This gives $\mathbb{Q}(i) = \{\alpha + \beta i \mid \alpha, \beta \in \mathbb{Q}\}$, with complex conjugation as its non-trivial field automorphism, so that

$$\mathrm{tr}(x) = x + \bar{x} = 2\,\mathrm{Re}(x) \quad \text{and} \quad \mathrm{nr}(x) = x\bar{x} = |x|^2$$

for $x \in \mathbb{Q}(i)$. The ring of integers in $\mathbb{Q}(i)$ is $\mathbb{Z}[i]$, which is a PID known as the ring of *Gaussian integers* [HW08]. The unit group is

$$\mathbb{Z}[i]^\times = \mathbb{Z}[i] \cap \mathbb{S}^1 = \{1, i, -1, -i\} \simeq C_4.$$

In the Euclidean plane, $\mathbb{Z}[i]$ is nothing but the square lattice $\mathbb{Z}^2$, see also Figure 3.1 below, and $\mathbb{Z}[i]^\times$ is its rotation symmetry group.

An analogous situation is met for the imaginary field $\mathbb{Q}(\rho)$ with the number $\rho = \frac{1}{2}(-1 + i\sqrt{3})$, which is a primitive 3rd root of unity. The non-trivial field automorphism is once again given by complex conjugation. Its ring of integers is $\mathbb{Z}[\rho]$, known as the ring of *Eisenstein integers* and again a PID. In the plane, it coincides with the triangular (or hexagonal) lattice of edge length 1, and the unit group is

$$\mathbb{Z}[\rho]^\times = \mathbb{Z}[\rho] \cap \mathbb{S}^1 = \{(-\rho)^m \mid 0 \leqslant m \leqslant 5\} \simeq C_6,$$

again with the geometric interpretation as a rotation symmetry group. $\Diamond$

Remark 2.8 (*Euclidean domains*). All quadratic fields which were explicitly discussed above are actually examples of *Euclidean domains*; see [HW08, Thms. 246 and 247]. Here, the property refers to the existence of a Euclidean algorithm, where the field norm provides the function v from the definition. The PID and UFD properties then follow from Eq. (2.13). $\Diamond$

2.5.2. Cyclotomic fields

The two fields of Example 2.15 are quadratic fields, but they are also the simplest non-trivial examples of cyclotomic fields. The latter play an important role in the theory of aperiodic order, both implicitly and explicitly.

Let $\xi_n \in \mathbb{C}$ be a primitive nth root of unity (which means $\xi_n^m = 1$ precisely when n divides m, written as $n|m$ from now on). The field $\mathbb{Q}(\xi_n)$ is a field extension of $\mathbb{Q}$ of degree $\phi(n)$, compare [Was97, Thm. 2.5], where ϕ is Euler's totient function. It is defined by

$$\phi(n) := \mathrm{card}\{1 \leqslant k \leqslant n \mid \gcd(k, n) = 1\}.$$

For $n \in \{1, 2\}$, one has $\mathbb{Q}(\xi_n) = \mathbb{Q}$, while $\mathbb{Q}(\xi_n)$ is a true extension and complex for all $n > 2$. In the latter case, $\mathbb{Q}(\xi_n)$ is referred to as a *cyclotomic field*; see [Was97] for background material and a comprehensive treatment.

The ring of integers in $\mathbb{Q}(\xi_n)$ is $\mathbb{Z}[\xi_n]$, see [Was97, Thm. 2.6], which is a $\mathbb{Z}$-module of rank $\phi(n)$. It is a PID for several important values of n (including

TABLE 2.1. Basic data for the three cyclotomic fields $\mathbb{Q}(\xi_n)$ with $\phi(n) = 4$.

cyclotomic field	$\mathbb{Q}(\xi_5) = \mathbb{Q}(\xi_{10})$	$\mathbb{Q}(\xi_8)$	$\mathbb{Q}(\xi_{12})$
defining polynomial	$x^4 + x^3 + x^2 + x + 1$	$x^4 + 1$	$x^4 - x^2 + 1$
ring of integers	$\mathbb{Z}[\xi_5]$	$\mathbb{Z}[\xi_8]$	$\mathbb{Z}[\xi_{12}]$
max. real subfield	$\mathbb{Q}(\sqrt{5})$	$\mathbb{Q}(\sqrt{2})$	$\mathbb{Q}(\sqrt{3})$
fundamental unit	$\tau = \tfrac{1}{2}(1 + \sqrt{5})$	$1 + \sqrt{2}$	$2 + \sqrt{3}$

all $n < 23$), though this is not true in general; compare [Was97, Ch. 11] and Remark 3.7 below. In the latter case, unique factorisation also breaks down. Note that $\mathbb{Q}(\xi_n) = \mathbb{Q}(\xi_{2n})$ for all odd n. Therefore, one usually restricts to integers $n \not\equiv 2 \bmod 4$ to avoid duplications. This is the standard convention in the mathematics literature, as it is most convenient from the point of view of systematic formulas for units, class numbers and zeta functions. Note that a different convention is sometimes used in the physics literature [RMW87], when putting more emphasis on the rotational symmetry.

The maximal real subfield of $\mathbb{Q}(\xi_n)$ is $\mathbb{Q}(\xi_n + \bar{\xi}_n)$, with relative degree 2 (for $n > 2$). Its ring of integers is $\mathbb{Z}[\xi_n + \bar{\xi}_n]$, see [Was97, Prop. 2.16], with unit group $\mathbb{Z}[\xi_n + \bar{\xi}_n]^\times$. Let us adopt the convention $n \not\equiv 2 \bmod 4$, and define $U_n = \{\pm \xi_n^\ell \mid \ell \in \mathbb{Z}\}$. This satisfies $U_n \simeq C_{\mathrm{lcm}(2,n)}$, where lcm is the least common multiple. If $n = p^r$ is a prime power, the unit group of $\mathbb{Z}[\xi_n]$ is

$$\mathbb{Z}[\xi_n]^\times = U_n \times \mathbb{Z}[\xi_n + \bar{\xi}_n]^\times,$$

which follows from [Was97, Thm. 4.12 and Cor. 4.13]. Whenever n is divisible by two distinct primes, the unit group $\mathbb{Z}[\xi_n]^\times$ is slightly larger, with index $[\mathbb{Z}[\xi_n]^\times : (U_n \times \mathbb{Z}[\xi_n + \bar{\xi}_n]^\times)] = 2$. One choice for the additional generator is $z = 1 - \xi_n$ in this case; compare [Was97, Prop. 2.8]. This works because $z/\bar{z} = -\xi_n$, but $-\xi_n$ is not a square in U_n.

Example 2.16 (*Cyclotomic fields with $\phi(n) \leqslant 4$*). The two imaginary quadratic fields of Example 2.15 are the cyclotomic fields with $\phi(n) = 2$. The Euler function ϕ takes the value 4 precisely for $n \in \{5, 8, 10, 12\}$, which corresponds to three distinct cyclotomic fields. These will show up in many of our later examples. Their basic information is summarised in Table 2.1; consult Example 2.14 for their maximal real subfields. In these cases, the unit group within the ring of integers of the maximal real subfield is always of the form $\{\pm u^m \mid m \in \mathbb{Z}\} \simeq C_2 \times C_\infty$, where u is a fundamental unit. For $n = 5$ and $n = 8$, the additional complex units come from the corresponding groups U_n from above, giving C_{10} and C_8 respectively. For $n = 12$, one has an additional generating element as mentioned earlier. One convenient (albeit somewhat surprising) choice is $z = \sqrt{2 + \sqrt{3}}\, \xi_{24}$. The unit group is isomorphic with $C_{12} \times C_{\infty}$, where the infinite cyclic group is generated by z, and not by a real unit as in the other two examples. ◇

Remark 2.9 (*Cyclotomic polynomials*). The defining polynomials in Table 2.1 are examples of *cyclotomic polynomials*, which emerge as follows. The polynomial $x^n - 1$ (with $n \geqslant 1$) has a unique factorisation (in $\mathbb{Q}[x]$) into integer polynomials that are irreducible over $\mathbb{Q}$,

$$x^n - 1 = \prod_{\ell \mid n} Q_\ell(x),$$

where $Q_\ell(x) \in \mathbb{Z}[x]$ has degree $\phi(\ell)$ and is called the ℓth cyclotomic polynomial. The polynomials are recursively defined this way, via the Euclidean algorithm. Explicitly, they are given by

$$Q_\ell(x) = \prod_{\xi}(x - \xi) = \prod_{k \mid \ell}(x^k - 1)^{\mu(\ell/k)},$$

where ξ (in the first product) runs over the $\phi(\ell)$ distinct primitive ℓth roots of unity (which are the ξ_n^ℓ with $1 \leqslant \ell \leqslant n$ and $\gcd(\ell, n) = 1$), and μ denotes the Möbius function from elementary number theory. The first few of them read $Q_1(x) = x - 1$, $Q_2(x) = x + 1$, $Q_3(x) = x^2 + x + 1$, $Q_4(x) = x^2 + 1$, $Q_6 = x^2 - x + 1$, while the $Q_\ell(x)$ for $\ell \in \{5, 8, 12\}$ are listed in Table 2.1. Note that the coefficients are not restricted to the set $\{-1, 0, 1\}$ in general, with the first counterexample emerging for $\ell = 105$. ◇

In our context, algebraic numbers and their conjugates enter via their consequences for harmonic analysis; compare [Mey72, BDG+92]. Let us thus recall some important concepts.

Definition 2.13. An algebraic integer $\alpha > 1$ (which is then real) is called a *Pisot–Vijayaraghavan number*, or *PV number* for short, if all its algebraic conjugates (apart from α itself) lie inside the open unit disk. A PV number that is also a unit is referred to as a *PV unit*.

The three (real) fundamental units of Table 2.1 are PV units. The set of PV numbers (called class S in [Sal63]) is closed and bounded from below, but it is not a perfect set. Every real algebraic number field of degree n over $\mathbb{Q}$ contains a PV number of degree n [Sal63, Thm. 2, p. 3]. The smallest PV number [Sie44], also known as the plastic number [SAH06], is the maximal root of $x^3 - x - 1$ (with approximate value 1.32472 and a complex conjugate pair $-0.66236 \pm 0.56228\mathrm{i}$ of algebraic conjugates), while the smallest PV number that is itself a non-trivial limit of a sequence of PV numbers is the golden mean τ; see [BDG+92, Chs. 6 and 7]. Both PV numbers are units.

Definition 2.14. An algebraic integer $\alpha > 1$ is called a *Salem number*, if all its algebraic conjugates (apart from α itself) lie inside the closed unit disk, with at least one conjugate on the unit circle.

Salem's theorem [Sal63, Ch. III.3, p. 26] states that the degree of a Salem number is always even and $\geqslant 4$, that α and $1/\alpha$ are the only real conjugates, and that all other conjugates come in complex conjugate pairs and lie on the unit circle (none of them a root of unity). A Salem number is always a unit. Every PV number is the limit of an increasing and of a decreasing sequence of Salem numbers; see [Sal63, Thm. IV, p. 30] or [BDG+92, Thm. 6.4.1].

Moreover, an algebraic integer $\alpha > 1$ is called a *Perron number* when all algebraic conjugates $\alpha' \neq \alpha$ satisfy $|\alpha'| < \alpha$, and it is called a *Lind number* [Lag99] when $|\alpha'| \leqslant \alpha$ with at least one conjugate satisfying $|\alpha'| = \alpha$. All PV and Salem numbers are Perron numbers. They all surface in the theory of Delone sets via inflation symmetries (which will be discussed in more detail in Chapter 6). Let us summarise the connection with a result of Lagarias, which will be useful later to distinguish between different types of Delone sets, in particular between Meyer and non-Meyer sets.

Theorem 2.4 ([Lag99, Thm. 4.1]). *If $\Lambda \subset \mathbb{R}^d$ is a Delone set with $\alpha\Lambda \subset \Lambda$ for some $\alpha > 1$, the following properties hold.*

(1) *If Λ is finitely generated, α is an algebraic integer of a degree that divides $\mathrm{rank}\langle \Lambda - \Lambda \rangle_{\mathbb{Z}}$.*

(2) *If Λ is FLC, α is a Perron or a Lind number.*

(3) *If Λ is Meyer, α is a PV or a Salem number.* $\square$

The last result and similar results for tilings (compare [KS10] and references therein) provide ample evidence that concepts from algebraic number theory are both important and natural in this context.

2.5.3. OTHER ALGEBRAIC NUMBER FIELDS

Beyond the cases discussed so far, which are perhaps the most relevant ones for physical applications, one encounters mixtures of real and complex algebraic conjugates. Recall that a general algebraic number field (over $\mathbb{Q}$) is of the form $\mathbb{Q}(\theta)$, where θ is an algebraic integer of degree n. As such, θ is a root of a monic polynomial P with integer coefficients that is irreducible over $\mathbb{Q}$ and of degree n. As a consequence of irreducibility, the n roots of P over $\mathbb{C}$ (θ and its algebraic conjugates) are distinct. They comprise n_1 real roots and n_2 pairs of complex conjugate ones, so that $n = n_1 + 2n_2$.

In the study of substitution rules, one quickly encounters such fields, for instance cubic or quartic fields. In general, the algebraic conjugates of θ define different, but isomorphic fields. Let $\sigma_1, \ldots, \sigma_{n_1}$ denote the real Galois isomorphisms, with σ_1 the identity, and $\tau_1, \ldots, \tau_{n_2}$, together with their complex conjugate mappings, the remaining complex isomorphisms. The trace and norm of the field $\mathbb{Q}(\theta)$ are now defined as

(2.14)
$$\text{tr}_{\mathbb{Q}(\theta)/\mathbb{Q}}(x) = \sum_{i=1}^{n_1} \sigma_i(x) + 2 \sum_{j=1}^{n_2} \text{Re}(\tau_j(x)),$$
$$\text{nr}_{\mathbb{Q}(\theta)/\mathbb{Q}}(x) = \prod_{i=1}^{n_1} \sigma_i(x) \prod_{j=1}^{n_2} |\tau_j(x)|^2.$$

As before, we will simply write tr and nr whenever the field is clear from the context. When the field is cubic, one has either three real conjugates, or one real and one pair of complex conjugate ones.

Example 2.17 ($\mathbb{Q}(\beta)$ *with* $\beta^3 - \beta - 1 = 0$). Consider the cubic number field generated by the minimal PV number β (the 'plastic' number) mentioned above. It has a complex conjugate pair $\alpha, \overline{\alpha}$ of algebraic conjugates, so that $n_1 = n_2 = 1$ in this case. The field discriminant is -23, as follows from the defining cubic polynomial. The ring of integers is $\mathbb{Z}[\beta]$, which is generated by 1, β and β^2. One has $\text{nr}(\beta) = \text{nr}(\beta^2) = 1$ (they are both units) as well as $\text{tr}(\beta) = 0$ and $\text{tr}(\beta^2) = 2$ (note that β^2 is a root of $x^3 - 2x^2 + x - 1 = 0$). Moreover, one finds $\text{Re}(\alpha) = -\beta/2$ and $\text{Re}(\alpha^2) = 1 - \beta^2/2$, while one has $\text{Im}(\alpha) = (4 + 9\beta - 6\beta^2)/2\sqrt{23}$ and $\text{Im}(\alpha^2) = (6 + 2\beta - 9\beta^2)/2\sqrt{23}$. $\Diamond$

Another example of the same kind is given by the real root of the polynomial $x^3 - 2x^2 - 1$, which occurs in the context of the Kolakoski sequence; see [BS04a] and Example 4.8 below. Further cases are studied in many places; consult [PF02, Sec. 7.4] and references therein. Cubic polynomials with three real roots are also possible, but encountered less frequently in this context; see [LGJJ93, Sec. 5.2] for a fully worked example.

2.5.4. QUATERNIONS

In view of their applications in crystallography and physics, structures in 3- and 4-space are important in the theory of aperiodic order. Here, it is often advantageous to employ quaternions for their description, in particular for the parametrisation of rotations. The following material is slightly more advanced, and only needed occasionally. However, the approach via quaternions is very natural and powerful, particularly for icosahedral structures and their analogues in 4-space. Since (to our knowledge) there is no coherent account of this material in a single source, we present a short but systematic summary. For a more abstract and complete treatment, we refer to [Vig80, Rei03]. Specific references will be mentioned as we go along.

The starting point is the associative *quaternion algebra*

$$\mathbb{H} = \mathbb{H}(\mathbb{R}) = \mathbb{R} + \mathbb{R}\mathrm{i} + \mathbb{R}\mathrm{j} + \mathbb{R}\mathrm{k}$$

together with Hamilton's defining relations

$$\mathrm{i}^2 = \mathrm{j}^2 = \mathrm{k}^2 = \mathrm{i}\mathrm{j}\mathrm{k} = -1,$$

where it is understood that 1 is in the centre of $\mathbb{H}$. The centre is isomorphic with $\mathbb{R}$, and hence canonically identified with it. The relations define a non-commutative multiplication between quaternions that turns $\mathbb{R}^4$ into a skew field. The elements $1, \mathrm{i}, \mathrm{j}, \mathrm{k}$ correspond to the standard Euclidean basis vectors. A quaternion $q \in \mathbb{H}$ is thus also written as $q = (q_0, q_1, q_2, q_3)$, where $q_0 = \mathrm{Re}(q)$ is the *real part* of q, while $(q_1, q_2, q_3) = \mathrm{Im}(q)$ is its *imaginary part*. The latter is important due to the relation of quaternions to the geometry of 3-space. In this context, one also defines $\mathbb{H}_0 = \{q \in \mathbb{H} \mid q_0 = 0\}$, and we use A_0 analogously for subsets $A \subset \mathbb{H}$.

Real and imaginary parts are connected via the *conjugation* $q \mapsto \bar{q}$, which is the unique mapping that fixes the elements of the centre of $\mathbb{H}$ and reverses the sign on its complement. Explicitly, this reads as $\bar{q} = (q_0, -q_1, -q_2, -q_3)$, and one has $\overline{pq} = \bar{q}\bar{p}$. For later use, we also introduce the *reduced trace* $\mathrm{tr}(q)$ and the *reduced norm* $\mathrm{nr}(q)$ by

$$\mathrm{tr}(q) := q + \bar{q} \quad \text{and} \quad \mathrm{nr}(q) := q\bar{q} = |q|^2,$$

where $|q|$ is the Euclidean length of q.

In 4-space, when viewing the elements of $\mathbb{R}^4$ as quaternions, any rotation can be written as a mapping $x \mapsto px\bar{q}$ with $p, q \in \mathbb{S}^3$, where

$$\mathbb{S}^3 = \{q \in \mathbb{H} \mid |q|^2 = 1\}.$$

Since -1 is in the centre, the pairs (p, q) and $(-p, -q)$ define the same rotation. It is easy to verify that this is the only ambiguity for $p, q \in \mathbb{S}^3$. This observation establishes $\mathbb{S}^3 \times \mathbb{S}^3$ as the standard double cover of the rotation group $\mathrm{SO}(4, \mathbb{R})$; see [KR91] for details. The elements of $\mathrm{O}(4, \mathbb{R}) \setminus \mathrm{SO}(4, \mathbb{R})$ are covered by mappings of the form $x \mapsto p\bar{x}\bar{q}$, again with $p, q \in \mathbb{S}^3$.

Moreover, when identifying 3-space with the subspace $\mathbb{H}_0$, all rotations (of 3-space) are of the form $x \mapsto qxq^{-1}$ with $0 \neq q \in \mathbb{H}$. Without loss of generality, one can choose $q \in \mathbb{S}^3$ as above, which makes the parametrisation unique up to multiplication of q with -1 and shows $\mathbb{S}^3$ to be the double cover of $\mathrm{SO}(3, \mathbb{R})$. Here, due to $x \in \mathbb{H}_0$, the reflections are parametrised as $x \mapsto q\bar{x}q^{-1} = -qxq^{-1}$ with $q \in \mathbb{S}^3$.

Quaternion algebras can also be defined over other fields than $\mathbb{R}$, in particular (without modifications to the defining relations) over any real algebraic number field; see [O'Mea73] for a more general approach. The simplest case is the field $\mathbb{Q}$ of rational numbers, where the quaternion algebra $\mathbb{H}(\mathbb{Q})$ is also known as the skew field of *Hurwitz quaternions* [Hur19]. Another field of interest in our context is the quadratic field $\mathbb{Q}(\sqrt{5})$ of Example 2.14, which leads to the quaternion algebra $\mathbb{H}(\mathbb{Q}(\sqrt{5}))$; see [Vig80] for background.

Example 2.18 (*Hurwitz ring*). A quaternion $q \in \mathbb{H}(\mathbb{Q})$ is called *integral* when both $\text{tr}(q)$ and $\text{nr}(q)$ are integers. An interesting subset is the ring

$$\mathbb{J} = \left\{ (q_0, q_1, q_2, q_3) \in \mathbb{H}(\mathbb{Q}) \mid \text{all } q_i \in \mathbb{Z} \text{ or all } q_i \in \frac{2\mathbb{Z}+1}{2} \right\},$$

known as the Hurwitz ring of *integer* quaternions. Note that, for instance, $(0, 0, \frac{3}{5}, \frac{4}{5})$ is an integral but not an integer quaternion in this setting. $\mathbb{J}$ is a *maximal order* in $\mathbb{H}(\mathbb{Q})$, meaning that $\mathbb{J}$ is a $\mathbb{Z}$-module with three properties: (1) the $\mathbb{Q}$-span of $\mathbb{J}$ is $\mathbb{H}(\mathbb{Q})$, (2) $\mathbb{J}$ is a ring which contains 1, and (3) $\mathbb{J}$ is not contained in a larger object of this kind. As such, it is a PID, which leads to the non-commutative analogue of unique factorisation [Hur19]. At the same time, $\mathbb{J}$ is also a lattice in $\mathbb{R}^4$. It is the weight lattice D_4^*, to be discussed in Example 3.2 below. The units within $\mathbb{J}$ form a finite group $\mathbb{J}^\times$ of order 24 which comprises the elements

$$(\pm 1, 0, 0, 0), \ (0, \pm 1, 0, 0), \ (0, 0, \pm 1, 0), \ (0, 0, 0, \pm 1) \ \text{and} \ \tfrac{1}{2}(\pm 1, \pm 1, \pm 1, \pm 1).$$

$\mathbb{J}^\times$ is a double cover of the tetrahedral group T, known as the binary tetrahedral group (which is not isomorphic with the tetrahedral group T_d). The 24 quaternions also form a crystallographic root system of type D_4, denoted by Δ_{D_4}; see Section 3.3 for more. The set $\{x \mapsto px\bar{q}, \ x \mapsto p\bar{x}\bar{q} \mid p, q \in \Delta_{D_4}\}$ comprises $24^2 = 576$ distinct orthogonal transformations (due to the double covering property), which thus constitute one half of the point group of D_4^*. The latter group is isomorphic with $\text{Aut}(\Delta_{D_4})$, which is of order 1152. Defining $r = (1 + \text{i})/\sqrt{2}$, the missing half is (doubly) covered by $\{x \mapsto prx\bar{r}\bar{q}, \ x \mapsto pr\bar{x}\bar{r}\bar{q} \mid p, q \in \Delta_{D_4}\}$. Note that $r \in \mathbb{S}^3$ is not an element of $\mathbb{J}$. The symmetry property follows from the relation $(1 + \text{i})\mathbb{J} = \mathbb{J}(1 + \text{i})$. ◇

Example 2.19 (*Icosian ring*). In $\mathbb{H}(\mathbb{Q}(\sqrt{5}))$, quaternions are integral when $\text{tr}(q)$ and $\text{nr}(q)$ are elements of $\mathbb{Z}[\tau]$, the ring of integers in $\mathbb{Q}(\sqrt{5})$. Within the integral quaternions lies the ring

$$\mathbb{I} = \left\langle (1, 0, 0, 0), (0, 1, 0, 0), \tfrac{1}{2}(1, 1, 1, 1), \tfrac{1}{2}(\tau', \tau, 0, 1) \right\rangle_{\mathbb{Z}[\tau]},$$

which has rank 4 over $\mathbb{Z}[\tau]$, but is also a $\mathbb{Z}$-module of rank 8. It is called the *icosian ring*, and satisfies $\mathbb{I} = \langle \Delta_{H_4} \rangle_{\mathbb{Z}}$. It is a maximal order of class number 1 (hence a PID) in $\mathbb{H}(\mathbb{Q}(\sqrt{5}))$; see [Vig80]. Componentwise algebraic conjugation $'$ from $\mathbb{Q}(\sqrt{5})$ leads to $\mathbb{I}'$, which is another maximal order.

The unit group $\mathbb{I}^\times = \{q \in \mathbb{I} \mid q^{-1} \in \mathbb{I}\}$ is an infinite group. It can be characterised [MW94] as

$$\mathbb{I}^\times = \{\pm \tau^m \varepsilon \mid m \in \mathbb{Z} \text{ and } \varepsilon \in \Delta_{H_4}\} \simeq \mathbb{Z}[\tau]^\times \times \Delta_{H_4},$$

where $\Delta_{H_4} = \mathbb{I}^\times \cap \mathbb{S}^3$ is the non-crystallographic root system of type H_4 encountered above in Eq. (2.10). Viewed as a subgroup of $\mathbb{I}^\times$ of order 120, Δ_{H_4} is a double cover of the icosahedral group Y, and sometimes referred to as the binary icosahedral group (which is not isomorphic with Y_h). The orthogonal transformations $\{x \mapsto pxq,\ x \mapsto p\bar{x}q \mid p, q \in \Delta_{H_4}\}$ form a representation of the Coxeter group H_4 of order 120^2. This is also the point symmetry group of the icosian ring $\mathbb{I}$, seen as a (dense) point set in $\mathbb{R}^4$. $\diamondsuit$

Another interesting object is the *cubian ring*, which is a maximal order (with class number 1) in $\mathbb{H}(\mathbb{Q}(\sqrt{2}))$. It can be obtained as the $\mathbb{Z}[\sqrt{2}]$-span of $\mathbb{J} \cup r\mathbb{J}$ with the r of Example 2.18; see [Vig80, BM99] for details.

Example 2.20 (*Icosahedral modules in 3-space*). Icosahedrally symmetric point sets in $\mathbb{R}^3$ are important for aperiodic order [KN84, RMW87]. In view of the structure of root systems [CMP98], one obvious possibility is

$$\mathcal{M}_\mathsf{F} := 2\langle \Delta_{H_3} \rangle_\mathbb{Z},$$

which is both a $\mathbb{Z}$-module of rank 6 and a $\mathbb{Z}[\tau]$-module of rank 3. The prefactor 2 is a matter of convenience; some authors prefer to work without it [CMP98]. The module is related to the icosian ring $\mathbb{I}$ of Example 2.19 by $\mathcal{M}_\mathsf{F} = 2\,\mathrm{Im}(\mathbb{I}_0)$, with $\mathbb{I}_0 = \{q \in \mathbb{I} \mid q_0 = 0\}$, and satisfies

$$\mathcal{M}_\mathsf{F} = \big\langle (2,0,0), (-1,-\tau',\tau), (0,2,0) \big\rangle_{\mathbb{Z}[\tau]},$$

where the choice of basis reflects the Coxeter diagram H_3 in Figure 2.5 (relative to the quadratic form $\frac{1}{2}\langle x|y\rangle$, with $\langle x|y\rangle$ as in Section 2.3.2).

Another icosahedrally symmetric module is obtained as

$$\mathcal{M}_\mathsf{B} := 2\,\mathrm{Im}(\mathbb{I}) = \big\langle (2,0,0), (1,1,1), (\tau,0,1) \big\rangle_{\mathbb{Z}[\tau]},$$

which is again a $\mathbb{Z}[\tau]$-module of rank 3 (and a $\mathbb{Z}$-module of rank 6). In fact, it is *dual* to $\mathcal{M}_\mathsf{F}$ in the sense that

$$\mathcal{M}_\mathsf{B} = \big\{ x \in \mathbb{H}(\mathbb{Q}(\sqrt{5})) \mid \tfrac{1}{2}\langle x|y\rangle \in \mathbb{Z}[\tau] \text{ for all } y \in \mathcal{M}_\mathsf{F} \big\}.$$

Moreover, one has $\mathcal{M}_\mathsf{F} \subset \mathcal{M}_\mathsf{B}$ with index $[\mathcal{M}_\mathsf{B} : \mathcal{M}_\mathsf{F}] = 4$. With $u = (1,1,1)$, the coset structure can be made explicit as

$$(2.15) \qquad \mathcal{M}_\mathsf{B} = \mathcal{M}_\mathsf{F} \,\dot\cup\, (\mathcal{M}_\mathsf{F} + u) \,\dot\cup\, (\mathcal{M}_\mathsf{F} + \tau u) \,\dot\cup\, (\mathcal{M}_\mathsf{F} + \tau^2 u).$$

The union of $\mathcal{M}_\mathsf{F}$ with any one of the other three cosets forms a $\mathbb{Z}$-module of rank 6. Note that these are $\mathbb{Z}[2\tau]$-modules, but *not* $\mathbb{Z}[\tau]$-modules. This follows from $\tau^3 = 2\tau + 1$ together with the observation that the vectors u, τu and $\tau^2 u$ form a 3-cycle under multiplication by τ modulo $\mathcal{M}_\mathsf{F}$. Nevertheless, icosahedral symmetry is preserved in all three cases [RMW87].

If $\mathcal{M}_\mathsf{P}$ denotes any of these intermediate modules, one has

$$\mathcal{M}_\mathsf{F} \overset{2}{\subset} \mathcal{M}_\mathsf{P} \overset{2}{\subset} \mathcal{M}_\mathsf{B},$$

where the integers are the corresponding indices. Let us specify a coordinati-sation of the three modules (as in [Baa97, Sec. 5.2] or [Huc09, Sec. 3.1]) in terms of the standard basis $\{e_1, e_2, e_3\}$ of 3-space. This is

$$\mathcal{M}_\mathsf{B} = \left\{ \textstyle\sum_{i=1}^{3} \alpha_i e_i \mid \tau^2 \alpha_1 + \tau \alpha_2 + \alpha_3 \equiv 0 \bmod 2 \right\},$$

$$\mathcal{M}_\mathsf{P} = \left\{ x \in \mathcal{M}_\mathsf{B} \mid \alpha_1 + \alpha_2 + \alpha_3 \equiv 0 \text{ or } \tau^k \bmod 2 \right\},$$

$$\mathcal{M}_\mathsf{F} = \left\{ x \in \mathcal{M}_\mathsf{B} \mid \alpha_1 + \alpha_2 + \alpha_3 \equiv 0 \bmod 2 \right\},$$

where $k \in \{0, 1, 2\}$ according to the three possible choices for $\mathcal{M}_\mathsf{P}$, which are similar copies of each other. In fact, up to similarity, $\mathcal{M}_\mathsf{B}$, $\mathcal{M}_\mathsf{P}$ and $\mathcal{M}_\mathsf{F}$ are the only icosahedrally symmetric $\mathbb{Z}$-modules of rank 6 in 3-space; see [RMW87] for a proof.

For later use and practical applications, it is advantageous to specify a $\mathbb{Z}$-basis of these modules that is adapted to the geometric structure of the icosahedron. To be explicit, we follow [RMW87] and choose

$$v_1 = (\tau, 0, 1), \qquad v_2 = (\tau, 0, -1), \qquad v_3 = (1, \tau, 0),$$
$$v_4 = (-1, \tau, 0), \qquad v_5 = (0, 1, \tau), \qquad v_6 = (0, -1, \tau).$$

Their $\mathbb{Z}$-span gives $\mathcal{M}_\mathsf{P}$ as $\mathcal{M}_\mathsf{P} = \mathcal{M}_\mathsf{F} \,\dot\cup\, (\mathcal{M}_\mathsf{F} + \tau^2 u)$, which corresponds to the choice $k = 2$, while

$$\mathcal{M}_\mathsf{B} = \left\langle v_1, v_2, v_3, v_4, v_5, \tfrac{1}{2}(v_1 + v_2 + v_3 + v_4 + v_5 + v_6) \right\rangle_\mathbb{Z},$$

$$\mathcal{M}_\mathsf{F} = \left\langle v_1 + v_2, v_2 + v_3, v_3 + v_4, v_4 + v_5, v_5 + v_6, v_6 - v_1 \right\rangle_\mathbb{Z}.$$

In this setting, $\mathcal{M}_\mathsf{B} = \mathcal{M}_\mathsf{P} \,\dot\cup\, (\mathcal{M}_\mathsf{P} + u)$. The vector u used here and in Eq. (2.15) is $u = \tfrac{1}{2}(v_1 - v_2 + v_3 - v_4 + v_5 - v_6) = (1, 1, 1)$. The above vectors (rewritten in column form) will reappear later in Example 3.8 and in Eq. (A.2) of Appendix A.   ◇

To understand the connection of these modules with lattices in 6-space (which will also explain the notation), we now turn our attention to classical lattice theory.

CHAPTER 3

Lattices and Crystals

Before we begin to develop and discuss systems without periodicity, we recall some basic notions and properties of lattices and perfect crystals. This is important because they will still play an essential role in the description of almost periodic systems. Moreover, Meyer sets are a natural generalisation of lattice periodic point sets. In this sense, many of our later examples will be closer to lattices than to random structures. In this chapter, our main focus lies on the crystallographic restriction as well as on a systematic method to generate the lattices that are needed later.

3.1. Periodicity and lattices

Point sets $\Lambda \subset \mathbb{R}^d$ are important objects for most of our considerations. An element $t \in \mathbb{R}^d$ is a *period* of Λ when $t + \Lambda = \Lambda$. The set

$$\mathrm{per}(\Lambda) := \{t \in \mathbb{R}^d \mid t + \Lambda = \Lambda\}$$

is called the *set of periods* of Λ. It always contains the trivial period 0, and it is a subgroup of $\mathbb{R}^d$. The structure of the group $\mathrm{per}(\Lambda)$ determines the type of periodicity of the set Λ. The $\mathbb{R}$-span of $\mathrm{per}(\Lambda)$ is a subspace of $\mathbb{R}^d$, and may in general be of lower dimension, called the *rank* of $\mathrm{per}(\Lambda)$. However, of particular importance is the case that $\mathrm{per}(\Lambda)$ is a lattice in $\mathbb{R}^d$.

Recall from Definition 2.4 and the ensuing discussion that a *lattice* Γ in $\mathbb{R}^d$ (or, more generally, in an LCAG G) is a co-compact discrete subgroup of it. In $\mathbb{R}^d$, $\Gamma = \mathbb{Z}b_1 \oplus \cdots \oplus \mathbb{Z}b_d$ is a free $\mathbb{Z}$-module of rank d with a $\mathbb{Z}$-basis $\{b_1, \ldots, b_d\}$ that spans $\mathbb{R}^d$ over the field $\mathbb{R}$.

Definition 3.1. A (discrete) point set $\Lambda \subset \mathbb{R}^d$ is called *periodic* (of *rank m*) when $\mathrm{per}(\Lambda) \subset \mathbb{R}^d$ is non-trivial (with $1 \leqslant m = \dim\langle\mathrm{per}(\Lambda)\rangle_{\mathbb{R}} \leqslant d$), and *non-periodic* when $\mathrm{per}(\Lambda) = \{0\}$. The set Λ is called *crystallographic* when $\mathrm{per}(\Lambda)$ is a lattice in $\mathbb{R}^d$, and *non-crystallographic* otherwise.

In this setting, a non-periodic point set is non-crystallographic, while a non-crystallographic set may still have non-trivial periods (a situation that is sometimes called sub-periodic). Some authors also use the term *fully periodic* instead of crystallographic. Though the term *aperiodic* is a rather common synonym for non-periodic, we prefer to reserve it for a slightly stronger property (to be introduced later), so that we have the implications

$$(3.1) \qquad \text{aperiodic} \implies \text{non-periodic} \implies \text{non-crystallographic},$$

none of which being reversible in general.

The co-compactness of a lattice $\Gamma \subset \mathbb{R}^d$ means that the factor group $\mathbb{R}^d/\Gamma$ is compact. One can select a set of representatives in $\mathbb{R}^d$ that is relatively compact and measurable. Such a set is called a *fundamental domain* of Γ, abbreviated as FD_Γ. If $\{b_1, \ldots, b_d\}$ is a basis of Γ, a natural choice is

$$(3.2) \qquad \text{FD}_\Gamma = \{\textstyle\sum_{i=1}^{d} \alpha_i b_i \mid 0 \leqslant \alpha_i < 1 \text{ for all } i\}.$$

Its volume is given by $\text{vol}(\text{FD}_\Gamma) = |\det(b_1, \ldots, b_d)|$, which is also the volume of $\mathbb{R}^d/\Gamma$ in the measure induced on it by Lebesgue measure λ on $\mathbb{R}^d$. The choice of FD_Γ is clearly not unique, but all (measurable) fundamental domains of a given lattice have the same volume. In this sense, $\text{vol}(\text{FD}_\Gamma)$ is an invariant of Γ; compare [Cas71] for general background.

Recall that the Minkowski sum $A + B$ of two sets is a set (and not a multiset). For instance, one has $\{0, \frac{1}{2}, 1\} + \mathbb{Z} = \mathbb{Z} \, \dot\cup \, (\mathbb{Z} + \frac{1}{2}) = \frac{1}{2}\mathbb{Z}$, so that possible multiplicities are disregarded. When each element of $A + B$ has a unique representation as a sum, one usually speaks of a *direct* Minkowski sum, denoted as $A \oplus B$. Note that $S = A \oplus B$ implies $S = A + B$.

Proposition 3.1. *A locally finite point set $\Lambda \subset \mathbb{R}^d$ is crystallographic if and only if there is a lattice $\Gamma \subset \mathbb{R}^d$ and a finite point set $F \subset \mathbb{R}^d$ such that $\Lambda = F + \Gamma$. In fact, F can then be chosen so that $\Lambda = F \oplus \Gamma$.*

PROOF. If $\Gamma \subset \mathbb{R}^d$ is a lattice and F a finite set, $\Lambda := F + \Gamma$ is locally finite, with $\Gamma \subset \text{per}(\Lambda)$. Clearly, $\text{per}(\Lambda)$ is a subgroup of $\mathbb{R}^d$ that is discrete, so $\text{per}(\Lambda)/\Gamma$ must be a finite group. Consequently, $\mathbb{R}^d/\text{per}(\Lambda)$ is still compact, so that $\text{per}(\Lambda)$ is a lattice, whence Λ is crystallographic.

Conversely, let Λ be crystallographic with $\Gamma := \text{per}(\Lambda)$ as lattice of periods. Let C be a fundamental domain of Γ that is relatively compact (one can choose it as in Eq. (3.2) after specifying a basis, or it can be constructed from a closed Voronoi cell of Γ by removing part of its boundary; see also the proof of Proposition 9.5 below). Then, one has $\mathbb{R}^d = \bigcup_{t \in \Gamma}(t + C)$, and $F := C \cap \Lambda$ is a finite set that satisfies $\Lambda = F \oplus \Gamma$ by construction. $\qquad \square$

The locally finite, crystallographic point sets are precisely the crystallographic point packings of Definition 2.5. Proposition 3.1 shows that they are only a 'mild extension' of a lattice. A subset Γ' of a lattice Γ that is itself a lattice is called a *sublattice* of Γ. When viewed as Abelian groups, the group-subgroup index $[\Gamma : \Gamma']$ is referred to as the *index* of Γ' in Γ.

Lemma 3.1. *If Γ' is a sublattice of the lattice $\Gamma \subset \mathbb{R}^d$ of index n, the lattice $n\Gamma$ is a sublattice of Γ' of index n^{d-1}.*

PROOF. By assumption, Γ/Γ' is an Abelian group with n elements, each thus having an order that divides n. Now, Γ is the disjoint union of n cosets of the form $u_i + \Gamma'$, with $1 \leqslant i \leqslant n$, where we may choose $u_1 = 0$ without loss of generality. The previous observation means that $nu_i \in \Gamma'$, hence $n\Gamma \subset \Gamma'$. Since $n^d = [\Gamma : n\Gamma] = [\Gamma : \Gamma'][\Gamma' : n\Gamma]$, the resulting index is clear. □

The index is an important quantity, which is related to the volumes of the fundamental domains via

$$(3.3) \qquad [\Gamma : \Gamma'] = \frac{\text{vol}(\text{FD}_{\Gamma'})}{\text{vol}(\text{FD}_\Gamma)}.$$

It can be calculated from a relative basis by a determinant.

If Γ is a lattice in $\mathbb{R}^d$, an important related lattice is the so-called *dual lattice* Γ^*, which (with the scalar product $\langle x|y\rangle$ as before) is defined as

$$(3.4) \qquad \Gamma^* = \{y \in \mathbb{R}^d \mid \langle x|y\rangle \in \mathbb{Z} \text{ for all } x \in \Gamma\}.$$

Let $\{b_1, \ldots, b_d\}$ be a lattice basis of Γ, and define vectors b_i^* (with $1 \leqslant i \leqslant d$) by the d^2 conditions $\langle b_i^*|b_j\rangle = \delta_{i,j}$, where $\delta_{i,j}$ is Kronecker's function. It is a standard exercise to verify that $\{b_1^*, \ldots, b_d^*\}$ is uniquely specified this way, that it is a basis of $\mathbb{R}^d$, and a $\mathbb{Z}$-basis of Γ^*. It is called the *dual basis* and implies that Γ^* is indeed a lattice. Note that this form of duality is an involution, so that $(\Gamma^*)^* = \Gamma$ and $(b_i^*)^* = b_i$.

Example 3.1 (*Dual and self-dual lattices*). The integer lattice $\mathbb{Z}^d$ of Example 2.7 satisfies $(\mathbb{Z}^d)^* = \mathbb{Z}^d$ and is thus an example of a *self-dual lattice*. If one chooses the standard Euclidean basis $\{e_1, \ldots, e_d\}$ as a lattice basis, it is self-dual as well.

For the standard triangular lattice in $\mathbb{R}^2$ with basis vectors $b_1 = (1, 0)$ and $b_2 = \frac{1}{2}(-1, \sqrt{3})$, the dual basis vectors are $b_1^* = (1, \frac{1}{3}\sqrt{3})$ and $b_2^* = (0, \frac{2}{3}\sqrt{3})$. A common alternative choice replaces b_2 by $\frac{1}{2}(1, \sqrt{3})$, which implies to change b_1^* into $(1, -\frac{1}{3}\sqrt{3})$. The dual lattice is again a triangular lattice, which is obtained from the original one by rotation through $\pi/6$, followed by an expansion by $\frac{2}{3}\sqrt{3}$. The two lattices are thus similar.

In general, if B is the basis matrix of a lattice Γ that contains the basis vectors columnwise, the dual basis matrix (for Γ^*) is given by $(B^{-1})^T$. ◇

Given a lattice $\Gamma \subset \mathbb{R}^d$, with a basis $\{b_1, \ldots, b_d\}$ say, a useful quantity is the corresponding *Gram matrix* $G = (G_{ij})_{1 \leqslant i,j \leqslant d}$, which is defined via $G_{ij} = \langle b_i|b_j\rangle$ and thus encodes all scalar (or inner) products between the basis vectors. Collecting the coordinates of the basis vectors columnwise in a basis matrix B as in Example 3.1, one has the relation

$$(3.5) \qquad\qquad G = B^T B.$$

Though G depends on the choice of basis, it is still useful for identifying lattices that are related by similarity transformations. Also, one has

$$\det(G) = \big(\det(B)\big)^2 = \big(\text{vol}(\text{FD}_\Gamma)\big)^2 > 0,$$

which is the *discriminant* of Γ and independent of the choice of basis.

A locally finite point set $\Lambda \subset \mathbb{R}^d$ defines a measure via the corresponding Dirac comb $\delta_\Lambda = \sum_{x \in \Lambda} \delta_x$, where δ_x is the normalised point measure at x; compare Example 8.6 on page 317. Since measures provide a useful unifying frame for many aspects of this book, let us briefly jump ahead and extend the above concepts to general measures on $\mathbb{R}^d$, written as $\mathcal{M}(\mathbb{R}^d)$; details will be explained later in Section 8.5. If $\mu \in \mathcal{M}(\mathbb{R}^d)$ and $*$ denotes the convolution of measures, the set

$$\text{per}(\mu) := \{ t \in \mathbb{R}^d \mid \delta_t * \mu = \mu \}$$

is called the set of *periods* of μ. It is again a subgroup of $\mathbb{R}^d$. Note that the convolution of an arbitrary measure μ with δ_t is well-defined.

Definition 3.2. A measure $\mu \in \mathcal{M}(\mathbb{R}^d)$ is called *periodic* when $\text{per}(\mu) \subset \mathbb{R}^d$ is non-trivial, and *non-periodic* when $\text{per}(\Lambda) = \{0\}$. Moreover, the measure μ is called *crystallographic* when $\text{per}(\mu)$ contains a lattice in $\mathbb{R}^d$.

This definition matches the previous one for point sets Λ via

$$\delta_t * \delta_\Lambda = \delta_{t+\Lambda},$$

which is a simple consequence of the convolution identity $\delta_x * \delta_y = \delta_{x+y}$ for point measures. Note that there is no obvious analogue of Proposition 3.1. If $\Gamma \subset \mathbb{R}^d$ is a lattice and ϱ a finite measure, the convolution $\mu = \varrho * \delta_\Gamma$ is well-defined (see Theorem 8.5 below) and a periodic measure. Due to $t + \Gamma = \Gamma$ for all $t \in \Gamma$, $\text{per}(\mu)$ certainly contains Γ. However, $\text{per}(\mu)$ need not be a lattice, but can be a much larger group (even all of $\mathbb{R}^d$). An example is

$$(3.6) \qquad \mu = \varrho * \delta_{\mathbb{Z}^2} \quad \text{with} \quad \varrho = 1_{[0,1) \times [0, \frac{1}{2})} \lambda,$$

where λ is Lebesgue measure and 1_A denotes the characteristic function of the set A. The periods are $\text{per}(\mu) = \mathbb{R} \times \mathbb{Z}$. It is thus reasonable to distinguish measures according to their groups of periods.

Remark 3.1 (*Translation symmetries of measures*). The term 'non-crystallographic' is tailored to uniformly discrete point sets and related structures (such as tilings). In the setting of measures, it is more sensible to distinguish the possible situations according to the nature of the factor group $\mathbb{R}^d / \text{per}(\mu)$. The latter can be compact, for instance if $\mu = \delta_\Lambda$ is crystallographic. One can even have $\mathrm{per}(\mu) = \mathbb{R}^d$, where the factor group is trivial, or various intermediate cases with finite factor group or finite components in it. When the factor group is non-compact, there may still be discrete translation symmetries in a subspace of $\mathbb{R}^d$. ◇

An important aspect of aperiodic order emerges from the compatibility between translation and rotation symmetries, which we discuss next.

3.2. The crystallographic restriction

Let us start with a formulation of the crystallographic restriction (CR) for point sets that are themselves lattices. Recall the we use (S)O(d) for the (special) orthogonal group of $\mathbb{R}^d$; see [Schw80] for general background.

Lemma 3.2 (CR for lattices). *Consider a lattice $\Gamma \subset \mathbb{R}^d$. If $R \in \mathrm{O}(d)$ satisfies $R\Gamma \subset \Gamma$, one has $R\Gamma = \Gamma$. The corresponding characteristic polynomial $P(\lambda) = \det(R - \lambda\mathbb{1})$ has integer coefficients only, so that $P(\lambda) \in \mathbb{Z}[\lambda]$.*

PROOF. Define $S_r := S \cap \overline{B_r(0)}$ for any subset $S \subset \mathbb{R}^d$. Since $R \in \mathrm{O}(d)$ is a linear isometry, $R\Gamma \subset \Gamma$ implies $(R\Gamma)_r \subset \Gamma_r$, for any $r \geqslant 0$, with both sets having the same cardinality, hence $(R\Gamma)_r = \Gamma_r$. This gives the first claim.

Let $\{b_1, \ldots, b_d\}$ be a basis for Γ, so that $\Gamma = \mathbb{Z}b_1 \oplus \cdots \oplus \mathbb{Z}b_d$. When $R\Gamma = \Gamma$, each b_i must be mapped to a lattice point, hence to an integer linear combination of the basis vectors. Consequently, $Rb_i = \sum_j b_j a_{ji}$, with all $a_{ji} \in \mathbb{Z}$. In matrix form, this means $RB = BA$, with B the basis matrix of Γ. Since the latter is invertible ($\det(B) \neq 0$), one has $R = BAB^{-1}$, where A is an integer matrix, hence $A \in \mathrm{GL}(d, \mathbb{Z})$. Clearly, R and A share the same characteristic polynomial, which then has integer coefficients only. □

It is obvious from the last proof that the appearance of a group as a symmetry of a lattice is really a statement about the existence of an integer representation of that group (namely relative to a lattice basis).

Corollary 3.1. *A lattice $\Gamma \subset \mathbb{R}^d$ with $d = 2$ or $d = 3$ can have n-fold rotational symmetry at most for $n \in \{1, 2, 3, 4, 6\}$.*

PROOF. For $d = 2$, a rotation by an angle φ is represented by the matrix

$$R_\varphi = \begin{pmatrix} \cos(\varphi) & -\sin(\varphi) \\ \sin(\varphi) & \cos(\varphi) \end{pmatrix}$$

with characteristic polynomial

$$P_2(\lambda) = \det(R_\varphi - \lambda\mathbb{1}) = \lambda^2 - \mathrm{tr}(R_\varphi)\lambda + \det(R_\varphi) = \lambda^2 - 2\cos(\varphi)\lambda + 1.$$

By the crystallographic restriction (Lemma 3.2), $R_\varphi\Gamma = \Gamma$ implies that $2\cos(\varphi) \in \mathbb{Z}$, hence $|\cos(\varphi)| \in \{0, \frac{1}{2}, 1\}$. Working out the corresponding rotation angles gives the claim for $d = 2$.

For $d = 3$, notice that the eigenvalues of $R_\varphi \in SO(3)$ must be real or come in complex conjugate pairs. As $\det(R_\varphi) = 1$, this implies that 1 always is an eigenvalue of R_φ, with any corresponding eigenvector v being a rotation axis (this is also known as Euler's theorem). When choosing the eigenvector v as the z-direction, and a suitable basis of the orthogonal complement, R_φ has the matrix representation

$$R_\varphi = \begin{pmatrix} \cos(\varphi) & -\sin(\varphi) & 0 \\ \sin(\varphi) & \cos(\varphi) & 0 \\ 0 & 0 & 1 \end{pmatrix}$$

with characteristic polynomial $P_3(\lambda) = (1 - \lambda) \cdot P_2(\lambda)$, wherefore we get the same conclusion as above. $\qquad\square$

To continue, we need a notation for *Euclidean motions*. We write (a, R) with $a \in \mathbb{R}^d$ and $R \in O(d)$ for the affine mapping $x \mapsto (a, R)x$ defined by

$$(a, R)x = Rx + a.$$

It is easy to check that this results in the multiplication rule

$$(a, R)(b, S) = (a + Rb, RS).$$

This also shows that the Euclidean motions form a group. It is the semi-direct product $\mathbb{R}^d \rtimes O(d)$, with neutral element $(0, \mathbb{1})$ and $(-R^{-1}a, R^{-1})$ as the inverse of (a, R). At this point, we also lift the restriction to lattices and consider locally finite crystallographic point sets. The latter are the crystallographic point packings of Definition 2.5, by an application of Proposition 3.1. They are simply called *packings* in [CS99].

Proposition 3.2 (CR for crystallographic point packings). *Let $\Lambda \subset \mathbb{R}^d$ be a locally finite point set that is crystallographic, with $\Gamma = \mathrm{per}(\Lambda)$ as its lattice of periods. If (a, R) is a Euclidean motion with $(a, R)\Lambda = \Lambda$, the linear isometry R must fix Γ, so that $R\Gamma = \Gamma$. In particular, $\det(R - \lambda\mathbb{1}) \in \mathbb{Z}[\lambda]$.*

PROOF. By Proposition 3.1, we may write $\Lambda = F \oplus \Gamma$ with F a finite set such that

$$\Lambda = \overset{\cdot}{\bigcup_{v \in F}} (v + \Gamma).$$

As $\Lambda = (a, R)\Lambda = \bigcup_{v \in F} (Rv + R\Gamma) + a$, consider

$$\Lambda' = \Lambda - a = \overset{\cdot}{\bigcup_{v \in F}} (Rv + R\Gamma),$$

where $R\Gamma$ is again a lattice. Since $\mathrm{per}(\Lambda') = \mathrm{per}(\Lambda)$, we may conclude that

$$R\Gamma \subset \mathrm{per}(\Lambda') = \mathrm{per}(\Lambda) = \Gamma,$$

hence $R\Gamma = \Gamma$ by Lemma 3.2. The last claim is then clear. $\qquad\square$

Remark 3.2 (*Choice of origin*). If $x \in \Lambda$, with Λ and (a, R) as in Proposition 3.2, the affine map (a', R) fixes $\Lambda - x$, with $a' := Rx - x + a$, since

$$(a', R)(\Lambda - x) = R\Lambda - Rx + a' = R\Lambda + a - x = (a, R)\Lambda - x = \Lambda - x.$$

We may thus assume that $0 \in \Lambda$, without loss of generality. In this case, $(a, R)0 = a$ shows that $a \in \Lambda$. When using the representation $\Lambda = F \oplus \Gamma$ with $0 \in \Lambda$, one can also arrange for $0 \in F$ without loss of generality. ◇

Corollary 3.2. *If the Euclidean motion (a, R) with $R \in \mathrm{O}(d)$, $d \in \{2, 3\}$, maps a locally finite crystallographic point set Λ onto itself, R generates a finite cyclic group C_n of order n, for some $n \in \{1, 2, 3, 4, 6\}$.*

PROOF. Since Λ is locally finite, it can be decomposed as $\Lambda = F \oplus \Gamma$ with a finite set F and $\Gamma = \mathrm{per}(\Lambda)$ a lattice in $\mathbb{R}^d$, by an application of Proposition 3.1. Now, Proposition 3.2 implies $R\Gamma = \Gamma$, and our claim for $R \in \mathrm{SO}(d)$ follows from Corollary 3.1.

For $d = 2$, any isometry $R \in \mathrm{O}(2) \setminus \mathrm{SO}(2)$ is itself an involution, and hence of order 2. When $d = 3$, any $R \in \mathrm{O}(3) \setminus \mathrm{SO}(3)$ can be written as the product of a rotation with $-\mathbb{1}$. Since the latter is in the centre of $\mathrm{O}(3)$ and an involution, no new orders are added to the list. ☐

Let us now take a look at the realisation of finite order symmetries in lattices of minimal dimension, where Euler's totient function ϕ (which was introduced in Example 2.16) and an additive variant of it will become important. First, we determine the minimal dimension of a lattice that displays n-fold rotational symmetry, without any further restriction. This question was considered by [Her49] and [Schw80] for the irreducible case where n is a prime power; see also [Vai28]. The result reads as follows.

Lemma 3.3. *Let Γ be a lattice with a point symmetry group that contains an element of order p^r, with p a prime and $r \geqslant 1$. Then, the minimal dimension of Γ is $d = \phi(p^r) = p^{r-1} \cdot (p - 1)$.*

PROOF. Let R be a (linear) isometry of order p^r, with $R\Gamma = \Gamma$. By the crystallographic restriction, $P(x) := \det(R - x\mathbb{1}) \in \mathbb{Z}[x]$. By assumption on the order of R, we must have at least one primitive p^rth root of unity, ξ_{p^r} say, among the roots of $P(x)$.

Recall from Remark 2.9 that the cyclotomic polynomial $Q_{p^r}(x)$ is irreducible over $\mathbb{Q}$ and an element of $\mathbb{Z}[x]$. We thus have $Q_{p^r}(x)|P(x)$, because $\mathbb{Z}[x]$ is factorial and at least one of the roots of Q_{p^r} must also be a root of P.

Clearly, $d := \deg(Q_{p^r}) = \phi(p^r) = p^{r-1} \cdot (p - 1)$ is then a lower bound for the dimension of Γ. The Minkowski embedding (see Section 3.4 below) of $\mathbb{Z}[\xi_{p^r}]$ as a lattice in $\mathbb{R}^d$, with $d = \phi(p^r)$, provides an example where the unit ξ_{p^r} results in a lattice rotation of order p^r. ☐

TABLE 3.1. Minimal embedding dimensions d_n of n-fold lattice symmetry.

n	1	2	3	4	5	6	7	8	9	10	11	12	13	14	15	16	17	18	19	20
d_n	0	1	2	2	4	2	6	4	6	4	10	4	12	6	6	8	16	6	18	6

The composite case requires some care, and was solved by [Hil85]; compare [Vai28] for a related result. For a formulation, it is advantageous to define the *additive* counterpart of Euler's totient function, denoted by ϕ_{a}. The latter is defined by $\phi_{\mathsf{a}}(1) = 0$, by

$$(3.7) \qquad \phi_{\mathsf{a}}(p^r) \,=\, \phi(p^r)$$

on all prime powers with $r \geqslant 1$,

$$(3.8) \qquad \phi_{\mathsf{a}}(2n) \,=\, \phi_{\mathsf{a}}(n)$$

for all odd $n > 1$, and

$$(3.9) \qquad \phi_{\mathsf{a}}(mn) \,=\, \phi_{\mathsf{a}}(m) + \phi_{\mathsf{a}}(n)$$

for all remaining coprime integers m and n (thus excluding $m = 2$ and $n = 2$). The values $d_n = \phi_{\mathsf{a}}(n)$ with $1 \leqslant n \leqslant 20$ are shown in Table 3.1.

Theorem 3.1. *Let $\varGamma$ be a lattice with a point symmetry of order n. Then, the minimal dimension of $\varGamma$ is $d_n = \phi_{\mathsf{a}}(n)$.*

PROOF. The case $n = 1$ is trivial (with $d = 0$), while $n = p^r$ follows from Lemma 3.3. If $R\varGamma = \varGamma$ with a symmetry R of odd order n, the matrix $-R$ is a point symmetry of $\varGamma$ of order $2n$, as taken care of by Eq. (3.8). The remaining cases precisely correspond to (3.9), because the minimal characteristic polynomial with integer coefficients is obtained as a product of cyclotomic polynomials (whose degrees have to be added).

Once again, the minimal dimension may occur, as each irreducible factor leads to a lattice realisation of the correct dimension, via the Minkowski embedding mentioned in the proof of Lemma 3.3. □

In general, Theorem 3.1 does not give the correct answer for the appearance of n-fold rotational symmetry in a planar quasicrystal (or model set). In view of the projection method (as discussed in Chapter 7), we need to search for a minimal embedding of n-fold symmetry with the additional constraint that a two-dimensional subspace of $\mathbb{R}^d$ is invariant under the action of the lattice symmetry; compare [BJS90, App. A].

Theorem 3.2. *Consider a locally finite planar point set with n-fold symmetry that is constructed from a lattice in $\mathbb{R}^d$ by a symmetry preserving (partial) projection. Then, $d \geqslant \phi(n)$, with the lower bound being sharp.*

TABLE 3.2. Differences between embedding dimensions d_n and d_n^{QC} for $n \leqslant 50$.

n	1	15	20	21	24	28	30	33	35	36	39	40	42	44	45	48
d_n	0	6	6	8	6	8	6	12	10	8	14	8	8	12	10	10
d_n^{QC}	1	8	8	12	8	12	8	20	24	12	24	16	12	20	24	16

PROOF. This time, in order to get a faithful representation of the n-fold lattice symmetry in the plane, the characteristic polynomial $P(x)$ has to be divisible by the cyclotomic polynomial $Q_n(x)$, as we must have at least one primitive n-th root of unity (and hence all, due to the irreducibility of $Q_n(x)$); see our earlier argument. As $\deg(Q_n(x)) = \phi(n)$, the claim follows.

As mentioned before, the ring $\mathbb{Z}[\xi_n]$ lifts to a lattice in $\mathbb{R}^d$ (with $d = \phi(n)$) that shows a point symmetry of order n; see Section 3.4 below for the details. The latter emerges as the linear transformation induced by the multiplication map $x \mapsto \xi_n x$ in $\mathbb{Z}[\xi_n]$. This shows constructively that (and how) the minimal dimension in Theorem 3.2 can be realised. □

Apart from the trivial case $n = 1$ (which is more of a convention [Hil85]), the first difference occurs at $n = 15$; see Table 3.2 for all further examples with $n \leqslant 50$. The cyclic group C_{15} can act as a lattice symmetry in $\mathbb{R}^6$ (for instance on the root lattice $A_2 \times A_4$; see below). This makes use of the factorisation $15 = 3 \cdot 5$. However, no two-dimensional subspace shows a faithful real representation of order 15. To this end, we need to go to $\mathbb{R}^8$, where a realisation is provided by the standard Minkowski (or Galois) embedding of $\mathbb{Z}[\xi_{15}]$ as a lattice. For a general treatment of symmetries for crystals and tilings, we refer to [CBG08] and references therein.

An important application concerns the minimal embedding of patterns such as the vertex set of the Penrose tiling. Though often constructed from a lattice in $\mathbb{R}^5$, Theorem 3.2 suggests that a lattice in $\mathbb{R}^4$ should suffice. We shall see later in Chapter 7 that this is indeed the case; compare Example 7.11 and Remark 7.8. A systematic approach to minimal embeddings naturally leads to the class of root and weight lattices, which we will now discuss.

3.3. Root lattices

The most commonly used lattice in $\mathbb{R}^d$ is the integer lattice $\mathbb{Z}^d$ of Example 2.7, which is also called the primitive (hyper)cubic lattice. It is spanned by the d standard Euclidean basis vectors of unit length,

$$\mathbb{Z}^d = \mathbb{Z}e_1 \oplus \cdots \oplus \mathbb{Z}e_d.$$

In our context, where metric aspects are important, one has to consider other lattices as well, preferably ones with many symmetries. In the Euclidean

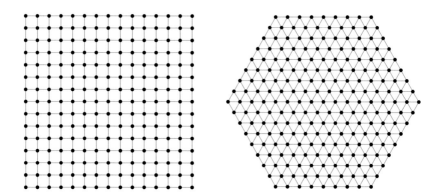

FIGURE 3.1. Finite patches of the square (left) and the triangular (right) lattice in the plane, shown with their Delone cell complexes. At the same time, when the edge length is 1, the points represent the rings of Gaussian (left) and Eisenstein (right) integers from Example 2.15.

plane, this is illustrated by the square versus the triangular (or hexagonal) lattice; see Figure 3.1. As Abelian groups alone, they are isomorphic (as are all planar lattices); as objects of (metric) geometry, they are not.

Viewed in an appropriate scale, the triangular lattice is an (indecomposable) *root lattice* [CS99], spanned by the vectors of the root system A_2. Root lattices are spanned by crystallographic root systems (as introduced in Section 2.3.2). They form a versatile class of lattices with nice properties, and feature in many branches of mathematics and physics; see [BJKS90, MP93, CMP98] for their appearance in the context of aperiodic order.

Definition 3.3. A lattice $\Gamma \subset \mathbb{R}^d$ is called a *root lattice* (in the strict sense) if it possesses a lattice basis $\{b_1, \ldots, b_d\}$ such that $\langle b_i | b_i \rangle = 2$ (for $1 \leqslant i \leqslant d$) and $\langle b_i | b_j \rangle \in \{0, -1\}$ (for all $i \neq j$).

More generally, any lattice that is similar to a root lattice in the strict sense is also called a root lattice. This is justified by the observation that one can then simply adjust the bilinear form for the scalar product to match the above requirements.

Based on Definition 3.3, one can use a graph-theoretic approach to come to a classification of root lattices as follows. A basis vector of squared length 2 is represented by a node (or vertex) of the graph. Two nodes are connected by a single line (or edge) if their scalar product is -1, while there is no edge if they are orthogonal to each other. All graphs that emerge this way are simple graphs, which means that they contain neither loops nor multiple edges. To avoid confusion with the use of the word 'simple' in the context of root

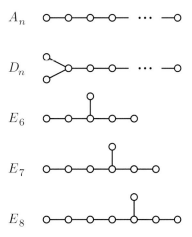

FIGURE 3.2. Diagrams of indecomposable root lattices, consisting of the two series A_n ($n \geqslant 1$) and D_n ($n \geqslant 4$), where n is the number of nodes, and the three exceptional lattices E_6, E_7 and E_8.

systems, we will use the term graph to mean simple graph for the remainder of this section. A graph is called *admissible* when it is a representing graph of a root lattice according to this rule. It is not difficult to see that different bases for the same root lattice (both in line with Definition 3.3) lead to isomorphic graphs, while non-isomorphic graphs correspond to different lattices. This gives a bijection between isomorphism classes of admissible graphs and finite-dimensional root lattices by standard arguments [Ebel02, CS99, Hum92].

A root lattice is called *indecomposable* when its representing graph is connected. This property is preserved under graph isomorphism. For a classification, it is thus sufficient to determine all admissible graphs that are connected (the remainder being an exercise in combinatorics). Important examples, which are actually exhaustive, are shown in Figure 3.2.

Theorem 3.3. *If Γ is an indecomposable root lattice, its representing graph is of type[1] A_n, $n \geqslant 1$, of type D_n, $n \geqslant 4$, or of type E_6, E_7 or E_8.*

SKETCH OF PROOF. The proof consists of four steps, which are spelt out in detail in [Ebel02, Sec. 1.4]. Let $\mathcal{G}$ be the graph of the indecomposable root lattice $\Gamma \subset \mathbb{R}^n$, and let $\{b_1, \ldots, b_n\}$ be the corresponding root lattice basis. Note that $\mathcal{G}$ is connected, and need only be considered up to graph isomorphisms.

---

[1]Note that we are using the symbol A_n in two different meanings, for root lattices and for alternating groups. Since the meaning will always be clear from the context, we prefer to follow this widely used notation. Later, we will distinguish a root lattice, D_n say, from its generating (finite) root system, by writing the latter as Δ_{D_n}.

Step 1: The graph $\mathcal{G}$ cannot contain any cycle. Indeed, the existence of a cycle would imply that of a minimal cycle, with nodes $x_1, \ldots, x_r$, where $2 \leqslant r \leqslant n$ and the x_i are distinct elements of the basis. Writing y^2 for the Euclidean scalar product $\langle y|y \rangle$, this would imply

$$0 < (x_1 + \cdots + x_r)^2 = 2r - 2r = 0,$$

which is impossible. Note that this argument also excludes the existence of a two-cycle; this is already clear from our definition of the graph $\mathcal{G}$.

Step 2: The graph $\mathcal{G}$ contains at most one vertex of degree (or valency) three, and none of higher degree. Otherwise, $\mathcal{G}$ would contain a subgraph of the form

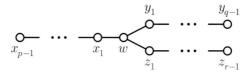

for some $1 \leqslant r \leqslant n - 4$. This leads to a contradiction via the calculation

$$0 < (2x_1 + \cdots + 2x_r + x_{r+1} + \cdots + x_{r+4})^2 = 8r + 8 - 8(r-1) - 16 = 0.$$

Step 3: Classification of remaining graphs. All that is left to consider are graphs of the form

with $1 \leqslant p \leqslant q \leqslant r$ (without loss of generality) and $p+q+r = n+2$. Working out some scalar products of slightly more complicated linear combinations (see [Ebel02] for the detailed expressions) yields the necessary condition

$$\frac{1}{p} + \frac{1}{q} + \frac{1}{r} > 1.$$

The solutions are $(1, q, r)$ for arbitrary $1 \leqslant q \leqslant r$ (which gives A_n with $n = q + r - 1$) and $(2, 2, r)$ for $r \geqslant 2$ (yielding D_n with $n = r + 2$), together with $(2, 3, 3)$, $(2, 3, 4)$ and $(2, 3, 5)$, which correspond to E_6, E_7 and E_8.

Step 4: The lattice realisation of each solution is shown by an explicit construction of a corresponding basis; see [CS99, Ebel02] for details.   □

Arbitrary root lattices are obtained by combining indecomposable ones, where the dimension is the total number of nodes. The integer lattice $\mathbb{Z}^d$ corresponds — up to scale — to the decomposable root lattice $(A_1)^d$, represented by the edgeless graph with d nodes. A detailed discussion of all root lattices, together with their symmetries, can be found in [MP92] and [CS99, Ch. 4]. The latter source also discusses the duals of the root lattices, known as *weight lattices*, which form another useful class of lattices.

Example 3.2 (*Hypercubic lattices*). In dimensions $d = 3$ and $d \geqslant 5$, there are three distinct lattices (up to similarity transformations) with point symmetry group $\mathrm{O}(d, \mathbb{Z}) \simeq C_2 \wr S_d \simeq C_2^d \rtimes S_d$ (up to isomorphism), where the symbol $\wr$ denotes the standard wreath product of group theory. These are the hypercubic lattices; see [Schw80, Sec. 5.6] for details and proofs. Their standard representatives are the integer lattice $\mathbb{Z}^d$ (primitive) of Example 2.7, the root lattice D_d (face-centred), with $D_3 = A_3$, and the weight lattice D_d^* (body-centred), again with $D_3^* = A_3^*$.

The body-centred lattice is $D_d^* = \mathbb{Z}^d \,\dot\cup\, (v + \mathbb{Z}^d)$, with $v = \frac{1}{2}(1, \dots, 1)$ being the centre of the unit cube $[0, 1]^d$. The face-centred lattice derives from $\mathbb{Z}^d$ as the smallest lattice that contains $\mathbb{Z}^d$ and all centres of the $(d-1)$-facets of $[0, 1]^d$, which gives D_d (up to scale); compare [Schw80, Sec. 5.6].

For $d = 1$ and $d = 2$, the three possibilities coincide via similarity transformations, while there are only two possibilities in dimension $d = 4$, because D_4 and D_4^* are again similar lattices [CS99, Sec. 4.7.4]. Moreover, the point symmetry of D_4 is larger than that of $\mathbb{Z}^4$ (the latter being an index 3 subgroup in this case). $\diamond$

Example 3.3 (*The lattices A_4 and A_4^**). The standard representation of the root lattice A_4 follows from the diagram

where the e_i denote the standard basis vectors of $\mathbb{R}^5$, and the lattice A_4 is embedded in the four-dimensional hyperplane $\left(\mathbb{R}(e_1 + \dots + e_5)\right)^\perp$. An explicit representation in 4-space (with the golden ratio τ) is provided by the basis

$$(3.10) \qquad \left\{(1, 0, 0, 0), \tfrac{1}{2}(-1, 1, 1, 1), (0, -1, 0, 0), \tfrac{1}{2}(0, 1, \tau - 1, -\tau)\right\},$$

which gives A_4 with its scale reduced by the factor $\sqrt{2}$. This representation is particularly useful when working with quaternions [BHM08, BGHZ08].

One way to construct the weight lattice A_4^* is via the corresponding dual basis, which (up to a scale factor $\frac{2}{5}$) reads

$$\left\{\tfrac{1}{2}(5, 0, 2+\tau, 3-\tau), (0, 0, 2+\tau, 3-\tau), \tfrac{1}{2}(0, -5, 1+3\tau, 4-3\tau), (0, 0, 2\tau-1, 1-2\tau)\right\}.$$

For many purposes, it is preferable to work with a symmetry-adapted basis. One convenient choice (also used later) leads to the two Gram matrices

$$G_{A_4} = \begin{pmatrix} 2 & -1 & 0 & 0 \\ -1 & 2 & -1 & 0 \\ 0 & -1 & 2 & -1 \\ 0 & 0 & -1 & 2 \end{pmatrix} \quad \text{and} \quad G_{A_4^*} = \begin{pmatrix} 4 & -1 & -1 & -1 \\ -1 & 4 & -1 & -1 \\ -1 & -1 & 4 & -1 \\ -1 & -1 & -1 & 4 \end{pmatrix};$$

see [CS99, Secs. 4.6.1 and 4.6.6] for details. $\diamond$

Remark 3.3 (*Crystallographic root systems*). In a more general context, root lattices are generated by crystallographic root systems, which are classified in terms of Dynkin diagrams. There are additional types beyond the simply laced diagrams in Theorem 3.3, namely B_n (with $n \geqslant 2$), C_n (with $n \geqslant 3$), G_2 and F_4. However, the corresponding root systems do not generate new lattices [CS99, Table 4.1]. Indeed, the root systems Δ_{A_2} and Δ_{G_2} both generate the root lattice A_2, see [BG97, Fig. 1], while Δ_{F_4} generates the weight lattice D_4^*. Moreover, Δ_{B_n} generates the integer lattice $\mathbb{Z}^n$, Δ_{C_2} again $\mathbb{Z}^2$, Δ_{C_3} the lattice A_3, and Δ_{C_n} (with $n \geqslant 4$) the lattice D_n. $\Diamond$

Incidentally, the proof of Theorem 3.3, with suitable modifications, also leads to the classification of the finite subgroups of $\mathrm{SO}(3)$ as follows; compare Section 2.3.1, and see [CM80, Ch. 4] or [Schw80, Sec. 4.2] for further details.

Theorem 3.4. *Up to isomorphism, the finite subgroups of* $\mathrm{SO}(3)$ *are the cyclic groups* C_n *of order* n, *with* $n \geqslant 1$, *the dihedral groups* D_n *of order* $2n$, *with* $n \geqslant 2$, *and the three rotation groups of the Platonic solids, namely the tetrahedral group* T *of order* 12, *the octahedral group* O *of order* 24, *and the icosahedral group* Y *of order* 60. $\square$

In this context, talking about groups actually means talking about faithful representations of the corresponding abstract groups by three-dimensional rotation matrices. The group D_1 is not listed because it is isomorphic to C_2. The Platonic solids were shown in Figure 2.4 on page 23. As mentioned earlier, the cube and the octahedron are dual to each other, as are the dodecahedron and the icosahedron, which is why each pair shares the same symmetry group [Cox73]. The corresponding groups T, O and Y were described in Section 2.3.1, and Y is looked at in more detail in Appendix A.

From our previous discussion of the crystallographic restriction, we know that not all groups of Theorem 3.4 are compatible with a lattice in 3-space. In particular, this applies to the icosahedral group Y, because it contains a fivefold element. This conclusion also applies to the larger class of crystallographic packings, as follows from Proposition 3.2.

Corollary 3.3. *Among the finite subgroups of* $\mathrm{SO}(3)$ *in Theorem 3.4, only finitely many can occur as rotation symmetry groups of lattices or crystallographic point packings in 3-space, namely* C_n *and* D_n, *for* $n \leqslant 6$ *except* $n = 5$, *as well as* T *and* O. $\square$

It is clear that no other groups are possible, due to the crystallographic restriction. Moreover, it is not difficult to construct examples of lattices or point packings which realise these cases, for instance by starting from the maximal point groups in 3-space (see [BBN+78, Fig. 6]) and the corresponding lattices.

3.4. Minkowski embedding

In the theory of aperiodic order, one often has to deal with dense point sets and their relations to lattices. A particularly useful instance is given by a dense $\mathbb{Z}$-module with an underlying algebraic structure, which permits a natural lattice embedding into a Euclidean space of higher dimension. Such an embedding is called a geometric image in [BS66]. However, we prefer to use the term *Minkowski embedding*, which seems justified in view of Minkowski's work on lattices and the geometry of numbers.

3.4.1. REAL FIELDS

Let us first look at the Minkowski embedding of rings of integers of real algebraic number fields. The latter are of the form $\mathbb{Q}(\theta)$ with θ a real algebraic integer. Note that its algebraic conjugates might be complex, as in Example 2.17. We begin with the simpler situation when all algebraic conjugates are real ($\mathbb{Q}(\theta)$ is then called totally real).

Example 3.4 (*Lattice embedding of* $\mathbb{Z}[\tau]$). Consider $\mathbb{Z}[\tau]$ with τ being the golden ratio, which is a PV unit of degree 2; see Example 2.14. Consequently,

$$\mathbb{Z}[\tau] = \{m + n\tau \mid m, n \in \mathbb{Z}\}.$$

This is a dense point set in $\mathbb{R}$, and thus not a lattice. It is both a $\mathbb{Z}[\tau]$-module of rank 1 (with $\tau\mathbb{Z}[\tau] = \mathbb{Z}[\tau]$) and a $\mathbb{Z}$-module of rank 2. It can be viewed as the projection of a two-dimensional lattice in various ways. Perhaps the most natural one is provided by the Galois structure of the underlying quadratic number field $\mathbb{Q}(\sqrt{5})$. Recall that there is one non-trivial algebraic conjugation, defined by $\sqrt{5} \mapsto -\sqrt{5}$ together with its unique extension to a field automorphism. The image of an element x is written as x'.

If we now define the *diagonal embedding* of $\mathbb{Z}[\tau]$ into $\mathbb{R}^2$ as

$$\mathcal{L} = \big\{(x, x') \mid x \in \mathbb{Z}[\tau]\big\},$$

one can check that $\mathcal{L}$ is both discrete and co-compact in $\mathbb{R}^2$, and thus a planar lattice, generated by the two basis vectors $(1,1)$ and (τ, τ'). This is the Minkowski embedding of the ring $\mathbb{Z}[\tau]$; see Figure 3.3. In the co-ordinatisation that is implicit to the construction (via the two basis vectors chosen), the invariance of $\mathbb{Z}[\tau]$ under multiplication by the unit τ corresponds to the invariance of $\mathcal{L}$ under the linear transformation defined by the matrix $M = \left(\begin{smallmatrix} 0 & 1 \\ 1 & 1 \end{smallmatrix}\right) \in \mathrm{GL}(2, \mathbb{Z})$. $\diamond$

Remark 3.4 (*Embedding of* $\mathbb{Z}[\tau]$ *with maximal symmetry*). In the construction of Example 3.4 and Figure 3.3, we used the basis matrix $B = \left(\begin{smallmatrix} 1 & \tau \\ 1 & 1-\tau \end{smallmatrix}\right)$ for the lattice $\mathcal{L}$. Note that, according to our convention, the basis vectors

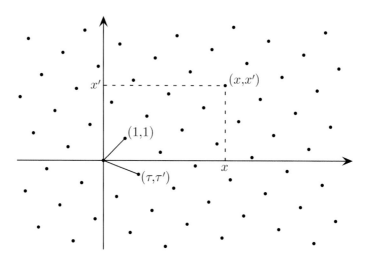

FIGURE 3.3. The Minkowski embedding of $\mathbb{Z}[\tau]$ as a planar lattice, where the projections are illustrated for the lattice point (x, x') with $x = 4 + \tau$.

enter as columns in B. Clearly, we can scale the two coordinates relative to each other by an arbitrary real number $\alpha > 0$, leading to the basis matrix

$$B(\alpha) = \begin{pmatrix} 1 & 0 \\ 0 & \alpha \end{pmatrix} B = \begin{pmatrix} 1 & \tau \\ \alpha & \alpha(1-\tau) \end{pmatrix},$$

which has the Gram matrix

$$G(\alpha) = \begin{pmatrix} 1 + \alpha^2 & \tau + \alpha^2(1-\tau) \\ \tau + \alpha^2(1-\tau) & (\tau+1) + \alpha^2(2-\tau) \end{pmatrix}.$$

Choosing $\alpha = \tau$ gives $G(\tau) = (\tau + 2)\mathbb{1}$ and thus a scaled square lattice. This is the standard embedding of $\mathbb{Z}[\tau]$ as a planar lattice (similar to $A_1 \times A_1$) with maximal symmetry, which is frequently used in the literature. $\quad\lozenge$

The construction of Example 3.4 can be generalised considerably, compare [BS66, Ple00], to provide a powerful tool for the description of mathematical quasicrystals. The extension of Example 3.4 to other quadratic fields such as $\mathbb{Q}(\sqrt{2})$ and $\mathbb{Q}(\sqrt{3})$ from Example 2.14 leads to the lattices

(3.11) $\qquad \big\langle (1,1), (\sqrt{2}, -\sqrt{2}) \big\rangle_{\mathbb{Z}}$ and $\big\langle (1,1), (\sqrt{3}, -\sqrt{3}) \big\rangle_{\mathbb{Z}}$

for $\mathbb{Z}[\sqrt{2}]$ and $\mathbb{Z}[\sqrt{3}]$, respectively. The choice for the generating elements is not unique. One natural alternative, in analogy to Example 3.4, employs a fundamental unit, which means replacing $\sqrt{2}$ by $1 + \sqrt{2}$ or $\sqrt{3}$ by $2 + \sqrt{3}$. Via suitable relative scalings (as in the previous example), one can arrange the embedding lattices to be similar to $\mathbb{Z}^2$ or to A_2, respectively, which are then again embeddings with maximal symmetry.

The Minkowski embedding for any totally real number field $\mathbb{Q}(\theta)$ (where all algebraic conjugates of θ are real) works analogously. The change needed in the presence of complex roots is best explained with an example.

Example 3.5 (*Embedding of* $\mathbb{Z}[\beta]$). Consider the ring $\mathbb{Z}[\beta]$ with the cubic PV unit β (the plastic number) of Example 2.17. The algebraic conjugates are $\beta, \alpha, \overline{\alpha}$, and the embedding of $\mathbb{Z}[\beta]$ is constructed from β and α alone as

$$\big\langle (1,1,0), (\beta, \mathrm{Re}(\alpha), \mathrm{Im}(\alpha)), (\beta^2, \mathrm{Re}(\alpha^2), \mathrm{Im}(\alpha^2)) \big\rangle_{\mathbb{Z}},$$

which defines a lattice in 3-space. Its fundamental domain has volume $\sqrt{23}/2$, which is the determinant of the three spanning vectors.   ◊

The construction for other real fields proceeds analogously; see [Ple00] for a general formulation in the same context.

3.4.2. COMPLEX FIELDS

When the field $\mathbb{Q}(\theta)$ is totally complex, it is advantageous to employ complex numbers. In particular, the lattices obtained from the Minkowski embeddings of rings of cyclotomic integers appear frequently. To expand on this, we consider the $\mathbb{Z}$-span of the solutions of $x^n - 1 = 0$, which are the nth roots of unity, for $n \geqslant 3$. This gives the ring $\mathbb{Z}[\xi_n]$, which is a planar lattice when $n = 4$ (the square lattice) or $n = 3$ (the triangular lattice, also obtained from $n = 6$); compare Example 2.15. All other values of n lead to dense point sets in the plane, in line with the crystallographic restriction of Lemma 3.2 and the argument from Figure 1.4. As discussed in Example 2.16, $\mathbb{Z}[\xi_n]$ is the ring of integers in the cyclotomic field $\mathbb{Q}(\xi_n)$, where ξ_n is a primitive nth root of unity. As a point set, $\mathbb{Z}[\xi_n]$ has N-fold rotational symmetry, where

$$N = N(n) = \mathrm{lcm}(2,n).$$

If n is odd, one has $\mathbb{Z}[\xi_n] = \mathbb{Z}[\xi_{2n}]$, whence we may (and will) restrict to integers $n \not\equiv 2 \bmod 4$ without loss of generality. This convention has various advantages algebraically (though the symmetry might suggest otherwise), as explained earlier in Section 2.5.2; see [Was97] for background.

The *Galois group* of the field extension $\mathbb{Q}(\xi_n)$ of $\mathbb{Q}$ is isomorphic with $(\mathbb{Z}/n\mathbb{Z})^\times$, with the latter viewed as a multiplicative group of order $\phi(n)$. Its elements correspond to field automorphisms of $\mathbb{Q}(\xi_n)$, specified by $\xi_n \mapsto \xi_n^\ell$ with ℓ, n coprime. For $n \geqslant 3$, they come in complex conjugate pairs (and $\phi(n)$ is even). Since we consider $\mathbb{Z}[\xi_n]$ as a subset of the complex plane, we may pick one representative out of each such pair, including the identity (rather than complex conjugation) for the first mapping. This gives $\frac{1}{2}\phi(n)$ distinct mappings σ_i, with σ_1 being the identity. The diagonal embedding

$$(3.12) \quad \mathcal{L}_n = \big\{ (x, \sigma_2(x), \dots, \sigma_{\frac{1}{2}\phi(n)}(x)) \mid x \in \mathbb{Z}[\xi_n] \big\} \subset \mathbb{C}^{\frac{1}{2}\phi(n)} \simeq \mathbb{R}^{\phi(n)}$$

defines a $\mathbb{Z}$-module of rank $\phi(n)$. As a point set, it is both uniformly discrete and relatively dense, so $\mathcal{L}_n$ is a lattice in $\mathbb{R}^{\phi(n)}$. This is the standard *Minkowski embedding* of $\mathbb{Z}[\xi_n]$. If π denotes the natural projection onto the first complex coordinate, one recovers $\mathbb{Z}[\xi_n]$ from $\mathcal{L}_n$ via $\mathbb{Z}[\xi_n] = \pi(\mathcal{L}_n)$.

Note that the rotation symmetry of $\mathbb{Z}[\xi_n]$, described by $x \mapsto \xi_n x$ with $\xi_n \in \mathbb{Z}[\xi_n]^\times$, lifts to a symmetry of the lattice $\mathcal{L}_n$, as mentioned above in Theorem 3.1. When n is odd, one can use $x \mapsto -\xi_n x$, which defines a symmetry of order $2n$. In all cases, this rotation symmetry has an integer representation in the lattice basis by construction.

Example 3.6 (*Coordinatisation of* $\mathbb{Z}[\xi_8]$ *and* $\mathbb{Z}[\xi_{12}]$). The cases $n = 8$ and $n = 12$ permit a simple explicit choice of four-dimensional coordinates, because $\sqrt{-1} = \mathrm{i} \in \mathbb{Z}[\xi_n]$. We use the Minkowski embedding of $\mathbb{Z}[\xi_n]$ with the explicit choice $\xi_n = \mathrm{e}^{2\pi \mathrm{i}/n}$ and the conjugation map defined by $\xi_n \mapsto \xi_n^{n/2-1}$. For $n = 8$, this leads to the lattice $\mathcal{L}_8 = \sqrt{2} R_8 \, \mathbb{Z}^4$, with the rotation matrix

$$
R_8 = \frac{1}{2}\begin{pmatrix} \sqrt{2} & 1 & 0 & -1 \\ 0 & 1 & \sqrt{2} & 1 \\ \sqrt{2} & -1 & 0 & 1 \\ 0 & 1 & -\sqrt{2} & 1 \end{pmatrix}.
$$

By the congruence $\mathcal{L}_8 \simeq (A_1)^4$ and Definition 3.3, $\mathcal{L}_8$ is a root lattice (in the strict sense). Since $(R\,\mathbb{Z}^4)^* = R\,(\mathbb{Z}^4)^* = R\,\mathbb{Z}^4$, the dual lattice is $\mathcal{L}_8^* = \frac{1}{2}\mathcal{L}_8$. With $\mathbb{Z}[\xi_8] = \pi(\mathcal{L}_8)$, this also implies that $\pi(\mathcal{L}_8^*) = \frac{1}{2}\mathbb{Z}[\xi_8]$.

For $n = 12$, the resulting basis and Gram matrices read

$$
B_{12} = \frac{1}{2}\begin{pmatrix} 2 & \sqrt{3} & 1 & 0 \\ 0 & 1 & \sqrt{3} & 2 \\ 2 & -\sqrt{3} & 1 & 0 \\ 0 & 1 & -\sqrt{3} & 2 \end{pmatrix} \quad \text{and} \quad G_{12} = \begin{pmatrix} 2 & 0 & 1 & 0 \\ 0 & 2 & 0 & 1 \\ 1 & 0 & 2 & 0 \\ 0 & 1 & 0 & 2 \end{pmatrix}.
$$

The shortest non-zero lattice vectors have squared length 2, and there are precisely 12 of them (the lifts of the 12th roots of unity). This shows that the lattice $\mathcal{L}_{12}$ is different from both the hypercubic lattice $\mathbb{Z}^4$ and the root lattice D_4; compare [CS99, Sec. 4.7.2]. Swapping the two middle columns of B_{12} and changing their signs leads to an alternative basis matrix for $\mathcal{L}_{12}$, with new Gram matrix $\left(\begin{smallmatrix} 2 & -1 \\ -1 & 2 \end{smallmatrix}\right) \oplus \left(\begin{smallmatrix} 2 & -1 \\ -1 & 2 \end{smallmatrix}\right)$. This shows that $\mathcal{L}_{12} \simeq A_2 \times A_2$, which is a root lattice with 12 vectors of squared length 2 and a point symmetry group of order 288; compare [BBN+78, Fig. 7].

Sticking with the original basis, the dual lattice $\mathcal{L}_{12}^*$ has basis matrix

$$
B_{12}^* = \frac{1}{2\sqrt{3}}\begin{pmatrix} \sqrt{3} & 2 & 0 & -1 \\ -1 & 0 & 2 & \sqrt{3} \\ \sqrt{3} & -2 & 0 & 1 \\ 1 & 0 & -2 & \sqrt{3} \end{pmatrix},
$$

which satisfies $B_{12} = B_{12}^* G_{12}$. Since the Gram matrix is integral, $\mathcal{L}_{12}$ is a sublattice of $\mathcal{L}_{12}^*$. In fact, it is a similar sublattice of index 9, and one can check explicitly that $\pi(\mathcal{L}_{12}^*) = \frac{1}{\sqrt{3}} \mathbb{Z}[\xi_{12}]$. This can be verified by comparing the top two rows of the basis matrices B_{12} and B_{12}^*. ◊

Remark 3.5 (*Relation to the root lattice D_4 and Schur rotation*). In a manner similar to Example 3.4 and Remark 3.4, one can shear the lattice embedding of $\mathbb{Z}[\xi_{12}]$ by defining

$$B_{12}(\alpha) = \mathrm{diag}(1, 1, \alpha, \alpha) B_{12}$$

with $\alpha > 0$. In particular, when choosing $\alpha = \sqrt{2 \pm \sqrt{3}}$, one obtains a four-dimensional lattice of higher symmetry. It is a scaled version of the root lattice D_4. This is a lattice embedding of $\mathbb{Z}[\xi_{12}]$ in 4-space with maximal symmetry (the point group has order 1152, which is the largest one for lattices in 4-space; see [BBN+78, Fig. 7]). This has been used for the construction of mathematical quasicrystals with twelvefold symmetry [BJS90].

Interestingly, also the embedding of $\mathbb{Z}[\xi_8]$ can be realised within the root lattice D_4, via $\alpha = 3 - \sqrt{2}$ in $B_8(\alpha) := \mathrm{diag}(1, 1, \alpha, \alpha) B_8$, with $B_8 = \sqrt{2} R_8$. Then, one can fix a common C_4-subgroup of 8- and 12-fold rotation that is compatible with a one-parameter rotation due to Schur's lemma. One may now relate the two symmetries via such a rotation in the embedding lattice [BJKS90, BJK91]. An analogous mechanism exists for the relation between cubic and icosahedral symmetry, via the common tetrahedral subgroup as described in Section 2.3.1; see [Kra87b, BJK91] for details. ◊

Example 3.7 (*Minkowski embedding of $\mathbb{Z}[\xi_5]$*). The structure of $\mathbb{Z}[\xi_5]$ is slightly more complex. The bilinear form $\mathrm{tr}(\bar{x}y) = (\bar{x}y + x\bar{y}) + (\bar{x}y + x\bar{y})'$ employs the explicit version of the trace for the number field $\mathbb{Q}(\xi_5)$. Using this for the $\mathbb{Z}$-basis $\{1, \xi_5, \xi_5^2, \xi_5^3\}$, one obtains the Gram matrix $G_{A_4^*}$ of Example 3.3. This shows that the Minkowski embedding of $\mathbb{Z}[\xi_5]$ is the weight lattice A_4^*, with (algebraic) dual A_4.

If one works with the diagonal embedding in $\mathbb{C}^2 \simeq \mathbb{R}^4$, the standard Euclidean scalar product is related to the trace via $\langle (x, x^\star) | (y, y^\star) \rangle = \frac{1}{2} \mathrm{tr}(\bar{x}y)$, where * denotes the algebraic conjugation in $\mathbb{Q}(\xi_5)$ defined by $\xi_5 \mapsto \xi_5^2$. This corresponds to a change of scale by a factor of $\sqrt{2}$, which will become relevant later. An explicit Cartesian coordinatisation can easily be derived, but is less useful than in the previous example, because $\mathrm{i} \notin \mathbb{Z}[\xi_5]$. ◊

Remark 3.6 (*Alternative choice of bilinear form*). The bilinear form $\mathrm{tr}(\bar{x}y)$ in Example 3.7 is a natural choice, but not the only one. Indeed, following [CMP98], consider

$$(x, y) := \mathrm{tr}\left(\frac{2\bar{x}y}{5 + \sqrt{5}}\right) = \frac{2(\bar{x}y + x\bar{y})}{5 + \sqrt{5}} + \frac{2(\bar{x}y + x\bar{y})'}{5 - \sqrt{5}}.$$

The Gram matrix of our $\mathbb{Z}$-basis $\{1, \xi_5, \xi_5^2, \xi_5^3\}$ relative to this new quadratic form immediately gives the Gram matrix G_{A_4} of Example 3.3, in line with [CMP98, Cor. 6.8]. Since $(5 + \sqrt{5})/(5 - \sqrt{5}) = \tau^2$, the geometric meaning of this new quadratic form is an overall change of scale together with a relative scaling of the internal coordinates by a factor τ. The latter plays the same role as the scaling factor α in Remarks 3.4 and 3.5. $\Diamond$

The possibility to scale the additional 'internal' space relative to the direct one in order to obtain an embedding with larger symmetry is also described in [Gäh88, Sec. 2.4], in a slightly different setting.

An important property of a point set is its behaviour under *inflation*. For lattices and modules, this leads to the question of which homotheties map such sets into or onto themselves. More generally, one is interested in affine mappings of such sets into themselves. For a lattice Γ, by an application of part (1) of Theorem 2.4, the equation $a\Gamma \subset \Gamma$ with $a \in \mathbb{R}$ implies $a \in \mathbb{Z}$, where $a = \pm 1$ are the only cases with equality. In the more general case of a $\mathbb{Z}$-module (such as $\mathbb{Z}[\xi_n]$), further possibilities might exist, because the underlying structure may have further units. This is certainly the case when the $\mathbb{Z}$-module under consideration is an R-module at the same time, with R some ring of integers with further units. Let us discuss some examples in this spirit, which will reappear in many of our later constructions.

Remark 3.7 (*Classification of planar modules*). Let $n \in \mathbb{N}$ with $n \not\equiv 2$ mod 4 and $n \geqslant 3$. As mentioned above, the ring of integers $\mathbb{Z}[\xi_n]$ is a $\mathbb{Z}$-module in the plane of rank $\phi(n)$ with N-fold rotational symmetry, where $N = \mathrm{lcm}(n, 2)$ and the rank is minimal for this property (in line with Theorem 3.2). It is important to know whether there are other rotation symmetric modules of the same rank that are inequivalent up to similarity. This question can be reformulated as a question about ideal classes in $\mathbb{Z}[\xi_n]$; compare [MRW87]. Whenever the class number h of $\mathbb{Q}(\xi_n)$ is $h = 1$ (which means that $\mathbb{Z}[\xi_n]$ is a PID), the planar, rank-$\phi(n)$ module with N-fold symmetry is unique up to similarity. This happens precisely for the 29 cases [Was97, Thm. 11.1]

$$n \in \{3, 4, 5, 7, 8, 9, 11, 12, 13, 15, 16, 17, 19, 20, 21, 24,$$
$$25, 27, 28, 32, 33, 35, 36, 40, 44, 45, 48, 60, 84\}.$$

Note that $n = 1$ refers to the field $\mathbb{Q}$, which also has class number 1, but is not considered here because its ring of integers is $\mathbb{Z}$ and does not span the plane (over $\mathbb{R}$).

The first case with $h > 1$ occurs for $n = 23$ (where $h = 3$), so that additional solutions exist. This connection was first highlighted in [MRW87], while its core is a classic result of algebraic number theory [Was97]. $\Diamond$

3.4.3. EMBEDDING OF $\mathbb{Z}[\tau]$-MODULES

Beyond the above examples, one also frequently encounters modules over rings of integers in the theory of aperiodic order. Since one also needs lattice embeddings for these, it is natural to build such an embedding on the Minkowski embedding of the underlying ring. We explain this for the most important class of $\mathbb{Z}[\tau]$-modules here; other cases can be treated similarly.

Example 3.8 (*Minkowski embedding of icosahedral modules*). Let us first look at the icosahedral modules of Example 2.20, using the explicit $\mathbb{Z}$-bases specified there. The standard Minkowski embedding maps a (row) vector $v = (a, b, c)$ with $a, b, c \in \mathbb{Z}[\tau]$ to the column vector $(a, b, c, a', b', c')^{T}$, again with $'$ denoting algebraic conjugation in $\mathbb{Q}(\sqrt{5})$. Here, we modify this by using $(a, b, c, \tau a', \tau b', \tau c')^{T}$ instead, in line with Remark 3.6. Applying this map to $\{v_1, \ldots, v_6\}$, we obtain, up to a scaling factor, the matrix B of Eq. (A.2) in Appendix A. This shows that the (modified) Minkowski embedding of $\mathcal{M}_\mathsf{P}$ is similar to $\mathbb{Z}^6$.

Similarly, observing $D_6 \overset{2}{\subset} \mathbb{Z}^6 \overset{2}{\subset} D_6^*$ and comparing this with the corresponding relation for the modules of Example 2.20, we see that $\mathcal{M}_\mathsf{F}$ lifts to a similar copy of the root lattice D_6, see also [CMP98, Cor. 6.8], and $\mathcal{M}_\mathsf{B}$ to one of the weight lattice D_6^*. This way, the three icosahedral modules in 3-space precisely correspond to the three hypercubic lattices in 6-space described in Example 3.2. $\diamond$

Example 3.9 (*Icosian ring and root lattice E_8*). Following [MP93, Eq. 3.6], we choose a $\mathbb{Z}$-basis for the icosian ring $\mathbb{I}$ of Example 2.19 as

$$\frac{1}{2}\begin{pmatrix} \tau-1 & 0 & 0 & 0 & 0 & 0 & 1 & 0 \\ -\tau & \tau-1 & 1 & -\tau & \tau & 1 & -\tau-1 & -1 \\ 0 & -\tau & \tau-1 & 1 & 1 & -\tau-1 & 0 & \tau-1 \\ -1 & 1 & -\tau & \tau+1 & -\tau-1 & \tau & -\tau & \tau \end{pmatrix},$$

which is to be read columnwise for the coordinates of the eight basis vectors. The order refers to the following enumeration of the E_8 Dynkin diagram

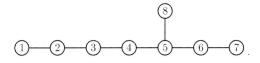

In analogy to our previous examples, we follow [MW94] and choose the bilinear form

$$(3.13) \qquad (x, y) := \mathrm{tr}_{\mathbb{Q}(\sqrt{5})}\left(\frac{4\langle x|y\rangle}{5+\sqrt{5}}\right) = \frac{4\langle x|y\rangle}{5+\sqrt{5}} + \frac{4\langle x|y\rangle'}{5-\sqrt{5}}.$$

As in Remark 3.6 and Example 3.8, this choice of bilinear form reflects an additional scaling factor τ for the added (internal) coordinates, relative to the direct Minkowski embedding via coordinate-wise algebraic conjugation in $\mathbb{Q}(\sqrt{5})$. For our choice of basis, and in line with [CMP98, Cor. 6.8], the Gram matrix (of inner products taken with (3.13)) reads

$$
G_{E_8} = \begin{pmatrix}
2 & -1 & 0 & 0 & 0 & 0 & 0 & 0 \\
-1 & 2 & -1 & 0 & 0 & 0 & 0 & 0 \\
0 & -1 & 2 & -1 & 0 & 0 & 0 & 0 \\
0 & 0 & -1 & 2 & -1 & 0 & 0 & 0 \\
0 & 0 & 0 & -1 & 2 & -1 & 0 & -1 \\
0 & 0 & 0 & 0 & -1 & 2 & -1 & 0 \\
0 & 0 & 0 & 0 & 0 & -1 & 2 & 0 \\
0 & 0 & 0 & 0 & -1 & 0 & 0 & 2
\end{pmatrix},
$$

which is the Cartan matrix of the E_8 root system [CS99]. In this sense, the modified Minkowski embedding of the icosian ring is the root lattice E_8. $\Diamond$

Remark 3.8 (*Fivefold hierarchy*). We have seen that a small variation of the standard Minkowski embedding can lead to a larger symmetry group of the embedding lattice. This is particularly striking for the modules connected with the golden ratio — and hence with fivefold symmetry. Indeed, the embeddings of $\mathbb{I}$ (Example 3.9), $\mathcal{M}_\mathsf{F}$ (Example 3.8), $\mathbb{Z}[\xi_5]$ (Example 3.7 and Remark 3.6) and $\mathbb{Z}[\tau]$ (Example 3.4 and Remark 3.4) all lead to the lattice sequence

$$
E_8 \searrow D_6 \searrow A_4 \searrow A_1 \times A_1
$$

which is not an accident. In fact, one can 'fold' the E_8 diagram in such a way as to derive this hierarchy (and its relation to the Coxeter groups H_4, H_3 and H_2) from first principles; see [CMP98] for details.

Geometrically, this means that $\mathbb{I}$ contains $\mathcal{M}_\mathsf{F}$ in a subspace of dimension 3, which contains $\mathbb{Z}[\xi_5]$ in a plane, which finally has $\mathbb{Z}[\tau]$ in a line. On the level of tilings and model sets, this hierarchy will appear in Example 7.14. $\Diamond$

At this point, we are prepared to analyse a large class of systems with aperiodic order via lattice embeddings. It is now time to turn to various construction methods for such systems. We shall return to lattices in Chapter 7.

CHAPTER 4

Symbolic Substitutions and Inflations

A powerful tool for the construction of ordered systems, particularly in one dimension, is provided by substitution rules on finite alphabets and their geometric counterparts, which we call inflation rules. We will start by reviewing some basic notions of symbolic dynamics, which is a natural setting in the presence of a $\mathbb{Z}$-action (such as the one defined by the shift operator). This is then gradually lined up with geometry, where we ultimately have an action of the group $\mathbb{R}$ by (continuous) translations. As we go along, we introduce several paradigmatic substitutions and discuss their properties.

4.1. Substitution rules

In view of several group-theoretic aspects, we prefer an approach to substitution systems based on groups rather than formal alphabets and their dictionaries. This automatically leads to the free group and its endomorphisms, and hence to various notions of combinatorial group theory; see [MKS76] for background material, and [PF02, AS03] for extensions and as thorough sources for topics and results that we cannot include here.

Let us thus start from a finite alphabet $\mathcal{A}_n = \{a_i \mid 1 \leqslant i \leqslant n\}$ and consider the *free group* $F_n := \langle a_1, \ldots, a_n \rangle$ generated by the letters of the alphabet. The elements of F_n consist of the empty word e and all possible finite words in the letters a_i and their formal inverses, up to equivalence according to the relations $a_i a_i^{-1} = a_i^{-1} a_i = e$. This turns F_n into an infinite group with composition of words as multiplication and e as the neutral element. The composition $u \circ v$ of two words is abbreviated as uv. The term 'free' refers to $uu^{-1} = e$, with $u \in F_n$ arbitrary, being the only relation available (observe that $e = u^{-1}u$ already is a consequence of this). Note that the same group F_n permits different sets of generators. Whenever we refer to the particular choice of generators from above, we continue to call them 'letters'.

Definition 4.1. A *general substitution rule* ϱ on a finite alphabet $\mathcal{A}_n$ with n letters is an endomorphism of the corresponding free group F_n.

The endomorphism property means that $\varrho(uv) = \varrho(u)\varrho(v)$ together with $\varrho(u^{-1}) = \big(\varrho(u)\big)^{-1}$ holds for any $u, v \in F_n$. A general substitution rule is thus completely specified by the images $\varrho(a_i)$ of the letters, which are themselves finite words in the letters and their inverses. In this setting, the multiplication of general substitution rules is well-defined as a composition of mappings, as the set $\mathrm{End}(F_n)$ of endomorphisms forms a monoid. A special role is played by the subgroup $\mathrm{Aut}(F_n)$ of *automorphisms* of the free group F_n. This group is finitely (but not freely) generated; see [MKS76] for details.

For the characterisation of general substitution rules, it is useful to introduce the corresponding substitution matrices through *Abelianisation*. The latter is induced by the group homomorphism $\alpha \colon F_n \longrightarrow \mathbb{Z}^n$ defined on the generators by $a_i \mapsto e_i$, where e_i is the conventional Euclidean basis vector of $\mathbb{R}^n$ for the ith coordinate. The image of an arbitrary word $w \in F_n$ is thus obtained by separately adding up the (positive and negative) powers of each a_i in w, denoted by $\mathrm{card}_{a_i}(w)$, which is an element of $\mathbb{Z}$ here. The map α induces a homomorphism $M \colon \mathrm{End}(F_n) \longrightarrow \mathrm{End}(\mathbb{Z}^n) \simeq \mathrm{Mat}(n, \mathbb{Z})$. Under M, the group $\mathrm{Aut}(F_n)$ is mapped onto the group $\mathrm{Aut}(\mathbb{Z}^n) \simeq \mathrm{GL}(n, \mathbb{Z})$. Given ϱ, $M(\varrho)$ is the unique mapping that makes the diagram

$$
\begin{array}{ccc}
F_n & \xrightarrow{\ \varrho\ } & F_n \\[4pt]
{\scriptstyle\alpha}\Big\downarrow & & \Big\downarrow{\scriptstyle\alpha} \\[4pt]
\mathbb{Z}^n & \xrightarrow{\ M(\varrho)\ } & \mathbb{Z}^n
\end{array}
$$

commutative. Since we mainly work with matrix representations in this context, we may specify the homomorphism M by

$$
\bigl(M(\varrho)\bigr)_{i,j} = \mathrm{card}_{a_i}\bigl(\varrho(a_j)\bigr).
$$

One can check that this definition matches with the usual matrix multiplication in the sense that $M(\varrho\sigma) = M(\varrho)M(\sigma)$ holds for any $\varrho, \sigma \in \mathrm{End}(F_n)$.

Definition 4.2. For a given general substitution rule $\varrho \in \mathrm{End}(F_n)$, the matrix $M_\varrho := M(\varrho) \in \mathrm{Mat}(n, \mathbb{Z})$ is called the *substitution matrix* of ϱ.

With this definition, one has $M_{\varrho\sigma} = M_\varrho M_\sigma$. Different general substitution rules can share the same substitution matrix, as the kernel $\mathrm{ker}(M)$ is non-trivial. In particular, this kernel contains all *inner* automorphisms of F_n, which are the mappings of the form $a_i \mapsto u a_i u^{-1}$ with some fixed $u \in F_n$.

For many examples and illustrations, it is sufficient to consider a two-letter alphabet, which we mostly write as $\mathcal{A}_2 = \{a, b\}$. The kernel of the restriction $M|_{\mathrm{Aut}(F_2)}$ consists precisely of the inner automorphisms [MKS76, Thm. 3.9], while there are further elements for $n \geqslant 3$. This simplifies the situation for $\mathcal{A}_2$ considerably, but one should bear in mind that this is deceptive in the sense that several obstacles emerge for larger alphabets.

From now on, we mainly consider rules ϱ where the images $\varrho(a_i)$ of the letters contain no negative powers of the letters. This is the standard setting of symbolic dynamics. Such rules are simply referred to as *substitution rules* from now on, to distinguish them from the general setting of Definition 4.1. If some emphasis seems helpful, such substitution rules are explicitly called *non-negative*. In this context, for a given alphabet $\mathcal{A}_n$, it is often useful to also consider the infinite set

$$\mathcal{A}^* := \{e\} \,\dot{\cup}\, \mathcal{A}_n \,\dot{\cup}\, \{a_i a_j \mid 1 \leqslant i, j \leqslant n\} \,\dot{\cup}\, \{a_i a_j a_k \mid 1 \leqslant i, j, k \leqslant n\} \,\dot{\cup}\, \dots$$

of all finite words in $\mathcal{A}_n$, which is referred to as the *dictionary* of the alphabet.

Definition 4.3. A (non-negative) substitution rule ϱ on a finite alphabet $\mathcal{A}_n$ is called *irreducible* when, for each index pair (i, j), there exists some $k \in \mathbb{N}$ such that a_j is a subword of $\varrho^k(a_i)$. Moreover, ϱ is *primitive* when some $k \in \mathbb{N}$ exists such that every a_j is a subword of each $\varrho^k(a_i)$.

Recalling the definitions of Section 2.4, the following characterisation of ϱ in terms of its substitution matrix is immediate.

Lemma 4.1. *A (non-negative) substitution rule ϱ is irreducible or primitive if and only if its substitution matrix M_ϱ is an irreducible or a primitive non-negative integer matrix, respectively.* □

This setting permits an application of Perron–Frobenius (PF) theory to derive various geometric and statistical properties of substitution rules.

Example 4.1 (*A simple two-letter substitution*). Consider the rule

$$\varrho : \begin{array}{l} a \mapsto abb \\ b \mapsto a \end{array}$$

which has the substitution matrix and associated directed graph

$$M_\varrho = \begin{pmatrix} 1 & 1 \\ 2 & 0 \end{pmatrix} \quad \text{and} \quad$$.

Here, we use the convention for the orientation that follows from the interpretation $M_{ij} = M_{i \leftarrow j}$. This reverses the orientation used earlier in Section 2.4, compare Lemma 2.4 and the discussion preceding it, as it fits better to the substitution picture. Note that sometimes also a directed multigraph (with multiple edges according to the matrix entries) is used. For our purposes, single edges are sufficient, which can be given weights according to the entries of M_ϱ. Here, the edge from a to b carries weight 2, while all others have unit weight. From now on, we indicate the weight by the number of arrows.

The matrix M_ϱ is primitive (because $M_\varrho^2 = M_\varrho + 2\mathbb{1} \gg 0$), so that ϱ is primitive by Lemma 4.1. Irreducibility and primitivity are also immediate from the associated graph and its cycle structure. The matrix M_ϱ keeps track of the power counting of a (finite) word u under ℓ-fold substitution via

$$\begin{pmatrix} \mathrm{card}_a(\varrho^\ell(u)) \\ \mathrm{card}_b(\varrho^\ell(u)) \end{pmatrix} = M_\varrho^\ell \begin{pmatrix} \mathrm{card}_a(u) \\ \mathrm{card}_b(u) \end{pmatrix},$$

which is easily checked by induction. Since the PF eigenvalue $\lambda = 2$ of M_ϱ is the leading eigenvalue and simple, a standard eigenvector decomposition gives access to the asymptotic behaviour of the cardinalities for large ℓ. The total number of letters is asymptotically doubled in each step, and the relative frequencies of the two letters are encoded in the right PF eigenvector of M_ϱ, which is $\nu_{\mathrm{PF}} = \frac{1}{2}(1,1)^T$ in statistically normalised form (where the entries sum to 1). Here, the letters are equally frequent asymptotically. In particular, this applies to any fixed point of the substitution (see below).   ◇

To develop the theory significantly further, one needs the concept of a limit, and hence a topology. The finite alphabet $\mathcal{A}_n$ naturally comes with the discrete topology, and is compact in it. We are ultimately interested in infinite or bi-infinite sequences, meaning elements $w = w_0 w_1 w_2 w_3 \ldots$ or $w = \ldots w_{-2} w_{-1} w_0 w_1 w_2 \ldots$ of the spaces $\mathcal{A}_n^{\mathbb{N}_0}$ or $\mathcal{A}_n^{\mathbb{Z}}$. Both are equipped with the product topology, and are then compact spaces by Tychonov's theorem [RS80, Thm. IV.5]. Let $w_{[k,\ell]}$ with $k,\ell \in \mathbb{Z}$ and $k \leqslant \ell$ be the finite subword of w from position k to ℓ (with $w_{[k,k]} = w_k$), and define cylinder sets for a finite word u of length $m \geqslant 1$ as $Z_k(u) = \{ w \in \mathcal{A}^{\mathbb{Z}} \mid w_{[k,k+m-1]} = u \}$ with $k \in \mathbb{Z}$. The family of all such cylinder sets forms a basis of the topology. For this reason, it is also called the *local topology*, and two sequences w, w' are close when they agree on a large region around index 0. When we talk about the convergence of a sequence of finite words (of increasing length), we implicitly consider them as embedded objects in $\mathcal{A}_n^{\mathbb{N}_0}$ or $\mathcal{A}_n^{\mathbb{Z}}$.

In fact, the above spaces are also *metric* spaces. To see this, recall that the *Hamming distance* of two finite words u, v of equal length is the number of positions in which u and v differ, so that $\mathrm{d}(u,v) := \mathrm{card}\{ i \mid u_i \neq v_i \}$. This is a metric on $\mathcal{A}_n^k$ for any $k \in \mathbb{N}$. To extend this to infinite $u, v \in \mathcal{A}_n^{\mathbb{Z}}$, one can employ the standard Fréchet construction

$$(4.1) \qquad \mathrm{d}_{\mathrm{F}}(u,v) := \sum_{m \in \mathbb{N}_0} \frac{\mathrm{d}\big(u_{[-m,m]}, v_{[-m,m]} \big)}{2^m}.$$

This defines a metric on $\mathcal{A}_n^{\mathbb{Z}}$ that generates the product topology mentioned before, and turns $\mathcal{A}_n^{\mathbb{Z}}$ into a compact metric space. The construction for $\mathcal{A}_n^{\mathbb{N}_0}$ is analogous, this time employing $\mathrm{d}\big(u_{[0,m]}, v_{[0,m]} \big)$ in the sum.

The *shift operator* S acts on both spaces, and is defined by $(Sw)_i := w_{i+1}$, where w is an infinite or bi-infinite word, written as a sequence that is indexed by $\mathbb{N}_0$ or by $\mathbb{Z}$, respectively. The shift is continuous in both cases, but possesses a continuous inverse only on $\mathcal{A}_n^{\mathbb{Z}}$, given by $(S^{-1}w)_i = w_{i-1}$.

Definition 4.4. An S-invariant closed subset $X \subset \mathcal{A}_n^{\mathbb{N}_0}$ or $X \subset \mathcal{A}_n^{\mathbb{Z}}$ is called a one-sided or a two-sided *shift space*.

Clearly, the empty set, $\mathcal{A}_n^{\mathbb{N}_0}$ and $\mathcal{A}_n^{\mathbb{Z}}$ are shift spaces, but some more interesting ones will now be constructed via primitive substitution rules. For an iteration based approach, we need to know admissible initial words.

Definition 4.5. Let ϱ be a substitution rule on a finite alphabet $\mathcal{A}_n$. A finite word is called *legal* for ϱ, if it occurs as a subword of $\varrho^k(a_i)$ for some $1 \leqslant i \leqslant n$ and some $k \in \mathbb{N}$.

Legal words have the property that they are mapped to legal words under the substitution. As an example, define a sequence $(w^{(i)})_{i \in \mathbb{N}}$ of finite words by starting with the legal word $w^{(1)} = a$ and iterating $w^{(i+1)} = \varrho(w^{(i)})$ for $i \geqslant 1$, with the substitution rule ϱ from Example 4.1. One obtains

$$a \overset{\varrho}{\longmapsto} abb \overset{\varrho}{\longmapsto} abbaa \overset{\varrho}{\longmapsto} abbaaabbabb \overset{\varrho}{\longmapsto} \cdots \overset{\varrho}{\longmapsto} w^{(i)} \overset{i \to \infty}{\longrightarrow} w = \varrho(w),$$

which is a sequence of words of increasing lengths that converges (in the product topology, after appropriate embedding) to an infinite word that is fixed under the substitution ϱ. Intuitively, the convergence means that an ever-growing initial part of the words becomes stable under the iteration. Each $w^{(i+1)}$ starts with $w^{(i)}$ but grows in length. In fact, for $i \geqslant 2$, one has the recursion

$$w^{(i+1)} = w^{(i)} w^{(i-1)} w^{(i-1)},$$

which follows from the initial relation $w^{(3)} = abbaa = w^{(2)} w^{(1)} w^{(1)}$ by induction. Define

$$\mathbb{X}_{\mathbb{N}_0}(w) := \overline{\{S^i w \mid i \geqslant 0\}},$$

where the closure is taken in the product topology. This gives a closed subspace of $\mathcal{A}_2^{\mathbb{N}_0}$ that is (one-sidedly) shift invariant by construction. It is also invariant under ϱ, because $\varrho(S^i w) = S^j w$ for some $j \geqslant i$ due to our construction of the fixed point w as a limit, and this argument transfers to all elements of $\mathbb{X}_{\mathbb{N}_0}(w)$ by the (obvious) continuity of ϱ. The latter is equivalent to sequential continuity, so that ϱ maps a converging sequence in $\mathbb{X}_{\mathbb{N}_0}(w)$ to a converging sequence again, whose limit is then in $\mathbb{X}_{\mathbb{N}_0}(w)$.

Although much of the theory in symbolic dynamics is formulated with one-sided sequences, their two-sided counterparts are better suited for our purposes. This is mainly a matter of convenience rather than a difference in principle between these approaches. The substitution ϱ of Example 4.1 does not have a bi-infinite fixed point, but its square does:

$$b|a \overset{\varrho^2}{\longmapsto} abb|abbaa \overset{\varrho^2}{\longmapsto} abbaaabbabb|abbaaabbabbabbaaabbaa \overset{\varrho^2}{\longmapsto} \cdots$$

where the vertical line marks the reference point (so that $w = \ldots w_{-1}|w_0 \ldots$ for $w \in \mathcal{A}_n^{\mathbb{Z}}$) and $b|a$ is a legal two-letter word (it occurs in $\varrho^2(a)$).

Definition 4.6. A bi-infinite word w is called a *fixed point* of a primitive substitution ϱ if $\varrho(w) = w$ and $w_{-1}|w_0$ is a legal two-letter word of ϱ.

A little later, we shall see that, for a primitive substitution, we can always define a bi-infinite word this way, possibly after replacing ϱ by a suitable power. Given such a bi-infinite word w, one defines

$$(4.2) \qquad\qquad \mathbb{X}(w) := \overline{\{S^i w \mid i \in \mathbb{Z}\}},$$

which is a two-sided shift space that is also ϱ-invariant. One thus has a *topological dynamical system* $(\mathbb{X}(w), \mathbb{Z})$ with the continuous $\mathbb{Z}$-action of the shift on the compact space $\mathbb{X}$, and the additional action of the substitution ϱ on $\mathbb{X}$, which is continuous as well. For general background on dynamical systems, we refer to [Pet83, Nad95, LM95, Wal00], and to [Que10] for a specific exposition in the context of substitution systems.

Definition 4.7. Given an element $w \in \mathcal{A}^{\mathbb{Z}}$, the shift space $\mathbb{X}(w)$ of Eq. (4.2) is called the (two-sided, symbolic or discrete) *hull* of w.

If w is a fixed point of a primitive substitution ϱ, the word $\varrho(w_0)$ starts with w_0 and $\varrho(w_{-1})$ ends with w_{-1} by definition. Consequently, w can be obtained from $w_{-1}|w_0$ as an iteration limit, because the length of both $\varrho^n(w_0)$ and $\varrho^n(w_{-1})$ is > 1 for some $n = n_0 \geqslant 1$ (and then for all $n \geqslant n_0$) due to the primitivity of ϱ. The hull $\mathbb{X}(w)$ is then also called the *hull of the substitution* ϱ; we shall later see (in Proposition 4.2 and Theorem 4.1) that this notion is consistent, also with taking powers of ϱ.

Example 4.2 (*A periodic example and the role of legality*). Consider the primitive two-letter substitution rule

$$\varrho : \quad \begin{array}{l} a \mapsto aba \\ b \mapsto bab \end{array}$$

which is an example of a substitution of constant length. Here, each two-letter seed leads to an iteration sequence which converges to a bi-infinite word that is fixed under ϱ. However, only $a|b$ and $b|a$ are legal and produce bi-infinite fixed points according to Definition 4.6, which are periodic sequences with fundamental word ab. The remaining two seeds $a|a$ and $b|b$ are illegal and result in distinct bi-infinite words without non-trivial periods. In fact, they can be obtained from the two fixed points by removing w_0. This is an example of a 'defect', which occurs just once, and not 'repetitively' as all other (finite) subwords of w in the complement of the defect. Legality is included in the definition of a fixed point to avoid such artifacts. $\qquad\Diamond$

Two-sided hulls constructed from a bi-infinite fixed point w of a substitution ϱ on $\mathcal{A}_n$ have various convenient advantages over their one-sided counterparts. An obvious one is the simpler notion of periodicity, which can be stated as $S^i w = w$ for some $i \in \mathbb{N}$. Another one (in the context of Proposition 4.6 below) is the compatibility with inner automorphisms of F_n as follows. Let u be an arbitrary element of F_n and define the conjugate (general) substitution ϱ_u by $\varrho_u(a_i) = u\varrho(a_i)u^{-1}$ on the generators of Γ_n. For an arbitrary $v \in F_n$, the homomorphism property implies $\varrho_u(v) = u\varrho(v)u^{-1}$, because the additional words arising from conjugation always cancel between two letters of the word v.

Under certain circumstances, ϱ and ϱ_u define the same hull. Conversely, the concept of inner automorphisms provides a useful tool for the comparison of distinct (but possibly equivalent) substitutions; see Proposition 4.6 below.

Remark 4.1 (*Hulls from one-sided fixed points*). It is possible to define a two-sided hull from a one-sided fixed point v of a primitive substitution ϱ on a finite alphabet $\mathcal{A}$. To this end, consider v formally as a bi-infinite word, whose letters to the left of the reference point are unknown. The set $\{S^j(v) \mid j \in \mathbb{N}_0\}$ is then a relatively compact subset of the (compact) two-sided shift space $\mathcal{A}^{\mathbb{Z}}$. This means that we can select a converging subsequence $\big(S^{j_i}(v)\big)_{i \in \mathbb{N}}$ with $j_{i+1} > j_i$. Let w be the limit of this subsequence in the product topology of $\mathcal{A}^{\mathbb{Z}}$. By construction, each finite subword of w also occurs in v (in its original one-sided version) and vice versa (due to repetitivity; compare Lemma 4.4 below). The two-sided hull $\mathbb{X}(w)$ is the correct object to consider, and any converging sequence will define the same hull (due to primitivity). It coincides with the hull that is constructed from any bi-infinite fixed point of ϱ or one of its powers. Other constructions (like reflections in the origin) used in the literature rarely produce the correct hull, because they tend to introduce patches that are not present in v and hence illegal. The two sequences of Example 4.2 with a 'defect' are examples of this problem.   ◇

The substitution matrix M_ϱ (as introduced in Definition 4.2) is a powerful tool to derive various statistical properties of a fixed point w of a primitive substitution ϱ (and then also of all elements of its hull $\mathbb{X}(w)$). On the one hand, it keeps track of the power counting for the various letters under an iteration of the substitution, as explained in Example 4.1. As a consequence, the right eigenvector ν_{PF} to the leading or PF eigenvalue λ of M_ϱ can be chosen strictly positive and normalised so that its entries are the relative frequencies of the letters. Note that, under the primitivity assumption for the matrix, it is an easy exercise to derive the existence of the frequencies.

In our Example 4.1, the letters a and b are equally frequent. It is not obvious so far whether the fixed point w has any periods or not. Observing that the letter b always comes in pairs, we can map w into a new word in the alphabet $\{a, B\}$ via $B = bb$. It is easy to check that this turns w into a fixed

point of the square of another substitution rule, namely the one defined by
the rule $a \mapsto aB$ and $B \mapsto aa$. This is the period doubling substitution rule,
which will be discussed below in Section 4.5.1. The absence of any periodic
element in the period doubling hull will then imply the same property for
our original Example 4.1; see Corollary 4.5. An alternative way to detect
non-periodicity will be discussed around Corollary 4.2.

On the other hand, the *left* eigenvector $v_{\mathrm{PF}} = (2, 1)$ paves the way for
a *geometric* realisation of the symbolic sequence as a tiling of the real line,
by two types of intervals (which form the *prototiles*) in this example. Indeed, turning a into an interval of length 2 and b into one of length 1, the
substitution rule can be interpreted as an *inflation rule* for two prototiles as

This rule consists of two steps. First, one 'inflates' the prototiles by the
inflation multiplier $\lambda = \lambda_{\mathrm{PF}}$; second, the inflated prototiles are dissected
into copies of the original prototiles, according to the substitution rule (in
the specified order of the letters). The first step is often combined with the
second and not shown explicitly. The use of the PF eigenvalue λ and left
eigenvector v_{PF} guarantees that the dissection into smaller copies is length
preserving. It is easy to check that this construction is consistent. Moreover,
it is clearly not restricted to our example.

Definition 4.8. Consider a primitive substitution rule ϱ on a finite alphabet
with substitution matrix M_ϱ and PF eigenvalue λ. The associated *geometric
inflation rule* with inflation multiplier λ is obtained by turning the letters a_i
into closed intervals (the *prototiles*) with lengths proportional to the entries
of the left PF eigenvector of M_ϱ, and by dissecting the λ-inflated prototiles
into copies of the original ones, respecting the order specified by ϱ.

Geometric inflation rules will be considered in more detail later. They
are particularly important for generalisations to higher dimensions. They also
provide a natural step in the suspension of the underlying discrete dynamical
system to a continuous one; see [KH95, CFS82, EW11] for background on
suspensions and special flows. Before we turn our attention to the inflation
picture, we continue our symbolic treatment of the one-dimensional case.

4.2. Hulls and their properties

It is important to analyse the hulls in more detail. To this end, let us
introduce an equivalence relation that captures the idea of a local comparison
of words, particularly infinite ones.

Definition 4.9. Two words u and v in the same alphabet are *locally indistinguishable* (or LI for short), and are denoted by $u \overset{\text{LI}}{\sim} v$, when each finite subword of u is also a subword of v and vice versa.

Two finite words are LI if and only if they are equal. The LI concept becomes more relevant for infinite words, and for bi-infinite ones in particular. Then, the LI *class* of a given word $w \in \mathcal{A}^{\mathbb{Z}}$ is defined as

$$\text{LI}(w) := \left\{ z \in \mathcal{A}^{\mathbb{Z}} \mid z \overset{\text{LI}}{\sim} w \right\}.$$

This definition can also be adapted to one-sided sequences.

Lemma 4.2. *If w is a bi-infinite word, its LI class is contained in the hull of w, and one has $\mathbb{X}(w) = \overline{\text{LI}(w)}$. In particular, $\mathbb{X}(u) = \mathbb{X}(v)$ holds for any two bi-infinite words u and v that are LI.*

PROOF. Let $z \in \text{LI}(w)$. For each $m \in \mathbb{N}$, the subword $z_{[-m,m]}$ of z (of length $2m+1$) occurs also as a subword of w. We can thus find an integer j_m such that $S^{j_m}w$ coincides with z on all positions i with $-m \leqslant i \leqslant m$, which means $(S^{j_m}w)_{[-m,m]} = z_{[-m,m]}$. The sequence $(S^{j_m}w)_{m \in \mathbb{N}}$ thus converges to z in the product topology, showing that $z \in \mathbb{X}(w)$. This proves $\text{LI}(w) \subset \mathbb{X}(w)$. Since $S^i w \in \text{LI}(w)$ for all $i \in \mathbb{Z}$, one has

$$\mathbb{X}(w) = \overline{\{S^i w \mid i \in \mathbb{Z}\}} \subset \overline{\text{LI}(w)} \subset \overline{\mathbb{X}(w)} = \mathbb{X}(w),$$

which establishes the first claim.

If u and v are LI, one clearly has $\text{LI}(u) = \text{LI}(v)$, because LI is an equivalence relation. Taking closures and using the first claim completes the proof. $\square$

Example 4.3 (*Hulls versus LI classes*). Let us consider the bi-infinite word $w \in \{a, b\}^{\mathbb{Z}}$ defined by $w_0 = b$ and $w_i = a$ for all $i \neq 0$. Its LI class is $\text{LI}(w) = \{S^i w \mid i \in \mathbb{Z}\}$, while its hull is given by

$$\mathbb{X}(w) = \overline{\text{LI}(w)} = \text{LI}(w) \cup \{w'\},$$

where w' is the bi-infinite word with $w'_i = a$ for all $i \in \mathbb{Z}$, which is periodic. This shows that LI classes need not be closed. Moreover, the hull (which is closed by definition) fails to be a perfect set because each element of the form $S^i w$ is isolated in $\mathbb{X}(w)$. $\diamond$

It is clearly of interest to look more closely at hulls that do *not* show the phenomenon of Example 4.3.

Definition 4.10. A two-sided shift space $\mathbb{X} \subset \mathcal{A}^{\mathbb{Z}}$ is called *minimal* when, for all $w \in \mathbb{X}$, the shift orbit $\{S^i w \mid i \in \mathbb{Z}\}$ is dense in $\mathbb{X}$.

In particular, we use this concept for hulls. The hull of Example 4.3 fails to be minimal, and we need a good criterion for minimality.

Proposition 4.1. *If w is a bi-infinite word in the finite alphabet $\mathcal{A}$, with LI class $\mathrm{LI}(w)$ and hull $\mathbb{X}(w)$, the following assertions are equivalent.*

(1) $\mathbb{X}(w)$ *is minimal;*

(2) $\mathrm{LI}(w)$ *is closed;*

(3) $\mathbb{X}(w) = \mathrm{LI}(w)$.

PROOF. The equivalence of (2) and (3) follows from Lemma 4.2, because $\mathbb{X}(w) = \overline{\mathrm{LI}(w)}$. It remains to show that $\mathbb{X}(w)$ is minimal if and only if $\mathrm{LI}(w) = \overline{\mathrm{LI}(w)}$.

When $\mathrm{LI}(w)$ is not closed, there is an element $z \in \overline{\mathrm{LI}(w)}$ that is not LI with w. Since, by the construction of the hull, z cannot contain any subword that does not occur in w, some finite subword u of w must be missing in z. Consequently, the orbit $\{S^i z \mid i \in \mathbb{Z}\}$ cannot be dense in $\mathbb{X}(w) = \overline{\mathrm{LI}(w)}$, so that the hull is not minimal.

Conversely, when $\mathbb{X}(w)$ is not minimal, there is an element $z \in \mathbb{X}(w)$ such that $\mathbb{X}(w) \setminus \mathbb{X}(z) \neq \varnothing$. In particular, $w \notin \mathbb{X}(z)$, which also implies that $\mathrm{LI}(w) \cap \mathbb{X}(z) = \varnothing$. Since we know that $\overline{\mathrm{LI}(w)} = \mathbb{X}(w) \supset \mathrm{LI}(w) \, \dot\cup \, \mathbb{X}(z)$, the set $\mathrm{LI}(w)$ is not closed. ☐

For the next step, we need the *length* of a finite word w, which we denote by $|w|$. It will only be used when no negative exponents of letters occur.

Lemma 4.3 (Existence of bi-infinite fixed points). *If ϱ is a primitive substitution rule on a finite alphabet $\mathcal{A}_n$ with $n \geqslant 2$, there exists some $k \in \mathbb{N}$ and some $w \in \mathcal{A}_n^{\mathbb{Z}}$ such that w is a fixed point of ϱ^k (which means that $w_{-1} w_0$ is legal and $\varrho^k(w) = w$).*

PROOF. Let $\mathcal{A}_n = \{a_1, \ldots, a_n\}$ as before. Assume that $|\varrho(a_i)| > 1$ for all $1 \leqslant i \leqslant n$. This is no restriction because ϱ is primitive, so that some power of it satisfies this requirement. Define the mapping $g \colon \mathcal{A}_n^2 \longrightarrow \mathcal{A}_n^2$ by $g(xy) = \varrho(x)_{|\varrho(x)|-1} \varrho(y)_0$. The image (of length 2) thus consists of the last letter of $\varrho(x)$ followed by the first of $\varrho(y)$. Note that g maps the non-empty subset of legal two-letter words of ϱ into itself.

Select any initial legal two-letter word and iterate it under g. Since there are only n^2 distinct words of length 2 over $\mathcal{A}_n$, Dirichlet's pigeonhole principle guarantees that at least one legal two-letter word, xy say (which need not be the initial word), must have reappeared after n^2 iterations of g. Consequently, $g^k(xy) = xy$ for some $1 \leqslant k \leqslant n^2$, where x and y need not be distinct. Using $w^{(1)} = x|y$ as a legal seed and iterating ϱ^k via $w^{(i+1)} = \varrho^k(w^{(i)})$ produces a sequence of words of increasing lengths that, by construction, converges towards a bi-infinite fixed point of ϱ^k in the product topology. ☐

Note that the existence of one-sided fixed points follows suit, for instance by choosing the right half of a bi-infinite fixed point.

Proposition 4.2. *Let ϱ be a primitive substitution rule on a finite alpha-bet. Then, any two bi-infinite fixed points u and v of ϱ are locally indistin-guishable. The same conclusion holds if u and v are fixed points of possibly different positive powers of ϱ.*

PROOF. It is sufficient to prove the first claim, as $\varrho^k(u) = u$ and $\varrho^\ell(v) = v$ with $k, \ell \in \mathbb{N}$ implies that u and v are fixed points of $\varrho^{\operatorname{lcm}(k,\ell)}$, which is again primitive.

Let u and v be bi-infinite fixed points of ϱ, and let a be any fixed letter of the alphabet. Any finite subword w of u is also a subword of $z = \varrho^p(u_{-1}|u_0)$ for some $p \in \mathbb{N}$. Since ϱ is primitive and $u_{-1}u_0$ is legal, the latter occurs itself in some $\varrho^q(a)$ with $q \in \mathbb{N}$, so that a translate of z is also a subword of $\varrho^{p+q}(a)$. Observe that a must be a subword of v, again by the primitivity of ϱ. This is then also true of $\varrho^{p+q}(a)$, so that a translate of z, and hence also of w, must be a subword of v. Since this argument is symmetric in u and v, the two fixed points are LI. $\qquad\square$

Proposition 4.2 together with Lemmas 4.2 and 4.3 justify our previous approach to the hull of a primitive substitution via two-sided fixed points, which we formalise as follows.

Definition 4.11. If ϱ is a primitive substitution and w a bi-infinite fixed point of ϱ^n for some $n \in \mathbb{N}$, the hull $\mathbb{X}(w)$ specified by Definition 4.7 is called the (symbolic, two-sided) *hull* of the substitution ϱ.

In particular, the hull of a primitive substitution does not depend on the choice of the fixed point, and it is thus uniquely determined. To use Proposition 4.1 to further characterise the hull, we need another important concept for bi-infinite words.

Definition 4.12. A bi-infinite word w (over a finite alphabet) is called *repet-itive* when every finite subword of w reappears in w with bounded gaps.

Let us expand on the meaning of repetitivity. The word $u = u_0 u_1 \cdots u_{n-1}$ of length n occurs in w if $(S^j w)_{[0,n-1]} = u$ for some $j \in \mathbb{Z}$, where we again use the shorthand $v_{[0,\ell]} = v_0 v_1 \cdots v_\ell$. Now, w is repetitive when the set

$$T_u := \left\{ j \in \mathbb{Z} \mid (S^j w)_{[0,|u|-1]} = u \right\}$$

is relatively dense in $\mathbb{Z}$ for all finite words u that occur in w. This shows that repetitivity can be considered as a special case of almost periodicity, which is a common notion in the theory of dynamical systems; see also Chapter 8. Note that repetitivity as defined above also implies that arbitrary finite patterns in w repeat with bounded gaps, as they can always be related to finite subwords.

Proposition 4.3. *If w is a bi-infinite word in the finite alphabet $\mathcal{A}$, the hull $\mathbb{X}(w)$ is minimal if and only if w is repetitive.*

PROOF. This is a variant of Gottschalk's theorem, see [Pet83, Thm. 4.1.2], which we spell out for our situation.

Assume first that $\mathbb{X}(w)$ is minimal, and let u be a finite subword of w, so that $(S^j w)_{[0,|u|-1]} = u$ for some $j \in \mathbb{Z}$. Define

$$U = \left\{ z \in \mathbb{X}(w) \mid z_{[0,|u|-1]} = u \right\},$$

which is an open neighbourhood of $S^j w$. We now have to show that the set $T_u = \{ m \in \mathbb{Z} \mid S^m w \in U \}$ is relatively dense. Since $\mathbb{X}(w)$ is minimal, the orbit of each element meets U, so that $\mathbb{X}(w) \subset \bigcup_{n \in \mathbb{Z}} S^{-n} U$. The compactness of $\mathbb{X}(w)$ implies the existence of a finite subcover, hence

$$\mathbb{X}(w) \subset \bigcup_{n \in I} S^{-n} U$$

for some finite set $I \subset \mathbb{Z}$. For any $m \in \mathbb{Z}$, there is thus an element $n_m \in I$ with $S^m w \in S^{-n_m} U$, hence $S^{m+n_m} w \in U$. This means $m + n_m \in T_u$, which shows that $T_u - I = \mathbb{Z}$, so T_u is relatively dense.

Conversely, assume w to be repetitive and let $V \subset \mathbb{X}(w)$ be an arbitrary compact neighbourhood of w. Define $R = \{ m \in \mathbb{Z} \mid S^m w \in V \}$, which is relatively dense. This implies $\mathbb{Z} = R + J$ for some finite set $J \subset \mathbb{Z}$, whence

$$\{ S^i w \mid i \in \mathbb{Z} \} = \bigcup_{j \in J} S^j \{ S^i w \mid i \in R \} \subset \bigcup_{j \in J} S^j V.$$

Since the last union is compact, we also have $\mathbb{X}(w) \subset \bigcup_{j \in J} S^j V$. This means that the orbit of any $z \in \mathbb{X}(w)$ has non-empty intersection with V. Since V was arbitrary, it follows that $w \in \mathbb{X}(z)$, and hence $\mathbb{X}(w) = \mathbb{X}(z)$. □

Let us now connect repetitivity with the fixed point property, thus expanding on a structural property that was briefly discussed in Example 4.2.

Lemma 4.4. *Any bi-infinite fixed point of a primitive substitution on a finite alphabet is repetitive.*

PROOF. Let ϱ be a primitive substitution rule on the finite alphabet $\mathcal{A} = \{ a_i \mid 1 \leqslant i \leqslant n \}$, and assume that w is a bi-infinite fixed point of ϱ. By primitivity, there is an integer $k \in \mathbb{N}$ such that a_1 is a subword of $\varrho^k(a_i)$ for all i. Consequently, a_1 occurs with bounded gaps in w.

On the other hand, any finite subword u of w must occur in the substitution word $\varrho^\ell(a_1)$ for some $\ell \in \mathbb{N}$. This implies that also the gaps between consecutive occurrences of u in w are bounded, and w is repetitive. □

As a power of a primitive substitution remains primitive, Lemma 4.4 also applies to fixed points of ϱ^m with $m \in \mathbb{N}$. Moreover, when w is repetitive, then so is any $w' \in \mathrm{LI}(w)$. Let us consider how this extends to the hull.

Theorem 4.1. *Every primitive substitution rule on a finite alphabet possesses a unique hull. This hull consists of a single, closed* LI *class.*

PROOF. Let ϱ be a primitive substitution rule. By Lemma 4.3, there is a bi-infinite fixed point w of ϱ^k for some $k \in \mathbb{N}$. By Definition 4.11, the hull of ϱ is $\mathbb{X}(w)$, which does not depend on the specific fixed point w by an application of Proposition 4.2. We also know that $\mathbb{X}(w) = \overline{\mathrm{LI}(w)}$ from Lemma 4.2. It thus remains to show that $\mathrm{LI}(w)$ is closed.

The primitivity of ϱ in conjunction with Lemma 4.4 implies that the fixed point w is repetitive, so that $\mathbb{X}(w)$ is minimal by Proposition 4.3. This gives $\mathrm{LI}(w) = \overline{\mathrm{LI}(w)}$ by an application of Proposition 4.1. $\square$

This theorem also tells us that the hull of a primitive substitution contains repetitive elements only. It is thus reasonable to call the hull itself *repetitive*, as we shall do from now on. More generally, whenever we are dealing with primitive substitutions, certain properties hold uniformly for all elements of the hull, so that we can view this as a property of the hull.

Remark 4.2 (*Characterisation of hulls via legal words*)**.** For a primitive substitution, each element of its unique (symbolic) hull shares the property that all finite subwords are legal, and are thus also subwords of $\varrho^n(a)$ for some $n \in \mathbb{N}$. Alternatively, the hull can also be defined by this subword property. This is an important alternative that can easily be adapted to situations beyond a single, primitive substitution. In particular, it is more useful when no meaningful fixed point concept is available, for instance when dealing with non-periodic sequences of distinct substitutions. $\Diamond$

We are particularly interested in systems without periodicity. In view of Theorem 4.1, we define a stronger notion of non-periodicity as follows.

Definition 4.13 (*Aperiodic sequences and substitution rules*)**.** A bi-infinite sequence w in a finite alphabet is called (topologically) *aperiodic* when its hull $\mathbb{X}(w)$ contains no periodic sequence. A primitive substitution rule ϱ is *aperiodic* when the unique hull defined by ϱ contains no periodic element.

A counterpart of this concept, namely measure-theoretic aperiodicity, will be introduced in Chapter 11. Clearly, any aperiodic sequence is non-periodic, but not necessarily vice versa. A non-periodic sequence that is also repetitive is aperiodic. However, the sequence $w = \ldots aaabaaa \ldots$ of Example 4.3 is a non-periodic sequence that fails to be aperiodic. The corresponding hull also fails to be minimal. The substitution rule ϱ of Example 4.1 is aperiodic, as will be shown in Corollary 4.5. Let us add another example for later use.

Example 4.4 (*Substitution with minimal PV inflation multiplier*)**.** Consider the primitive three-letter substitution ϱ defined by $a \mapsto b \mapsto c \mapsto ab$, with substitution matrix and graph

$$
\begin{pmatrix} 0 & 0 & 1 \\ 1 & 0 & 1 \\ 0 & 1 & 0 \end{pmatrix} \quad \text{and} \quad
$$

The inflation multiplier is

$$
\beta = \frac{(9+\sqrt{69}\,)^{\frac{1}{3}} + (9-\sqrt{69}\,)^{\frac{1}{3}}}{18^{\frac{1}{3}}} \approx 1.32472,
$$

which is the smallest PV number (mentioned on page 38, and also known as the plastic number). It satisfies the relations $\beta^3 = \beta + 1$ and $\beta^{-1} = \beta^2 - 1$. The corresponding ring of integers $\mathbb{Z}[\beta]$ has rank three. One-sided fixed points can be constructed via ϱ^3, starting from any letter, and two-sided ones via ϱ^6 and the seed $b|a$ (for instance), which is legal. The resulting hull is aperiodic (see Theorem 4.6 below).

A geometric realisation is possible with intervals of lengths 1, β and β^2 for types a, b and c. In any element of the hull, they appear with frequencies $2 - \beta^2$, $\beta^2 - \beta$ and $\beta - 1$, respectively.   ◇

A primitive substitution which is aperiodic leads to a hull that has many interesting properties. We can only mention some, and, for others, have to refer to the extensive literature on the subject; see [PF02, Que10] and references therein. One important feature is *recognisability*, which is also called the *unique composition property* and means that one can locally identify the words that have emerged by the substitution of a single letter. Such words are often called level-1 superwords. This procedure can then be repeated to come to level-2 superwords and so on. Consequently, one can 'invert' the action of the substitution here, which is useful for many theoretical arguments; compare [BFS12]. This property cannot hold for substitutions that define a periodic hull, as one can quickly see from Example 4.2; see [Que10, Sec. 5.5.2] and [Sol98a] for more. Here, one needs non-local information to invert the action of the substitution. Note that subtly different versions of this concept are in use, so care is needed when comparing results.

4.3. Symmetries, invariant measures and ergodicity

For any finite word $w = w_0 w_1 \cdots w_m$, the reflected (or reversed) word is defined as $\widetilde{w} = w_m \cdots w_1 w_0$. When $\widetilde{w} = w$, the word w is called a *palindrome*. The empty word is also considered to be a palindrome for convenience. For a bi-infinite word w, the reflected word (relative to the marker $|$) is specified by $\widetilde{w}_i = w_{-i-1}$. Here, w is an infinite palindrome when $\widetilde{w} = w$ or $\widetilde{w} = Sw$. The former (latter) case takes care of the possibility that the reflection symmetry is relative to the marker (to the centre of the letter w_0). More generally, a bi-infinite word w is also called a palindrome when $\tilde{w} = S^n w$ for some $n \in \mathbb{Z}$. Which version is used will always be clear from the context.

The reflection symmetry of an individual (bi-infinite) word is generally not preserved under local indistinguishability. It is thus reasonable to employ the symmetry concept of dynamical systems instead. A (closed) shift space $\mathbb{X} \subset \mathcal{A}^{\mathbb{Z}}$ is called *reflection symmetric* when $w \in \mathbb{X}$ implies $\tilde{w} \in \mathbb{X}$; compare Definition 5.14 below. In particular, this applies to hulls of the form $\mathbb{X}(w)$.

Definition 4.14. A primitive substitution ϱ on a finite alphabet is called *reflection symmetric* if its hull $\mathbb{X}$ is reflection symmetric. If $\mathbb{X}$ contains at least one infinite palindrome, ϱ is called *palindromic*.

Recall that a primitive substitution ϱ defines a unique hull, which coincides with the LI class of w, via a fixed point w of ϱ^n for some $n \in \mathbb{N}$. If ϱ is palindromic, w (and any element of LI(w)) must contain arbitrarily long palindromic subwords. Such an LI class (and analogously a hull) is then called *palindromic*. Conversely, if an element u of a hull contains palindromic subwords of arbitrary length, one can find a sequence $(S^{k_i} u)_{i \in \mathbb{N}}$ with growing palindromic core, which contains a converging subsequence due to the compactness of the hull. The limit is then an infinite palindrome within the hull, so that these two points of view are equivalent for primitive substitutions.

Let us recall the following result from [HKS95, Lemma 3.1].

Lemma 4.5. *Let ϱ be a primitive substitution on $\mathcal{A} = \{a_1, a_2, \ldots, a_n\}$, with $\varrho(a_i) = pq_i$ for all $1 \leqslant i \leqslant n$, where p and all q_i are palindromes. Then, ϱ is palindromic.*

PROOF. Observe first that the image of a (finite) palindrome w under ϱ has the form $\varrho(w) = pu$, where u is also a palindrome. It is easy to see by induction that $\varrho^k(a_i) = p^{(k)} q_i^{(k)}$ for all $k \in \mathbb{N}$ and all $1 \leqslant i \leqslant n$, where all $p^{(k)}$ and $q_i^{(k)}$ are again palindromes.

Since $\varrho^k(a_1) = p^{(k)} q_1^{(k)}$ contains arbitrarily long palindromic subwords as $k \to \infty$, the hull is palindromic. □

Palindromic primitive substitutions have reflection symmetric hulls. At least for two-letter substitutions, the converse is also true [Tan07].

Remark 4.3 (*Permutation symmetries*). Given an alphabet $\mathcal{A}$ of cardinality n, the symmetric group S_n acts via permutations of the letters. Consequently, one may also consider permutation symmetries of primitive substitutions via the invariance of its hull under a subgroup $G \subset S_n$. For instance, the substitution $a \mapsto aba$, $b \mapsto bab$ of Example 4.2 is obviously S_2-symmetric. Other examples will be the Thue–Morse substitution of Section 4.6 and the binary Rudin–Shapiro sequence of Section 4.7.1. ◇

To further our discussion, we need the standard concept of a positive *measure* on the (compact) hull $\mathbb{X}$ of a substitution; see [RS80, Secs. I.4 and IV.4] for background material. More precisely, we consider regular Borel measures on $\mathbb{X}$. These can be identified with the (continuous) positive linear functionals on $C(\mathbb{X})$, the space of continuous functions on $\mathbb{X}$, the latter equipped with the supremum norm topology. The identification is justified by the Riesz–Markov representation theorem [RS80, Thm. IV.14].

For now, it is sufficient to consider probability measures, which simplifies the task because the set $\mathbb{P}(\mathbb{X})$ of all probability measures on $\mathbb{X}$ is weak-$*$ compact (which follows, for instance, from an application of the Banach–Alaoglu theorem [RS80, Thm. IV.21]). More on measures, in particular unbounded ones in the context of non-compact LCAGs, follows in Section 8.5.

Consider a two-sided shift space $\mathbb{X} \subset \mathcal{A}^{\mathbb{Z}}$ for a finite alphabet $\mathcal{A}$. For any $w \in \mathbb{X}$, the point (or Dirac) measure δ_w is an element of $\mathbb{P}(\mathbb{X})$. It is defined by $\delta_w(A) = 1$ if w is an element of $A \subset \mathcal{A}^{\mathbb{Z}}$, and $\delta_w(A) = 0$ otherwise. Clearly,

$$(4.3) \qquad \mu_N := \frac{1}{2N+1} \sum_{i=-N}^{N} \delta_{S^i w},$$

where S again is the shift, defines a sequence in $\mathbb{P}(\mathbb{X})$. Since $\mathbb{P}(\mathbb{X})$ is compact, this sequence has a converging subsequence, whose limit, μ say, is then a *shift invariant* (or simply *invariant*) element of $\mathbb{P}(\mathbb{X})$, which means that $S.\mu(A) := \mu(S^{-1}(A)) = \mu(A)$ for all Borel sets A, giving the following result.

Lemma 4.6. *If $\mathbb{X}$ is a two-sided shift space over a finite alphabet, the subset $\mathbb{P}_{\mathbb{Z}}(\mathbb{X})$ of shift invariant measures on $\mathbb{X}$ is non-empty.* $\square$

The set $\mathbb{P}_{\mathbb{Z}}(\mathbb{X})$ is convex (since any convex combination of two invariant measures is again invariant) and obviously closed, hence compact again. Note that the elements of this convex set may be distinguished by their behaviour under further symmetry transformations (such as reflections or permutations). An important property is the existence of *extremal* measures, which are measures that cannot be written as the convex combination of two distinct invariant measures.

To highlight the role of extremal measures, we introduce the concept of ergodicity. For its formulation, we need Borel sets A that satisfy $S^{-1}(A) = A$. Sets with this property are called *invariant sets*.

Definition 4.15. Let $\mathbb{X} \subset \mathcal{A}^{\mathbb{Z}}$ be a two-sided shift space. An invariant probability measure μ on $\mathbb{X}$ is called *ergodic* (with respect to the $\mathbb{Z}$-action of the shift) if the measure $\mu(A)$ of any invariant Borel set A is either 0 or 1.

In view of the identification of regular Borel measures with linear functionals on $C(\mathbb{X})$, it is natural to also reformulate ergodicity via (measurable) functions. With $S.f(w) := f(S^{-1}w)$, a function $f \in L^1(\mathbb{X}, \mu)$ is *invariant* if $S.f = f$, which means that $S.f(w) = f(w)$ for μ-almost all $w \in \mathbb{X}$. Now, the measure μ is ergodic if and only if the only invariant functions are constant μ-almost everywhere. An invariant set A gives the invariant function 1_A, while an invariant function f leads to $A_c = \{w \in \mathbb{X} \mid |f(w)| \leqslant c\}$, which (for any c) is invariant up to a null set; see [DGS76] for background.

Lemma 4.7. *Let $\mathbb{X} \subset \mathcal{A}^{\mathbb{Z}}$ be a two-sided shift space. A measure $\mu \in \mathbb{P}_{\mathbb{Z}}(\mathbb{X})$ is ergodic if and only if it is extremal.*

PROOF. To show that extremal implies ergodic, assume that μ fails to be ergodic. Then, an invariant Borel set A exists with $0 < \mu(A) < 1$, and hence also $0 < \mu(A^c) < 1$, as $\mu(A^c) = 1 - \mu(A)$. Note that $A^c = \mathbb{X} \setminus A$ is an invariant set, too. Define two distinct measures ν_1 and ν_2 via

$$\nu_1(B) = \frac{1}{\mu(A)} \mu(B \cap A) \quad \text{and} \quad \nu_2(B) = \frac{1}{\mu(A^c)} \mu(B \cap A^c)$$

for arbitrary Borel sets B. These measures are clearly invariant and normalised. It is immediate that $\mu(B) = \mu(A)\nu_1(B) + (1 - \mu(A))\nu_2(B)$ represents μ as a convex combination, whence it is not extremal.

Conversely, if μ is not extremal, we have $\mu = \alpha\nu_1 + (1-\alpha)\nu_2$ with $0 < \alpha < 1$ and $\nu_1, \nu_2 \in \mathbb{P}_{\mathbb{Z}}(\mathbb{X})$ distinct. It follows that ν_1 is absolutely continuous with respect to μ, since $\mu(A) = 0$ implies $\nu_1(A) = 0$ for all Borel sets A. We thus have a representation as $\nu_1 = h\mu$ with $h \in L^1(\mathbb{X}, \mu)$ by the Radon–Nikodym theorem [RS80, Thm. I.19]. Since $0 \neq \nu_1 \neq \nu_2$, the density h cannot be constant almost everywhere. Since h is an invariant function by construction, μ cannot be ergodic. □

Let us now recall Birkhoff's ergodic theorem from [Wal00, Thm. 1.14 and Sec. 1.6], in our present setting of two-sided shift spaces of the form $\mathcal{A}^{\mathbb{Z}}$; see also [KP06] and references therein for a simplified approach. The ergodic theorem is an important tool to relate orbit and space averages.

Theorem 4.2 (Ergodic theorem for shift spaces). *Let $\mathbb{X} \subset \mathcal{A}^{\mathbb{Z}}$ be a two-sided shift space and let μ be a regular Borel probability measure on $\mathbb{X}$ that is invariant under the shift S. If $f \in L^1(\mathbb{X}, \mu)$, the sequence $\left(\frac{1}{n}\sum_{i=0}^{n-1} f(S^i x)\right)_{n \in \mathbb{N}}$ converges, for μ-almost every $x \in \mathbb{X}$, to a function $F \in L^1(\mathbb{X}, \mu)$ that is S-invariant and satisfies $\int_{\mathbb{X}} F \, \mathrm{d}\mu = \int_{\mathbb{X}} f \, \mathrm{d}\mu$.*

Moreover, if μ is an ergodic probability measure, the function F is constant μ-almost everywhere, and

$$\lim_{n \to \infty} \frac{1}{n} \sum_{i=0}^{n-1} f(S^i x) = \int_{\mathbb{X}} f \, \mathrm{d}\mu$$

holds for μ-almost all $x \in \mathbb{X}$. □

A powerful situation emerges when the convex set of invariant probability measures on $\mathbb{X}$ consists of a single point only (which is then extremal and hence ergodic). The system is then called *uniquely ergodic*. If a uniquely ergodic system is also minimal, it is called *strictly ergodic*. In our context of symbolic dynamics, minimality is equivalent to repetitivity by Proposition 4.3. Let us complete the picture by considering unique ergodicity.

Remark 4.4 (*Refined repetitivity*). One often employs a stronger property than mere repetitivity of $w \in \mathcal{A}^{\mathbb{Z}}$. Due to the finiteness of $\mathcal{A}$, repetitivity of w according to Definition 4.12 implies that, for each $n \in \mathbb{N}$, there is a number N_n such that every subword of w of length at least N_n contains *all* legal words of length n. One can now distinguish different growth bounds for N_n as a function of n. If $N_n = \mathcal{O}(n)$, w is called *linearly repetitive* (in general, other functions for the bounds are possible as well).

Primitive substitutions are linearly repetitive; see [DL06a] for a proof in a slightly more general setting, and for some history of this question. Its geometric counterpart for tilings is proved in [Sol98a]; see also Proposition 5.3 below. Linear repetitivity has strong consequences on the hull $\mathbb{X}(w)$ as a dynamical system. In particular, it implies unique (and hence strict) ergodicity; see [Dur00, Len02] for proofs and [LP03] for generalisations.   ◇

Theorem 4.3. *Let ϱ be a primitive substitution on a finite alphabet. Its hull $\mathbb{X}$ is then strictly ergodic under the $\mathbb{Z}$-action of the shift.*

SKETCH OF PROOF. By Theorem 4.1, the hull $\mathbb{X}$ of ϱ is unique and consists of a single LI class. Proposition 4.1 then implies its minimality. As discussed in Remark 4.4, fixed points of primitive substitutions are linearly repetitive, with uniquely ergodic hull under the shift.   □

An alternative strategy for the last step of the proof can be sketched as follows. Here, unique ergodicity follows from proving that all ergodic measures coincide on the cylinder sets defined by finite words. By the ergodic theorem, the measure of such a cylinder set almost surely coincides with the frequency of the defining word. For a primitive substitution, in the light of Proposition 4.1, the frequencies can be determined from a fixed point (and then apply to any member of the hull). These frequencies are unique by an application of the Perron–Frobenius theorem to the induced substitution matrices (see Section 4.8.3 below), which are all primitive as well [Que10, Prop. 5.10]. This completely specifies the ergodic measures which thus form a singleton set in $\mathbb{P}_{\mathbb{Z}}(\mathbb{X})$. Indeed, the connection between frequencies and unique ergodicity is more general. We follow [Que10, Cor. 4.2], which is a consequence of Oxtoby's theorem; compare [Oxt52] and [Que10, Thm. 4.3]. Note that the orbit averages of continuous functions are determined from those of continuous functions with a finite range (or window) on the sequences, as the latter are dense in $C(\mathbb{X})$ by the Stone–Weierstrass theorem.

Proposition 4.4 (Oxtoby). *Let $\mathcal{A}$ be a finite alphabet and let $\mathbb{X} \subset \mathcal{A}^{\mathbb{Z}}$ be a two-sided shift space. Then, the dynamical system of $\mathbb{X}$ under the shift action is uniquely ergodic if and only if the frequencies of all finite words exist uniformly, for each element of $\mathbb{X}$. Moreover, it is strictly ergodic if and only if all frequencies exist uniformly and are positive.* □

Remark 4.5 (*Unique ergodicity versus minimality*). Let us point out that minimality is not required for unique ergodicity. For instance, consider the hull of the periodic binary sequence w with one defect from Example 4.3. Here, $\mathbb{X}(w) = \{S^i w \mid i \in \mathbb{Z}\} \cup \{w'\}$, with w' the periodic sequence. Clearly, $\delta_{w'}$ is an invariant probability measure on $\mathbb{X}(w)$, which gives no weight to the translation orbit of w. If ν is any invariant probability measure on $\mathbb{X}(w)$, it must satisfy $\nu(\{w\}) = 0$, as σ-additivity otherwise leads to a contradiction. Consequently, $\delta_{w'}$ is the only invariant probability measure here, and $(\mathbb{X}(w), \mathbb{Z})$ is uniquely ergodic, but not minimal.

The reason behind this structure is the restriction to invariant probability measures. Under certain circumstances, it is reasonable to widen the class of invariant measures to also include unbounded ones, in particular in the context of non-primitive substitutions. For a systematic treatment, we refer the reader to [BKMS00, CS11] and references therein. ◇

For the uniquely ergodic setting, we recall a refinement of the ergodic theorem (Theorem 4.2); see [Wal00, Thm. 6.19 and Sec. 6.5] for a proof and further details. It uses the fact that the shift action on $\mathcal{A}^{\mathbb{Z}}$ is continuous.

Theorem 4.4 (Birkhoff's theorem for unique ergodicity). *If, in the situation of Theorem 4.2, μ is the only shift-invariant probability measure on $\mathbb{X}$ and the orbit average is taken for a continuous function f on $\mathbb{X}$, the limit*

$$\lim_{n \to \infty} \frac{1}{n} \sum_{i=k}^{k+n-1} f(S^i x) = \int_{\mathbb{X}} f \, d\mu$$

exists for all $x \in \mathbb{X}$ and $k \in \mathbb{Z}$, and the convergence is uniform in k. □

Let us continue with a result on the combinatorial richness of the hull of symbolic sequences. If w is a periodic sequence, the hull $\mathbb{X}(w)$ consists of one translation orbit, which is closed. In particular, $\mathbb{X}(w)$ contains only finitely many elements, and any invariant measure induces the uniform distribution on the representatives. If w is the bi-infinite sequence of Example 4.3, which is non-periodic (but not repetitive), $\mathbb{X}(w)$ comprises two translation orbits, but altogether still only countably many elements. In view of this, the following result is perhaps surprising.

Theorem 4.5. *Let w be a bi-infinite word in a finite alphabet that is repetitive and non-periodic. Then, the hull $\mathbb{X}(w)$ is uncountable, and even contains uncountably many pairwise disjoint translation orbits.*

PROOF. Note first that the hull $\mathbb{X} = \mathbb{X}(w)$ is minimal by Proposition 4.3, so that w is actually aperiodic ($\mathbb{X}$ contains no periodic element at all; see Definition 4.13). Let μ be an invariant probability measure on $\mathbb{X}$ (so that $\mu(\mathbb{X}) = 1$), which exists by Lemma 4.6. Observe that $\mathbb{X} = \bigcup_{u \in \mathbb{X}}\{u\}$, where the singleton sets $\{u\}$ are closed and measurable, and $\mu(\{u\}) = \mu(\{S^i u\})$ for all $i \in \mathbb{Z}$ by the shift invariance of μ.

Select an arbitrary $u \in \mathbb{X}$. By our assumption, which implies minimality, the orbit of u is dense in $\mathbb{X}$, while the infinitely many singleton sets $\{S^i u\}$ are distinct (as u is non-periodic). So, we must have $\mu(\{u\}) = 0$ (by the σ-additivity of μ). This holds for all $u \in \mathbb{X}$. As the countable union of null sets is a null set, $\mathbb{X}$ must be composed of uncountably many elements. Since any translation orbit is countable, $\mathbb{X}$ must also partition into uncountably many translation orbits. $\qquad\square$

An alternative argument will be mentioned below after Proposition 4.5. Theorem 4.5 in particular means that, for a fixed point w of a primitive, aperiodic substitution, uncountably many new elements (and even translation orbits) arise by taking the closure of the (countable) shift orbit $\{S^j w \mid j \in \mathbb{Z}\}$. In terms of translation orbits, one has the following dichotomy.

Corollary 4.1. *The symbolic hull of a repetitive word w over a finite alphabet consists of either one (periodic case) or uncountably many (non-periodic case) pairwise disjoint $\mathbb{Z}$-orbits under the shift action. In the periodic case, the hull itself is a finite set.* $\qquad\square$

Another important dichotomy is the existence or non-existence of *proximal* elements. In a periodic hull, two elements u, v are either equal or substantially different, in the sense that the distance $d_{\mathrm{F}}(S^n u, S^n v)$ is bounded away from 0 for all $n \in \mathbb{Z}$. In an aperiodic hull, this is not the case. In fact, one then always has a pair of (asymptotically) proximal elements, meaning $u \neq v$ with $d_{\mathrm{F}}(S^n u, S^n v) \longrightarrow 0$ for $n \to \infty$ or $n \to -\infty$; see [BDH03, BD07] and [BO14, Prop. 1] for details. We shall often meet this phenomenon in the slightly more special situation of a *singular* element of a hull, which agrees with another element on almost all positions in a way that is compatible with proximality. This means that the positions of mismatches have upper density 0 (and hence density 0, both with respect to an averaging sequence of centred intervals of growing length), and that the sequences satisfy the proximality condition. The existence of singular elements is the standard situation for hulls of sequences that can also be described by the projection method, while the hull of the Thue–Morse sequence in Section 4.6 will provide an example of proximal elements of a different kind. Here, the proximal pair consists of two sequences that agree on all non-negative positions, and differ on all other ones. This type of structure is also well-known as a relevant tool in Kotani theory of random Schrödinger operators; compare [CL90, Prop. VII.5.1] and its proof. We will revisit this in the setting of point sets in Section 5.5.

Corollary 4.2. *If w is a repetitive, bi-infinite word in a finite alphabet, the hull $\mathbb{X}(w)$ is aperiodic if and only if it contains a proximal pair.*

PROOF. Aperiodicity implies the existence of distinct elements u, v in the hull that are proximal by [BO14, Prop. 1]; see also Proposition 5.6 below. Since periodic hulls cannot contain a proximal pair, the claim is clear.   □

This innocently looking criterion provides a powerful method to decide on aperiodicity of primitive substitutions, particularly in situations where other simple criteria (such as the irrationality of the PF eigenvalue of the substitution matrix in Theorem 4.6 below) fail.

Proposition 4.5. *The symbolic hull $\mathbb{X}$ of a repetitive, bi-infinite word w over a finite alphabet $\mathcal{A}$ is either finite or a Cantor set.*

PROOF. Due to repetitivity, w is either periodic (with finite hull due to Corollary 4.1) or aperiodic in the sense of Definition 4.13, then with uncountable hull. The hull is compact and non-empty by construction. It is also metrisable, for instance via the metric d_F of Eq. (4.1). Since $\mathbb{X} = \mathbb{X}(w)$ is a subset of the totally disconnected space $\mathcal{A}^{\mathbb{Z}}$, it is totally disconnected as well.

It remains to show that an aperiodic $\mathbb{X}$ is perfect. Assume that $u \in \mathbb{X}$ is isolated, which means that $\{u\}$ is an open set in $\mathbb{X}$, in the topology induced by the product topology of $\mathcal{A}^{\mathbb{Z}}$. This implies that some open set $O \subset \mathcal{A}^{\mathbb{Z}}$ exists such that $\{u\} = \mathbb{X} \cap O$, hence also $\{u\} = u_{\mathbb{Z}} \cap O$, where $u_{\mathbb{Z}} := \{S^i u \mid i \in \mathbb{Z}\}$. However, a basis for the open sets in $\mathcal{A}^{\mathbb{Z}}$ is given by the cylinder sets that specify a finite patch. No such set can single out the element u from $u_{\mathbb{Z}}$ because u is repetitive, wherefore $\{u\}$ cannot be isolated.   □

The Cantor property is important in the theory of aperiodic order. It also provides an independent proof of the uncountability of aperiodic hulls, because any Cantor set is homeomorphic to the classic middle thirds Cantor set, which is uncountable; compare Definition 2.8 and Figure 8.5.

4.4. Metallic means sequences

The golden mean τ belongs to the family of so-called *metallic means*, defined as the roots of the polynomial $x^2 - px - q$ with $p, q \in \mathbb{N}$; see [dSp99] for a summary. Such a polynomial appears as the characteristic polynomial of the substitution matrix $\left(\begin{smallmatrix} p & q \\ 1 & 0 \end{smallmatrix}\right)$. The polynomial is reducible over $\mathbb{Q}$ if and only if $p^2 + 4q$ is a square. Both eigenvalues are then integers, as in Example 4.1, which corresponds to $p = 1$, $q = 2$ and is sometimes referred to as the 'copper mean' case [dSp99]. The positive root for $q = 1$ with $p \in \mathbb{N}$ is a quadratic algebraic integer (in fact, a unit) with periodic continued fraction expansion $[p; p, p, p, \ldots]$. These numbers are often referred to as the *noble means*.

Let us recapitulate our various notions with another simple example.

Example 4.5 (*Silver mean substitution*). Consider the substitution

$$(4.4) \qquad\qquad \varrho = \varrho_{\mathrm{sm}} : \begin{array}{l} a \mapsto aba \\ b \mapsto a \end{array}$$

which allows the construction of a bi-infinite (and reflection symmetric) fixed point as follows. Starting from the legal seed $w^{(1)} = a|a$, with the vertical line denoting the reference point, and defining $w^{(i+1)} = \varrho(w^{(i)})$ for $i \geqslant 1$, one obtains the iteration sequence

$$a|a \xrightarrow{\varrho} aba|aba \xrightarrow{\varrho} abaaaba|abaaaba \xrightarrow{\varrho} \cdots \xrightarrow{\varrho} w^{(i)} \xrightarrow{i \to \infty} w = \varrho(w)$$

where w is a bi-infinite word in the alphabet $\{a, b\}$ and the limit of the iteration sequence is taken (as usual) in the product topology. The seed $a|a$ leads to a fixed point that is an infinite palindrome. Consequently, the possible finite subwords of w are either palindromes or come in reversed pairs. The same property is inherited by the entire hull $\mathbb{X}_{\mathrm{sm}} = \mathbb{X}(w)$, in the sense that $v \in \mathbb{X}_{\mathrm{sm}}$ implies $\tilde{v} \in \mathbb{X}_{\mathrm{sm}}$, so that $\mathbb{X}_{\mathrm{sm}}$ is reflection symmetric and palindromic according to Definition 4.14. Note that the reflection symmetry also implies that two subwords which are reversed copies of each other are equally frequent.

The corresponding substitution matrix and graph read

$$M_{\mathrm{sm}} = \begin{pmatrix} 2 & 1 \\ 1 & 0 \end{pmatrix} \quad \text{and} \quad$$

where the loop at a has double weight. The symmetric matrix is primitive, with PF eigenvalue $\lambda_{\mathrm{sm}} = 1 + \sqrt{2}$. This is a PV unit, with defining polynomial $x^2 - 2x - 1$ and algebraic conjugate $1 - \sqrt{2} \approx -0.4142$. This is why ϱ is a so-called binary *PV substitution*. It is often referred to as the *silver mean* substitution, due to the continued fraction expansion $\lambda_{\mathrm{sm}} = [2; 2, 2, 2, \ldots]$, in comparison to $\tau = [1; 1, 1, 1, \ldots]$ for the golden mean. One convenient choice for the geometric inflation rule according to Definition 4.8 represents a by an interval of length $1 + \sqrt{2}$, and b by one of length 1, thus giving

The fixed point of the substitution rule (with the reference marker as its 'origin') is thus turned into a fixed point of the (induced) geometric inflation rule (with the marker being mapped to the point 0). This gives a face-to-face tiling of $\mathbb{R}$ by two types of intervals (the prototiles). We also refer to this structure as the *silver mean chain*. In a second step, we extract two point sets $\Lambda_a, \Lambda_b \subset \mathbb{R}$ as the left endpoints of the two types of intervals. The *silver mean point set* is then given by $\Lambda_{\mathrm{sm}} = \Lambda_a \dot{\cup} \Lambda_b$.

The relative letter (or interval) frequencies of the fixed point w are $\frac{1}{2}\sqrt{2}$ and $\frac{1}{2}(2 - \sqrt{2})$, which are irrational numbers. An immediate consequence is that the silver mean chain cannot have any non-trivial period, because pattern frequencies in periodic words are always rational numbers.

Let us finally mention that the matrix M_{sm} is also compatible with two further (non-negative) substitutions. As will be discussed in Remark 4.7, they define the same hull. A closely related hull will also appear in the context of the octagonal tiling in Remark 6.1. $\Diamond$

More generally, one has the following criterion for the aperiodicity of a fixed point (and hence the hull) of a primitive substitution.

Theorem 4.6. *Let ϱ be a primitive substitution rule on a finite alphabet with substitution matrix M_ϱ, and let w be a bi-infinite fixed point of ϱ. If the PF eigenvalue of M_ϱ is irrational, the sequence w is aperiodic.*

PROOF. Assume, to the contrary, that w itself has a non-trivial period and hence a periodically repeated fundamental word of finite length. This forces all (relative) letter frequencies of w to be rational numbers, which are the entries of the right PF eigenvector of M_ϱ. Since the latter is an integer matrix, its PF eigenvalue must then be rational, which is a contradiction. So, w is non-periodic.

Since $\mathbb{X}(w) = \mathrm{LI}(w)$ by Theorem 4.1, the hull cannot contain any periodic element, so that w (and hence ϱ) is aperiodic. $\square$

Recall that any primitive substitution rule ϱ permits the construction of a bi-infinite fixed point, possibly by employing a suitable power of ϱ. Since $M_{\varrho^i} = M_\varrho^i$ and the PF eigenvector is the same for M_ϱ and M_ϱ^i, Theorem 4.6 extends to any such fixed point as well (even when λ_{PF}^i becomes rational, along with a degeneracy of the largest eigenvalue). Primitivity is also preserved by taking powers, so that again all elements of the hull $\mathbb{X}(w)$ must be non-periodic.

Corollary 4.3. *A primitive substitution rule with irrational PF eigenvalue (inflation multiplier) is aperiodic. The same conclusion applies to the induced geometric inflation rule.* $\square$

Example 4.6 (*Fibonacci substitution*). Arguably, the best-studied aperiodic, primitive substitution rule is

$$\varrho = \varrho_{\mathrm{F}} : \begin{array}{c} a \mapsto ab \\ b \mapsto a \end{array}$$

with substitution matrix $M_{\mathrm{F}} = \left(\begin{smallmatrix} 1 & 1 \\ 1 & 0 \end{smallmatrix}\right)$, PF eigenvalue $\lambda = \tau$ and corresponding eigenvectors with entries of ratio τ. The associated graph is essentially that of Examples 4.1 and 4.5, but with all edges having unit weight.

Starting an iteration of ϱ on a (one-sided) seed $w^{(0)} = a$ produces a sequence of words $w^{(i)}$ of length f_{i+2}, where the f_i are the *Fibonacci numbers*[1] defined by the initial condition $f_0 = 0$ and $f_1 = 1$ together with the recursion

$$f_{i+1} = f_i + f_{i-1} \quad \text{and} \quad f_{i-1} = f_{i+1} - f_i$$

for $i \geqslant 1$ (forward direction) and $i \leqslant 0$ (backward direction). With this convention, one has

$$M_{\mathrm{F}}^i = \begin{pmatrix} f_{i+1} & f_i \\ f_i & f_{i-1} \end{pmatrix}$$

for $i \in \mathbb{Z}$, and many statistical properties of words and their behaviour under iterations of ϱ can be expressed by the Fibonacci numbers and their variants. Let us note that

$$(4.5) \qquad \lim_{i \to \pm\infty} \frac{f_{i+1}}{f_i} = \frac{1 \pm \sqrt{5}}{2} = \begin{cases} \tau \\ \tau' \end{cases}$$

which shows how the two algebraic conjugates τ and τ' are related to the ratio limits in the two different directions of the $\mathbb{Z}$-indexed Fibonacci numbers. More generally, one has Binet's formula,

$$(4.6) \qquad f_m = \frac{1}{\sqrt{5}}\left(\tau^m - (\tau')^m\right) = \frac{1}{\sqrt{5}}\left(\tau^m - \frac{(-1)^m}{\tau^m}\right),$$

which holds for all $m \in \mathbb{Z}$. An efficient calculation of the f_m is possible via $f_0 = 0$, the symmetry relation $f_{-m} = (-1)^{m+1} f_m$, and the observation that, for all $m \in \mathbb{N}$, the number f_m is the integer closest to $\tau^m/\sqrt{5}$.

A two-sided hull $\mathbb{X}$ can once again be constructed via a fixed point of ϱ^2, obtained from the legal seed $a|a$ as a limiting 2-cycle

$$a|a \xmapsto{\varrho} \underline{\underline{ab}}|ab \xmapsto{\varrho} \underline{aba}|aba \xmapsto{\varrho} abaab|abaab \xmapsto{\varrho} abaababa|abaababa$$

$$\xmapsto{\varrho} abaababaabaab|abaababaabaab \xmapsto{\varrho} \cdots$$

---

[1] Fibonacci numbers are named after Leonardo of Pisa, known as Fibonacci. His book *Liber Abaci* (1202) introduced the sequence, which apparently was previously described in Indian mathematics [Sin85].

The two members of the 2-cycle differ only at the underlined positions. Nevertheless, they are locally indistinguishable and define the same hull $\mathbb{X}$. Accordingly, there are precisely two fixed points of ϱ^2, distinguished by the two seeds $a|a$ and $b|a$. The Fibonacci substitution is aperiodic due to Corollary 4.3. Since the 2-cycle alternates between two elements that are singular and hence proximal, aperiodicity also follows from Corollary 4.2.

A natural choice for the geometric representation, compare Definition 4.8, employs intervals of lengths τ and 1 for the letters a and b. Considering the left endpoints of all intervals and writing $\nu_{\mathrm{PF}} = (\nu_a, \nu_b)^T$, the average distance between neighbouring points is

$$\bar{\ell} = \nu_a \tau + \nu_b 1 = \frac{\tau^2}{\tau + 1} + \frac{1}{\tau + 1} = \frac{\tau + 2}{\tau + 1} = 3 - \tau,$$

which implies that the resulting point set has density $(\tau + 2)/5$. $\diamond$

Let us investigate some further properties of the Fibonacci substitution.

Corollary 4.4. *The Fibonacci substitution of Example* 4.6 *is aperiodic and palindromic. Its hull is thus reflection symmetric.*

PROOF. The aperiodicity follows from Corollary 4.3 (or from Corollary 4.2), while Lemma 4.5 applies with $p = a$, $q_a = b$ and q_b the empty word. Since palindromicity implies reflection symmetry, the last claim is clear. $\square$

Remark 4.6 (*An alternative substitution rule*). Quite frequently, one also finds the rule

$$\varrho' : \begin{array}{l} a \mapsto ba \\ b \mapsto a \end{array}$$

instead of the Fibonacci substitution rule ϱ from Example 4.6. This rule has the same substitution matrix M_{F}. In fact, ϱ' defines the same hull as ϱ, as one can show explicitly in this case. This rests upon the conjugation relation

$$\varrho(v) = a\varrho'(v)a^{-1},$$

which is valid for all finite words v in the alphabet $\{a, b\}$. In particular, the iterations of the squares of the rules from the seed aa, combined with suitable shifts, produce converging sequences with growing equal cores, which then define the same hull.

Alternatively, one can observe that suitably shifted copies of $\varrho^{3n}(aa)$ produce a sequence with growing *palindromic* core, hence showing the invariance of the hull under reflection. Since the iterates of $a|a$ under ϱ' are precisely the reflected copies of the iterates under ϱ, the two hulls are reflections of one another, and thus equal by the symmetry property. $\diamond$

The alternative rule of Remark 4.6 reveals an interesting reflection property of the Fibonacci substitution.

Lemma 4.8. *If $\varrho = \varrho_{\mathrm{F}}$ is the Fibonacci substitution of Example 4.6 and w is any finite word in the alphabet $\{a, b\}$, one has $\varrho(\widetilde{w})\, a = a\, \widetilde{\varrho(w)}$.*

PROOF. Recall that $\widetilde{w}$ denotes the reversal of w, and assume $w = w_0 \cdots w_n$. Using ϱ' from Remark 4.6, one finds

$$\widetilde{\varrho(w)} = \widetilde{\varrho(w_n)} \cdots \widetilde{\varrho(w_0)} = \varrho'(w_n) \cdots \varrho'(w_0)$$
$$= a^{-1}\varrho(w_n)\, a \, \cdots \, a^{-1}\varrho(w_0)\, a = a^{-1}\varrho(\widetilde{w})\, a.$$

Multiplying with a from the left establishes the claim.   □

It is also possible to consider substitutions that consist of arbitrary (finite) concatenations of ϱ and ϱ'. All of them define the same hull, which is perhaps surprising (because this is certainly not a general property for distinct substitutions). A proof of this claim can be formulated via the Sturmian property of the Fibonacci rules that we discuss later in Example 4.11.

Let us expand on the role of inner automorphisms for the comparison of substitution rules. Recall that ϱ_u, with u a finite word, denotes the conjugate substitution defined by $\varrho_u(a_i) = u\varrho(a_i)u^{-1}$ on the alphabet $\mathcal{A} = \{a_1, \ldots, a_n\}$.

Proposition 4.6. *Let ϱ be a primitive substitution on the finite alphabet $\mathcal{A}$, and let u be a finite word such that ϱ_u is a (non-negative) substitution as well. Then, ϱ_u is primitive, and ϱ and ϱ_u define the same hull.*

PROOF. The first claim is obvious, because ϱ and ϱ_u possess the same substitution matrix, and hence the same directed substitution graph. Since ϱ is primitive, Lemma 4.3 guarantees that there is some $k \in \mathbb{N}$ and some bi-infinite word w such that $\varrho^k(w) = w$, where the two-letter core (or seed) $w_{-1}|w_0$ of w is legal for ϱ. The hull of ϱ is $\mathbb{X}(w) = \mathrm{LI}(w)$ by Theorem 4.1, and w is repetitive by Lemma 4.4.

With $w_{[-n,n]} = w_{-n} \cdots w_{-1}|w_0 \cdots w_n$ as above, one has

$$\varrho_u^k(w_{[-n,n]}) = v\varrho^k(w_{[-n,n]})v^{-1},$$

where $v = u\varrho(u)\varrho^2(u) \cdots \varrho^{k-1}(u)$, by an iterated application of the conjugation relation. Observe that, for fixed $k \in \mathbb{N}$, $\varrho_u^k(w_{[-n,n]}) \xrightarrow{n \to \infty} \varrho_u^k(w)$ in the product topology, as well as

$$v\varrho^k(w_{[-n,n]})v^{-1} \xrightarrow{n \to \infty} S^{\ell}\varrho^k(w) = S^{\ell}w \in \mathbb{X}(w)$$

for some $\ell \in \mathbb{Z}$. The shift by ℓ positions is a result of the conjugation by v. Repeated application of ϱ_u^k shows that $\varrho_u^{mk}(w) = S^{\ell_m}w$ with $\ell_m \in \mathbb{Z}$.

Recalling the proof of Lemma 4.3, we know that, after at most $|\mathcal{A}|^2$ steps, one seed $\alpha|\beta$ must have occurred twice, with r steps in between. Then, the iteration of ϱ_u^{rk} on a shifted version of w with seed $\alpha|\beta$ converges to some bi-infinite word z with $\varrho_u^{rk}(z) = z$. By construction, z is an element of $\mathbb{X}(w)$, and thus LI with w, wherefore z is repetitive. The subword $\alpha|\beta$ must then reoccur in z on the right hand side of β, so that it must be a subword of $\varrho_u^{sk}(\beta)$ for some $s \in \mathbb{N}$. This means that $\alpha|\beta$ is legal for ϱ_u, so that z is a fixed point of ϱ_u^{rk}. The hull of ϱ_u is then $\mathbb{X}(z) = \mathbb{X}(w)$. $\qquad\square$

Remark 4.7 (*Noble means substitutions*). Consider the substitution matrix $M = \left(\begin{smallmatrix} p & 1 \\ 1 & 0 \end{smallmatrix}\right)$ with $p \in \mathbb{N}$. It is compatible with the noble mean substitution

$$\varrho : \begin{array}{l} a \mapsto a^p b \\ b \mapsto a \end{array}$$

with PF eigenvalue $\lambda = \frac{1}{2}\left(p + \sqrt{p^2 + 4}\right) = [p; p, p, p, \ldots]$, which is always a PV unit. Its algebraic conjugate is $\lambda' = p - \lambda$. For $p = 1$, we see that ϱ is the Fibonacci substitution of Example 4.6, while $p = 2$ differs from the silver mean substitution used in Example 4.5. However, these two versions are related by conjugation with the word a, and hence define the same hull by Proposition 4.6, analogously to the situation met in Remark 4.6.

For a general $p \in \mathbb{N}$, we have the freedom to place the letter b anywhere in the word $\varrho(a)$, which results in $p + 1$ distinct choices. They can be conjugated into each other by suitable powers of a, so that all (non-negative) substitution rules for M with fixed p define the same (symbolic) hull. The relative frequencies of the letters a and b are $\frac{\lambda}{\lambda+1}$ and $\frac{1}{\lambda+1}$. A natural geometric realisation, via the inflation rule according to Definition 4.8, works with intervals of length λ (for a) and 1 (for b). $\qquad\Diamond$

The existence of distinct substitutions that define the same hull opens an interesting possibility for the (local) mixture of such rules. We shall briefly come back to this point of view in Section 11.2.3.

Example 4.7 (*Platinum mean substitution rules*). The algebraic integer $\lambda_{\mathrm{pm}} = 2 + \sqrt{3}$ (the fundamental unit of $\mathbb{Z}[\sqrt{3}]$) does not appear among the metallic means discussed above, but is sometimes referred to as the platinum number, alluding to its regular continued fraction expansion $\lambda_{\mathrm{pm}} = [3; 1, 2, 1, 2, \ldots]$, although 'electrum number' would seem a more logical choice. It is a PV unit with defining polynomial $x^2 - 4x + 1$. The latter is the characteristic polynomial of six distinct non-negative integer matrices, namely

$$M_1 = \begin{pmatrix} 3 & 2 \\ 1 & 1 \end{pmatrix}, \quad M_2 = \begin{pmatrix} 1 & 2 \\ 1 & 3 \end{pmatrix}, \quad M_3 = \begin{pmatrix} 2 & 3 \\ 1 & 2 \end{pmatrix},$$

together with their transposed versions. Each matrix is compatible with a variety of different (non-negative) substitution rules, all of which are primitive and aperiodic, with the standard consequences for the corresponding hulls.

Let us discuss the rules that are compatible with M_1 in more detail, while leaving all PF exercises to the reader. The possible images of a are $aaab$, $aaba$, $abaa$ and $baaa$, while those of b are aab, aba and baa, thus giving 12 distinct substitution rules. Nevertheless, in the light of Proposition 4.6, some may still define the same hull.

The 12 substitution rules partition into five equivalence classes relative to inner automorphisms as follows. The largest class contains six rules and is represented by $a \mapsto aaab$, $b \mapsto aab$. Its hull is reflection symmetric and distinguished from the others in that it contains neither the subword bb nor bab. Next, there are two pairs of equivalent rules, represented by $a \mapsto aaab$, $b \mapsto aba$ and by $a \mapsto baaa$, $b \mapsto aba$. Their hulls are the only ones to contain the subword bab. They differ in the word $aaaabaa$ which is only a subword of the first hull, while its reflected version $aabaaaa$ shows up only in the second (the hulls are images of one another under reflection). Two rules remain, namely $a \mapsto aaab$, $b \mapsto baa$ and $a \mapsto baaa$, $b \mapsto aab$. Their hulls, which are mapped onto each other by reflection again, are singled out by containing the subword bb; they differ in the words $baaba$ and $abaab$ which only occur as subwords in the first and the second hull, respectively.

It is clear that the reflection symmetric hull is different from the others. The (topological) distinction of the other examples is less obvious. However, on the basis of the Anderson–Putnam complex [AP98], one can determine their topological invariants[2] and conclude that a hull with subword bb cannot be homeomorphic with any of the hulls that contain bab.   ◊

At this point, we turn our attention to another important class of examples, namely substitutions with an integer inflation multiplier (or factor).

4.5. Period doubling and paper folding

The period doubling sequence has its origin in chaotic dynamics. More precisely, when following the symbolically coded itinerary of the unique critical point under the iteration of a unimodal map, one encounters certain periodic itineraries along a period doubling route to chaos. Here, the fundamental block is doubled in each step and follows the period doubling substitution to be described below. This property (the genealogy of periodic points) is part of the kneading theory in one-dimensional dynamics; see [Dev89, Sec. 1.19] or [CE80, Sec. II.3] for details.

The structure of a repeatedly folded sheet of paper leads to an interesting combination of a four-letter substitution rule with a subsequent reduction to a two-letter alphabet; see [AMF95, Ch. I.3 and Part II] and [AS03, Ch. 6], as

[2]We thank Franz Gähler for discussions; see [Gäh13] for further details and examples.

well as references therein, and [B-AQS13] for some higher-dimensional analogues. Both sequences provide examples of aperiodic substitutions with an integer inflation multiplier, which have an underlying limit-periodic structure (in the sense of [GK97]); compare the proof of Proposition 4.7 below as well as Chapter 8. Since this mechanism is rather different from what we have seen above and will also play an important role in higher dimensions, we discuss these two sequences in some detail.

Many important examples in this spirit have the property that the images of all letters under the substitution have equal length. They are called *substitutions of constant length*. Here, the left PF eigenvector is always a multiple of $(1, 1, \ldots, 1)$, so that the geometric picture leads to a realisation with intervals of equal length and hence does not add anything new to the picture. Note, however, that the frequencies of the letters in a fixed point need not be equal, as we shall see shortly.

4.5.1. THE PERIOD DOUBLING SUBSTITUTION

The period doubling substitution (of constant length) is given by

$$(4.7) \qquad \varrho = \varrho_{\mathrm{pd}} : \quad \begin{matrix} a \mapsto ab \\ b \mapsto aa \end{matrix}$$

which was mentioned earlier on page 74 in connection with the substitution rule of Example 4.1. This is a special case of the class of *Toeplitz sequences* as introduced in [JK69]. The substitution matrix and graph of ϱ are

$$M_{\mathrm{pd}} = \begin{pmatrix} 1 & 2 \\ 1 & 0 \end{pmatrix} \quad \text{and} \quad$$

where the edge from b to a has weight 2. The leading eigenvalue of M_{pd} is $\lambda = 2$, which is the constant word length of the substitution. Starting from the legal seed $a|a$, an iteration of ϱ gives

$$\underline{a}|a \xmapsto{\varrho} a\underline{b}|ab \xmapsto{\varrho} aba\underline{a}|abaa \xmapsto{\varrho} abaaaba\underline{b}|abaaaabab$$
$$\xmapsto{\varrho} abaaabababaaaba\underline{a}|abaaabababaaaabaa$$
$$\xmapsto{\varrho} \ldots abaaabababaaaba\underline{b}|abaaabababaaaabaa \ldots \xmapsto{\varrho} \cdots$$

which converges on all positions except the underlined one. The latter position alternates between a and b, so that we have created a 2-cycle under ϱ, and hence two bi-infinite fixed points of ϱ^2. These two fixed points are LI and differ at one position ($i = -1$) only. Let w be the fixed point of ϱ^2 obtained from the seed $a|a$, and $w' = \varrho(w)$. They form a proximal pair (so Corollary 4.2 applies). Both w and w' are palindromes, with $w_{-n-2} = w_n$ for all $n \in \mathbb{Z}$, and analogously for w'.

An application of PF theory reveals that the letter a is twice as frequent as b, while a geometric representation requires intervals of equal lengths (which may then be distinguished by colour). Consider w, choose intervals of length 1, and take the left endpoints as their characteristic (or control) points, denoted by Λ_a and Λ_b. This gives $\Lambda_a \,\dot\cup\, \Lambda_b = \mathbb{Z}$ with

$$\Lambda_a = \{i \mid w_i = a\} \quad \text{and} \quad \Lambda_b = \{i \mid w_i = b\}.$$

The geometric fixed point equation for ϱ^2 now implies the identities

$$
\begin{aligned}
\Lambda_a &= 4\Lambda_a \,\dot\cup\, (4\Lambda_a + 2) \,\dot\cup\, (4\Lambda_a + 3) \,\dot\cup\, 4\Lambda_b \,\dot\cup\, (4\Lambda_b + 2), \\
\Lambda_b &= (4\Lambda_a + 1) \,\dot\cup\, (4\Lambda_b + 1) \,\dot\cup\, (4\Lambda_b + 3),
\end{aligned}
\tag{4.8}
$$

as can be deduced from $\varrho^2(a) = abaa$ and $\varrho^2(b) = abab$. These equations can be decoupled to obtain

$$\Lambda_a = 2\mathbb{Z} \,\dot\cup\, (4\Lambda_a + 3) \quad \text{and} \quad \Lambda_b = (4\mathbb{Z} + 1) \,\dot\cup\, (4\Lambda_b + 3).$$

By iteration, these lead to the solutions

$$
\begin{aligned}
\Lambda_a &= \dot{\bigcup_{i \geqslant 0}} \left(2{\cdot}4^i\mathbb{Z} + (4^i - 1)\right) \,\dot\cup\, \{-1\}, \\
\Lambda_b &= \dot{\bigcup_{i \geqslant 1}} \left(4^i\mathbb{Z} + (2{\cdot}4^{i-1} - 1)\right),
\end{aligned}
\tag{4.9}
$$

where the right-hand sides are unions of disjoint sets. The role of the singleton set $\{-1\}$ is exceptional in the sense that it emerges from the iteration only as a limit (in the topology of the 2-adic numbers of Example 2.10). Adding it to Λ_a, as in Eq. (4.9), corresponds to the fixed point w, while moving it to Λ_b would give w'. Due to the conditions that $\Lambda_a \cap \Lambda_b = \varnothing$ and $\Lambda_a \cup \Lambda_b = \mathbb{Z}$, one can show that (4.9) and its partner for w' are the *only* solutions of Eq. (4.8). In general, finding all solutions of such equations is a difficult, and still unsolved, problem.

Proposition 4.7. *The period doubling sequence is aperiodic.*

PROOF. Since we already identified a pair of proximal elements in the hull of the primitive substitution ϱ_{pd}, the claim follows from Theorem 4.3 and Corollary 4.2. Let us nevertheless present an alternative argument, as it highlights another important mechanism for aperiodicity.

It is clear from Eq. (4.9) that the sequence w cannot have any nontrivial period because Λ_a is a disjoint union of translated lattices of increasing periods (all powers of 2) with increasing translations (all odd). The only compatible period is thus 0. Since ϱ is a primitive substitution, the hull $\mathbb{X}(w)$ coincides with $\mathrm{LI}(w)$ by Proposition 4.1, so that all elements of the hull must be non-periodic as well, which was our definition of aperiodicity. □

Recall that the primitive substitution rule of Example 4.1 could be turned into the period doubling substitution by a simple morphism. In the converse direction, doubling each letter b in a period doubling sequence (meaning any member of the hull) turns it into a sequence of the hull of Example 4.1. Proposition 4.7 thus has the following immediate consequence.

Corollary 4.5. *The primitive substitution rule $a \mapsto abb$, $b \mapsto a$ of Example 4.1 is aperiodic as well.* $\qquad\Box$

Aperiodic systems cannot have any non-trivial translation symmetries. However, they can show other interesting symmetries. In our present setting, the most obvious question is that of the behaviour under reflection. As before, this can be dealt with systematically on the level of the hull.

Proposition 4.8. *The hull of the period doubling substitution $\varrho = \varrho_{\mathrm{pd}}$ from Eq. (4.7) is reflection symmetric.*

PROOF. Let w be a two-sided fixed point of ϱ^2, and set $\varrho'\colon a \mapsto ba, b \mapsto aa$. Then, $\tilde{w} = (\varrho')^2(\tilde{w})$, where $\tilde{w}$ is the reflected version of w and $a|a$ is legal for $(\varrho')^2$. Since $\varrho'(v) = a^{-1}\varrho(v)a$, Proposition 4.6 implies that ϱ and ϱ' define the same (minimal) hull. The latter is thus reflection symmetric. $\qquad\Box$

Note that Proposition 4.8 also follows from the earlier observation that the defining fixed point w of the period doubling substitution is palindromic.

4.5.2. THE PAPER FOLDING SEQUENCE

Following [AMF95, Ch. I.3], the primitive substitution rule that underlies the classic paper folding sequence can be given as

$$(4.10) \qquad \varrho = \varrho_{\mathrm{pf}} : \begin{array}{l} a \mapsto ab \\ b \mapsto cb \\ c \mapsto ad \\ d \mapsto cd \end{array}$$

which is of constant length and has a coincidence in the sense of Dekking [Dek78]; see also [Que95, Sec. 6.2]. This means that, for some power of the substitution, the images of the letters share the same symbol at least at one position. The period doubling substitution showed a coincidence at the level of ϱ itself. Here, a coincidence occurs on the level of ϱ^2 (the first letter in the images is always a, the third is c). Bi-infinite fixed points are obtained by iterations starting from the seeds $b|a$ or $d|a$, which are both legal. They read

$$\ldots abcbadcdabcdadc{\textstyle{b \atop d}}|abcbadcbabcdadcb \ldots$$

and differ only at position -1, whence they are singular and proximal. Consequently, Theorem 4.3 and Corollary 4.2 imply aperiodicity.

Realising the letters as unit intervals with left endpoints of four different types, one obtains point sets $\Lambda_a, \ldots, \Lambda_d$ that satisfy the set-valued equations

$$
\begin{aligned}
\Lambda_a &= 4\Lambda_a \mathbin{\dot{\cup}} 4\Lambda_b \mathbin{\dot{\cup}} 4\Lambda_c \mathbin{\dot{\cup}} 4\Lambda_d, \\
\Lambda_b &= (4\Lambda_a + 1) \mathbin{\dot{\cup}} (4\Lambda_a + 3) \mathbin{\dot{\cup}} (4\Lambda_b + 3) \mathbin{\dot{\cup}} (4\Lambda_c + 1), \\
\Lambda_c &= (4\Lambda_a + 2) \mathbin{\dot{\cup}} (4\Lambda_b + 2) \mathbin{\dot{\cup}} (4\Lambda_c + 2) \mathbin{\dot{\cup}} (4\Lambda_d + 2), \\
\Lambda_d &= (4\Lambda_b + 1) \mathbin{\dot{\cup}} (4\Lambda_c + 3) \mathbin{\dot{\cup}} (4\Lambda_d + 1) \mathbin{\dot{\cup}} (4\Lambda_d + 3),
\end{aligned}
$$
(4.11)

as a consequence of the fixed point property under ϱ^2. Since we obviously have $\Lambda_a \mathbin{\dot{\cup}} \Lambda_b \mathbin{\dot{\cup}} \Lambda_c \mathbin{\dot{\cup}} \Lambda_d = \mathbb{Z}$, it follows that

$$
\Lambda_a = 4\mathbb{Z}, \quad \Lambda_c = 4\mathbb{Z} + 2 \quad \text{and} \quad \Lambda_b \mathbin{\dot{\cup}} \Lambda_d = 2\mathbb{Z} + 1,
$$

which manifests the Toeplitz structure in this case. Observing the relation $(4\Lambda_b + 1) \mathbin{\dot{\cup}} (4\Lambda_d + 1) = 8\mathbb{Z} + 5$, the remaining equations of (4.11) simplify to

$$
\begin{aligned}
\Lambda_b &= (8\mathbb{Z} + 1) \mathbin{\dot{\cup}} (16\mathbb{Z} + 3) \mathbin{\dot{\cup}} (4\Lambda_b + 3), \\
\Lambda_d &= (8\mathbb{Z} + 5) \mathbin{\dot{\cup}} (16\mathbb{Z} + 11) \mathbin{\dot{\cup}} (4\Lambda_d + 3).
\end{aligned}
$$

By formal iteration of the decoupled equations, one finds the infinite unions

$$
U_b = \dot{\bigcup_{i \geqslant 1}} (2^{i+2}\mathbb{Z} + 2^i - 1) \quad \text{and} \quad U_d = \dot{\bigcup_{i \geqslant 1}} (2^{i+2}\mathbb{Z} + 3 \cdot 2^i - 1).
$$

As in the period doubling case, the point -1 is not contained in either union, but is a 2-adic limit point of both. More precisely, one has

$$
U_b \mathbin{\dot{\cup}} U_d = (2\mathbb{Z} + 1) \setminus \{-1\},
$$

and the point sets that correspond to the two fixed points with seeds $b|a$ and $d|a$ (which both lead to the same set of set-valued equations) are obtained as $\Lambda_b = U_b \mathbin{\dot{\cup}} \{-1\}$, $\Lambda_d = U_d$ and $\Lambda_b = U_b$, $\Lambda_d = U_d \mathbin{\dot{\cup}} \{-1\}$, respectively.

If we now subject our four-letter alphabet to the (continuous) morphism $\varphi \colon \{a, b, c, d\} \longrightarrow \{0, 1\}$ defined by $\varphi(a) = \varphi(b) = 1$ and $\varphi(c) = \varphi(d) = 0$, we obtain (for the seed $b|a$)

$$
\begin{aligned}
\Lambda_1 &= \Lambda_a \mathbin{\dot{\cup}} \Lambda_b = \dot{\bigcup_{i \geqslant 0}} (2^{i+2}\mathbb{Z} + 2^i - 1) \mathbin{\dot{\cup}} \{-1\}, \\
\Lambda_0 &= \Lambda_c \mathbin{\dot{\cup}} \Lambda_d = \dot{\bigcup_{i \geqslant 0}} (2^{i+2}\mathbb{Z} + 3 \cdot 2^i - 1),
\end{aligned}
$$
(4.12)

so that $\Lambda_1 \mathbin{\dot{\cup}} \Lambda_0 = \mathbb{Z}$. For the seed $d|a$, the point -1 has to be moved from Λ_1 to Λ_0. The one-sided paper folding sequence leads to one-sided point sets that are obtained from (4.12) upon removing $\{-1\}$ and replacing $\mathbb{Z}$ by $\mathbb{N}_0$.

By arguments used before, the (binary) paper folding sequence has a limit-periodic structure, wherefore its aperiodicity is of the same kind as in the period doubling system. Moreover, φ is locally invertible, because 111 has the unique preimage bab and occurs at relatively dense positions. This suffices since the letters a, c and b, d occupy different cosets of $\mathbb{Z}$ modulo 2.

Corollary 4.6. *The hull of the quaternary paper folding substitution is aperiodic, as is the hull of its binary image under φ. Moreover, the two hulls are locally equivalent.* □

The equivalence alluded to here will later be introduced in more detail as mutual local derivability; see Definition 5.7.

4.6. Thue–Morse substitution

Here, we consider the two letter substitution rule

$$(4.13) \qquad \varrho = \varrho_{\mathrm{TM}} : \quad \begin{array}{l} a \mapsto ab \\ b \mapsto ba \end{array}$$

which is known as the Thue–Morse substitution (also referred to as the Prouhet–Thue–Morse or simply the Morse substitution); see [AS99] and references therein for background, as well as [Mię93]. This substitution is permutation invariant in the sense of Remark 4.3, and simplest example of a primitive bijective substitution. Bijectivity refers to the property that ϱ defines a permutation of the alphabet in each column of the letter images, which excludes any coincidence; compare [Que10, Sec. 9.1] and [Fra05, Fra09].

The Thue–Morse (TM) substitution graph and matrix read

$$ \text{and} \quad M_{\mathrm{TM}} = \begin{pmatrix} 1 & 1 \\ 1 & 1 \end{pmatrix}. $$

Two distinct bi-infinite fixed points w and w' are obtained as limits of the iteration of ϱ^2, starting from the legal seeds $a|a$ and $b|a$, as

$$ a|a \xmapsto{\varrho^2} abba|abba \xmapsto{\varrho^2} abbabaabbaababba|abbabaabbaababba \xmapsto{\varrho^2} \cdots \longrightarrow w $$

$$ b|a \xmapsto{\varrho^2} baab|abba \xmapsto{\varrho^2} baababbaabbabaab|abbabaabbaababba \xmapsto{\varrho^2} \cdots \longrightarrow w' $$

where w and w' constitute a 2-cycle of ϱ, meaning $w \xmapsto{\varrho} w' \xmapsto{\varrho} w$. Due to the construction from a primitive substitution, w and w' are locally indistinguishable. Nevertheless, while w and w' are identical to the right of the reference point, their left sides are mapped onto each other by $a \leftrightarrow b$. Consequently, w and w' are proximal (in one direction). Also, w is an infinite palindrome, while w' is anti-symmetric. For our further discussion, it is useful to write $w = u|v$ and $w' = \bar{u}|v$, with the convention $\bar{a} = b$ and $\bar{b} = a$.

Remark 4.8 (*Alternative approaches to the TM sequence*). The one-sided word v is a fixed point of ϱ (in $\{a,b\}^{\mathbb{N}_0}$) and satisfies the identities

$$(4.14) \qquad v_{2i} = v_i \quad \text{and} \quad v_{2i+1} = \overline{v_i},$$

which hold for all $i \in \mathbb{N}_0$. This follows immediately from the observation that $\varrho(x) = x\bar{x}$ for $x \in \{a,b\}$. Eq. (4.14) also implies that

$$(4.15) \qquad v = v_0 v_2 v_4 \ldots \quad \text{and} \quad \bar{v} = v_1 v_3 v_5 \ldots$$

which is clear from the observation that the inflation can be *locally* reversed by recognising the words ab and ba (which always begin at even positions) as $\varrho(a)$ and $\varrho(b)$.

The rule of Eq. (4.14) is often used to define the Thue–Morse sequence recursively, starting from $v_0 = a$ (or from $v_0 = 1$ with $\bar{1} = -1$ for a common numerical realisation). One can also consider the recursive concatenation $v^{(n+1)} = v^{(n)} \overline{v^{(n)}}$ with $n \in \mathbb{N}_0$ and $v^{(0)} = v_0$, which gives a sequence of words of length 2^n that converges to v. Another characterisation is

$$v_i = \begin{cases} a, & \text{if the binary digits of } i \text{ sum to an even number,} \\ b, & \text{otherwise.} \end{cases}$$

The symmetric bi-infinite fixed point w from above is related to v by

$$(4.16) \qquad w_i = \begin{cases} v_i, & \text{for } i \geqslant 0, \\ v_{-i-1}, & \text{for } i < 0. \end{cases}$$

We refer to [AS99] for more on this sequence. ◇

The letters a and b are equally frequent in any of the Thue–Morse fixed points (both one-sided or two-sided), and a natural geometric realisation once again uses intervals of length 1 for both letters, which may be distinguished by a colouring. This follows from the substitution matrix M_{TM} given above via the usual PF arguments. The symmetry properties immediately transfer to this geometric setting, in the sense of colour symmetries. The interplay between w and w' has important consequences and will later exclude the Thue–Morse chain from the class of model sets. As the PF eigenvalue is an integer, we cannot decide on the aperiodicity of the Thue–Morse sequence on the basis of M_{TM}. Nevertheless, this sequence shows a strong form of aperiodicity. Thue constructed the sequence to be cubefree, which means that it does not contain any subword of the form zzz. What is more, it is *overlap-free* or *strongly cube-free* in the following sense.

Proposition 4.9. *The one-sided fixed point v of the Thue–Morse substitution of Eq. (4.13) does not contain any subword of the form zzz_0 with a non-empty finite word z.*

PROOF. Our exposition follows [Off08]. Using Eq. (4.14) of Remark 4.8, the identity $v_i = v_{i+1}$ implies i odd, so that no three consecutive letters in v can be equal. With Eq. (4.15), this also implies that the words *abab* and *babab* cannot occur in v. Any subword of v with five or more letters thus contains either *aa* or *bb* as a subword.

Assume, contrary to the claim, that a non-empty word z of length ℓ exists such that zzz_0 is a subword of v, where $\ell > 1$ due to our previous argument. Without loss of generality, we consider the minimal possible ℓ only.

If ℓ is odd, zzz_0 contains more than five letters, and thus either *aa* or *bb*. In fact, this subword must then appear at least twice, each starting at an odd position. One distance between the two appearances has to be ℓ, which must then be even by our previous observation, resulting in a contradiction.

If $\ell \geqslant 2$ is even, consider the word $z' = z_0 z_2 \cdots z_{\ell-2}$ of length $\ell/2$. By Eq. (4.15) once again, either $z'z'z_0'$ or $\overline{z'z'z_0'}$ must be a subword of v as well, which contradicts our minimality assumption on ℓ. $\qquad\square$

Corollary 4.7. *The Thue–Morse substitution is aperiodic and palindromic.*

PROOF. The two-sided hull is defined as the orbit closure of the bi-infinite fixed point w from above, which is an infinite palindrome. By Proposition 4.9, w cannot be periodic. Since ϱ is a primitive substitution, the hull of w satisfies $\mathbb{X}_{\mathrm{TM}} = \mathbb{X}(w) = \mathrm{LI}(w)$. None of its elements can be periodic, because this would imply the existence of a cube also in w which is impossible. $\qquad\square$

Once again, aperiodicity already follows from Corollary 4.2, but that tells us very little about the type (or mechanism) of aperiodicity.

Let us now define a specific (sliding) block map ψ that sends two-letter words to single letters by $\psi(ab) = \psi(ba) = a$ and $\psi(aa) = \psi(bb) = b$; compare [LM95, Ch. 6] for general properties of such mappings. The action of ψ on the palindromic Thue–Morse fixed point (under ϱ_{TM}^2) with seed $a|a$ is

$$(4.17) \qquad \begin{aligned} &\ldots abbabaabb\overrightarrow{aa}babba|abbabaabbaababba \ldots \\ &\ldots abaaababab aaababab|abaaababababaaabaa \ldots \end{aligned}$$

where the shading indicates the sliding block. This induced mapping is also called ψ. It is easy to check that the image is the fixed point of the squared period doubling substitution ϱ_{pd}^2 with seed $b|a$ (which is yet another way to infer aperiodicity, via Proposition 4.7). This induces the following relation between the corresponding hulls, which will be generalised and further explained in Remark 4.9 below.

Theorem 4.7. *The hull $\mathbb{X}_{\mathrm{TM}}$ of the Thue–Morse substitution is a double cover of the period doubling hull $\mathbb{X}_{\mathrm{pd}}$ under the continuous block map ψ, so* *that*

$$\mathbb{X}_{\mathrm{TM}} \xrightarrow{\ \varrho_{\mathrm{TM}}\ } \mathbb{X}_{\mathrm{TM}}$$

$$\psi \downarrow \qquad\qquad \downarrow \psi$$

$$\mathbb{X}_{\mathrm{pd}} \xrightarrow{\ \varrho_{\mathrm{pd}}\ } \mathbb{X}_{\mathrm{pd}}$$

is a commutative diagram. Moreover, the map ψ also commutes with the action of $\mathbb{Z}$ defined by the shift.

PROOF. A simple calculation with the two substitution rules acting on finite subwords shows that ϱ_{TM} induces ϱ_{pd} via the block map ψ. This extends to the fixed points of ϱ_{TM}^2 and ϱ_{pd}^2 as explained above. Since the two hulls are the LI classes of the fixed points by Theorem 4.1, ψ is well-defined on all of $\mathbb{X}_{\mathrm{TM}}$, with $\psi(\mathbb{X}_{\mathrm{TM}}) = \mathbb{X}_{\mathrm{pd}}$ due to minimality. The continuity of ψ is obvious.

Clearly, ψ is at least $2:1$, because $\psi(w) = \psi(\bar{w})$ when $\bar{w}$ is the image of w under the letter exchange $a \leftrightarrow b$, where always $w \neq \bar{w}$. Conversely, starting from an arbitrary element $u \in \mathbb{X}_{\mathrm{pd}}$, the binary choice of a single letter at position $i = 0$ completely and consistently specifies one compatible preimage of u under ψ. Consequently, $\psi^{-1}(u)$ is a subset of $\mathbb{X}_{\mathrm{TM}}$ with precisely two elements. Commutativity of the diagram is clear by construction, which is also true of the commutativity of ψ with the shift. □

The importance of Theorem 4.7 stems from the fact that the dynamical system defined by the action of the shift on $\mathbb{X}_{\mathrm{pd}}$ is a topological factor (via the mapping ψ) of $(\mathbb{X}_{\mathrm{TM}}, \mathbb{Z})$ with maximal pure point spectrum (see Section 9.4.4 and Appendix B). This maximal factor is an almost everywhere $1:1$ cover of the *Kronecker factor* of the Thue–Morse system; see [Que10, Sec. 3.4.2] for more. The latter is defined as the maximal equicontinuous factor (for the group action), which leads to the dyadic solenoid Σ_2 of Example 2.11.

In general, the Kronecker factor is *not* a substitution system. The special situation here is that a factor with maximal pure point spectrum can itself be realised via a primitive substitution. An analogous property holds for the following generalisation.

Remark 4.9 (*Generalised TM sequences and their images*). A natural generalisation of the TM substitution, in the spirit of [Kea68], is

$$\varrho_{\mathrm{gTM}}^{(k,\ell)} : \quad \begin{aligned} a &\mapsto a^k b^\ell \\ b &\mapsto b^k a^\ell \end{aligned}$$

with $k, \ell \in \mathbb{N}$, where $k = \ell = 1$ is the classic TM substitution. The substitution matrix reads $\left(\begin{smallmatrix} k & \ell \\ \ell & k \end{smallmatrix} \right)$, with eigenvalues $k \pm \ell$. This two-parameter family of constant length substitutions shares many properties with its classic ancestor.

Fixed points are constructed as above via the squares of the substitution rules, starting from any binary seed, which are all legal here. The structure of the underlying 2-cycles implies aperiodicity via Corollary 4.2. In analogy to Remark 4.8, the one-sided fixed point v satisfies

$$v_{(k+\ell)m+r} = \begin{cases} v_m, & \text{if } 0 \leqslant r \leqslant k-1, \\ \overline{v_m}, & \text{if } k \leqslant r \leqslant k+\ell-1, \end{cases}$$

where $m \in \mathbb{N}_0$ and $0 \leqslant r \leqslant k+\ell-1$.

Moreover, the block map ψ from Eq. (4.17) above is well-defined, and induces a generalisation of the period doubling substitution, namely

$$\varrho_{\mathrm{gpd}}^{(k,\ell)} : \begin{array}{l} a \mapsto ub \\ b \mapsto ua \end{array}$$

with the finite word $u = b^{k-1}ab^{\ell-1}$. This is another constant length substitution, with substitution matrix $\left(\begin{smallmatrix} 1 & k+\ell-1 \\ 2 & k+\ell-2 \end{smallmatrix}\right)$. The PF eigenvalue is again $k + \ell$, but, unlike the gTM sequences, this family always has a coincidence in the sense of Dekking [Dek78]. The analogue of Theorem 4.7 remains true for arbitrary $k, \ell \in \mathbb{N}$. Moreover, the gpd systems define topological factors with maximal pure point spectrum for the gTM systems. They are covers of the Kronecker factors, which are solenoids; see Example 2.11.   ◇

Let us mention in passing that the properties of the gTM sequences may change dramatically under simultaneous permutations of positions in the words $\varrho(a)$ and $\varrho(b)$. Consider, for instance, the case $k = 2$, $\ell = 1$, with $\{\varrho(a), \varrho(b)\} = \{aab, bba\}$, in obvious shorthand notation. This defines an aperiodic gTM sequence. A simultaneous cyclic permutation by one position leads to $\{aba, bab\}$, which obviously defines the hull of the periodic sequence $\ldots ababab \ldots$; compare Example 4.2. Such observations lead to interesting spectral consequences, as will be discussed later.

Remark 4.10 (*Fixed points under iterated inflation*). Given a primitive substitution rule ϱ, it is an interesting and important question to determine its periodic points, and, in particular, to count all solutions of $\varrho^n(w) = w$ for $n \in \mathbb{N}$. A systematic approach was introduced in [AP98], which also permits the calculation of the Artin–Mazur (or dynamical) zeta function. The approach is effectively based on the corresponding geometric inflation rule (as explained in Example 4.5) and refers to the elements of the *continuous* hull under the translation action of the group $\mathbb{R}$. If different letters lead to intervals of the same length in this setting, as in any substitution rule with an integer inflation factor, they need to be distinguished by colours.   ◇

4.7. Rudin–Shapiro and Kolakoski sequences

Let us continue with two important examples with a less direct (or obvious) connection to the substitution approach.

4.7.1. The Rudin–Shapiro sequence

An interesting sequence was independently constructed by Rudin and Shapiro, answering a question from the classical theory of Fourier series [Sha51, Rud59]; see also [Alz95] and references therein, as well as [Gol49] for a predecessor. A discussion of its spectral properties appears in [MN83, Fra03, Que10]. Its modern reformulation involves a four-letter substitution rule of constant length in the first step. Following [Que10], we start with the quaternary alphabet $\mathcal{A} = \{0, 1, 2, 3\}$ and define the substitution rule by

$$(4.18) \qquad \varrho_{\mathrm{RS}}: \quad 0 \mapsto 02, \quad 1 \mapsto 32, \quad 2 \mapsto 01, \quad 3 \mapsto 31,$$

with the symmetric substitution matrix

$$M_{\mathrm{RS}} = \begin{pmatrix} 1 & 0 & 1 & 0 \\ 0 & 0 & 1 & 1 \\ 1 & 1 & 0 & 0 \\ 0 & 1 & 0 & 1 \end{pmatrix}$$

and the directed substitution graph

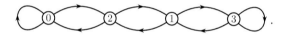

.

The PF eigenvalue is 2, and each letter will be equally frequent in a fixed point. Bi-infinite fixed points can be constructed from any of the legal seeds $2|0$, $2|3$, $1|0$ or $1|3$ under the square of the substitution from Eq. (4.18). Each of these fixed points (which are locally indistinguishable) defines the unique *quaternary* Rudin–Shapiro hull as a closed subset of $\mathcal{A}^{\mathbb{Z}}$ via Eq. (4.2).

The second step of the construction applies the morphism $\varphi \colon \mathcal{A} \longrightarrow \{a, b\}$ defined by $\varphi(0) = \varphi(2) = a$ and $\varphi(1) = \varphi(3) = b$, and its extension to $\mathcal{A}^{\mathbb{Z}}$, also called φ. The four fixed points mentioned above are turned into four distinct bi-infinite *binary* words this way, with a and b being equally frequent. Moreover, these four sequences are locally indistinguishable, and read

$$\ldots bbbabbabbbbaaabaaaabaabababbbaaaba | aaabaabaaaabbbabaaabaababbbaaaba \ldots$$
$$\ldots bbbabbabbbbaaabaaaabaabababbbaaaba | bbbabbabbbbaaababbbabbabaaabbbab \ldots$$
$$\ldots aaabaabaaaabbbabbbbabbabaaabbbab | aaabaabaaaabbbabaaabaabababbbaaaba \ldots$$
$$\ldots aaabaabaaaabbbabbbbabbabaaabbbab | bbbabbabbbbaaababbbabbabaaabbbab \ldots$$

The construction of these four sequences involved the substitution rule of Eq. (4.18) and the morphism φ. Any of these, in turn, defines the *binary* Rudin–Shapiro (RS) hull, now as a closed subset of $\{a, b\}^{\mathbb{Z}}$. Note that various pairs are proximal, but not singular. Nevertheless, Corollary 4.2 implies aperiodicity. Due to the importance of the RS system, it seems advantageous to augment this approach with some alternative descriptions.

Lemma 4.9. *The four bi-infinite sequences constructed above are fixed under the staggered two-letter substitution rules*

$$\varrho_{\text{even}} : \begin{array}{l} a \mapsto aaab, \\ b \mapsto bbba, \end{array} \quad and \quad \varrho_{\text{odd}} : \begin{array}{l} a \mapsto aaba, \\ b \mapsto bbab, \end{array}$$

where ϱ_{even} and ϱ_{odd} have to be applied to letters at even and odd positions, respectively.

PROOF. Any bi-infinite fixed point w of the square of the rule in Eq. (4.18) satisfies $w_{2i} \in \{0, 3\}$ and $w_{2i+1} \in \{1, 2\}$ for $i \in \mathbb{Z}$. At even and odd positions, one thus obtains the rules

$$0 \mapsto 0201 \xmapsto{\varphi} aaab, \qquad\qquad 2 \mapsto 0232 \xmapsto{\varphi} aaba,$$

$$3 \mapsto 3132 \xmapsto{\varphi} bbba, \qquad\qquad 1 \mapsto 3101 \xmapsto{\varphi} bbab,$$

to be read columnwise. These consistently give ϱ_{even} and ϱ_{odd} by observing that $\varphi(0) = \varphi(2) = a$ and $\varphi(3) = \varphi(1) = b$.   □

For one-sided fixed points, it is possible to use the simpler staggered rule $a \mapsto aa$, $b \mapsto bb$ (even) and $a \mapsto ab$, $b \mapsto ba$ (odd). This follows from Eq. (4.18) by the same argument, which shows the commutativity of the diagram

$$\begin{array}{ccc}
\{0, 1, 2, 3\}^{\mathbb{Z}} & \xrightarrow{\ \text{Eq. (4.18)}\ } & \{0, 1, 2, 3\}^{\mathbb{Z}} \\
\varphi \downarrow & & \downarrow \varphi \\
\{a, b\}^{\mathbb{Z}} & \xrightarrow[\text{rule}]{\text{staggered}} & \{a, b\}^{\mathbb{Z}}
\end{array}$$

for both situations. Note that either version of the staggered rule is *non-local*, as the knowledge of the reference point (index 0) is required to apply the rule. In particular, the rule does not commute with the shift S (although it does with S^2).

Corollary 4.8. *Let $w \in \{a, b\}^{\mathbb{N}_0}$ be the one-sided fixed point of the staggered substitution rule with $w_0 = a$. Under the map $a \mapsto 1$, $b \mapsto -1$, the fixed point is mapped to the recursively defined sequence $r = (r_i)_{i \in \mathbb{N}_0}$ with $r_0 = 1$, $r_{2i} = r_i$ and $r_{2i+1} = (-1)^i r_i$ for $i \in \mathbb{N}_0$.*

PROOF. The initial condition $r_0 = 1$ is clear, while the transportation of values to even and odd sites precisely reflects the action of the staggered substitution rule.   □

Remark 4.11 (*Binary versus quaternary hull*). Let $\mathbb{X}_{\text{RS}}^{(4)}$ denote the quaternary Rudin–Shapiro hull and define $\mathbb{X}_{\text{RS}}^{(2)} = \varphi(\mathbb{X}_{\text{RS}}^{(4)})$, where the morphism φ from above is continuous and commutes with the shift action. It is perhaps surprising to note that φ is a *bijective* mapping between the two hulls, wherefore the dynamical systems $(\mathbb{X}_{\text{RS}}^{(4)}, \mathbb{Z})$ and $(\mathbb{X}_{\text{RS}}^{(2)}, \mathbb{Z})$ are topologically conjugate. To see this, observe that *bbbb* is the longest *a*-free subword of (any element of) $\mathbb{X}_{\mathrm{RS}}^{(2)}$, and that its unique preimage under φ is 1313. The entire preimage of any element of $\mathbb{X}_{\mathrm{RS}}^{(2)}$ is thus uniquely determined. Due to repetitivity, we know that *bbbb* must occur with bounded gaps, so that the topological conjugation rule is local. $\Diamond$

An extra benefit of the sequence r comes from its arithmetic interpretation. If q_i is the number of occurrences of digit pairs 11 in the binary representation of $i \in \mathbb{N}_0$ (so that k successive 1s count as $k-1$ pairs), one has $r_i = (-1)^{q_i}$. It is a straightforward exercise (in multiplying by 2) to check that this relation satisfies the recursion of r. In fact, this arithmetic characterisation is often the starting point for a discussion of the Rudin–Shapiro sequence [Alz95], with the substitution picture being derived as a property; compare [Que10] for details. It can be used to show that the one-sided Rudin–Shapiro sequence is not ultimately periodic, in the following sense.

Lemma 4.10. *For the sequence* r *of Corollary* 4.8, *there is no* $p \in \mathbb{N}$ *and no* $i_0 \in \mathbb{N}_0$ *such that* $r_{i+p} = r_i$ *holds for all* $i \geqslant i_0$.

PROOF. Assume to the contrary that $r_{i+p} = r_i$ is satisfied for some $p \in \mathbb{N}$ and all $i \geqslant i_0$, with some $i_0 \in \mathbb{N}_0$. Then, there are $k, \ell \in \mathbb{N}$ such that $2^{k-1} \leqslant p < 2^k$ and $i_0 < 2^\ell$. If q_p is odd, $i = 2^{\min(k+1,\ell)}$ would give $r_i = 1$ and $r_{i+p} = -1$, in contradiction to the assumption. If q_p is even, the choice $i = 2^{\min(k+2,\ell)} + 2^k$ leads to $q_{i+p} = 1 + q_p$ and hence to the same contradiction. $\square$

Theorem 4.8. *The Rudin–Shapiro substitution of Eq.* (4.18) *is primitive and aperiodic. Aperiodicity is inherited by the binary Rudin–Shapiro sequence, which also defines a minimal hull.*

PROOF. Primitivity follows from the fact that $M_{\mathrm{RS}}^3 \gg 0$, or from an inspection of the graph. The one-sided binary version of the Rudin–Shapiro sequence is not ultimately periodic by Lemma 4.10. Since this forms the right half of a bi-infinite fixed point of the staggered substitution rule, the latter cannot have any periods, and neither can any of its preimages under φ. Since the hulls on both levels are unique, the claims follow, with the last one being a consequence of the continuity of φ. $\square$

As before, aperiodicity can be concluded from proximality, but the structure of the sequence r is more revealing.

Remark 4.12 (*Palindromes in Rudin–Shapiro sequences*)**.** Both the binary Rudin–Shapiro sequence and its underlying 4-letter sequence fail to be palindromic [All97, Baa99]. Starting from all possible words of length n, and selecting the legal ones that are palindromic, one can compute the counts for the quaternary $(p_4(n))$ and the binary $(p_2(n))$ version. Note that, whenever the palindrome complexity satisfies $p(k) = p(k+1) = 0$ for some $k \in \mathbb{N}$, one obviously has $p(n) = 0$ for all $n \geqslant k$; for a more general exposition, we refer to [ABCD03] and references therein. The numbers read as follows.

n	1	2	3	4	5	6	7	8	9	10	11	12	13	14	15	16
$p_4(n)$	4	0	8	0	8	0	4	0	0	0	0	0	0	0	0	0
$p_2(n)$	2	2	4	4	4	4	4	4	0	4	0	2	0	2	0	0

Moreover, the binary chain contains the 6-letter words *aaaabb*, *abbaba*, *baabab* and *bbbbaa*, but none of their reverses. This implies that the hull of the binary chain (and hence also that of the quaternary one) fails to be inversion symmetric. Nevertheless, the binary hull is symmetric under the exchange mapping $a \leftrightarrow b$. ◊

It is interesting to note that the idea of the sliding block map used for the Thue–Morse sequence leads to valuable insight also for the Rudin–Shapiro sequence. We define a block map χ by its action on the legal two-letter words of the quaternary Rudin–Shapiro sequence as

$$\chi(01) = \chi(32) = A, \qquad \chi(02) = \chi(31) = B,$$
$$\chi(10) = \chi(23) = C, \qquad \chi(13) = \chi(20) = D,$$

thus using the alphabet $\{A, B, C, D\}$ for the image. If $\mathbb{Y}^{(4)} = \chi(\mathbb{X}_{\mathrm{RS}}^{(4)})$, we obtain the commutative diagram

$$(4.19) \qquad \begin{array}{ccc} \mathbb{X}_{\mathrm{RS}}^{(4)} & \xrightarrow{\varrho_{\mathrm{RS}}} & \mathbb{X}_{\mathrm{RS}}^{(4)} \\ \chi \downarrow & & \downarrow \chi \\ \mathbb{Y}^{(4)} & \xrightarrow{\varrho} & \mathbb{Y}^{(4)} \end{array}$$

where ϱ_{RS} is the substitution rule of Eq. (4.18), while ϱ turns out to be

$$\varrho: \quad A \mapsto BC, \quad B \mapsto BD, \quad C \mapsto AD, \quad D \mapsto AC.$$

This induced substitution resembles the paper folding substitution, and can be treated along similar lines. Bi-infinite fixed points of ϱ^2 are obtained by iteration from the legal seeds $C|B$ or $D|B$. These two fixed points differ only at position -1 and form a 2-cycle under ϱ. In particular, they are proximal.

The resulting set-valued equations for the left endpoint sets lead to

$$\Lambda_A = 4\mathbb{Z}+2, \quad \Lambda_B = 4\mathbb{Z} \quad \text{and} \quad \Lambda_C \dot\cup \Lambda_D = 2\mathbb{Z}+1,$$

hence to $\Lambda_A \dot\cup \Lambda_B = 2\mathbb{Z}$, and to the decoupled equations

$$\Lambda_C = (8\mathbb{Z}+5) \dot\cup (16\mathbb{Z}+3) \dot\cup (4\Lambda_C+3),$$
$$\Lambda_D = (8\mathbb{Z}+1) \dot\cup (16\mathbb{Z}+11) \dot\cup (4\Lambda_D+3).$$

Formal iteration of these equations produces the unions

$$U_C = \dot{\bigcup_{n \geqslant 1}} \big((2^{2n+1}\mathbb{Z} + 3 \cdot 2^{2n-1} - 1) \dot{\cup} (2^{2n+2}\mathbb{Z} + 2^{2n} - 1) \big),$$

(4.20)

$$U_D = \dot{\bigcup_{n \geqslant 1}} \big((2^{2n+1}\mathbb{Z} + 2^{2n-1} - 1) \dot{\cup} (2^{2n+2}\mathbb{Z} + 3 \cdot 2^{2n} - 1) \big).$$

As before, $U_C \dot{\cup} U_D = (2\mathbb{Z} + 1) \setminus \{-1\}$, and the point -1 has to be added to one of these sets, depending on the fixed point chosen. The aperiodicity of the RS sequence can alternatively be derived from here.

The block map χ on $\mathbb{X}_{\mathrm{RS}}^{(4)}$ also gives rise to a block map on the binary hull $\mathbb{X}_{\mathrm{RS}}^{(2)}$; this is exactly the block map ψ used in the Thue–Morse case in Eq. (4.17). Starting from $\mathbb{Y}^{(4)}$, this amounts to $A, C \mapsto a$ and $B, D \mapsto b$, which maps $\mathbb{Y}^{(4)}$ to a compact subset $\mathbb{Y}^{(2)} \subset \{a, b\}^{\mathbb{Z}}$. This mapping is locally invertible as a consequence of Remark 4.11 (which can also be verified directly by observing that bbb has the unique preimage DBD).

The hull $\mathbb{Y}^{(2)} = \psi\big(\mathbb{X}_{\mathrm{RS}}^{(2)}\big)$ reveals the corresponding limit-periodic sub-structure of $\mathbb{X}_{\mathrm{RS}}^{(2)}$. In fact, the new point sets for the sequence with seed $a|b$ or $b|a$ read

$$\Lambda_a = \{-1\} \dot{\cup} (4\mathbb{Z} + 2) \dot{\cup} U_C \quad \text{and} \quad \Lambda_b = 4\mathbb{Z} \dot{\cup} U_D,$$

where the point -1 has to be moved to Λ_b for the seed $a|a$ or $b|b$. We will revisit this structure of Toeplitz type in our later chapter on diffraction. In fact, the comparison with the known pure point part of the dynamical spectrum of the RS system [Fra03] then shows that $\big(\mathbb{Y}^{(2)}, \mathbb{Z}\big)$ defines a topological factor with maximal pure point spectrum, both for the quaternary and for the binary RS sequence. It is again a cover of the Kronecker factor of the system, the latter usually realised as a dyadic solenoid; compare Example 2.11.

4.7.2. KOLAKOSKI SEQUENCES

The discussion of the binary Rudin–Shapiro sequence indicates that matters may become more difficult rather quickly when one leaves the realm of primitive substitution rules. Particularly enigmatic examples of this kind are the classic Kolakoski sequence, introduced as a problem in [Kol65] (though it was discussed earlier in [Old39]), and some of its relatives.

A one-sided infinite sequence w over the alphabet $\mathcal{A} = \{1, 2\}$ is called a (classic) *Kolakoski sequence* if it equals the sequence defined by its 'run lengths', for instance

(4.21)
$$w = 22\ 11\ 2\ 1\ 22\ 1\ 22\ 11\ 2\ 11\ 22\ 1\ 2\ 11\ 2\ 1\ 22\ 11 \dots$$
$$ 2\quad 2\ 1\ 1\ 2\ 1\ 2\quad 2\ 1\ 2\quad 2\ 1\ 1\ 2\ 1\ 1\ 2\quad 2 \dots = w.$$

Here, a *run* is a maximal subword consisting of identical letters. The sequence $w' = 1w$ is the only other sequence with this property (over $\mathcal{A}$). An alternative way to obtain w iterates the staggered substitution rule

$$\varrho_{\text{even}} : \begin{array}{c} 1 \mapsto 2 \\ 2 \mapsto 22 \end{array} \quad \text{and} \quad \varrho_{\text{odd}} : \begin{array}{c} 1 \mapsto 1 \\ 2 \mapsto 11 \end{array}$$

on the seed 2 (which is at position 0). This gives

$$2 \mapsto 22 \mapsto 2211 \mapsto 221121 \mapsto 221121221 \mapsto 22112122122112 \mapsto \cdots$$

The iterates converge to the Kolakoski sequence w in the product topology, and w is the unique (one-sided) fixed point of this staggered substitution. The reasoning behind this rule becomes transparent if one reads Eq. (4.21) from the bottom to the top line. Although this looks like a mild variation of the staggered Rudin–Shapiro rule, the violation of the constant length property has severe consequences. In fact, many questions about the Kolakoski sequence, even elementary ones such as the symbol frequencies, are still unanswered or unproved; see [Dek97] for a review, and [Sin11] and references therein for some more recent developments. The extensive numerical calculation of the Kolakoski sequence in [Nil12b] supports the conjecture that the two symbols (1 and 2) are equally frequent.

Some progress is possible for the generalised Kolakoski sequences on the binary alphabet $\mathcal{A} = \{p, q\}$, where p and q are distinct positive integers. The sequences are again defined by the run length property as above, and permit a characterisation by the staggered substitution

$$(4.22) \qquad \varrho_{\text{even}} : \begin{array}{c} p \mapsto p^p \\ q \mapsto p^q \end{array} \quad \text{and} \quad \varrho_{\text{odd}} : \begin{array}{c} p \mapsto q^p \\ q \mapsto q^q \end{array}$$

with starting letter p (at position 0), so that the order of p and q becomes relevant. Here, as before, x^n is a shorthand for the word $x \cdots x$ of length n. The corresponding fixed point is called the Kolakoski-(p, q) sequence or $\mathrm{Kol}(p, q)$ for short. The classic Kolakoski sequence w of Eq. (4.21) is then given as $w = \mathrm{Kol}(2, 1)$, while $w' = \mathrm{Kol}(1, 2)$.

Remark 4.13 (*Bi-infinite Kolakoski sequences*). Every one-sided sequence $\mathrm{Kol}(p, q)$ can be extended to a bi-infinite sequence in a unique way, by complementing it with the reversed copy of $\mathrm{Kol}(q, p)$ to the left of the reference point. For the classic case, this gives

$$\ldots 11221221211221 | 22112122122112 \ldots$$

By this construction, the bi-infinite sequence still equals the sequence of its run lengths, when counting commences at the reference point. Moreover, if $q = 1$ (or $p = 1$), the bi-infinite sequence is reflection symmetric in the first position to the left (right) of the reference point. The staggered substitution rule for (p, q) will produce these two-sided sequence as fixed points by iteration from the seed $q|p$.   ◇

Example 4.8 (*Kolakoski-$(3, 1)$ as a primitive substitution*). The (one-sided) fixed point $\mathrm{Kol}(3, 1)$ starts as

$$3331113331313331113331333133111333\dots$$

A careful inspection reveals that it is uniquely composed of three pairs, namely $a = 33$, $b = 31$ and $c = 11$. Moreover, this is compatible with the primitive three-letter substitution rule

$$\varrho: \begin{array}{l} a \mapsto abc \\ b \mapsto ab \\ c \mapsto b \end{array} \quad \text{with} \quad M_\varrho = \begin{pmatrix} 1 & 1 & 0 \\ 1 & 1 & 1 \\ 1 & 0 & 0 \end{pmatrix},$$

see [BS04a] and references therein for details. A bi-infinite fixed point of ϱ can be obtained by iteration from the legal seed $b|a$, giving

$$b|a \overset{\varrho}{\longmapsto} ab|abc \overset{\varrho}{\longmapsto} abcab|abcabb \overset{\varrho}{\longmapsto} abcabbabcab|abcabbabcabab \overset{\varrho}{\longmapsto} \cdots$$

The limit corresponds to the unique bi-infinite word derived from $\mathrm{Kol}(3, 1)$, as explained in Remark 4.13.

The matrix M_ϱ is primitive (because $M^3 \gg 0$). Its characteristic polynomial is $\det(x\mathbb{1} - M_\varrho) = x^3 - 2x^2 - 1$, which is irreducible over $\mathbb{Q}$, with discriminant -59. The irrational PF eigenvalue $\lambda \approx 2.206$ is a real algebraic unit of degree 3 whose algebraic conjugates form a complex conjugate pair with approximate values $-0.103 \pm 0.665\mathrm{i}$. This also shows that λ is a PV unit. A possible choice for the left PF eigenvector is

$$v_{\mathrm{PF}} = (\ell_a, \ell_b, \ell_c) = (\lambda^2 - \lambda, \lambda, 1) \approx (2.659, 2.206, 1),$$

while the normalised letter frequencies are $\nu_a = \frac{1}{2}(-\lambda^2 + 3\lambda - 1) \approx 0.376$, $\nu_b = \lambda^2 - 2\lambda \approx 0.453$ and $\nu_c = \frac{1}{2}(-\lambda^2 + \lambda + 3) \approx 0.171$. The average interval length in the geometric representation thus is

$$\overline{\ell} = \nu_a \ell_a + \nu_b \ell_b + \nu_c \ell_c = \frac{1}{2}(-\lambda^2 + \lambda + 7) \approx 2.171.$$

Also the symbol frequencies in $\mathrm{Kol}(3, 1)$ can now be calculated, resulting in $\nu_3 = \frac{1}{2}(\lambda - 1) \approx 0.603$ and $\nu_1 = \frac{1}{2}(3 - \lambda) \approx 0.397$.   ◇

Remark 4.14 (*Other Kolakoski sequences*). As mentioned before, little is known about the classic Kolakoski sequence $\mathrm{Kol}(2, 1)$. The same is true of all $\mathrm{Kol}(p, q)$ with $p + q$ odd. Example 4.8 demonstrates that the situation might be more favourable for other parameters. Indeed, if both p and q are even $(p \neq q)$, one can rewrite the staggered substitution (4.22) as a substitution of constant length, based on 4-letter blocks [Sin03]. One case with p and q odd was explained above; the other cases can be treated by substitution rules in a similar fashion. Here, one gets substitutions with a PV unit multiplier precisely when $|p - q| = 2$. More generally, PV numbers show up if and only if $2(p + q) \geqslant (p - q)^2$. Otherwise, all eigenvalues of the substitution matrix lie outside the unit circle; see [Sin02, Sin11] for details. $\Diamond$

4.8. Complexity and further directions

Symbolic substitution rules are structurally very clear and elegant, with many applications in mathematics, physics and computer science. Due to the linear structure of $\mathbb{R}$, it is always possible to relate symbolic sequences to geometric tilings of the line. The purpose of this section is to briefly introduce simple generalisations of substitution rules and to explain how the symbolic structure can be used to derive various higher-order combinatorial properties of the sequences.

4.8.1. AFFINE SUBSTITUTIONS

In several of our previous examples, we needed to use powers of the substitution rule to define a fixed point and consecutively the hull of the substitution. An alternative approach (particularly useful in the bi-infinite setting) employs the combination of a substitution with the shift S. Let us explain this for a variant of the silver mean substitution of Example 4.5.

Example 4.9 (*Affine substitution for the silver mean sequence*)**.** Consider the primitive substitution

$$\varrho : \begin{array}{l} a \mapsto baa \\ b \mapsto a \end{array}$$

which is conjugate to ϱ_{sm} of Example 4.5, but has no fixed point (while ϱ^2 does). Keeping track of our reference marker (the origin), the iteration of the *affine substitution* $\sigma := S \circ \varrho$ leads to

$$|a \xrightarrow{\sigma} b|aa \xrightarrow{\sigma} ab|aabaa \xrightarrow{\sigma} baaab|aabaaaabaabaa \xrightarrow{\sigma} \cdots \longrightarrow w = S(\varrho(w)).$$

The bi-infinite limit word w is thus fixed by ϱ up to a shift by one position. In this example, it was even sufficient to start from a single letter. It is obvious that w is locally indistinguishable from any fixed point of ϱ^2, so that the orbit closure of w defines the silver mean hull $\mathbb{X}_{\mathrm{sm}}$. $\Diamond$

Reviewing (and slightly modifying) the proof of Lemma 4.3, it is clear that fixed points exist also for suitable powers of any mapping of the form $\sigma = S^k \circ \varrho^\ell$ with ϱ a primitive substitution, $k \in \mathbb{Z}$ and $\ell \in \mathbb{N}$. Note that the definition of a legal seed requires some care, in particular concerning its length and its initial position with respect to the marker. The use of σ is then consistent in the following sense.

Proposition 4.10. *Let ϱ be a primitive substitution with hull $\mathbb{X}$, defined by a bi-infinite fixed point of some (positive) power of ϱ. Let $\sigma = S^k \circ \varrho^\ell$ with $k \in \mathbb{Z}$ and $\ell \in \mathbb{N}$ be an affine substitution with the same ϱ, and w be a bi-infinite fixed point of σ^m for some $m \in \mathbb{N}$, with legal core. Then, the orbit closure of w coincides with $\mathbb{X}$. In particular, ϱ and σ define the same hull.*

PROOF. The hull $\mathbb{X}$ of ϱ is unique and minimal by Theorem 4.1. We are thus done as soon as we know that $w \in \mathbb{X}$. By our notion of a fixed point, we know that w has a core that emerges from the non-empty overlap between a finite legal word $u = w_{[p,q]}$ (for suitable $p, q \in \mathbb{Z}$) and its image $\sigma^m(u)$. The latter is legal, too, because $\mathbb{X}$ is closed both under shifts and under substitution. So, w is the limit of a sequence of legal words of growing length and must then be an element of the closed set $\mathbb{X}$. □

This observation, which we have formulated in the symbolic setting, possesses a counterpart for geometric inflation rules (then called affine inflations). Moreover, its usefulness increases when considering higher dimensions, where one can exploit the full structure of the Euclidean group. We shall meet examples later. Let us briefly mention that one could also employ anti-homomorphisms (meaning $\varrho(ab) = \varrho(b)\varrho(a)$) to define a hull. Since ϱ^2 is then a homomorphism again, this does not significantly extend the theory.

4.8.2. SUBWORD COMPLEXITY AND STURMIAN SEQUENCES

Given an arbitrary sequence w in a finite alphabet $\mathcal{A}_n$, let $W_m = W_m(w)$ be the set of all subwords (or factors) of w of length m and define

$$p_w(m) := \operatorname{card}(W_m) \qquad \text{for } m \geqslant 0.$$

This is called the *complexity function* of w. By convention, W_0 contains only the empty word, hence $p_w(0) = 1$. As $W_m \subset \mathcal{A}_n^m$, one has $0 \leqslant p_w(m) \leqslant n^m$.

The function p_w is an interesting measure for the complexity of w. For infinite sequences, the quantity

$$(4.23) \qquad\qquad h := \lim_{m \to \infty} \frac{\log\bigl(p_w(m)\bigr)}{m}$$

is called the (combinatorial) *entropy* of w. Note that the limit always exists. This follows from $p_w(k+\ell) \leqslant p_w(k)p_w(\ell)$ for arbitrary $k, \ell \in \mathbb{N}$ by a standard subadditivity argument known as Fekete's lemma [Fek23]; see also [Gri99, App. II] and references therein, or [BGE97, Lemma 1].

Example 4.10 (*Complexity of periodic sequences*). Let w be a bi-infinite sequence that is periodic, with minimal period s. This means that w is the periodic repetition of a fundamental word u of length s. It is thus immediate that $p_w(m) \leqslant s$ for all $m \geqslant 0$. Moreover, one has $p_w(m) = s$ for all $m \geqslant s$

as a consequence of the minimality of the period s. This can be seen by 'wrapping' the word u on a circle.

Conversely, any bi-infinite sequence w with bounded complexity function is periodic, with period

$$\inf\{s \in \mathbb{N} \mid p_w(m) \leqslant s \text{ for all } m \in \mathbb{N}\}.$$

All periodic sequences have entropy $h = 0$. ◇

Interestingly, the periodicity of w already follows if $p_w(m) \leqslant m$ holds for a single $m \in \mathbb{N}$. More generally, one has the following result, which builds on the pioneering work of Morse and Hedlund [MH38, MH40, CH73]. We formulate it for binary sequences; see [Lot02, LP92] for extensions.

Proposition 4.11. *Let w be a bi-infinite binary sequence with complexity function p_w. Then, the following statements are equivalent.*

(1) $p_w(m + 1) = p_w(m)$ *for some* $m \in \mathbb{N}$*;*

(2) $k := \sup_{n \in \mathbb{N}} \big(p_w(n) \big) < \infty$*;*

(3) w *is periodic, with minimal period k from* (2)*;*

(4) $p_w(m) \leqslant m$ *for some* $m \in \mathbb{N}$*;*

(5) *there exists $k \in \mathbb{N}$ such that $p_w(n) \geqslant n + 1$ for $1 \leqslant n \leqslant k - 1$ and $p_w(n) = k$ for all $n \geqslant k$.*

PROOF. The implications (3) $\Longrightarrow$ (2) $\Longrightarrow$ (1) are obvious, so let us first show that (1) $\Longrightarrow$ (3). If $p_w(m + 1) = p_w(m) = n$ for some m, we know that each subword of w of length m has precisely one extension (by one letter) to the right, and also precisely one to the left. Locate and fix one particular subword $u = u_0 \cdots u_{m-1}$ of w of length m. The letter following u is fixed, so that the m-letter word starting with u_1 is determined. Continuing this process, the entire sequence of m-letter words (to the right) is fixed. After n steps, one word must have reappeared, which defines a cycle. This cycle must periodically repeat to the right, and thus also to the left. Consequently, this cycle must be primitive and of length n, as otherwise some words would be missing. This shows that the sequence is n-periodic. The characterisation of $k = n$ follows from Example 4.10.

Since $p_w(k) = k$ from Example 4.10 with k the (minimal) period of w, the implications (3) $\Longrightarrow$ (5) $\Longrightarrow$ (4) are clear. It now suffices to show (4) $\Longrightarrow$ (1). Since w is a binary sequence, we know $p_w(1) = 2$ and $p_w(n + 1) \geqslant p_w(n)$ for all $n \in \mathbb{N}$. If $p_w(m) \leqslant m$, there is a minimal m' with $2 \leqslant m' \leqslant m$ that satisfies the same inequality. But this implies $p_w(m' - 1) = p_w(m')$. □

Similar results hold for one-sided sequences (compare [AS03, Thm. 10.2.6] and [Lot02, Thm. 1.3.13]). Various refinements can be obtained by taking the alphabet size into account; see [CH73, AS03] and references therein. As a consequence of Proposition 4.11, the complexity of any non-periodic bi-infinite word w must satisfy $p_w(m) \geqslant m+1$. This phenomenon is called the *complexity gap*, because there is no bi-infinite sequence with a complexity between ultimately constant and linear behaviour. A similar phenomenon exists for tilings in higher dimensions [LP02], although the detailed structure is more involved and not yet fully understood; see also [Jul10].

An example of a non-periodic bi-infinite sequence of minimal complexity (then necessarily over a binary alphabet) is provided by $w = \ldots aaaa|bbbb\ldots$, which gives $p_w(m) = m+1$ for all $m \in \mathbb{N}$. There is an entire family of similarly constructed words that are fully classified in [CH73, Thm. 4.12]. They all emerge from two periodic words (with coprime periods) that are glued together with a joining block that occurs precisely once. All these words are non-periodic, but *not* aperiodic in the sense of Definition 4.13; compare Example 4.3. In fact, none of these special words can be repetitive, wherefore this family is of limited interest in our context. More interestingly, there are many other examples of minimal complexity, all of which turn out to be repetitive (albeit only few of them being linearly repetitive).

Definition 4.16. A repetitive, bi-infinite, binary sequence w with complexity $p_w(m) = m+1$ for all $m \in \mathbb{N}$ is called a *Sturmian sequence*.

For a comprehensive exposition of the widely used one-sided analogue, we refer to [Lot02, Ch. 2]. Sturmian sequences are the aperiodic sequences of minimal complexity. They can be constructed via circle maps with an irrational number α or via concatenation rules derived from the continued fraction expansion of α; see [BIST89, LP92, DL99] and references therein. All Sturmian sequences have entropy 0, so that entropy does not distinguish periodic from Sturmian sequences. Here, the quotient $p_w(m)/m$ converges, as $m \to \infty$, to 0 in the periodic and to 1 in the Sturmian case. Nevertheless, $p_w(m)/m$ is not a good measure in general, due to lack of convergence. Instead, one could consider the sequence defined by

$$(4.24) \qquad b_m = \frac{1}{\log(m)} \sum_{k=1}^{m} \frac{d_w(k)}{k}, \quad \text{with } d_w(k) = p_w(k+1) - p_w(k),$$

which always converges[3]. In particular, the limit is 0 for any periodic and 1 for any Sturmian sequence, which shows that it can distinguish different cases with vanishing entropy. In general, one might need further concepts to 'measure' the complexity of sequences. Their choice might depend on the application at hand; compare [Dam01, DL06b] for examples in the context of Schrödinger operators.

---

[3]We thank Michael Boshernitzan for suggesting this approach.

Proposition 4.12. *The fixed point w, with seed $a|a$, of the primitive Fibonacci substitution ϱ_F^2 is Sturmian. Moreover, all elements of the Fibonacci hull are Sturmian.*

PROOF. We prove by an inductive argument that w has complexity function $p_w(n) = n + 1$ for $n \geqslant 1$, which is true for $n = 1$. Since w is aperiodic by Theorem 4.6, Proposition 4.11 implies $p_w(n) \geqslant n + 1$. It is clear that we only need to argue with extensions to one direction (to the right, say). Since any finite subword of w has at least one continuation to the right, we know that $p_w(n + 1) \geqslant p_w(n)$. In fact, since ϱ_F is aperiodic, we already know that $p_w(n + 1) \geqslant p_w(n) + 1$ for all $n \in \mathbb{N}_0$. Our result follows if we can show that there is at most one subword of length n with two extensions to the right.

To this end, we first observe that, for any legal word u, aua or bub must be illegal. This claim can easily be checked for any legal u with $0 \leqslant |u| \leqslant 2$. Now, assume to the contrary that a legal word u exists such that both aua and bub are legal as well. Without loss of generality, assume that the length of u is minimal (which must be at least 3). As bb is illegal, we must have $u = ava$, where v is a non-empty legal word. Inspecting the legal word $bavab$, we see that av must be a superword for $\varrho = \varrho_F$, meaning that $av = \varrho(z)$ for some legal word z. This is the recognisability property mentioned on page 80.

This observation implies that $aavaa$ is an extension of $\varrho(bzb)$, while the legality of $bavab$ forces also $abavab = \varrho(aza)$ to be legal. Since the preimages aza and bzb are unique (by recognisability), they must be legal as well. This is a contradiction, because z has smaller length than u.

Now assume that two legal words $u \neq v$ of length $n \geqslant 2$ exist such that all four words ua, ub, va and vb are legal. Let z be the longest common suffix of u and v, which is a word of length $< n$ (it might be the empty word). Since both possible extensions of z to the left are now realised, we see that aza and bzb would both be legal, which is a contradiction.

The second claim follows from $\mathbb{X}(w) = \mathrm{LI}(w)$ via Theorem 4.1. □

Let us briefly mention that there is a construction for substitution words which have precisely two extensions to the right. This construction (which we omit here) is based on the observation of Lemma 4.8. For the details, we refer to [Lot02, Ex. 2.1.1].

Remark 4.15 (*Fibonacci substitutions and general Sturmian words*). Let us highlight another interesting feature of the Fibonacci substitution. Consider $\varrho = \varrho_F$ and ϱ' from Remark 4.6 (defined by $a \mapsto ba$, $b \mapsto a$) together with σ, the latter defined by $a \leftrightarrow b$. All combinations of non-negative powers of these substitutions generate the monoid of positive substitutions within $\mathrm{Aut}(F_2)$; see [Lot02, Sec. 2.3.5] for details. They are precisely the automorphisms that map any Sturmian sequence to a Sturmian sequence again

[Lot02, Thm. 2.3.23]. Moreover, all Sturmian hulls can be constructed via suitable combinations of these maps; see [DL99] and references therein. ◇

More generally, one has the following result [Que10, Prop. 5.12].

Proposition 4.13. *If ϱ is a primitive substitution rule on a finite alphabet, its unique hull has at most linear complexity.* □

This property will emerge again in a more general setting in Chapter 6.

4.8.3. INDUCED SUBSTITUTIONS AND WORD FREQUENCIES

The relative letter frequencies of primitive substitution rules follow immediately from the substitution matrix by PF theory, as used above. More generally, one is interested in frequencies of arbitrary finite words. If a finite word, of length $m \geqslant 1$ say, occurs in the fixed point of a primitive substitution (meaning that it is legal), its frequency exists uniformly, by the unique ergodicity of the substitution system, and is positive due to repetitivity (otherwise, the frequency is clearly zero), via Theorem 4.4. The remaining task is to calculate the frequency. One possibility consists of deriving an induced substitution rule for all legal m-letter words, and applying PF theory to the corresponding substitution matrix (which also gives the existence of the limits involved, without reference to the ergodic theorem as formulated in Theorems 4.2 and 4.4).

Let ϱ be a primitive substitution on the alphabet $\mathcal{A}_n = \{a_1, \ldots, a_n\}$. If $w = w_0 w_1 \cdots w_{k-1}$ is a word of length k, and $\varrho(w) = w'_0 w'_1 \cdots w'_{k'-1}$, the *induced substitution* ϱ_m, with $1 \leqslant m \leqslant k$, acts on w as

$$(4.25) \quad \varrho_m(w) := (w'_0 w'_1 \cdots w'_{m-1})(w'_1 w'_2 \cdots w'_m) \ldots (w'_{\ell-1} w'_\ell \cdots w'_{\ell+m-2}),$$

where $\ell = |\varrho(w_0)|$ is the total length of $\varrho(w_0)$. This rule ensures that words of length m are neither over- nor undercounted relative to each other. Note that $k' \geqslant \ell + m - 1$ by construction, so that the action of ϱ_m is always well-defined. In particular, $\varrho_1(w) = \varrho(w)$ for $w \in \mathcal{A}$ by dismissing the parentheses.

Proposition 4.14. *Let ϱ be a primitive substitution on the alphabet $\mathcal{A}_n$, and ϱ_m the induced substitution according to Eq. (4.25). When interpreted as a substitution on the alphabet that consists of the ϱ-legal m-letter words, ϱ_m is again a primitive substitution.*

PROOF. Let $\{u^{(i)} \mid i \in I_m\}$ denote the (finite) set of ϱ-legal m-letter words. Due to the primitivity of ϱ, there is some $k \in \mathbb{N}$ such that $u^{(i)}$ is a subword of $v^{(j)} := \varrho^k(u_0^{(j)})$ for all $i, j \in I_m$. This implies that the image of $u^{(j)}$ under the induced substitution $(\varrho^k)_m$ contains $u^{(i)}$. The claim now follows from the observation that $(\varrho^k)_m = (\varrho_m)^k$, which can be verified from the definition; compare [Que10, Lemma 5.2] and its proof for further details. □

Example 4.11 (*Frequencies of Fibonacci words*). Consider the Fibonacci substitution $\varrho = \varrho_{\mathrm{F}}$ of Example 4.6, on the alphabet $\{a, b\}$. The legal 2- and 3-letter words are $\{aa, ab, ba\}$ and $\{aab, aba, baa, bab\}$. Since $|\varrho(a)| = 2$ and $|\varrho(b)| = 1$, Eq. (4.25) gives the following induced substitutions

$$
\varrho_2 : \begin{array}{l} (aa) \mapsto (ab)(ba) \\ (ab) \mapsto (ab)(ba) \\ (ba) \mapsto (aa) \end{array}
\quad \text{and} \quad
\varrho_3 : \begin{array}{l} (aab) \mapsto (aba)(bab) \\ (aba) \mapsto (aba)(baa) \\ (baa) \mapsto (aab) \\ (bab) \mapsto (aab) \end{array}
$$

for $m = 2$ and $m = 3$. Viewed as substitutions on 3- and 4-letter alphabets, ϱ_2 and ϱ_3 are primitive, with substitution matrices

$$
M_2 = \begin{pmatrix} 0 & 0 & 1 \\ 1 & 1 & 0 \\ 1 & 1 & 0 \end{pmatrix}
\quad \text{and} \quad
M_3 = \begin{pmatrix} 0 & 0 & 1 & 1 \\ 1 & 1 & 0 & 0 \\ 0 & 1 & 0 & 0 \\ 1 & 0 & 0 & 0 \end{pmatrix} .
$$

The eigenvalues are $\{\tau, \tau', 0\}$ and $\{\tau, \tau', 0, 0\}$, with normalised right PF eigenvectors $(\tau^{-3}, \tau^{-2}, \tau^{-2})^T$ and $(\tau^{-3}, \tau^{-2}, \tau^{-3}, \tau^{-4})^T$, respectively. They contain the relative frequencies of the corresponding 2- and 3-letter words.

In general, as the Fibonacci sequence is Sturmian by Proposition 4.12, there are $m+1$ legal words of length m, so that M_m is $(m+1)$-dimensional. All matrices M_m have the same PF eigenvalue [Que10, Prop. 5.10]. In fact, their eigenvalues are τ, τ' and 0, the latter with multiplicity $m - 1$, by an application of [Que10, Cor. 5.5]. All relative frequencies of legal words can be expressed by simple inverse powers of τ (as in the cases above). This can be seen from the following tree structure of subwords and their extensions,

$\underline{a}$			b		
aa		ab	$\underline{ba}$		
aab		$\underline{aba}$	baa		bab
$\underline{aaba}$		$abaa$	$abab$	$baab$	$baba$
$aabaa$	$aabab$	$abaab$	$ababa$	$\underline{baaba}$	$babaa$
$aabaab$	$aababa$	$\underline{abaaba}$	$ababaa$	$baabaa$	$baabab$	$babaab$

where the underlined words are the unique words on each level with two extensions to the right; compare Proposition 4.12 and the subsequent comment.

The corresponding tree structure of relative frequencies reads

τ^{-1}			τ^{-2}		
τ^{-3}		τ^{-2}	τ^{-2}		
τ^{-3}		τ^{-2}	τ^{-3}		τ^{-4}
τ^{-3}		τ^{-3}	τ^{-4}	τ^{-3}	τ^{-4}
τ^{-5}	τ^{-4}	τ^{-3}	τ^{-4}	τ^{-3}	τ^{-4}
τ^{-5}	τ^{-4}	τ^{-3}	τ^{-4}	τ^{-5}	τ^{-4}	τ^{-4}

where each row sums up to 1. This relies on the identity $\tau^{-1} + \tau^{-2} = 1$. In each vertical step down, precisely one word has two legal extensions, with a splitting of a frequency according to this identity. Uniquely extended words inherit the frequency from the parent word. Recall that each word and its reversed version occur with equal frequency due to the palindromicity of the hull. Since, in any extension pair, at most one word can be palindromic, all frequencies are recursively determined by this tree structure. This informal discussion can be turned into a formal inductive proof. Interestingly, there are two or three different frequencies in each row, but never more [Lot02, Thm. 2.2.37]. This property is common to all Sturmian words.

Since cylinder sets generate the σ-algebra of the hull, the above frequencies completely determine the unique invariant measure of the Fibonacci substitution. Moreover, it is now an easy exercise to determine the $\mathbb{Z}$-module spanned by the set of all word frequencies. This turns out to be $\mathbb{Z}[\tau]$ and is called the (symbolic) *frequency module* of the Fibonacci hull. $\Diamond$

It is clear that the frequencies of finite subwords carry important information about the system. Note that frequencies of more complex (finite) patterns can be expressed as integer linear combinations of word frequencies. This motivates the following concept for systems of symbolic dynamics (which will later be modified in a more geometric setting).

Definition 4.17. Let $\mathbb{X}$ be a uniquely ergodic shift space over a finite alphabet (whence the frequencies of all finite subwords exist uniformly by Proposition 4.4). The minimal $\mathbb{Z}$-module $\mathcal{F}$ that contains all these frequencies is called the (symbolic) *frequency module* of $\mathbb{X}$.

When $\mathbb{X}$ is defined by a (primitive) substitution ϱ, one speaks of the (symbolic) frequency module of ϱ.

In Example 4.11, we saw explicitly that the Fibonacci frequency module is $\mathbb{Z}[\tau]$. Let us describe how it can be determined for a general primitive substitution. To this end, one attaches a substitution matrix M_m to the induced substitution ϱ_m as in Example 4.11, which is then primitive as a consequence of Proposition 4.14; see also [BGJ93] for further references and examples. Its spectrum is related to that of $M = M_1$ as follows. First, the PF eigenvalue is the same for all matrices M_m; see [Que10, Lemma 5.4 and Prop. 5.10]. Furthermore, the matrix M_2 (which may possess additional non-zero eigenvalues) is representative in the sense that its spectrum is a subset of the spectrum of all M_m with $m > 2$, and only the eigenvalue 0 is added [Que10, Cor. 5.5]. When M_1 itself is insufficient, the matrix M_2 thus carries the complete information about the possible frequencies. Let us substantiate this fact with another example, where one indeed needs M_2 (and its PF eigenvector) to calculate the frequency module.

Example 4.12 (*Frequencies of Thue–Morse words*). Consider the Thue–Morse substitution rule $\varrho = \varrho_{\mathrm{TM}}$ of Eq. (4.13), which is a substitution of constant length. The number t_m of legal subwords of length m is given by $t_0 = 1$, $t_1 = 2$, $t_2 = 4$, $t_3 = 6$ together with the recursions $t_{2\ell} = l_\ell + t_{\ell+1}$ and $t_{2\ell+1} = 2t_{\ell+1}$ for $\ell > 1$; see [Brl89, Prop. 4.3] and [Sloane, A005942].

For $m = 2$ and $m = 3$, the induced substitution rules are

$$\varrho_2: \begin{aligned} (aa) &\mapsto (ab)(ba) \\ (ab) &\mapsto (ab)(bb) \\ (ba) &\mapsto (ba)(aa) \\ (bb) &\mapsto (ba)(ab) \end{aligned} \quad \text{and} \quad \varrho_3: \begin{aligned} (aab) &\mapsto (aba)(bab) \\ (aba) &\mapsto (abb)(bba) \\ (abb) &\mapsto (abb)(bba) \\ (baa) &\mapsto (baa)(aab) \\ (bab) &\mapsto (baa)(aab) \\ (bba) &\mapsto (bab)(aba) \end{aligned}$$

which are again of constant length. The corresponding substitution matrices M_m have spectra $\{2, 1, -1, 0\}$ and $\{2, 1, -1, 0, 0, 0\}$, and the normalised right PF eigenvectors are $\frac{1}{6}(1, 2, 2, 1)^T$ and $\frac{1}{6}(1, 1, 1, 1, 1, 1)^T$. As in our previous example, one can again identify a tree-like structure, and frequencies turn out to be of the form $\frac{k}{3 \cdot 2^n}$ with positive integers k and n.

Let us point out that, unlike in some other examples such as the noble mean chains, the PF eigenvector $\frac{1}{2}(1, 1)^T$ of M_1 alone is insufficient to determine the frequency module, so that M_2 is vital here, while the higher order matrices do not add any further generating element. $\diamond$

Let us explain why the frequency module of a primitive substitution can be calculated from the statistically normalised PF eigenvectors of M_1 and M_2, hence actually from M_2 alone. The reason is that an arbitrary legal word of a primitive substitution ϱ occurs either in $\varrho^n(a)$ for some (minimal) n or emerges across the boundary between two neighbouring superwords under substitution (again chosen minimally). Therefore, the calculation of its frequency is possible when one knows the frequencies of all letters and all legal words of length 2, together with the PF eigenvalue of course. The precise result is part of [Bel95, Thm. 18] and reads as follows.

Theorem 4.9. *Let ϱ be a primitive substitution, with induced substitution matrices $M_1 = M$ and M_2, and with PF eigenvalue $\lambda = \lambda_{\mathrm{PF}}$. Then, the frequency module $\mathcal{F}$ of ϱ is the $\mathbb{Z}$-module generated by the numbers a/λ^n, where $n \in \mathbb{N}_0$ and a runs through the entries of the statistically normalised PF eigenvectors of M_2.* $\square$

Alternatively, one can characterise the frequency module of a primitive substitution as the $\mathbb{Z}[1/\lambda_{\mathrm{PF}}]$-module generated by the entries of the statistically normalised PF eigenvectors of M_2. Let us close this paragraph with a brief discussion of a slightly more involved type of word pattern.

Example 4.13 (*Frequencies of Rudin–Shapiro words*). Let us once more consider the binary Rudin–Shapiro sequence of Section 4.7.1. Its subword complexity is given by the sequence

$$1, 2, 4, 8, 16, 24, 36, 46, 56, 64, 72, 80, 88, 96, 104, \ldots$$

which is sequence [Sloane, A005943]. For words of length $n \geqslant 8$, the subword complexity is $8n - 8$, which holds both for the binary and for the quaternary Rudin–Shapiro sequence [AS94]. The invariant measure defined by the word frequencies has a number of interesting properties. Employing the technique of the induced substitution rule for the underlying four-letter substitution, it is easy to verify that, up to length 3, all subwords of the same length have the same frequency. For length 4, the result is

$aaaa$	$aaab$	$aaba$	$aabb$	$abaa$	$abab$	$abba$	$abbb$
$\frac{1}{32}$	$\frac{3}{32}$	$\frac{3}{32}$	$\frac{1}{32}$	$\frac{3}{32}$	$\frac{1}{32}$	$\frac{1}{32}$	$\frac{3}{32}$
$bbbb$	$bbba$	$bbab$	$bbaa$	$babb$	$baba$	$baab$	$baaa$

where the frequencies apply both to the word above and below. In fact, one has S_2-permutation symmetry of the hull under the exchange $a \leftrightarrow b$ (in the sense of Remark 4.3). This follows from the existence of two locally indistinguishable bi-infinite Rudin–Shapiro words that are related by this mapping. Since our shift-invariant measure is unique, we inherit the stronger property that a word and its image under this permutation are equally frequent.

Interesting cluster frequencies in connection with spectral properties are those of words of given length $n \geqslant 2$ with specified first and last letter (but summed over everything in between), denoted by $\nu_{a.a}^{(n)}$, $\nu_{a.b}^{(n)}$, $\nu_{b.a}^{(n)}$ and $\nu_{b.b}^{(n)}$. It is immediate that they sum up to 1, and that they satisfy $\nu_{a.a}^{(n)} = \nu_{b.b}^{(n)}$ and $\nu_{a.b}^{(n)} = \nu_{b.a}^{(n)}$. In fact, one even has

$$(4.26) \qquad \nu_{a.a}^{(n)} = \nu_{a.b}^{(n)} = \nu_{b.a}^{(n)} = \nu_{b.b}^{(n)} = \frac{1}{4}$$

for all $n \geqslant 2$. To see this, we invoke Proposition 10.1, which is proved later by different (and completely independent) arguments, together with Remark 10.4. The latter states that the autocorrelation coefficients of the signed representation satisfy $\eta(m) = \delta_{m,0}$. The claim then follows from the observation that $\nu_{a.a}^{(n)} - \nu_{a.b}^{(n)} = \eta(n-1)/2$. $\Diamond$

Remark 4.16 (*Symbolic versus geometric picture*). It is essentially clear how these properties determine corresponding properties of the geometric counterparts to the symbolic sequences. However, one should be aware of certain subtleties that emerge from the differences between the $\mathbb{Z}$-action of the shift on the sequences and the $\mathbb{R}$-action on the tilings by translation. This is usually studied by the suspension of a discrete dynamical system (with $\mathbb{Z}$-action) into a flow ($\mathbb{R}$-action), which was briefly mentioned in the context of counting inflation fixed points in Remark 4.10. Concretely, in the terminology of [CFS82, Ch. 11.1], the $\mathbb{R}$-action on the tiling for a constant length substitution is equivalent to the special flow obtained from the shift action via the constant function $f \equiv 1$ on the discrete hull; see also [EW11, Exs. 2.9.1 and Lemmas 9.23 and 9.24]. Note that one needs other functions when the underlying substitution is not of constant length. $\Diamond$

Let us now turn our attention to higher dimensions, by generalising the substitution concept to block (or lattice) substitutions on $\mathbb{Z}^2$.

4.9. Block substitutions

Substitution rules on symbolic alphabets can also be used to define objects in higher dimensions; see [Rob99, Lin04, Fra05, Fra08] and references therein. Let us illustrate this with a planar example. Consider the *block substitution rule* on the alphabet $\{0, 1, 2, 3\}$ defined by

$$(4.27) \qquad 0 \mapsto \begin{smallmatrix} 1 & 0 \\ 0 & 3 \end{smallmatrix} \qquad 1 \mapsto \begin{smallmatrix} 1 & 2 \\ 0 & 1 \end{smallmatrix} \qquad 2 \mapsto \begin{smallmatrix} 1 & 2 \\ 2 & 3 \end{smallmatrix} \qquad 3 \mapsto \begin{smallmatrix} 3 & 2 \\ 0 & 3 \end{smallmatrix}$$

which can be iterated in an obvious way. In particular, when starting with the image of 0 as a (legal) seed, one obtains

$$
0 \longmapsto \frac{1 \,|\, 0}{0 \,|\, 3} \longmapsto
\begin{array}{cc|cc}
1 & 2 & 1 & 0 \\
0 & 1 & 0 & 3 \\
\hline
1 & 0 & 3 & 2 \\
0 & 3 & 0 & 3
\end{array}
\longmapsto
\begin{array}{cccc|cccc}
1 & 2 & 1 & 2 & 1 & 2 & 1 & 0 \\
0 & 1 & 2 & 3 & 0 & 1 & 0 & 3 \\
1 & 0 & 1 & 2 & 1 & 0 & 3 & 2 \\
0 & 3 & 0 & 1 & 0 & 3 & 0 & 3 \\
\hline
1 & 2 & 1 & 0 & 3 & 2 & 1 & 2 \\
0 & 1 & 0 & 3 & 0 & 3 & 2 & 3 \\
1 & 0 & 3 & 2 & 1 & 0 & 3 & 2 \\
0 & 3 & 0 & 3 & 0 & 3 & 0 & 3
\end{array}
\longmapsto \cdots .
$$

Choosing the centre as the reference point, this gives a sequence of square shaped blocks which grow in all directions and converge in the standard product topology. One can imagine this fixed point as a labelled (or coloured) square lattice, where each square (Delone cell) carries one of the four digits (or colours). The analogous property holds for the iteration sequences that start from the other three letters. Yet another fixed point, less obvious but ultimately perhaps more interesting, is obtained from the iteration

$$
(4.28) \qquad \frac{3 \,|\, 0}{2 \,|\, 1} \longmapsto
\begin{array}{cc|cc}
3 & 2 & 1 & 0 \\
0 & 3 & 0 & 3 \\
\hline
1 & 2 & 1 & 2 \\
2 & 3 & 0 & 1
\end{array}
\longmapsto
\begin{array}{cccc|cccc}
3 & 2 & 1 & 2 & 1 & 2 & 1 & 0 \\
0 & 3 & 2 & 3 & 0 & 1 & 0 & 3 \\
1 & 0 & 3 & 2 & 1 & 0 & 3 & 2 \\
0 & 3 & 0 & 3 & 0 & 3 & 0 & 3 \\
\hline
1 & 2 & 1 & 2 & 1 & 2 & 1 & 2 \\
0 & 1 & 2 & 3 & 0 & 1 & 2 & 3 \\
1 & 2 & 3 & 2 & 1 & 0 & 1 & 2 \\
2 & 3 & 0 & 3 & 0 & 3 & 0 & 1
\end{array}
\longmapsto \cdots .
$$

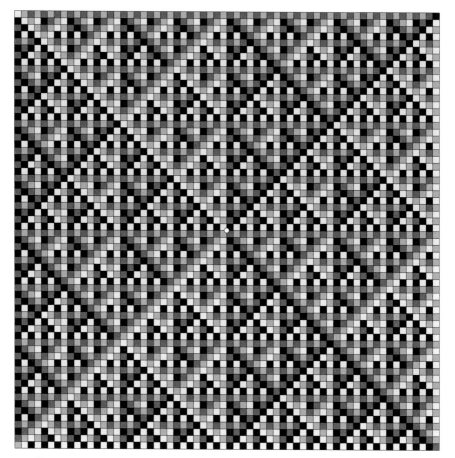

FIGURE 4.1. Patch of the square lattice, shaded according to the fifth iteration of the block substitution (4.27) with the seed of Eq. (4.28). The white disk in the centre marks the origin.

The legality of the seed follows from localising it in the previous iteration sequence. One interesting property of this fixed point is its D_4 *colour symmetry*. An anti-clockwise rotation through $\pi/2$ corresponds to the cyclic permutation (0123), while a reflection in the central horizontal gives the permutation (01)(23). The pattern is thus invariant under any D_4 transformation followed by the inverse of the corresponding permutation. Note that the other fixed points mentioned above agree with this one, except on the two diagonals (and thus show 'almost' symmetries).

One can extend the notion of proximality, and use it to assess the non-periodicity (and also aperiodicity) of the block substitution; see Section 5.5 below for the general method. Here, we alternatively follow a constructive route that is reminiscent of our approach to the period doubling sequence, because it emphasises the Toeplitz structure as the underlying mechanism.

In analogy to the situation with the previous substitution rules, one can assign a counting matrix to the block substitution, which keeps track of the number of blocks of each type. This matrix reads

$$M = \begin{pmatrix} 2 & 1 & 0 & 1 \\ 1 & 2 & 1 & 0 \\ 0 & 1 & 2 & 1 \\ 1 & 0 & 1 & 2 \end{pmatrix}$$

for the block substitution (4.27). The matrix M is symmetric and primitive ($M^2 \gg 0$), with PF eigenvalue $\lambda = 4$ and corresponding right eigenvector $\frac{1}{4}(1,1,1,1)^T$. Each letter of the alphabet is thus equally frequent in a fixed point. Since the left PF eigenvector is just the transpose, a natural geometric representation employs four square-shaped prototiles of equal area, distinguished by four colours. A patch of the resulting square lattice colouring is shown in Figure 4.1. We present an alternative geometric approach (via the so-called 'chair' inflation) in Section 6.4 on page 202.

Let us expand on this example by giving an algebraic formulation which resembles the treatment of the period doubling chain in Section 4.5.1. We choose a representation with four different unit squares and take the lower left corners as their reference points. The fixed point defined by the iteration in Eq. (4.28) then results in a partition of $\mathbb{Z}^2$,

$$\mathbb{Z}^2 = \Lambda_0 \,\dot\cup\, \Lambda_1 \,\dot\cup\, \Lambda_2 \,\dot\cup\, \Lambda_3,$$

into four point sets of equal density $\frac{1}{4}$. They satisfy the fixed point equations

(4.29)
$$\begin{aligned} \Lambda_0 &= 2\Lambda_0 \,\dot\cup\, (2\Lambda_0 + u + v) \,\dot\cup\, 2\Lambda_1 \,\dot\cup\, 2\Lambda_3, \\ \Lambda_1 &= (2\Lambda_0 + v) \,\dot\cup\, (2\Lambda_1 + u) \,\dot\cup\, (2\Lambda_1 + v) \,\dot\cup\, (2\Lambda_2 + v), \\ \Lambda_2 &= (2\Lambda_1 + u + v) \,\dot\cup\, 2\Lambda_2 \,\dot\cup\, (2\Lambda_2 + u + v) \,\dot\cup\, (2\Lambda_3 + u + v), \\ \Lambda_3 &= (2\Lambda_0 + u) \,\dot\cup\, (2\Lambda_2 + u) \,\dot\cup\, (2\Lambda_3 + u) \,\dot\cup\, (2\Lambda_3 + v), \end{aligned}$$

with $u = (1,0)^T$ and $v = (0,1)^T$. These equations derive from the block substitution rule of Eq. (4.27).

Defining $\Gamma_+ = \big\{ (x_1, x_2)^T \in \mathbb{Z}^2 \mid x_1 + x_2 \equiv 0 \bmod 2 \big\}$ as the even sublattice of $\mathbb{Z}^2$ and $\Gamma_- = \Gamma_+ + u$, one clearly has $\mathbb{Z}^2 = \Gamma_+ \cup \Gamma_-$, but also

$$\Lambda_0 \,\dot\cup\, \Lambda_2 = \Gamma_+ \quad \text{and} \quad \Lambda_1 \,\dot\cup\, \Lambda_3 = \Gamma_-,$$

as readily follows from the above fixed point equations, if one observes the relation $2\mathbb{Z}^2 \mathbin{\dot{\cup}} (2\mathbb{Z}^2 + u + v) = \Gamma_+$. This leads to the decoupled equations

$$\Lambda_0 = 2\Gamma_- \mathbin{\dot{\cup}} \left(2\Lambda_0 + \{0, u + v\}\right),$$
$$\Lambda_1 = \left(2\Gamma_+ + v\right) \mathbin{\dot{\cup}} \left(2\Lambda_1 + \{u, v\}\right),$$

together with $\Lambda_2 = \Gamma_+ \backslash \Lambda_0$ and $\Lambda_3 = \Gamma_- \backslash \Lambda_1$. These can be solved iteratively, which yields the relations (with the shorthands $Sx := \{sx \mid s \in S\}$ for sets $S \subset \mathbb{Z}$ and $S_r := \{0, 1, \ldots, 2^r - 1\}$ for $r \geqslant 0$)

(4.30)

$$\Lambda_0 = \mathbb{N}_0(u + v) \mathbin{\dot{\cup}} \bigcup_{r \geqslant 0}^{\textstyle\cdot} \left(2^{r+1}\Gamma_- + S_r(u + v)\right),$$

$$\Lambda_1 + v = \mathbb{N}_0(u - v) \mathbin{\dot{\cup}} \bigcup_{r \geqslant 0}^{\textstyle\cdot} \left(2^{r+1}\Gamma_- + S_r(u - v)\right),$$

$$\Lambda_2 + u + v = -\mathbb{N}_0(u + v) \mathbin{\dot{\cup}} \bigcup_{r \geqslant 0}^{\textstyle\cdot} \left(2^{r+1}\Gamma_- - S_r(u + v)\right),$$

$$\Lambda_3 + u = -\mathbb{N}_0(u - v) \mathbin{\dot{\cup}} \bigcup_{r \geqslant 0}^{\textstyle\cdot} \left(2^{r+1}\Gamma_- - S_r(u - v)\right),$$

where we used $\Gamma_+ + u = \Gamma_+ + v = \Gamma_- = \Gamma_- + u + v$. It is easy to confirm that each of the infinite unions on the right-hand side defines a point set of density $\frac{1}{4}$. Note that the extra sets (of density 0) of points along the diagonals have to be added for our fixed point from Eq. (4.28), because they are not contained in any of the lattice translates (but are covered by the 2-adic completion of the infinite unions, similar to the situation of the period doubling chain). One can also explicitly check that these solutions obey the relations $\Lambda_0 \mathbin{\dot{\cup}} \Lambda_2 = \Gamma_+$ and $\Lambda_1 \mathbin{\dot{\cup}} \Lambda_3 = \Gamma_-$. The other fixed points described earlier differ from the solution in Eq. (4.30) only along the diagonals. This situation corresponds to the special role of the point -1 in the period doubling case. In particular, Eq. (4.30) is the solution of Eq. (4.29) that belongs to the patch of Figure 4.1, but it is not the only solution of Eq. (4.29). Later, we will discuss a similar example in more detail in Section 6.4.

The equations in Eq. (4.30) will become crucial for the diffraction theory of the chair tiling in Section 9.4.5 on page 381. The system shows a structure of Toeplitz type, and can indeed be viewed as a planar generalisation of the period doubling substitution. Our explicit solution in Eq. (4.30) is non-periodic. The primitivity of the block substitution (4.27) thus implies the aperiodicity of this example, while the various properties of the discrete hull (under the two-dimensional shift) follow as before.

Corollary 4.9. *The block substitution of Eq. (4.27) defines a dynamical system under the action of $\mathbb{Z}^2$ that is strictly ergodic and aperiodic.* $\square$

Example 4.14 (*Block substitution for the table tiling*). An interesting related block substitution is given by

$$(4.31) \qquad 0 \mapsto \begin{smallmatrix} 1 & 0 \\ 3 & 0 \end{smallmatrix} \qquad 1 \mapsto \begin{smallmatrix} 0 & 2 \\ 1 & 1 \end{smallmatrix} \qquad 2 \mapsto \begin{smallmatrix} 2 & 1 \\ 2 & 3 \end{smallmatrix} \qquad 3 \mapsto \begin{smallmatrix} 3 & 3 \\ 0 & 2 \end{smallmatrix}$$

which was studied by Robinson [Rob99] in connection with the above example. A geometric counterpart, known as the 'table' tiling, is constructed with one rectangular prototile that occurs in two orientations; see Example 6.2 below for its definition and Figure 6.47 for a (coloured) patch.

On first sight, one is tempted to search for a relation between the two block substitutions ('table' and 'chair') similar to the one between Thue–Morse and period doubling. A closer inspection of the singular (and hence proximal) members of the two hulls, however, shows that the chair cannot be a topological factor of the table. Nevertheless, the chair defines a dynamical system that is isomorphic to the Kronecker factor of the table in the measure-theoretic sense; see [Rob99] for details. ◇

In view of the relation between period doubling and Thue–Morse, it is tempting to search for a suitable generalisation in the case of the table tiling.

Remark 4.17 (*Pure point factor for the table*). The block substitution of Example 4.14 defines a unique (discrete) hull, which is aperiodic. Each element contains the following 24 legal patches of size 2×2:

0 2	0 2	0 2	1 0	1 1	1 3	2 0	2 1
0 2	1 0	2 1	3 1	3 3	3 0	1 1	1 3
2 3	2 3	2 3	3 0	3 0	3 0	3 1	3 3
0 2	1 0	2 1	0 2	1 0	2 1	2 3	2 0
0 2	0 2	1 0	1 3	2 0	2 1	3 1	3 3
1 1	2 0	3 0	3 1	0 2	2 3	1 3	0 2

Mapping the 16 patches of the first two rows to 0 and the remaining eight patches from the third row to 1 defines a sliding block map to the binary alphabet $\{0,1\}$. If the image replaces the lower left corner of each patch, the map induces the block substitution

$$\ell \longmapsto \begin{smallmatrix} 0 & \ell \\ 1 & 0 \end{smallmatrix}$$

with $\ell \in \{0,1\}$. The new rule is primitive and admits two fixed points (with legal seed $\begin{smallmatrix} 0 & 1 \\ 1 & 0 \end{smallmatrix}$ or $\begin{smallmatrix} 0 & 1 \\ 0 & 0 \end{smallmatrix}$ and reference point in its centre, so that the fixed point covers $\mathbb{Z}^2$). The fixed point equations for the corresponding point sets Λ_0 and Λ_1 decouple and read

$$\Lambda_\ell = \left(2\mathbb{Z}^2 + \{u,v\} \right) \dot{\cup} \left(2\Lambda_\ell + (u+v) \right)$$

with u and v as before. Since $\Lambda_0 \cap \Lambda_1 = \varnothing$ and $\Lambda_0 \cup \Lambda_1 = \mathbb{Z}^2$, the only two solutions are

$$\Lambda_\ell = \bigcup_{n \geqslant 1} \left(2^n \mathbb{Z}^2 + 2^{n-1}\{u, v\} + (2^{n-1} - 1)(u + v) \right) \dot{\cup} A_\ell,$$

where $A_1 = \{-(u + v)\}$ and $A_0 = \varnothing$, or vice versa, which corresponds to the two seeds mentioned above. This shows a Toeplitz structure on $\mathbb{Z}^2$ that is reminiscent of that of the period doubling chain. As we shall see later in similar cases, this defines a topological factor with maximal pure point spectrum, and hence a cover of the Kronecker factor of the table tiling, which is formulated via Σ_2^2 with the solenoid from Example 2.11.

Note that this factor is not unique in the sense that various other factors exist. More precisely, one can construct a system of factors [BGG13] with (identical) maximal pure point spectrum that unveil topological details of the table tiling, viewed as a fibred extension of the dyadic solenoid. ◊

Many more examples in this spirit can be constructed; see [Lin04, Fra08] for an overview and further references. We shall meet another one in Example 6.7 below. It is clear that this approach works in Euclidean spaces of arbitrary dimension, although substitution factors with maximal pure point spectrum that are covers of the Kronecker factor need not always exist; see [Her13] for concrete examples.

As mentioned at the beginning of this section, block substitutions provide a natural generalisation of one-dimensional substitution rules. However, despite many open questions in this area, their scope is somewhat limited because the intricate geometric constraints of tile arrangements in dimensions $d \geqslant 2$ cannot be captured in sufficient generality. For this reason, we now turn our attention to geometric inflation rules for (finite) prototile sets.

CHAPTER 5

Patterns and Tilings

Although we formulate many properties primarily for point sets, other discrete structures, such as tilings (or tessellations), possibly with additional markers or colours, are important and useful alternatives. They provide different points of views that are related by natural equivalence concepts. These relations enable an efficient progress by selecting the most appropriate representative from the equivalence class for the question at hand.

5.1. Patterns and local indistinguishability

In our context, we need a sufficiently general concept for a (discrete or discrete-related) structure in d-space.

Definition 5.1. A *pattern* $\mathcal{T}$ in Euclidean space $\mathbb{R}^d$ is a non-empty set of non-empty subsets of $\mathbb{R}^d$. We refer to the elements of $\mathcal{T}$ as the *fragments* of the pattern $\mathcal{T}$.

A pattern can emerge from a point set $\Lambda \subset \mathbb{R}^d$, or from a tiling, such as the Voronoi or Delone complex of a locally finite point set. For a pattern $\mathcal{T}$ in $\mathbb{R}^d$, which may be finite, we will use the self-explanatory notation $\mathcal{T} \sqsubset \mathbb{R}^d$. Below, we will mainly be interested in infinite patterns that are 'distributed' over $\mathbb{R}^d$ in a more or less homogeneous manner. One can generalise the term pattern to allow for decorations or colourings, which amounts to replacing $\mathbb{R}^d$ by some product-type configuration space. Since this would introduce an additional complexity of the notation, we prefer to stick to $\mathbb{R}^d$. There is no difficulty to translate the concepts and results below to the more general situation. Also, colours and markers can usually be represented by suitable sets, as we shall see below in various examples, so that patterns according to Definition 5.1 are general enough for most purposes.

Example 5.1 (*Point sets as patterns*)**.** Let $\Lambda \subset \mathbb{R}^d$ be a locally finite point set. It is naturally turned into a pattern as

$$\mathcal{T} = \mathcal{T}_\Lambda = \big\{\{x\} \mid x \in \Lambda\big\}$$

(and not as $\{\Lambda\}$, for obvious reasons). Below, we tacitly identify Λ and $\mathcal{T}_\Lambda$ for simplicity. ◊

Besides point sets, another important special case of a pattern is a tiling. This term is used in a variety of meanings in the literature. We start from a fairly general notion.

Definition 5.2. A *tiling* in d-space is a pattern $\mathcal{T} = \{T_i \mid i \in I\} \sqsubset \mathbb{R}^d$, with countable index set I and non-empty closed sets $T_i \subset \mathbb{R}^d$, subject to the conditions $\bigcup_{i \in I} T_i = \mathbb{R}^d$ and $T_i^\circ \cap T_j^\circ = \varnothing$ for all $i \neq j$.

The fragments T_i of $\mathcal{T}$ are called the *tiles* of the tiling, and their equivalence classes up to translations (or, alternatively, up to congruence) are called *prototiles*. It will be clear from the context which version is used. For additional material with many examples, we refer to [Sad08].

To avoid pathologies, one often assumes that all T_i have non-empty interior and a boundary of Lebesgue measure 0. Moreover, the most relevant case is the one where all T_i are compact and satisfy $T_i = \overline{T_i^\circ}$, a property that is sometimes called *regular*. In particular, regularity avoids tiles in $\mathbb{R}^d$ that possess lower-dimensional components, such as an additional line segment of a polygon in the plane or an extra facet of a polyhedron in 3-space. In most cases we have in mind, the T_i will also be connected. In fact, they will often be simple polytopes. In this case, when all neighbouring tiles meet in complete facets, we call the corresponding tiling *face to face*.

Apart from very simple examples, one often needs additional markers or *decorations* of the prototiles, for instance to distinguish different types of the same shape (by set-valued labels), to introduce an orientation (by arrows) or to impose adjacency rules for neighbouring tiles (by facet markers). All these extensions are examples of patterns (or can be represented as such). Nevertheless, following the widespread abuse of language in this context, we will frequently call such patterns *decorated tilings*, or simply tilings when misunderstandings are unlikely.

Without the condition on the intersection of the fragments in Definition 5.2, a pattern is called a *covering*. This is an important extension of the tiling concept, which has found applications in crystallography. In Section 6.2.1 below, we shall meet one decorated example of this kind, namely Gummelt's covering of the classic Penrose tiling; for further cases and a thorough survey, we refer to [KP03] and references therein.

5.1.1. LOCAL INDISTINGUISHABILITY

If $\mathcal{T}$ is a pattern in $\mathbb{R}^d$, denoted as $\mathcal{T} \sqsubset \mathbb{R}^d$, and $A \subset \mathbb{R}^d$, we define $\mathcal{T} \sqcap A$ to be the subset of $\mathcal{T}$ that consists of all fragments of $\mathcal{T}$ which intersect A, so $\mathcal{T} \sqcap A = \{T \in \mathcal{T} \mid T \cap A \neq \varnothing\}$.

Definition 5.3. A pattern $\mathcal{T} \sqsubset \mathbb{R}^d$ is called *locally finite* if $\mathcal{T} \sqcap K$ has finite cardinality, for all compact $K \subset \mathbb{R}^d$.

From now on, we will almost exclusively deal with locally finite patterns.

Definition 5.4. Let $\mathcal{T} \sqsubset \mathbb{R}^d$ be a locally finite pattern. When $K \subset \mathbb{R}^d$ is compact, the pattern $\mathcal{T} \sqcap K$ is called a *cluster* of $\mathcal{T}$. We also speak of a *patch* when K is convex.

Remark 5.1 (*Shape of clusters*). For some aspects, it is easier to operate with $\mathcal{T} \sqcap K$ rather than with the fragments of $\mathcal{T}$ that are contained in K. If all fragments of the pattern are uniformly bounded in size, as is the case for many tilings and for locally finite point sets, this distinction is irrelevant. $\Diamond$

So far, we have used the notation $t + \Lambda = \{t + x \mid x \in \Lambda\}$ for the translate of a point set $\Lambda \subset \mathbb{R}^d$. This extends to a pattern $\mathcal{T} = \{T_i \mid i \in I\}$, where $T_i \subset \mathbb{R}^d$ are the I-indexed fragments of the pattern, via

$$t + \mathcal{T} := \{t + T_i \mid i \in I\}.$$

In this setting, since $t + \mathcal{T} = \mathcal{T}$ has a clear meaning, the notion of a *period* extends from point sets to patterns in an obvious way. The same applies to the concepts 'periodic', 'non-periodic' and 'crystallographic' of Definition 3.1, while the term 'non-crystallographic' is useful only under some discreteness assumption on $\mathcal{T}$, such as the local finiteness from Definition 5.3; compare the discussion in Remark 3.1.

Definition 5.4, as well as the following one, generalises notions introduced earlier in the context of symbolic sequences; compare Definition 4.9.

Definition 5.5. Two (locally finite) patterns $\mathcal{T}$ and $\mathcal{T}'$ in $\mathbb{R}^d$ are *locally indistinguishable*, or LI for short and written as $\mathcal{T} \overset{\text{LI}}{\sim} \mathcal{T}'$, when any cluster of $\mathcal{T}$ occurs also in $\mathcal{T}'$ and vice versa. This means that, for any compact $K \subset \mathbb{R}^d$, there are translations $t, t' \in \mathbb{R}^d$ such that $\mathcal{T} \sqcap K = (-t' + \mathcal{T}') \sqcap K$ together with $\mathcal{T}' \sqcap K = (-t + \mathcal{T}) \sqcap K$.

Local indistinguishability is an equivalence relation on the class of patterns, partitioning them into LI classes. The LI class of a pattern $\mathcal{T} \sqsubset \mathbb{R}^d$ is written as $\mathrm{LI}(\mathcal{T})$. Other terms in use are *locally isomorphic* and *locally isometric*. Since the former has a different meaning in dynamical systems theory, and the latter has no natural generalisation beyond the Euclidean setting, we prefer the term local indistinguishability, as in Chapter 4; see also the discussion in [Lif96, Lif97]. Fortunately, the shorthand LI is oblivious to the personal preference.

For the formulation of the next result, we need the concept of the *local topology*. Two patterns $\mathcal{T}, \mathcal{T}' \sqsubset \mathbb{R}^d$ are ε-close in the local topology when

(5.1) $$\mathcal{T} \sqcap \overline{B_{1/\varepsilon}(0)} = (-t + \mathcal{T}') \sqcap \overline{B_{1/\varepsilon}(0)}$$

holds for some $t \in B_\varepsilon(0)$. We will say more about the underlying topology in Section 5.4.

Lemma 5.1. *If $t \neq 0$ is a period of a locally finite pattern $\mathcal{T} \sqsubset \mathbb{R}^d$, any $\mathcal{T}' \in \mathrm{LI}(\mathcal{T})$ is t-periodic as well. The group of periods is thus an invariant of an LI class. Moreover, if $\Gamma = \mathrm{per}(\mathcal{T})$ is a lattice in $\mathbb{R}^d$ (whence $\mathcal{T}$ is crystallographic), one has*

$$\mathrm{LI}(\mathcal{T}) = \{t + \mathcal{T} \mid t \in \mathrm{FD}(\Gamma)\} = \overline{\{x + \mathcal{T} \mid x \in \mathbb{R}^d\}},$$

where $\mathrm{FD}(\Gamma)$ is a fundamental domain of Γ, and where the closure is taken in the local topology. In particular, $\mathrm{LI}(\mathcal{T})$ is compact in the local topology, and one has $\mathrm{LI}(\mathcal{T}) \simeq \mathbb{R}^d/\Gamma$ as topological spaces.

PROOF. By assumption, we have $(-t + \mathcal{T}) \sqcap K = \mathcal{T} \sqcap K$ for any $t \in \Gamma$ and any compact $K \subset \mathbb{R}^d$. Assume that, contrary to the claim, there is some $\mathcal{T}' \in \mathrm{LI}(\mathcal{T})$ which fails to be t-periodic. Consequently, there exists a compact set $K' \subset \mathbb{R}^d$ with $(-t + \mathcal{T}') \sqcap K' \neq \mathcal{T}' \sqcap K'$. Clearly, there is a compact $K \subset \mathbb{R}^d$ that contains both K' and $t + K'$. Since $\mathcal{T}' \overset{\mathrm{LI}}{\sim} \mathcal{T}$, we have $\mathcal{T}' \sqcap K = (-x + \mathcal{T}) \sqcap K$ for some $x \in \mathbb{R}^d$, and hence

$$(-t - x + \mathcal{T}) \sqcap K' = (-t + \mathcal{T}') \sqcap K' \neq \mathcal{T}' \sqcap K' = (-x + \mathcal{T}) \sqcap K',$$

where $\mathcal{T} \sqcap (t + K') = t + \left((-t + \mathcal{T}) \sqcap K'\right)$ was used. This contradicts the periodicity of $-x + \mathcal{T}$ and hence that of $\mathcal{T}$ itself. It is thus clear that all elements of an LI class share the same group of periods.

One obviously has $\mathcal{T} \overset{\mathrm{LI}}{\sim} (x + \mathcal{T})$ for all $x \in \mathbb{R}^d$, so $\mathrm{LI}(\mathcal{T})$ contains all translates of $\mathcal{T}$. There can be no further element, as can be seen by choosing a compact set K that contains a fundamental domain of Γ and using the LI property, together with the preserved periodicity. Since $\mathcal{T} = t + \mathcal{T}$ for all $t \in \Gamma$, the LI class it exhausted by the patterns $x + \mathcal{T}$ with x running through a fundamental domain of Γ.

Due to the crystallographic structure, it is clear that we can employ the local topology. For sufficiently small $\varepsilon > 0$, such that $B_{1/\varepsilon}(0)$ contains a fundamental domain $\mathrm{FD}(\Gamma)$, the condition in Eq. (5.1) clearly simplifies to $\mathcal{T} = (-t + \mathcal{T}')$ for some $t \in B_\varepsilon(0)$. Let $(x_i + \mathcal{T})_{i \in \mathbb{N}}$ be a sequence of patterns that converges in the local topology, where we may assume $x_i \in \mathrm{FD}(\Gamma)$ without loss of generality, with $\mathrm{FD}(\Gamma)$ relatively compact. Consequently, $(x_i)_{i \in \mathbb{N}}$ contains a subsequence that converges (in $\mathbb{R}^d$) to some $x \in \overline{\mathrm{FD}(\Gamma)}$, wherefore the corresponding subsequence of patterns converges to $x + \mathcal{T}$. Any other accumulation point x' of $(x_i)_{i \in \mathbb{N}}$ can only be a lattice translate of x, due to the assumed convergence of $(x_i + \mathcal{T})_{i \in \mathbb{N}}$, so that any limit point of translates of $\mathcal{T}$ is indeed a translate of $\mathcal{T}$ and thus in $\mathrm{LI}(\mathcal{T})$. Consequently, as topological spaces, we have $\mathrm{LI}(\mathcal{T}) \simeq \mathbb{R}^d/\Gamma$ which is compact. $\square$

Remark 5.2 (*Other notions of equivalence*). There are important alternative equivalence concepts, including topological conjugacy (as dynamical systems), bi-Lipschitz equivalence or wobbling equivalence. The dynamical systems point of view will naturally arise again when we discuss the notion of a (continuous) hull; see also Section 4.2 above. Bi-Lipschitz equivalence ignores the translation action, but asks for an invertible map between two locally finite point sets which is Lipschitz in both directions. It is known [BK98, McM98] that there are Delone sets in $\mathbb{R}^d$ (with $d \geqslant 2$) that are *not* bi-Lipschitz equivalent to $\mathbb{Z}^d$; compare [Sol11, A-PCG13] for positive results on subclasses of Delone sets.

The stronger notion of wobbling equivalence of two locally finite point sets requires an invertible mapping between the two sets such that the distance between any point and its image is uniformly bounded; see [DSS95, FG13] for a detailed discussion. Distinct lattices (or crystallographic packings) in $\mathbb{R}^d$ are wobbling equivalent if and only if they have the same density. More generally, however, having the same density is not sufficient for wobbling equivalence of two point sets.   ◇

Let us turn our attention to a generalisation of Lemma 5.1 to patterns without periods, but nevertheless significant repetitions of subpatterns.

5.1.2. THE LIMIT TRANSLATION MODULE

Given a (not necessarily locally finite) pattern $\mathcal{T} \sqsubset \mathbb{R}^d$ and a compact set $K \subset \mathbb{R}^d$, one can define a $\mathbb{Z}$-module $\Delta_K(\mathcal{T})$ as

$$\Delta_K(\mathcal{T}) := \big\langle\, t \mid \mathcal{T} \sqcap (x + K) = (-t + \mathcal{T}) \sqcap (x + K) \text{ for some } x \in \mathbb{R}^d \big\rangle_{\mathbb{Z}}.$$

This is the $\mathbb{Z}$-module generated by all translations between occurrences of some K-cluster in $\mathcal{T}$. When $K \subset K'$, one clearly has $\Delta_{K'}(\mathcal{T}) \subset \Delta_K(\mathcal{T})$. Also, for any two compact sets K and K', $\Delta_{K \cup K'}(\mathcal{T}) \subset \Delta_K(\mathcal{T}) \cap \Delta_{K'}(\mathcal{T})$, where $K \cup K'$ is again compact. The *limit translation module* (LTM) $\Delta(\mathcal{T})$ is then defined as the inductive limit [Lan02] of the $\Delta_K(\mathcal{T})$ over all compact subsets $K \subset \mathbb{R}^d$, ordered according to inclusion. It is a well-defined $\mathbb{Z}$-module embedded in $\mathbb{R}^d$. Generically, it will be the trivial module $\{0\}$. Nevertheless, we shall see that the LTM is an interesting and useful quantity for certain systems with some degree of order, in particular for Euclidean model sets. When the LTM is trivial, as in limit-periodic examples such as the period doubling chain, one can define a dual structure instead which carries important information; see Chapter 9 for more. Also, the support of the autocorrelation carries vital information of the system, for instance about the cut and project scheme, as discussed in Chapter 7 and Remark 9.21; see [BM04] for a general approach to reconstruct the embedding space.

Proposition 5.1. *The limit translation module of a (locally finite) pattern* $\mathcal{T} \sqsubset \mathbb{R}^d$ *is an invariant of* $\mathrm{LI}(\mathcal{T})$.

SKETCH OF PROOF. An argument similar to the one used in the proof of Lemma 5.1 can be employed to show that, for any compact $K \subset \mathbb{R}^d$, every generating element of $\Delta_K(\mathcal{T})$ is also contained in $\Delta_K(\mathcal{T}')$, which holds for arbitrary $\mathcal{T}' \in \mathrm{LI}(\mathcal{T})$. □

In general, the explicit determination of the LTM seems difficult, with the exception of certain primitive substitution rules (discussed in Chapter 4 and in Example 5.3 below) and Euclidean model sets (to be introduced in Chapter 7). In the latter case, one obtains the LTM as the projection of a lattice. This can be viewed as a generalisation of the crystallographic case. Conversely, the LTM for a pattern reveals important information about a potential embedding of the structure via a lattice in some 'superspace'.

Example 5.2 (*LTM of crystallographic patterns*). If a locally finite pattern $\mathcal{T} \sqsubset \mathbb{R}^d$ is crystallographic, its LTM coincides with its lattice of periods Γ. This can be seen as follows. It is clear that Γ forms a submodule of $\Delta(\mathcal{T})$. Conversely, if there is some $t \in \Delta(\mathcal{T}) \setminus \Gamma$, we can use the Γ-periodicity of $\mathcal{T}$ to see that t must translate arbitrarily large clusters $K \sqcap \mathcal{T}$ to identical copies. Since $\mathcal{T} = \Gamma + \mathcal{F}$ with $\mathcal{F}$ a bounded pattern in complete analogy to Proposition 3.1, t must then be a period of $\mathcal{T}$, in contradiction to the assumption. Consequently, $\Delta(\mathcal{T}) \subset \Gamma$, and hence $\Delta(\mathcal{T}) = \Gamma$. ◊

Example 5.3 (*LTM of the silver mean chain*). The silver mean point set Λ_{sm} introduced in Example 4.5 on page 88 is a subset of $\mathbb{Z}[\sqrt{2}]$ by construction, as is the difference set $\Lambda_{\mathrm{sm}} - \Lambda_{\mathrm{sm}}$. Consequently, all translation modules, and the limit translation module in particular, are contained in $\mathbb{Z}[\sqrt{2}]$.

Recall that 1 and λ_{sm} (with $\lambda_{\mathrm{sm}} = 1 + \sqrt{2}$) are the lengths of the two prototiles of the silver mean chain and hence possible translations between neighbouring points of Λ_{sm}. This implies that $\Delta_{[-\varepsilon,\varepsilon]}(\Lambda_{\mathrm{sm}}) = \mathbb{Z}[\sqrt{2}]$ for all sufficiently small $\varepsilon > 0$. We now need an argument why $\Delta_K(\Lambda_{\mathrm{sm}})$ does not shrink as $K \subset \mathbb{R}$ grows.

Due to the inflation construction, the nth inflations of the prototiles correspond to patches of Λ_{sm} that are now translated to shifted copies by λ_{sm}^n and $\lambda_{\mathrm{sm}}^{n+1}$. Their $\mathbb{Z}$-span is again $\mathbb{Z}[\sqrt{2}]$ because λ_{sm} is a unit, so the corresponding translation module is again $\mathbb{Z}[\sqrt{2}]$. Now recall that any finite patch of Λ_{sm} is contained in the patch obtained from an a-type interval under n-fold inflation for some n. Since $n \in \mathbb{N}$ in the previous argument is arbitrary, the translation modules are thus stable, so that the inductive limit is $\Delta(\Lambda_{\mathrm{sm}}) = \mathbb{Z}[\sqrt{2}]$. The same argument obviously works for any point set $\Lambda \in \mathrm{LI}(\Lambda_{\mathrm{sm}})$, in line with Proposition 5.1. ◊

As soon as we have generalised our notion of an inflation rule to tilings of $\mathbb{R}^d$, the same kind of argument can be applied to tilings obtained as fixed points of primitive inflation rules; see Chapter 6 below.

5.2. Local derivability

Beyond LI, we need to formalise relations between point sets and tilings, or, more generally, between distinct but related patterns (not necessarily locally finite). A useful concept, called local derivability, was introduced in [BSJ91]; see also [Baa02a].

Definition 5.6. A pattern $\mathcal{T}' \sqsubset \mathbb{R}^d$ is said to be *locally derivable* from a pattern $\mathcal{T} \sqsubset \mathbb{R}^d$, written as $\mathcal{T} \overset{\mathrm{LD}}{\leadsto} \mathcal{T}'$, when a compact neighbourhood $K \subset \mathbb{R}^d$ of 0 exists such that, whenever $(-x + \mathcal{T}) \sqcap K = (-y + \mathcal{T}) \sqcap K$ holds for $x, y \in \mathbb{R}^d$, one also has $(-x + \mathcal{T}') \sqcap \{0\} = (-y + \mathcal{T}') \sqcap \{0\}$.

Local derivability is a necessary and sufficient condition for the possibility to devise a (formal) rule to construct the part of $\mathcal{T}'$ around a given point from the sole knowledge of the K-neighbourhood of that point in $\mathcal{T}$. It is immediate that local derivability is reflexive and transitive.

Lemma 5.2. *Let the pattern $\mathcal{T}_1'$ be locally derivable from $\mathcal{T}_1 \sqsubset \mathbb{R}^d$, and let $\mathcal{T}_2 \in \mathrm{LI}(\mathcal{T}_1)$. Then, there exists some $\mathcal{T}_2' \in \mathrm{LI}(\mathcal{T}_1')$ which is locally derivable from $\mathcal{T}_2$.*

PROOF. Let K be a compact set for the derivation $\mathcal{T}_1 \overset{\mathrm{LD}}{\leadsto} \mathcal{T}_1'$ according to Definition 5.6, and assume that a derivation rule has been specified. This rule can now be applied to $\mathcal{T}_2$ because the latter cannot contain any new clusters in comparison with $\mathcal{T}_1$, leading to the locally derived pattern $\mathcal{T}_2'$. If $K' \subset \mathbb{R}^d$ is compact, the cluster $\mathcal{T}_2' \sqcap K'$ is completely specified by the cluster $\mathcal{T}_2 \sqcap (K + K')$. Since the latter also occurs in $\mathcal{T}_1$, meaning that $\mathcal{T}_2 \sqcap (K + K') = (-t + \mathcal{T}_1) \sqcap (K + K')$ for some $t \in \mathbb{R}^d$, the derivation rule implies that $(-t + \mathcal{T}_1') \sqcap K' = \mathcal{T}_2' \sqcap K'$. The same argument works in the reverse direction, thus establishing that $\mathcal{T}_1' \overset{\mathrm{LI}}{\sim} \mathcal{T}_2'$. $\qquad\square$

This result allows the extension of local derivability from single patterns to entire LI classes of patterns. A class $\mathrm{LI}(\mathcal{T}')$ is thus called *locally derivable* from $\mathrm{LI}(\mathcal{T})$, written as $\mathrm{LI}(\mathcal{T}) \overset{\mathrm{LD}}{\leadsto} \mathrm{LI}(\mathcal{T}')$, when patterns $\mathcal{T}_1 \in \mathrm{LI}(\mathcal{T})$ and $\mathcal{T}_1' \in \mathrm{LI}(\mathcal{T}')$ exist such that $\mathcal{T}_1 \overset{\mathrm{LD}}{\leadsto} \mathcal{T}_1'$.

Definition 5.7. Two patterns $\mathcal{T}_1, \mathcal{T}_2 \sqsubset \mathbb{R}^d$ are called *mutually locally derivable* (MLD) from each other when $\mathcal{T}_1 \overset{\mathrm{LD}}{\leadsto} \mathcal{T}_2$ and $\mathcal{T}_2 \overset{\mathrm{LD}}{\leadsto} \mathcal{T}_1$. Similarly, two LI classes are MLD when they are locally derivable from each other.

MLD is an equivalence relation, both on patterns and on LI classes, and is denoted by $\mathcal{T}_1 \overset{\text{MLD}}{\longleftrightarrow} \mathcal{T}_2$ and $\text{LI}(\mathcal{T}_1) \overset{\text{MLD}}{\longleftrightarrow} \text{LI}(\mathcal{T}_2)$, respectively. MLD is a special case of topological conjugacy [HRS05] that is extremely useful, though it does not cover general deformations; compare [CS06].

Remark 5.3 (*MLD classes and local finiteness*). MLD classes are generally huge; in particular, they comprise uncountably many LI classes. As a simple example, consider the tiling of the line by unit intervals (the Delone cells of $\mathbb{Z}$). Adding a suitable motif that is representable as a union of subsets of $(0,1)$ to each interval (via suitable translates) produces an MLD representative. In particular, replacing the interval $[0,1]$ by $[0,1] \cup \left\{ \left\{ \frac{1}{n+1} \right\} \mid n \in \mathbb{N} \right\}$ leads to a representative that fails to be locally finite, while the point set $\mathbb{Z}$ is another representative that is locally finite. Consequently, local finiteness is *not* an invariant of an MLD class. For obvious reasons, whenever possible, we will prefer to work with a locally finite representative. Whenever such a representative exists, we call the MLD class *locally finite*.   ◊

Example 5.4 (*Sublattices and local derivability*). Consider the two point sets $2\mathbb{Z}$ and $\mathbb{Z}$ as patterns in $\mathbb{R}$. Using the compact interval $K = [0,2]$ (or any larger interval), it is clear that $2\mathbb{Z} \overset{\text{LD}}{\leadsto} \mathbb{Z}$, where the local derivation rule consists of adding the middle point between any two neighbouring points of $2\mathbb{Z}$. However, the reverse direction cannot be local, because there is no local rule (in the sense of Definition 5.6) to remove every second point. Consequently, $\mathbb{Z}$ and $2\mathbb{Z}$ are *not* MLD. Intuitively speaking, no local rule can guarantee that two 'deriving agents', who start far apart from each other to remove points, choose matching sublattices (even or odd) without knowing the location of the origin. The latter, however, is a non-local information.

More generally, when Γ and Γ' are lattices in $\mathbb{R}^d$ with $\Gamma \subset \Gamma'$, one always has $\Gamma \overset{\text{LD}}{\leadsto} \Gamma'$, while mutual local derivability means $\Gamma = \Gamma'$.   ◊

Proposition 5.2. *For* $\mathcal{T}, \mathcal{T}' \sqsubset \mathbb{R}^d$ *with* $\mathcal{T} \overset{\text{LD}}{\leadsto} \mathcal{T}'$, *one has* $\Delta(\mathcal{T}) \subset \Delta(\mathcal{T}')$.

PROOF. Let $K \subset \mathbb{R}^d$ be a compact set for the derivation $\mathcal{T} \overset{\text{LD}}{\leadsto} \mathcal{T}'$ according to Definition 5.6. Assuming that an explicit derivation rule has been fixed, it is clear that

$$\Delta_{K+K'}(\mathcal{T}) \subset \Delta_{K'}(\mathcal{T}')$$

for all compact $K' \subset \mathbb{R}^d$, because every generating translation of a $(K + K')$-cluster in $\mathcal{T}$ derives down to a translation of a K'-cluster in $\mathcal{T}'$. The claim follows from the fact that the inclusion relation is preserved by the inductive limit.   □

One consequence of this result is that $\mathcal{T} \overset{\text{MLD}}{\longleftrightarrow} \mathcal{T}'$ implies $\Delta(\mathcal{T}) = \Delta(\mathcal{T}')$. In view of Proposition 5.1, Lemma 5.2 has the following consequence.

Corollary 5.1. *The LTM $\Delta(\mathcal{T})$ of a pattern $\mathcal{T} \sqsubset \mathbb{R}^d$ is an invariant of the entire MLD class of $\mathrm{LI}(\mathcal{T})$.* □

Specialising this invariance property to a crystallographic pattern (represented by a point set) leads to the following result.

Corollary 5.2. *Two crystallographic, locally finite point sets $\Lambda, \Lambda' \subset \mathbb{R}^d$ are MLD if and only if they have the same lattice of periods.*

PROOF. According to Proposition 3.1, we write $\Lambda = F \oplus \Gamma$ and $\Lambda' = F' \oplus \Gamma'$ with finite point sets $F, F' \subset \mathbb{R}^d$ and $\Gamma, \Gamma' \subset \mathbb{R}^d$ being the corresponding lattices of periods (which means that both lattices are maximal).
When $\Lambda \overset{\mathrm{MLD}}{\rightsquigarrow} \Lambda'$, we know that

$$\Gamma = \Delta(\Lambda) = \Delta(\Lambda') = \Gamma'$$

by Example 5.2 and Corollary 5.1. Conversely, if $\Gamma = \Gamma'$, we choose a relatively compact fundamental domain D of Γ. The representation of the point sets Λ and Λ' holds with $F = \Lambda \cap D$ and $F' = \Lambda' \cap D$, which simultaneously realises the derivation rule. □

5.3. Repetitivity and finite local complexity

Given a pattern, it is necessary and useful to formulate adequate generalisations of periods. A natural approach, in view of the local topology, replaces the global coincidence by a local one, and is thus asking for the re-appearance of finite clusters in the pattern.

Definition 5.8. A pattern $\mathcal{T} \sqsubset \mathbb{R}^d$ is called (translationally) *repetitive* when, for every compact $K \subset \mathbb{R}^d$, there is a compact $K' \subset \mathbb{R}^d$ such that, for every $x, y \in \mathbb{R}^d$, the relation $\mathcal{T} \sqcap (x + K) = (-t + \mathcal{T}) \sqcap (y + K)$ holds for some $t \in K'$.

The set K' quantifies the local 'search space' to locate arbitrary K-clusters of $\mathcal{T}$. In particular, in a repetitive pattern, one can see all K-clusters (essentially) within the compact set $x + K'$, for any $x \in \mathbb{R}^d$; compare Remark 4.4 for the symbolic case. As a consequence, $\{t \mid \mathcal{T} \sqcap (t + K) = \mathcal{T} \sqcap K\}$ is relatively dense in $\mathbb{R}^d$. Definition 5.8 deviates from the formulation of Definition 4.12 for reasons that will become clear from Proposition 5.5 below.

Example 5.5 (*Lattice periodicity and repetitivity*). Every crystallographic pattern is repetitive. Indeed, if Γ is the lattice of periods of a pattern $\mathcal{T} \sqsubset \mathbb{R}^d$, one can choose any compact K' that contains a fundamental domain of Γ to verify the property from Definition 5.8. ◊

In contrast to this example, the non-periodic set $\mathbb{Z} \setminus \{0\}$, although close to a lattice in some sense, fails to be repetitive, because no finite patch that contains the origin ever re-occurs; compare the related case of Example 4.3. Nevertheless, there are interesting non-periodic examples that are repetitive.

Example 5.6 (*Repetitivity of the silver mean chain*). Consider the silver mean chain of Example 4.5, in its formulation as an inflation tiling of $\mathbb{R}$, with prototiles a and b of lengths $\lambda_{\mathrm{sm}} = 1 + \sqrt{2}$ and 1, with their left endpoints as reference points. Clearly, each single tile reoccurs with bounded gaps (namely $1 + \lambda_{\mathrm{sm}}$ for type a and $1 + 2\lambda_{\mathrm{sm}}$ for type b). Due to primitivity, any finite cluster occurs in the nth inflation of a single tile of type a for some $n \in \mathbb{N}$, which is the level-n supertile for a. Such a cluster thus repeats with gaps that are bounded by $\lambda_{\mathrm{sm}}^n(1 + \lambda_{\mathrm{sm}})$. The FLC property of the silver mean chain implies that, for any compact interval I, one can find a matching interval I' to satisfy the repetitivity condition of Definition 5.8.   ◇

An analogous reasoning applies to all face to face tilings of finite local complexity based on primitive inflation rules. Another large class of repetitive point sets will emerge via projection in Chapter 7.

Lemma 5.3. *Let* $T, T' \sqsubset \mathbb{R}^d$ *be two patterns with* $T \stackrel{LD}{\leadsto} T'$. *If* T *is repetitive, then so is* T'.

PROOF. This follows from a minor variation of the arguments used in the proofs of Proposition 5.2 and Lemma 5.2.   □

Both repetitivity and the LTM of a pattern T quantify the re-occurrence of fragments or clusters. In fact, the LTM was originally introduced with repetitive patterns in mind. Although it exists more generally, it is significant mainly in the presence of repetitivity.

The concept of the repetitivity of a pattern $T \sqsubset \mathbb{R}^d$ can be refined by relating the compact sets K and K' of Definition 5.8. Let $K_r = \overline{B_r(0)}$ be the closed ball of radius $r > 0$ around 0, and choose an appropriate $K' = K_R$, where $R \geqslant r$. In fact, by choosing the minimal R, this will result in a function $R = R(r)$, often referred to as the *repetitivity function*; compare Remark 4.4.

Definition 5.9. A repetitive pattern $T \sqsubset \mathbb{R}^d$ is called *linearly repetitive* when its repetitivity function satisfies $R(r) = \mathcal{O}(r)$ as $r \to \infty$. More generally, T is *g-repetitive* when $R(r) = \mathcal{O}(g(r))$ as $r \to \infty$, for some positive function g on $\mathbb{R}_+$.

As is clear from Example 5.5, crystallographic patterns are repetitive with g a constant function. The converse is also true [LP03].

Proposition 5.3. *Let* $T \sqsubset \mathbb{R}$ *be a tiling that emerges via the geometric interpretation of a primitive substitution rule on a finite alphabet. Then,* T *is linearly repetitive.*

PROOF. If $\mathcal{T}$ is periodic (and hence crystallographic), there is nothing to show in view of g being constant in this case. In general, repetitivity is clear by the argument used in Example 5.6. To establish our claim, we essentially have to refine this argument by establishing that any legal finite patch of length L can be found in the nth inflation of a single prototile with $n = \mathcal{O}(\log(L))$, where the implied constant is universal.

To overcome the complication that a legal patch may first occur either inside a supertile (the geometric counterpart of a superword) or across two neighbouring ones, it is best to follow [Sol98a, Lemma 2.3] and to employ 'collared' tiles (or 1-coronae), of which there are finitely many. We call them the level-0 coronae. Let $\mathcal{T}$ be a fixed point tiling of the inflation, and $\mathrm{LI}(\mathcal{T})$ its LI class. If we establish linear repetitivity of $\mathcal{T}$, it extends to all elements of the LI class. Due to primitivity, and hence repetitivity, there is a length $\ell > 0$ such that any interval of the form $[x, x + \ell]$ with $x \in \mathbb{R}$ contains all level-0 coronae at least once, as a consequence of Proposition 4.3. Due to the fixed point property, any interval $[y, y + \lambda^k \ell]$ with $y \in \mathbb{R}$ contains all level-k super-coronae at least once, where $\lambda > 1$ is the inflation multiplier and the term super-corona refers to a supertile that is collared by two supertiles of the same level.

Now, if δ is the smallest distance between any two consecutive tiles, we know that two level-0 coronae in the tiling at distance less than δ must overlap in a tile. Similarly, any two level-k super-coronae at distance less than $\lambda^k \delta$ must then overlap in a level-k supertile.

Let P be a legal (finite) patch of $\mathcal{T}$. If $k \in \mathbb{N}$ is the unique integer such that $\lambda^{k-1} \delta \leqslant L < \lambda^k \delta$, where $L = \mathrm{diam}(P)$ is the diameter (or length) of P, we know that P is completely contained in some level-k super-corona, and hence in any translate of it within $\mathcal{T}$. By our previous argument, any interval of length $\lambda^k \ell$ contains one such super-corona. Since

$$\frac{\lambda^k \ell}{L} \leqslant \frac{\lambda^k \ell}{\lambda^{k-1} \delta} = \frac{\lambda \ell}{\delta} =: C,$$

we may conclude that any interval of length CL contains a translate of P, which proves the claim. $\square$

Previously, the notion of finite local complexity (FLC) was defined for point sets in Definition 2.3. Recall from Proposition 2.1 that a Delone set Λ is FLC if and only if the Minkowski difference $\Lambda - \Lambda$ is locally finite. A tiling $\mathcal{T}$ (with bounded tiles) is FLC when, for each $r > 0$, the set of r-patches of $\mathcal{T}$ (up to translations) is finite. The two notions are consistent in the sense that each FLC point set defines an FLC tiling via the Delone cell construction, while each FLC tiling gives rise to an FLC point set in the sense of local derivability. This can consistently be extended as follows.

Definition 5.10. A pattern $\mathcal{T} \sqsubset \mathbb{R}^d$ is *FLC* when, for every compact set $K \subset \mathbb{R}^d$, the set of K-clusters $\{(t+K) \sqcap \mathcal{T} \mid t \in \mathbb{R}^d\}$ consists of finitely many equivalence classes up to $\mathbb{R}^d$-translations.

In particular, any FLC pattern is locally finite. If $\mathcal{T} \sqsubset \mathbb{R}^d$ is an FLC pattern, it follows by standard arguments (compare the proof of [Schl00, Prop. 2.3]) that, for any compact $K \subset \mathbb{R}^d$, there is a compact $R \subset \mathbb{R}^d$ such that, for every $t \in \mathbb{R}^d$, there is some translation $r \in R$ with $(t + \mathcal{T}) \sqcap K = (r + \mathcal{T}) \sqcap K$. This means (as before) that FLC patterns explore all their K-clusters within a fixed compact (but K-dependent) region. In contrast to the case of Delone sets, this property alone does no longer imply the finiteness of the clusters of a given size, so that FLC does not follow.

Remark 5.4 (*FLC versus MLD*). Let us point out that the FLC property is *not* an invariant of an MLD class, as can be seen from the following example, which resembles the comment on local finiteness in Remark 5.3. Consider $\mathbb{Z}$ as a point set, which is FLC, and compare it with the Minkowski sum $\mathbb{Z} + S$, where $S = \{\frac{1}{n+1} \mid n \in \mathbb{N}\}$. It is obvious that $\mathbb{Z}$ and $\mathbb{Z} + S$ are MLD, but the latter has infinitely many points in any open neighbourhood of $0 \in \mathbb{Z}$ (or of any other point of $\mathbb{Z}$), hence fails to be FLC. Nevertheless, this type of FLC violation is non-essential precisely in the sense that one can choose an FLC representative of the MLD class. Note also that the non-FLC set $\mathbb{Z} + S$ still has a well-defined orbit closure in the local topology, which is compact. $\Diamond$

Below, except for Example 5.8, we will mainly work with FLC patterns, in particular with point sets and tilings, because the FLC case presently seems more relevant for applications. Later, we shall also consider examples that are not FLC with respect to translations, but satisfy the analogous property for Euclidean motions. Patterns with this property are frequently also called FLC (by slight abuse of nomenclature), or FLC with respect to Euclidean motions. For further results on the more complex general situation of genuine non-FLC structures, we refer to [FS09, FS14].

5.4. Geometric hull

For simplicity, we restrict the formulation to (locally finite) point sets in $\mathbb{R}^d$, though everything can be generalised to patterns. This restriction implies no loss of generality because, in essence, it means that we start with a point set representative of a general MLD class, which always exists. To simplify things a little further, we begin by considering FLC sets. For such sets, it is adequate to use the *local topology*, which was introduced for patterns in Eq. (5.1). Here, two FLC sets Λ and Λ' are called ε-close when one has $\Lambda \cap B_{1/\varepsilon}(0) = (-t + \Lambda') \cap B_{1/\varepsilon}(0)$ for some $t \in B_\varepsilon(0)$. The topology is generated by the possible neighbourhoods with all $\varepsilon > 0$ sufficiently small. It turns out to be a metric topology; see [Sol98b] for details. A perhaps more natural point of view is that of a uniform structure [vQu79, Schl00, BM04], which is generalisable to the setting of LCAGs.

This topology permits the concept of a continuous hull [RW92, BHZ00], which is the natural extension of the corresponding concepts from Chapter 4. We phrase them in the terminology of dynamical systems.

Definition 5.11. If $\Lambda \subset \mathbb{R}^d$ is an FLC set, its geometric or *continuous hull* is $\mathbb{X}(\Lambda) = \overline{\{t + \Lambda \mid t \in \mathbb{R}^d\}}$, where the closure is taken in the local topology. If the $\mathbb{R}^d$-orbit of every element $\Lambda' \in \mathbb{X}(\Lambda)$ is dense, the hull $\mathbb{X}(\Lambda)$ is called *minimal.*

The motivation for this definition becomes clear from the following classic compactness result [RW92, Schl00].

Lemma 5.4. *A point set $\Lambda \subset \mathbb{R}^d$ is FLC if and only if its translation orbit $\{t + \Lambda \mid t \in \mathbb{R}^d\}$ is precompact in the local topology.*

SKETCH OF PROOF. For any compact set $K \subset \mathbb{R}^d$, we have only finitely many clusters of the form $\Lambda \cap (t + K)$ up to translation (due to the FLC property). The corresponding cylinder sets generate the local topology. It is now clear that the translation orbit can always be covered by finitely many cylinder sets 'of a given size' (in the sense of a uniform structure), which implies precompactness [vQu79, Ch. 13].

Conversely, precompactness in the local topology means that, for any open set U, the translation orbit has a finite cover by open sets that are 'comparable' in size to U. Defining such open sets via patches of a given size around the origin, up to a small global translation, shows that Λ can only contain finitely many patches of this size (up to translations).

For the detailed (formal) argument via the uniform structure on locally finite point sets, we refer to [Schl00, Sec. 2 and Prop. 2.3]. □

As in the discrete case, compare Proposition 4.3, repetitivity of Λ and minimality of $\mathbb{X}(\Lambda)$ are connected via Gottschalk's theorem, compare [Pet83, Thm. 4.1.2], while Proposition 4.1 remains literally true.

Proposition 5.4. *Let $\Lambda \subset \mathbb{R}^d$ be an FLC set. In the local topology, the following assertions are equivalent.*

(1) *Λ is repetitive;*
(2) *$\mathbb{X}(\Lambda)$ is minimal;*
(3) *$\mathrm{LI}(\Lambda)$ is closed;*
(4) *$\mathbb{X}(\Lambda) = \mathrm{LI}(\Lambda)$.*

The continuous hull of an FLC set is compact in the local topology. □

Definition 5.12. A point set $\Lambda \subset \mathbb{R}^d$ is *non-periodic* when per(Λ) = {0}. An FLC set $\Lambda \subset \mathbb{R}^d$ is called *topologically aperiodic*, or *aperiodic* for short, if all elements of its hull $\mathbb{X}(\Lambda)$ are non-periodic.

As before in Chapter 4, this distinction is designed to rule out 'trivially' non-periodic point sets (such as $\mathbb{Z}^d \setminus \{0\}$) from being called aperiodic. One connection emerges via repetitivity as follows.

Proposition 5.5. *Let* $\Lambda \subset \mathbb{R}^d$ *be a locally finite point set that is repetitive. Then,* Λ *is also FLC, and non-periodicity of* Λ *implies its aperiodicity.*

PROOF. The repetitivity according to Definition 5.8 means that all K-clusters can be seen in the K'-neighbourhood of $0 \in \mathbb{R}^d$, whence there are only finitely many of them up to translations. This implies the FLC property.

If Λ is non-periodic, then so is every other element of LI(Λ). By assumption, Λ is repetitive, wherefore LI(Λ) is closed by Proposition 5.4. Consequently, no element of $\mathbb{X}(\Lambda)$ can have a non-trivial period, which means aperiodicity of Λ by Definition 5.12. $\square$

Remark 5.5 (*MLD versus topological conjugacy*). Clearly, the MLD relation between dynamical systems is a special case of *topological conjugacy*. In general, however, a topological conjugacy need not originate from a local derivation rule [CS03, HRS05, CS06]. Within the realm of FLC structures, where we can work with the local topology, the two equivalence notions coincide. In the context of symbolic dynamics over a finite alphabet, this is a special case of the Curtis–Lyndon–Hedlund theorem [LM95, Thm. 6.2.9]. ◊

Consider the continuous hull $\mathbb{X}(\Lambda)$ of an FLC set $\Lambda \subset \mathbb{R}^d$. It contains the subset

$$\mathbb{X}_0(\Lambda) = \{\Lambda' \in \mathbb{X}(\Lambda) \mid 0 \in \Lambda'\},$$

which is sometimes called the *discrete hull* of Λ, or the *transversal*. In general, there is no $\mathbb{Z}^d$-action defined on it. In one dimension, however, there may be a bijection of $\mathbb{X}_0(\Lambda)$ with the hull of a symbolic sequence, which then allows to define a $\mathbb{Z}$-action. In such cases, the connection between the hulls can be studied via suspension. Moreover, when Λ is aperiodic, the explicit embedding of $\mathbb{X}_0(\Lambda)$ in $\mathbb{X}(\Lambda)$ is the origin of the rich structure of $\mathbb{X}(\Lambda)$ as a topological space. In fact, the latter is *locally* of the form $\mathbb{R}^d \times \mathcal{C}$ with $\mathcal{C}$ a Cantor set [SW03]. Also, $\mathbb{X}_0(\Lambda)$ is a Cantor set in this case.

Let us mention that the concept of an invariant measure can be generalised to the dynamical system $(\mathbb{X}(\Lambda), \mathbb{R}^d)$. The difference to Chapter 4 is the appearance of a continuous translation group. Nevertheless, for FLC systems, the invariant measure is often defined via the discrete hull $\mathbb{X}_0(\Lambda)$, and then extended to $\mathbb{X}(\Lambda)$.

Remark 5.6 (*Local rubber topology and Hausdorff metric*). When working with non-FLC structures, a generalisation of the local topology is required if one wants to preserve compactness. For Delone sets, a useful concept is the *local rubber topology* [BL04]. Here, two Delone sets Λ and Λ' are ε-close when they agree on a central ball of radius $1/\varepsilon$, possibly after moving the individual points of one set (Λ, say) by at most a distance ε. This means that each individual point $x \in \Lambda \cap B_{1/\varepsilon}(0)$ may be displaced independently to match $\Lambda' \cap B_{1/\varepsilon}(0)$, but only within $B_{\varepsilon}(x)$. This topology is induced by the Hausdorff metric [RW92, Sol98b].

In what follows, we will rarely use this topology, because there is a yet more general approach via translation bounded measures together with the vague topology, which gives compact orbit closures by an application of [BL04, Thm. 2]. Delone sets in the rubber topology (as well as FLC sets in the local topology) can be viewed as special cases of measures via representing them as Dirac combs in the sense of Example 8.6 below. $\diamond$

5.5. Proximality

The concept of proximality, which was briefly introduced at the end of Section 4.3, also proves useful in more than one dimension. Here, we begin with a formulation for FLC Delone sets (which extends to more general patterns via local derivability). We can thus work in the local topology, which we know to be a metric topology. Let $d(\cdot, \cdot)$ be a (fixed) metric that generates the local topology, and let $\mathbb{S}^d$ again denote the unit sphere in $\mathbb{R}^d$.

Definition 5.13. Two distinct FLC Delone sets $\Lambda, \Lambda' \subset \mathbb{R}^d$ are called *proximal in the direction of* $u \in \mathbb{S}^d$ if $\lim_{s \to \infty} d(\Lambda - su, \Lambda' - su) = 0$. They are simply called *proximal* when they are proximal for some $u \in \mathbb{S}^d$.

For an example, consider the block substitution of Eq. (4.27) in Section 4.9. The iteration of each of the four letters in the alphabet $\{0, 1, 2, 3\}$ converges towards a fixed point, as shown for the 'letter' 0 after Eq. (4.27). Any two of them (interpreted as four-colour Delone sets, as in Figure 4.1) are identical in two quadrants, either two diagonally opposite or two adjacent ones (thus forming a half-space), while they differ in each of the remaining two quadrants (there are further coincidences in those quadrants, but they are not relevant for our present discussion). This coincidence structure is a stronger property than mere proximality. To expand on this, let us consider distinct elements of the hull $\mathbb{X}(\Lambda)$ (for some FLC Delone set Λ) that agree on an entire half-space, adapting [BO14, Prop. 1] to this situation.

Proposition 5.6. *Let $\Lambda \subset \mathbb{R}^d$ be a repetitive, non-periodic Delone set, and let $\mathbb{X}(\Lambda) = \overline{\{t + \Lambda \mid t \in \mathbb{R}^d\}}$ be the continuous hull of Λ. Then, Λ is FLC* *and aperiodic. Moreover, there are distinct Delone sets $\Lambda_1, \Lambda_2 \in \mathbb{X}(\Lambda)$ that nevertheless agree on an open half-space, and are thus proximal.*

PROOF. The first claim is a consequence of Proposition 5.5.

Let $B_r = B_r(0)$ be the centred open ball as usual. Now, select a sequence $(t_n)_{n \in \mathbb{N}}$ of non-zero translations with $\|t_n\| \xrightarrow{n \to \infty} \infty$ subject to the condition $\Lambda \cap B_n = (\Lambda - t_n) \cap B_n$. Such a sequence exists due to the repetitivity of Λ. Clearly, $\Lambda \neq \Lambda - t_n$, as Λ is non-periodic, so that

$$n \leqslant r_n := \sup\{r > 0 \mid \Lambda \cap B_r = (\Lambda - t_n) \cap B_r\} < \infty$$

holds for all $n \in \mathbb{N}$.

Let $R_c = \inf\{r > 0 \mid \Lambda + \overline{B_r} = \mathbb{R}^d\}$ be the covering radius of Λ as in Eq. (2.1), and let $B = B_{4R_c}$ be fixed. We can now pick some $s_n \in B_{r_n}$ such that $(\Lambda - t_n - s_n) \cap B \neq (\Lambda - s_n) \cap B$. By construction, we know that

$$\big((\Lambda - t_n - s_n) \cap B\big) \cap \big((\Lambda - s_n) \cap B\big) \neq \varnothing,$$

while the FLC property of Λ implies that there are only finitely many pairs $\big((\Lambda - t_n - s_n) \cap B, (\Lambda - s_n) \cap B\big)$ up to (simultaneous) translations. Possibly after passing to a subsequence, we may thus assume that

$$\big((\Lambda - t_n - s_n) \cap B, (\Lambda - s_n) \cap B\big) = (P_1 - z_n, P_2 - z_n)$$

with fixed distinct patches P_1 and P_2 and with $z_n \xrightarrow{n \to \infty} 0$.

By the compactness of $\mathbb{X}(\Lambda)$ and of the unit sphere $\mathbb{S}^d$, there exist a subsequence $(n_i)_{i \in \mathbb{N}}$, elements $\Lambda_1, \Lambda_2 \in \mathbb{X}(\Lambda)$, and a point $u \in \mathbb{S}^d$ such that

$$\Lambda - t_{n_i} - s_{n_i} \xrightarrow{i \to \infty} \Lambda_1, \quad \Lambda - s_{n_i} \xrightarrow{i \to \infty} \Lambda_2, \quad \text{and} \quad \frac{s_{n_i}}{|s_{n_i}|} \xrightarrow{i \to \infty} u.$$

Then, $\Lambda_1 \neq \Lambda_2$ (since $\Lambda_1 \supset P_1 \neq P_2 \subset \Lambda_2$), while Λ_1 and Λ_2 agree on the half-space $\{v \in \mathbb{R}^d \mid \langle u|v \rangle < 0\}$ by construction. $\square$

Observe that two distinct Delone sets $\Lambda, \Lambda' \in \mathbb{R}^d$ that agree on a half-space are proximal in many directions. More precisely, if Λ and Λ' agree on $\{x \in \mathbb{R}^d \mid \langle v|x \rangle > \alpha\}$ with $v \in \mathbb{S}^d$ for some $\alpha \in \mathbb{R}$, the two Delone sets are proximal in the direction of u for all $u \in \mathbb{S}^d$ with $\langle v|u \rangle > 0$.

Consider the continuous hull $\mathbb{X}(\Lambda)$ of a repetitive Delone set $\Lambda \in \mathbb{R}^d$. We then know that $\mathbb{X}(\Lambda)$ is a compact space that is minimal for the translation action of $\mathbb{R}^d$. Let us assume that some element $\Lambda' \in \mathbb{X}(\Lambda)$ has a non-trivial period, which means $t + \Lambda' = \Lambda'$ for some $0 \neq t \in \mathbb{R}^d$. Due to minimality, any other element of $\mathbb{X}(\Lambda)$ must share this period, including Λ itself. As a consequence, $\mathbb{X}(\Lambda)$ cannot contain a pair of distinct Delone sets that are proximal in the direction of t. If we now assume that $\Lambda_1, \Lambda_2 \in \mathbb{X}(\Lambda)$ are distinct elements that agree on a half-space $\{\langle v|x \rangle > \alpha\}$ for some $v \in \mathbb{S}^d$ and $\alpha \in \mathbb{R}$, we may conclude that $\langle v|t \rangle = 0$, which shows the following result.

Lemma 5.5. *Let $\Lambda \subset \mathbb{R}^d$ be a repetitive Delone set and $\mathbb{X}(\Lambda)$ the corresponding continuous hull. If any element of $\mathbb{X}(\Lambda)$ possesses a non-trivial period $t \in \mathbb{R}^d$, the entire hull is t-periodic. Moreover, $\mathbb{X}(\Lambda)$ cannot have proximality in the direction of t. In particular, if two distinct elements agree on a half-space, its normal vector must be orthogonal to t.* □

This result provides a tool to assess aperiodicity as follows.

Theorem 5.1. *Let $\mathbb{X}(\Lambda)$ be the continuous hull of a repetitive Delone set $\Lambda \subset \mathbb{R}^d$. Let $\{u_i \in \mathbb{S}^d \mid 1 \leqslant i \leqslant d\}$ be a basis of $\mathbb{R}^d$ with the property that there are, for each $1 \leqslant i \leqslant d$, two distinct elements of $\mathbb{X}(\Lambda)$ which agree on the half-space $\{\langle u_i|x\rangle > \alpha_i\}$ for some $\alpha_i \in \mathbb{R}$. Then, $\mathbb{X}(\Lambda)$ is aperiodic.*

PROOF. Since Λ is also FLC by Proposition 5.5, the hull $\mathbb{X}(\Lambda)$ is minimal due to repetitivity by Proposition 5.4. Aperiodicity means that no element of $\mathbb{X}(\Lambda)$ possesses a non-trivial period. If we had an element $\Lambda' \in \mathbb{X}(\Lambda)$ with $t + \Lambda' = \Lambda'$ for some $t \neq 0$, Lemma 5.5 would imply that $u_i \perp t$ for all $1 \leqslant i \leqslant d$, which means that all u_i lie in the hyperplane $\{x \in \mathbb{R}^d \mid \langle t|x\rangle = 0\}$, in contradiction to the basis property. □

The straightforward extension to FLC patterns reads as follows.

Corollary 5.3. *Let $\mathcal{T} \sqsubset \mathbb{R}^d$ be a pattern that is MLD with a repetitive Delone set, and let $\mathbb{X}(\mathcal{T})$ be its continuous hull. Then, the corresponding claims of Theorem 5.1 apply to $\mathbb{X}(\mathcal{T})$ as well.* □

As we shall see later, this criterion is applicable in many cases, particularly for substitutions and inflation tilings. For instance, for the block substitution of Eq. (4.27), this approach provides an aperiodicity proof that is independent of the Toeplitz structure of Eq. (4.30).

Remark 5.7 (*Coinciding wedges*). Our formulation above was based on Delone sets (or locally equivalent patterns) that agree on half-spaces. Clearly, this can be extended to situations where the exact agreement is on wedges or sectors. Given a repetitive (and hence minimal) hull $\mathbb{X}$ as before, aperiodicity can then be concluded when one finds enough proximal pairs with exact coincidence in a sector such that the (open) sectors cover the space (we leave further details to the reader). ◊

Let us turn our attention to other geometric properties of patterns.

5.6. Symmetry and inflation

Let us begin with a discussion of exact symmetries of individual point sets. In one dimension, the reflection $r_x : \mathbb{R} \longrightarrow \mathbb{R}$ in the point x is defined by $r_x(y) = 2x - y$. Clearly, $(r_x)^2$ is the identity on $\mathbb{R}$. A point set $\Lambda \subset \mathbb{R}$ is called *reflection symmetric* in the point x if

$$r_x(\Lambda) = \Lambda.$$

Observing $(r_x \circ r_y)(z) = 2(x - y) + z$, which means that the product of two reflections in $\mathbb{R}$ is a translation, the existence of more than one reflection centre has the following consequence.

Lemma 5.6. *Let $\Lambda \subset \mathbb{R}$ be a point set that is reflection symmetric in the distinct points x and y. Then, Λ is periodic with period $2|x - y|$.*   □

If $\Lambda \subset \mathbb{R}$ is locally finite, the existence of two distinct reflection centres implies that there are also two of minimal distance, x' and y' say. In this case, Λ is crystallographic, with $\mathrm{per}(\Lambda) = 2\,|x' - y'|\,\mathbb{Z}$ as lattice of periods. Moreover, the set $\{x', y'\} + \mathrm{per}(\Lambda)$ comprises all reflection centres of Λ.

In dimension d, one inherits this structure when the point set Λ is reflection symmetric in two parallel, but distinct, co-dimension one hyperplanes; see [Hum92] for background material, including a classification of finite and affine reflection groups.

Let us now take a closer look at individual rotation symmetries in the plane. Our arguments will revolve around the consequences of multiple symmetry centres. For convenience, we use complex numbers. Consider a uniformly discrete point set $\Lambda \subset \mathbb{C}$ and assume that it has two distinct fivefold rotation centres, z and z' say, which means that

$$\xi_5(\Lambda - z) = \Lambda - z \quad \text{and} \quad \xi_5(\Lambda - z') = \Lambda - z',$$

where ξ_5 is a primitive fifth root of unity. Note that either rotation centre may be, but need not be, an element of Λ. Still, the uniform discreteness of Λ implies that there is such a pair $z \neq z'$ with $|z' - z|$ minimal. Figure 5.1 (which is a refinement of Figure 1.4 from page 6) shows how each rotation centre produces five centres of the other type, namely $\{z + \xi_5^\ell(z' - z)\}$ and $\{z' + \xi_5^\ell(z - z')\}$ with $0 \leqslant \ell \leqslant 4$. Here, pairs of opposite centres have distance

(5.2) $$\left|\xi_5^\ell + \xi_5^{-\ell} - 1\right| |z' - z|,$$

where the first factor takes a non-zero value < 1 for some ℓ. This contradicts the assumption on the position of the rotation centres, wherefore we conclude that two rotation centres are impossible. The analogous argument applies to all non-crystallographic rotation symmetries ($n = 5$ or $n \geqslant 7$) of the plane (compare Corollary 3.1 on page 49). In contrast, the crystallographic symmetries 'escape' this argument, see also [Hum92, Sec. 6.6], because the prefactor in Eq. (5.2) is either always $\geqslant 1$ (for $1 \leqslant n \leqslant 4$) or the only value < 1 is zero (which happens for $n = 6$). This gives the following result.

Proposition 5.7. *Let $\Lambda \subset \mathbb{C} \simeq \mathbb{R}^2$ be a uniformly discrete point set with an exact n-fold rotational symmetry. If n is non-crystallographic, which*

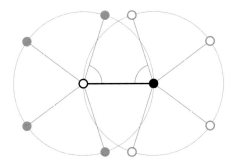

FIGURE 5.1. Interaction of two fivefold rotation centres.

means $n = 5$ or $n \geqslant 7$, there can only be one such rotation centre. When $n \in \{3, 4, 6\}$, the existence of more than one rotation centre is possible, and then implies lattice periodicity of Λ. When $n = 2$, the existence of another rotation centre means that Λ is at least rank-1 periodic.

PROOF. The claim about the impossibility of more than one rotation centre for non-crystallographic values of n follows from our above arguments. The case $n = 4$ leads to rotation centres on the lattice $\Gamma = |z' - z| \mathbb{Z}^2$, which results in $2\Gamma \subset \mathrm{per}(\Lambda)$. Similarly, $n = 6$ produces a triangular lattice, while the choice $n = 3$ generates rotation centres on the vertices of a honeycomb (or hexagonal) packing. Finally, $n = 2$ produces rotation centres along a line. This is (via an embedding) a consequence of Lemma 5.6.   □

For a systematic discussion of non-crystallographic root systems and their reflections in higher dimensions, we refer to [CMP98] and references therein. Let us just mention that a point set in 3-space can at most have one symmetry centre for icosahedral symmetry.

The exact symmetry of an individual point set or pattern is an important concept, but has the disadvantage that two patterns from the same LI class may (and generally will) have different symmetries in this sense. We have seen this phenomenon for symmetric versus non-symmetric fixed points of the Thue–Morse substitution rule in Section 4.6. One way to overcome this undesired feature is to employ local indistinguishability and to borrow a concept from the theory of dynamical systems [BHP97, HRB97, Baa02a, Kwa11].

Definition 5.14. Let R be a linear or affine transformation of $\mathbb{R}^d$. A pattern $\mathcal{T} \sqsubset \mathbb{R}^d$ is *symmetric* under the action of R when $R(\mathcal{T}) \overset{\mathrm{LI}}{\sim} \mathcal{T}$. Moreover, the hull $\mathbb{X}(\mathcal{T})$ is *symmetric* under the action of R when $R(\mathbb{X}(\mathcal{T})) \subset \mathbb{X}(\mathcal{T})$.

This is compatible with our earlier concept of reflection symmetry of primitive substitutions in Definition 4.14, because they define a unique hull which consists of a single LI class.

Remark 5.8 (*MLD and symmetries*). When comparing two hulls via local derivation rules, one will often meet the situation that rotated patches are treated with the rotated rule. To take care of this special case, we say that a certain local derivation rule preserves the symmetry of a structure if derivation rule and symmetry operation commute. This way, one defines *SMLD classes* (for symmetry-preserving MLD), which obviously form pairwise disjoint subclasses of MLD classes; compare [Baa02a] and references therein.

Note that two periodic structures belong to the same SMLD class if and only if they have the same space group. In this sense, SMLD is a refinement of MLD that is motivated by classic crystallographic notions.   ◇

Beyond 'classic' symmetries, one needs an extension of other invariance properties to discrete structures. If $\mathcal{T}$ has an LTM $\Delta(\mathcal{T})$ with $\lambda\Delta(\mathcal{T}) \subset \Delta(\mathcal{T})$ for some $0 \neq \lambda \in \mathbb{R}$ (or, more generally, for some invertible linear map instead of the homothety $x \mapsto \lambda x$), it is natural to ask how this is reflected in the pattern $\mathcal{T}$ itself, where a subset relation of this kind need not hold (nor even make sense).

Definition 5.15. A discrete structure $\mathcal{T}$ in $\mathbb{R}^d$ is said to have a *local scaling property* with respect to the homothety $x \mapsto \lambda x$ for some $0 \neq \lambda \in \mathbb{R}$, if $\mathcal{T}$ is locally derivable from $\lambda\mathcal{T}$, or $\lambda\mathcal{T} \overset{\text{LD}}{\rightsquigarrow} \mathcal{T}$ for short. When $M\mathcal{T} \overset{\text{LD}}{\rightsquigarrow} \mathcal{T}$ for some $M \in \mathrm{GL}(d)$, one speaks of a local scaling property relative to the linear map defined by M.

Note that the definition does not require $|\lambda| > 1$, though expanding versus contracting homotheties have rather different consequences for discrete structures. The difference might best be explained by an example.

Example 5.7 (*Local scalings for $\mathbb{Z}^d$*). Consider the discrete structure $\mathcal{T}$ defined by the integer lattice $\mathbb{Z}^d$, where the LTM is $\Delta(\mathcal{T}) = \mathbb{Z}^d$. We know that $\lambda\mathbb{Z}^d \subset \mathbb{Z}^d$ holds precisely for $\lambda \in \mathbb{Z}$. It is also clear that $n\mathbb{Z}^d \overset{\text{LD}}{\rightsquigarrow} \mathbb{Z}^d$ for all $n \in \mathbb{N}$, as the derivation rule here is simply the subdivision of an elementary (hyper)cube into smaller copies, which is clearly local. Note that the converse direction fails to be local for $n > 1$, by an obvious generalisation of the argument used in Example 5.4. This shows that the local scaling property does not hold in both directions here.   ◇

It is a fundamental and perhaps somewhat surprising insight that a large class of aperiodic structures cannot have a local scaling property for $M = \lambda\mathbb{1}$ with $\lambda > 1$ without automatically also having one for M^{-1}. This property is known as *local recognisability*, and usually formulated for inflation tilings; see [Sol98a] for details and a precise formulation.

By an application of Proposition 5.2, it is clear that $\Delta(M\mathcal{T}) \subset \Delta(\mathcal{T})$ is a necessary (but generally not a sufficient) condition for $M \in \mathrm{GL}(d)$ to define a local scaling property of $\mathcal{T}$. In view of Example 5.7, a dense module as LTM is thus a necessary requirement for the existence of local scaling rules in both 'directions'.

Definition 5.16. A discrete structure $\mathcal{T}$ in $\mathbb{R}^d$ is said to have a *local inflation deflation symmetry* (LIDS) relative to the linear map L if $\mathcal{T}$ and $L(\mathcal{T})$ are MLD, or $\mathcal{T} \overset{\text{MLD}}{\leftrightsquigarrow} L(\mathcal{T})$ for short. When $L(x) = \lambda x$, or when $L(x) = \lambda R x$ with $R \in \mathrm{O}(d, \mathbb{R})$, the number λ is called the *inflation multiplier* of the LIDS.

A necessary condition for L to define an LIDS of $\mathcal{T}$ is $\Delta(L(\mathcal{T})) = \Delta(\mathcal{T})$, which is a rather strong condition. The silver mean point set (realised as a subset of $\mathbb{Z}[\sqrt{2}\,]$) has an LIDS with inflation multiplier $\lambda = 1 + \sqrt{2}$, which is a unit. In one direction, this follows from the local inflation rule for the two interval types. In the other, it is a consequence of the local recognisability; compare the discussion on page 80.

The existence of an LIDS is usually easier to investigate in the tiling picture. Since we shall use this extensively below, we now introduce the notion of inflation rules for tilings; see [LW03] for the corresponding concepts for Delone sets.

Definition 5.17. Consider a finite set $\{T_1, T_2, \ldots, T_n\}$ of tiles, where each $T_i \subset \mathbb{R}^d$ is a compact set with non-empty interior and $\overline{T_i^{\circ}} = T_i$, so that we also have $0 < \mathrm{vol}(T_i) < \infty$. An *inflation rule* with inflation multiplier $\lambda > 1$ (and an extension map $x \mapsto \lambda x$) consists of the mappings

$$(5.3) \qquad \lambda T_i \longmapsto \bigcup_{j=1}^{n} T_j + A_{ji}$$

with finite sets $A_{ji} \subset \mathbb{R}^d$, subject to the mutual disjointness of the interiors of the sets on the right hand side and to the (individual) volume consistency conditions $\mathrm{vol}(\lambda T_i) = \sum_{j=1}^{n} \mathrm{vol}(T_j)\,\mathrm{card}(A_{ji})$, both for each $1 \leqslant i \leqslant n$.

More generally, one can equally well work with an extension map of the form $x \mapsto \lambda R x$ with $R \in \mathrm{O}(d, \mathbb{R})$, or with an expanding linear map.

Recall that we use the convention $T_i + \varnothing = \varnothing$, so that the right hand side of Eq. (5.3) need not (and generally will not) comprise translates of all tiles. Definition 5.17 assumes a *fixed* set of tiles, such as the prototile set of a tiling. Each of these tiles might be considered as having a specified control point (such as 0, say), in analogy to the one-dimensional situation in Chapter 4. Since $\lambda(t + T_i) = \lambda t + \lambda T_i$, it is clear how to extend Eq. (5.3) to translated copies of the tiles. An inflation rule is *consistent* when essentially disjoint tiles (meaning disjoint interiors) are mapped to essentially disjoint images this way. Below, we shall only consider consistent examples of inflation rules, without further mentioning this property explicitly.

In analogy to the superwords of a (primitive) substitution, see page 80, one can now introduce *supertiles*, and attach the corresponding level of inflation depth to it. Local recognisability (called unique composition property in [Sol98a]) amounts to the unique identification of level-1 supertiles (and then of any higher level as well) from strictly local information only.

The integer matrix M defined by $M_{k\ell} = \mathrm{card}(A_{k\ell})$ is a non-negative matrix that is called (in analogy to Chapter 4) the *inflation matrix* of the rule. The consistency conditions mean that λ^d is the leading eigenvalue of M and that $\big(\mathrm{vol}(T_1), \ldots, \mathrm{vol}(T_n)\big)$ is a corresponding left eigenvector of M. The terms *irreducible*, *cyclically primitive* and *primitive* can be defined as before in Section 2.4, and PF theory applies accordingly.

A tiling that is constructed by such an inflation rule is often called a *self-similar* tiling. We will not use this term (nor its extension to self-affine tilings; compare [LW96]) because we prefer to look at inflation rules from the more systematic point of view of local equivalence; compare Definition 5.16. One further advantage is that we need not distinguish between self-similar and pseudo self-similar tilings; compare [FSo01] and references therein for the concept and [Sol07] for some extensions to higher dimensions.

When an inflation rule satisfies equality in Eq. (5.3) between the left and the right hand sides, we call it a *stone inflation*, which is a notion introduced by Danzer. A stone inflation can also be extended, in various ways, beyond the setting of Eq. (5.3), for instance via replacing the homothety $x \mapsto \lambda x$ by an expansive linear mapping L. We refer to this extension as a stone inflation rule with *expansion map L*. Note that the leading eigenvalue of L is now given by $\det(L)$ or by $|\det(L)|$, and the notion of an inflation multiplier might no longer make sense. An important property of the linear structure is that translated images are replaced by translated patches, so that we can build a tiling by iterated applications of the inflation rule in a *local* fashion.

Example 5.8 (*A non-FLC stone inflation*). Though most of our later examples of stone inflations produce FLC tilings, the Frank–Robinson tiling [FR08], which is also known as the Priebe Frank non-PV tiling in the Tilings Encyclopedia [HFonl], shows that this is not the case in general (its local complexity is unbounded, also called infinite or ILC). Starting from two square-shaped and two rectangular prototiles with side lengths either 1 or $\alpha = (1 + \sqrt{13})/2 \approx 2.30278$, where $\alpha^2 = \alpha + 3$, one gets a primitive stone inflation with 'fault lines' in the inflation image of the larger square as follows.

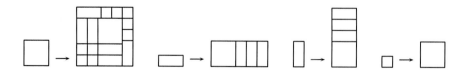

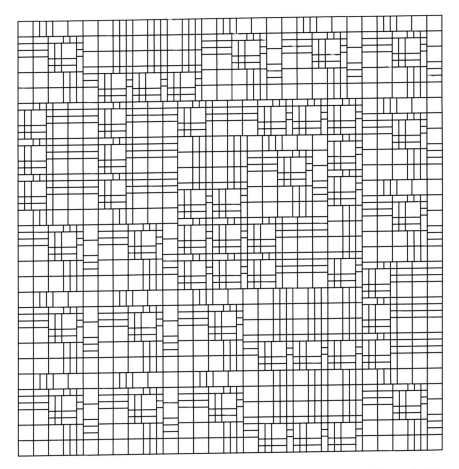

FIGURE 5.2. Square-shaped patch of the non-FLC stone inflation of Example 5.8.

The inflation multiplier is α, which is a quadratic non-PV number. This rule has the (diagonalisable) inflation matrix

$$M = \begin{pmatrix} 1 & 1 & 1 & 1 \\ 3 & 0 & 3 & 0 \\ 3 & 3 & 0 & 0 \\ 9 & 0 & 0 & 0 \end{pmatrix}$$

with eigenvalues $\alpha + 3$, $4 - \alpha$ and -3 (twice). The left PF eigenvector is $(\alpha^2, \alpha, \alpha, 1)$, in line with the prototile areas, while the statistically normalised right eigenvector with the prototile frequencies reads

$$\frac{1}{9}(4 - \alpha, 4\alpha - 7, 4\alpha - 7, 19 - 7\alpha)^T \approx (0.189, 0.246, 0.246, 0.320)^T.$$

The approximate area fractions covered by the 4 prototiles in a fixed point tiling are then given by $\frac{1}{26}(7+\sqrt{13}, 6, 6, 7-\sqrt{13}) \approx (0.408, 0.231, 0.231, 0.131)$.

Starting from the (legal) 2×2 block of 4 large squares, an iteration of the inflation rule leads to a sequence that converges (in the local rubber topology) towards a 2-cycle. The level-4 supertile emerging from the large square is illustrated in Figure 5.2. The resulting tiling fails to be FLC, which can, for instance, be seen from the fact that the number of vertex or pseudo vertex positions on the bottom edge of the large square is $3n + 2$ in the corresponding level-n supertile, hence unbounded; see [FR08] for details and a formal proof. This is also the reason why we need the local rubber topology of Remark 5.6 to define the fixed point tiling as a limit.

For interesting other properties of the Frank–Robinson tiling and its relatives, we refer to [FR08], while an extension in the framework of fusion tilings is discussed in [FS14].   ◊

Let us next illustrate why it is generally not sufficient to restrict Definitions 5.15–5.17 to homotheties.

Example 5.9 (*Product inflation*). Let us consider a prototile set with two rectangles (T_1 and T_2, both of length τ and height 1, distinguished by colour) and two squares (T_3 and T_4, both of side length 1, again with different colours). We now define the (extended) stone inflation

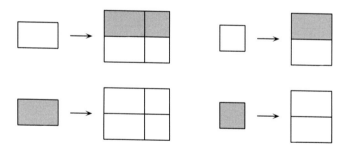

which is clearly primitive. One may consider each tile to be equipped with a reference (or control) point, for instance the mid point of the left vertical edge. This defines an example of a planar inflation tiling where the linear expansion is given by the matrix $L = \mathrm{diag}(\tau, 2)$, which is not a homothety.

This aperiodic inflation induces the inflation matrix

$$M = \begin{pmatrix} 1 & 2 & 1 & 2 \\ 1 & 0 & 1 & 0 \\ 1 & 2 & 0 & 0 \\ 1 & 0 & 0 & 0 \end{pmatrix} = \begin{pmatrix} 1 & 1 \\ 1 & 0 \end{pmatrix} \otimes \begin{pmatrix} 1 & 2 \\ 1 & 0 \end{pmatrix}$$

where $\otimes$ denotes the Kronecker (or tensor) product of matrices. Our example is thus the inflation induced by the direct product of the Fibonacci substitution of Example 4.6 (in its geometric version) and the period doubling substitution of Section 4.5.1, the latter having constant length.

The spectrum of M is $\{2\tau, 2\tau', -\tau, -\tau'\}$, with PF eigenvalue $\lambda_{\mathrm{PF}} = \det(L) = 2\tau$ and left and right PF eigenvectors $(\tau, \tau, 1, 1) = (\tau, 1) \otimes (1, 1)$ and $\frac{1}{3\tau^2}(2\tau, \tau, 2, 1)^T = (\tau^{-1}, \tau^{-2})^T \otimes (\frac{2}{3}, \frac{1}{3})^T$, which reflect the tensor product structure. The right eigenvector is statistically normalised and encodes the relative frequencies of the four prototiles in any fixed point tiling (or in any element of the corresponding hull). $\diamond$

It is clear that one can construct a plethora of similar examples, also multi-dimensional ones. Let us mention that there is another important extension of Definition 5.17. When working with a finite set of prototiles, one frequently encounters the situation that no subdivision as in a stone inflation is possible, and that one has to work with protruding and recessing parts. When this is possible in a consistent way that preserves the total volume, but possibly not the individual volume conditions for each prototile, one ends up with a minor variant of a tile inflation. Sometimes, however, one has to work with volume-increasing covers of the scaled tiles by copies of the original ones. When this leads to a consistent rule where overlapping tiles always exactly match, we speak of a *pseudo inflation rule*. We will meet several examples in Chapter 6.

Proposition 5.3 has the following generalisation; see [Sol98a, Lemma 2.3] or [FR14, Prop. 2.10] for a proof, which is a fairly straightforward extension of the previous one-dimensional argument, formulated via legal coronae.

Proposition 5.8. *Let $\mathcal{T} \sqsubset \mathbb{R}^d$ be an FLC tiling that is defined by a primitive stone inflation rule. Then, $\mathcal{T}$ is linearly repetitive.* $\square$

For an in-depth study of inflation tilings and their properties, in particular topological ones, further technical notions may be necessary. An important one was introduced by Kellendonk [Kel95], who calls an inflation rule *border forcing* when, for some $n \in \mathbb{N}$, all level-n supertiles completely determine ('force') the surrounding corona of tiles. The border forcing property is not an MLD invariant, but nevertheless important for the practical calculation of topological invariants [Sad08].

Of a similar nature is the *complexity* of a tiling. While subtle variants of this notion are in use, the underlying idea is to quantify the number of patches of a given size, which is a generalisation of the word complexity of Section 4.8.2. Since we will rarely use these concepts, we omit further details and refer to [Kel95, Jul10] and references therein for more.

5.7. Local rules

Consider a locally finite pattern $\mathcal{T} \sqsubset \mathbb{R}^d$. The collection of patches $\mathcal{T} \sqcap \overline{B_r(x)}$ with $r \geqslant 0$ and $x \in \mathbb{R}^d$ contains important information on $\mathcal{T}$. Their translation classes (which means considering patches up to translations) constitute the translational *atlas* of $\mathcal{T}$. For some patterns, it is useful to reduce such an atlas by considering patches up to Euclidean motions. Whether one considers an atlas up to translations or up to similarity will always be clear from the context. In general, such an atlas will be huge. In particular, not even its restriction to a finite radius needs to be finite.

Let us next consider the important case of an FLC pattern $\mathcal{T} \sqsubset \mathbb{R}^d$ with hull $\mathbb{X}(\mathcal{T})$. We denote the (translational) atlas of $\mathcal{T}$ by $\mathcal{A}(\mathcal{T})$, in analogy to the dictionary of sequences in Chapter 4. A pattern $\mathcal{T}' \sqsubset \mathbb{R}^d$ is called *compatible* with $\mathcal{A}(\mathcal{T})$ when $\mathcal{A}(\mathcal{T}') \subset \mathcal{A}(\mathcal{T})$.

Definition 5.18. The atlas $\mathcal{A}(\mathcal{T})$ of an FLC pattern $\mathcal{T} \sqsubset \mathbb{R}^d$ with hull $\mathbb{X}(\mathcal{T})$ is called a *defining atlas*, when the set of all patterns in $\mathbb{R}^d$ that are compatible with $\mathcal{A}(\mathcal{T})$ are precisely the elements of the hull $\mathbb{X}(\mathcal{T})$.

So far, an atlas comprises the list of all r-patches of $\mathcal{T}$ (up to translations), for all $r > 0$. This is generally not an economic way to describe a pattern, as smaller patches must re-occur inside larger ones. Conversely, some larger patches may be 'forced' by the available smaller patches. In the case of a defining atlas, one might thus want to reduce its size. A systematic way of doing this consists of first restricting to a set $\mathcal{R} = \{r_i \geqslant 0 \mid i \in I\}$ of radii (for some index set I) and then collecting (up to translations) all r_i-patches of the FLC pattern $\mathcal{T}$ into the *reduced atlas* $\mathcal{A}_\mathcal{R}(\mathcal{T})$. A pattern $\mathcal{T}' \sqsubset \mathbb{R}^d$ is *compatible* with the reduced atlas $\mathcal{A}_\mathcal{R}(\mathcal{T})$ when $\mathcal{A}_\mathcal{R}(\mathcal{T}') \subset \mathcal{A}_\mathcal{R}(\mathcal{T})$, which means that we have to use the same set $\mathcal{R}$. A reduced atlas $\mathcal{A}_\mathcal{R}(\mathcal{T})$ is then called a *defining atlas* when the set of all patterns in $\mathbb{R}^d$ that are compatible with $\mathcal{A}_\mathcal{R}(\mathcal{T})$ are precisely the elements of the hull $\mathbb{X}(\mathcal{T})$. In such as case, the elements of a reduced atlas are also referred to as a set of *rules* for $\mathcal{T}$.

A particularly relevant and interesting case emerges when an FLC pattern $\mathcal{T}$ possesses a defining reduced atlas that is a finite set. This occurs if a finite indexing set I for $\mathcal{R}$ suffices.

Definition 5.19. Let $\mathcal{T} \sqsubset \mathbb{R}^d$ be an FLC pattern with a reduced atlas that is both defining and finite. Then, $\mathcal{T}$ is said to possess a set of *local rules*.

Example 5.10 (*Local rules for crystallographic patterns*). It is clear that crystallographic patterns can be described in this way. For instance, the periodic tiling $\mathcal{T} = \{[n, n+1] \mid n \in \mathbb{Z}\}$ of the line by unit intervals is completely specified by the defining atlas of 0-patches, which consists of precisely two elements, namely the unit interval and two adjacent intervals. In particular, it does not contain the empty set, wherefore no finite pattern is compatible with the atlas. Effectively, the underlying structure is also a local growth rule, which uniquely specifies how to continue the tiling once an initial tile is chosen. $\diamond$

It is one of the surprising and deep results in the theory of aperiodic order that crystallographic or periodic systems are not the only examples which fit Definition 5.19. For instance, in Figure 1.2, we saw the example of the rhombic Penrose tiling, which possesses local rules of a very special kind.

Definition 5.20. A set of local rules is called *perfect* when it defines a unique LI class. A set of local rules is called *aperiodic* when all patterns compatible with it are non-periodic.

When one starts with a pattern with local rules, and applies a local derivation rule, one ends up with a new set of rules for the derived pattern, wherefore the following result is obvious.

Corollary 5.4. *If an FLC pattern* $\mathcal{T} \sqsubset \mathbb{R}^d$ *possesses local rules, the corresponding property holds for any member of its MLD class.* $\square$

Tilings in one dimension with finitely many intervals as prototiles are special in the sense that they cannot have aperiodic local rules. Since there is no geometric constraint beyond the consecutive positioning in this case, it is sufficient for an understanding of this claim to consider the purely symbolic case of a finite alphabet, whose letters represent the intervals. A finite atlas would then simply be a finite list of allowed words (of finite length), which must be subword-closed for consistency.

A simple example over $\mathcal{A} = \{a, b\}$ results from the atlas $\mathcal{Y} = \{a, b, ab, ba\}$, which is subword-closed. Here, $\mathcal{Y}$ is compatible with the periodic bi-infinite words $\dots abab|abab\dots$ and $\dots baba|baba\dots$ only. In the attempt to construct more complicated (in particular, non-periodic) situations, one encounters the following obstruction.

Lemma 5.7. *Let* $\mathcal{A}$ *be an alphabet of cardinality* n, *and let* $\mathcal{Y}$ *be a finite list of words in* $\mathcal{A}$, *with maximal length* r. *Assume further that at least one bi-infinite word* w *over* $\mathcal{A}$ *exists with the property that all subwords of* w *of length* $\leqslant r$ *are elements of* $\mathcal{Y}$. *Then, also a periodic bi-infinite word with this property exists.*

PROOF. There are at most n^r distinct words of length r over $\mathcal{A}$, not all of which need be in $\mathcal{Y}$. By the Dirichlet pigeonhole principle, any word of length $r(n^r + 1)$ over $\mathcal{A}$ must contain one of the possible words twice, without mutual overlap. Due to the assumed existence of w, the analogous situation must occur within w, with all occurring words of length r being in

$\mathcal{Y}$. Consequently, there is some $u \in \mathcal{Y}$ of length r such that uvu with v finite is a subword of w. Define the bi-infinite word w' as the repetition of uv, so that $w' = \ldots uvuv|uvuv\ldots$, which is clearly periodic. By construction, all subwords of w' of length $\leqslant r$ are $\mathcal{Y}$-legal, whence w' itself is compatible with $\mathcal{Y}$. This establishes our claim.   □

In fact, it suffices to assume the existence of a finite $\mathcal{Y}$-legal word of length $r(k+1)$ where k is the number of words of length r in $\mathcal{Y}$. In dimensions $d \geqslant 2$, the situation is rather different, and this discovery has contributed considerably to the development of the field of aperiodic order.

The scenario changes in higher dimensions, due to the additional geometric constraints. We already saw the example of the rhombic Penrose tiling in Figure 1.2, where two marked rhombuses (and their rotations through multiples of $\pi/5$) are only compatible with non-periodic tilings of the plane. More generally, this leads to the following concept.

Definition 5.21. A finite set of prototiles in $\mathbb{R}^d$ is called *aperiodic* (in the strict sense) when there exists at least one space-filling tiling with these tiles and when no tiling of $\mathbb{R}^d$ with these tiles possesses a non-trivial period.

More generally, a finite set of marked prototiles, together with a finite atlas of local configurations, is called aperiodic when the atlas constitutes aperiodic local rules in the sense of Definition 5.20.

Aperiodicity in the strict sense is thus a geometric property, defined via possible neighbourhoods of tiles. This can be reformulated via a suitable atlas, though not every aperiodic atlas can be realised as a purely geometric matching condition. For example, the arrow matching rules of the Penrose rhombuses (which only constrain common edges of two adjacent tiles) can be incorporated into the tile shape by using non-convex tiles such as

which can be used as shapes for a jigsaw puzzle.

As follows easily from Lemma 5.7, there are no aperiodic prototile sets for $d = 1$, while it turns out that there are such sets for all $d \geqslant 2$. We shall illustrate these concepts by a number of important examples in Sections 5.7.3–5.7.7 below. Let us first explain some important mechanisms for local rules, and how to verify them.

5.7.1. THE COMPOSITION-DECOMPOSITION METHOD

In general, it can be quite difficult to show that a given finite set of (marked) prototiles, together with a finite atlas of local configurations, is aperiodic. The composition-decomposition method is one approach that applies to aperiodic prototile sets which lead to tilings with an LIDS, so that one can employ the inflation as well as the corresponding deflation rule, the latter being a consequence of recognisability; compare [Sol98a]. The general procedure, which has often been used implicitly, was described in detail in [AGS92, Gäh93] and can be outlined as follows; see also Section 5.7.5 below.

Assume that, for our prototile set, we have a decomposition rule which respects the local rules. This means that, to each (marked) prototile, we associate a patch of scaled prototiles (by a factor $\lambda < 1$), which are compatible with the finite atlas (on the reduced scale) of allowed configurations. Rescaling by λ^{-1} gives rise to an inflation rule, which shows that a space-filling tiling compatible with the atlas exist, such as an inflation fixed point tiling.

Now, consider any space-filling tiling that is compatible with this atlas. The idea is to show that the decomposition rule is invertible in the sense that, for any admissible tiling, there exists a corresponding composition rule that provides a *unique* tiling of supertiles. The latter refer to prototiles scaled by λ^{-1} such that the original tiling corresponds to its decomposition. To prove this, we have to ensure that each tile in an admissible tiling can be composed, together with part of its neighbourhood, to a unique supertile, and that the supertiles inherit the markings and rules imposed by the atlas, such that the admissible configurations for the supertile tiling are equivalent to those of the original tiling.

If it is possible to show that these conditions hold, it follows that any admissible tiling has to be non-periodic [AGS92]. Indeed, assuming to the contrary that an admissible tiling possesses a non-trivial period $t \neq 0$ leads to a contradiction, because t is also a period of tilings obtained after applying the composition rule, which can be applied repeatedly. However, at some stage, the composed tiles become larger than the period t, which is impossible. This argument hence shows that all admissible tilings are non-periodic, and hence that the corresponding prototile set is aperiodic in the sense of Definition 5.21 [AGS92]. In fact, it shows more, because it implies that all admissible tilings are LI with an inflation tiling, so these are perfect aperiodic rules in the sense of Definition 5.20 [Gäh93].

This method was applied in [AGS92] to show that the prototile sets A2, A3, A4 and A5 due to Ammann [GS87] are aperiodic. In combination with an argument based on the projection method (see Chapter 7 below), it is used to prove quasiperiodicity of the Ammann–Beenker tiling (which corresponds to Ammann's aperiodic prototile set A5) and the dodecagonal shield tiling (see Example 7.12 below). Further examples are briefly discussed in [KG94]; see also [Sen95].

5.7.2. LOCAL RULES FOR PROJECTION TILINGS

There has been some effort to construct local rules for tilings obtained by the projection method, independently of the existence of an LIDS (which is the crucial ingredient in the composition-decomposition method outlined above). As we will only describe the projection method in Chapter 7 below, we can but sketch the idea at this point, and refer to the relevant literature for details, as well as to later contributions to this book series.

The basic ideas go back to Katz [Kat88] and Levitov [Lev88], and work for certain cut and project tilings with particular windows. In particular, the ideas apply to planar rhombus tilings with N-fold symmetry, where the tiles are then viewed as projections of 2-faces from a hypercubic Euclidean cell complex in n-space. Usually, one works with $n = N$ for N odd and $n = N/2$ otherwise, which often does not correspond to the minimal embedding discussed in Theorem 3.1. The tiling now lifts to a 'stepped surface' whose vertices essentially comprise all hypercubic lattice points that, when projected into the perpendicular (or 'internal') space, fall within a compact set (called the 'window'). Note that this window may comprise several parts, as illustrated in Figure 7.8 below for the case of the Penrose rhombus tiling.

The local rules are devised such that they prevent the stepped surface from crossing a collection of hyperplanes which are determined by the boundary faces of the window. Obviously, such an approach can only produce finite sets of local rules for windows that lead to a finite number of such constraints. This includes standard examples such as the Penrose or the Ammann–Beenker tilings, which have polygonal windows; see [Le95] for applications to pentagonal tilings. A detailed description of the Ammann–Beenker case can be found in [KG94, Kat95]; see also [BF13a, BF13b].

The analogous approach also works for higher-dimensional (rhombohedral) tilings. The local rules for the primitive icosahedral tiling (see Section 7.4 below), where the prototiles are decorated rhombohedra of two shapes, are derived in [Kat88]. Moreover, this method can also be adapted to Danzer's ABCK tiling, based on its projection description from Example 7.13 below.

While this approach leads to perfect local rules (as they completely determine the window or windows, and hence the LI class), Levitov also introduced the notion of *weak local rules* [Lev88], which only enforce that the projection into the internal (perpendicular) subspace remains bounded; see also [Soc90]. They are often weaker than 'imperfect' local rules in the sense that they are compatible with an ensemble of rhombus tilings of positive entropy, but may

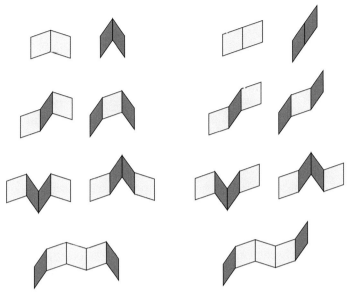

FIGURE 5.3. Sketch of the alternation condition for the Penrose rhombuses, via examples of allowed (left) and forbidden (right) patches. Along any 'worm' (or 'street'), rhombuses of any type have to alternate.

still enforce non-periodicity. In our terminology, they are then still aperiodic local rules. In some cases, such rules can take a rather simple form, as the next remark shows.

Remark 5.9 (*Alternation condition*). A simple local rule for rhombus (or rhombohedral) tilings is the alternation condition, investigated in detail in [Soc90]. This rule demands that, along any 'worm' (or 'street') formed by adjacent rhombuses (rhombohedra) joined at parallel edges (faces), rhombuses (rhombohedra) of the same type have to alternate; see Figure 5.3 for an illustration of the planar case (with Penrose rhombuses). The Penrose rhombus tiling of Figure 1.2, which satisfies the alternation condition, displays only a small set of the principally allowed local configurations. The 'anti-Penrose' tiling of Figure 7.19 is also allowed (and non-periodic), but contains a different subset of configurations.

Note that n-fold symmetry yields the crucial parameter for a more detailed discussion. As shown in [Soc90] for certain values of n (including $n = p$ and $n = 2p$ for $p \geqslant 5$ prime), the alternation rule determines the orientation of the lifted surface in the corresponding embedding in a hypercubic lattice, with bounded fluctuations. The rule then also enforces the Meyer set property together with non-periodicity. However, no perfect local rules are obtained, as several distinct LI classes are compatible with the alternation condition (as in the above example of the fivefold rhombus tilings). A classification of all LI classes compatible with the alternation condition seems difficult. In fact, the resulting local rules can be 'weak' in the previously mentioned sense that the ensemble of all compatible tilings may have entropy [Soc90]. In the fivefold case, however, it has been conjectured that the alternation condition specifies the class of generalised Penrose rhombus tilings[1] (which have no entropy), but we are not aware of a complete proof; see [Le97, Sec. 4.4] for a more detailed discussion.

For $n = 3$ or $n = 6$, there is only one rhombus, and an essentially unique periodic tiling that satisfies the alternation condition. For $n = 4$, one obtains the square lattice, as the square is degenerate as a rhombus for the alternation condition. Similarly, if n is a multiple of 4, this degeneracy persists, so that the alternation condition still enforces some form of quasiperiodicity, but is compatible with periodic tilings. Thus, one does not obtain weak local rules in this case. This is connected with the Schur rotation phenomenon; compare Remark 3.5.

The alternation condition also applies to the primitive icosahedral tiling of Section 7.4, which is built from the two Kepler rhombohedra. It may be that one actually obtains perfect local rules for this tiling LI class, compare [Soc90, Sec. 8], but we are not aware of a proof. A considerably more complex set of local rules (based on the projection method) is derived in [Kat88]. ◇

The question of which tilings possess local rules (with or without decorations) has been addressed by Le; see [Le97] and references therein. Even though he shows that there are infinitely many planar quasiperiodic tilings which admit local rules without decoration [Le97, Thm. 7.6], already the mere existence of weak local rules imposes severe restrictions on the cut and project scheme, such as the algebraicity result of [Le97, Thm. 8.3]. In the projection context, certain topological constraints are discussed in [Kal05]. Furthermore, recent results link the existence of local rules for planar tilings to the computability of the corresponding 'embedding slope', as represented by a basis of spanning vectors; see [FeS12, BF13a] for details.

5.7.3. Wang tiles

Historically, a very important class of aperiodic prototile sets consists of finite collections of squares together with markings and local rules how to put them together. More precisely, *Wang tiles* are unit squares with marked edges that are supposed to be packed face to face, respecting the markings. An interesting question (of historic origin) concerns the existence of a finite

---

[1]These tilings are briefly described below, in the context of de Bruijn's grid method in Section 7.5.2; see [PK87, Ing99] for more.

set of such tiles that is aperiodic. The first (but still huge) prototile set of this type was found by Berger [Ber66]; see [GS87, Ch. 10] for a detailed account. The cardinality of aperiodic Wang prototile sets has since gradually been reduced. We will describe a small prototile set in Section 5.7.4 below.

An interesting observation in this context is that there is no finite set of Wang tiles that permits non-trivial periods in only one direction.

Proposition 5.9. *Let* $\mathcal{W}$ *be a finite set of Wang tiles that admits a complete tiling* $\mathcal{T}$ *of the plane with a non-trivial period. Then, the prototile set* $\mathcal{W}$ *also admits a crystallographic tiling.*

PROOF. By assumption, $\mathcal{T} + u = \mathcal{T}$ for some $0 \neq u \in \mathbb{Z}^2$, since Wang tiles are unit squares. Let v be the image of u under a rotation by $\pi/2$, which is again a non-zero element of $\mathbb{Z}^2$, with $\|u\| = \|v\|$. The tiling $\mathcal{T}$ thus consists of the periodic repetition of an infinite v-strip (meaning a strip in the v-direction) of finite width. This v-strip is the (non-compact) closure of a fundamental region for the u translation, which is not unique. Let us construct one specific choice for a fundamental domain (with only horizontal and vertical edges) of the (square) lattice $\langle u, v \rangle_{\mathbb{Z}}$. For this, one can use a polyomino shape P (made from $\|u\|^2$ unit squares, where $\|u\|^2 \in \mathbb{N}$ by construction) that is either a square or the union of two squares (in line with the two integer coordinates of u). Then, the v-translates of P form a v-strip (with connected interior) whose $\mathbb{Z}u$-translates cover the plane. Note that the corresponding Wang tile configuration is periodic in the u-direction, but not necessarily along the v-strip.

At the same time, the u-translates of P form a u-strip with connected interior and u-periodic filling. The $\mathbb{Z}v$-translates of this u-strip again cover the plane, but we do not know anything about periodicity. However, among these v-translates, only finitely many different configurations of Wang tiles in a u-strip can occur, since P is a finite patch (which is periodically repeated along the u-strip). These configurations can now be labelled by letters of a finite alphabet, and we can read off the possible pairings of consecutive letters that occur in the v-direction. We thus get a (geometrically fitting) stacking sequence in the v-direction that observes all matching conditions of the Wang tiles by construction. Since the interior of the u-strip is connected, this can only lead to nearest neighbour restrictions (which is some kind of Markov property). This means that the induced atlas $\mathcal{Y}$ contains the letters of our alphabet and a certain list of two-letter words, but no longer ones (in the sense that no longer-range interaction can be enforced here).

We are thus in the situation of a symbolic sequence over a finite alphabet with a finite, subword-closed atlas of allowed subwords. By Lemma 5.7, there must be a bi-infinite periodic word over the alphabet which respects the atlas. Clearly, this corresponds (by proper stacking of the u-strips along the v-direction) to a Wang tiling with period u as before, and some period in the v-direction (in fact, an integer multiple of v), hence this new tiling is crystallographic.   □

Let us note in passing that Proposition 5.9 also means that any finite set of Wang tiles that permits a tiling with a non-trivial period must also permit $n\mathbb{Z}^2$-periodicity for some $n \in \mathbb{N}$, by an application of Lemma 3.1.

Example 5.11 (*Robinson's aperiodic tiling*)**.** In 1971, Raphael Robinson constructed an aperiodic set of six square-shaped prototiles [Rob71], up to rotations and reflections. Altogether, this adds up to 32 prototiles. Note that this version requires more than simple edge decorations. One way to recast the prototiles as Wang tiles led to 56 prototiles. At the time, this was the simplest set of Wang tiles. The original prototile set, a patch and a large piece of the resulting tiling are shown in Figure 6.50 on page 242. The local rules require that the squares have to be face to face, with the coloured line decorations matching in colour, position and direction at any common edge of two tiles. For an alternative description, see also [JM97].

Any resulting tiling of the plane inherits a hierarchical structure with interlocking squares of growing (and unbounded) size, indicated by different colours in Figure 6.50. Starting with the red squares, each square centre hosts a corner of a square of the next larger size. Since this procedure never terminates, the resulting tiling cannot have any non–trivial period, by a mechanism (a Toeplitz structure) that is reminiscent of the limit periodicity of the period doubling chain. This holds for the entire hull, which shows aperiodicity (as defined earlier). We shall meet a similar structure based on (half-) hexagons in Example 6.4 below.

The hierarchical structure suggests that at least a subset of repetitive Robinson tilings can alternatively be generated by an inflation rule. This is indeed the case, as was recently shown in [GJS12], with 208 decorated prototiles (up to translations). An alternative substitution approach was suggested by Taylor, which shifts the reference points to the tile centres in each step, and works with a pseudo inflation; see [Gäh13] for details.   ◊

5.7.4. THE KARI–CULIK PROTOTILES

An example of an apparently non-hierarchical set of aperiodic prototiles was recently constructed[2] by Kari [Kar96] with a subsequent simplification by Culik [Cul96]; see also [ENP07]. The simplified set consists of thirteen Wang tiles with edges distinguished by five different colours. One way to see

---

[2]We thank Arthur (Robbie) Robinson for bringing this work to our attention through a series of enlightening talks, and Joan Taylor for various helpful comments.

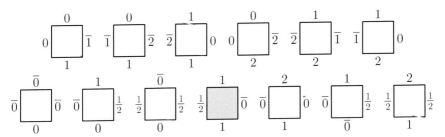

FIGURE 5.4. The Kari–Culik set of aperiodic tiles. The top row shows the six tiles in $\mathcal{W}_3$, the bottom row the seven tiles in $\mathcal{W}_{\frac{1}{2}}$.

that the system does not originate from a substitution rule is based on the positive entropy of the hull [NHL13].

Since the underlying structure is arithmetic, it is useful to properly label the edges for the proof that this prototile set is aperiodic. A convenient set of symbols is $\{0, \overline{0}, \frac{1}{2}, 1, \overline{1}, 2, \overline{2}\}$, where 0 and $\overline{0}$ both correspond to the numerical value 0, but are considered as different 'colours', and $\overline{1} = -1$, $\overline{2} = -2$. Only edges with equal labels are allowed to be adjacent. The set of labelled squares is shown in Figure 5.4. They fall into two classes $\mathcal{W}_3$ and $\mathcal{W}_{\frac{1}{2}}$, where $\mathcal{W}_3$ comprises the six tiles with labels $\{0, 1, \overline{1}, 2, \overline{2}\}$ on the vertical edges, and $\mathcal{W}_{\frac{1}{2}}$ the seven tiles with labels $\{\overline{0}, \frac{1}{2}\}$ on those edges. Denoting the labels of a square tile by a (top), b (left), c (bottom) and d (right), the labels of the Kari–Culik prototiles belonging to $\mathcal{W}_q$ satisfy the relation

$$(5.4) \qquad\qquad qa + b = c + d,$$

with $q \in \{\frac{1}{2}, 3\}$.

Clearly, any row of tiles has to consist entirely of tiles from either $\mathcal{W}_3$ or $\mathcal{W}_{\frac{1}{2}}$. Also, it is not possible to build a tiling with a single type of row. This can be seen as follows. Consider a (hypothetical) legal tiling of the plane that contains a row comprising tiles from $\mathcal{W}_3$. It has to contain at least one of the three tiles with top label 0, as the tiles with top label 1 cannot form rows of more than two tiles. Hence the row above has to consist of tiles of type $\mathcal{W}_{\frac{1}{2}}$. This shows that it is not possible that the tiling only contains tiles from $\mathcal{W}_3$. In fact, this argument also shows that, in any legal tiling, rows of type $\mathcal{W}_3$ must be sandwiched between rows of type $\mathcal{W}_{\frac{1}{2}}$.

Next, in order to show that any (still hypothetical) legal tiling has to contain tiles from $\mathcal{W}_3$, let us see how far we can get with rows of type $\mathcal{W}_{\frac{1}{2}}$ alone. To this end, consider three consecutive rows with tiles from $\mathcal{W}_{\frac{1}{2}}$, which is possible. However, the middle row is then restricted to only consist of the two tiles in $\mathcal{W}_{\frac{1}{2}}$ where both top and bottom labels are in $\{\overline{0}, 1\}$, because a lower label 0 or an upper label 2 would contradict the assumption. This, in turn, forces the top of the middle row to have labels 1 everywhere, and hence the top row cannot avoid a tile (from $\mathcal{W}_{\frac{1}{2}}$) with top label 2 (while the bottom row then must be restricted to bottom labels 0 and 1 to allow a continuation of the tiling by another row).

This shows that, after at most three consecutive rows with tiles from $\mathcal{W}_{\frac{1}{2}}$, a row of type $\mathcal{W}_3$ must follow in both directions, and hence a legal tiling has to contain rows of both types. In fact, rows of type $\mathcal{W}_{\frac{1}{2}}$ can at most occur in pairs, because the special triple considered above inevitably has bottom label 0001 in a periodically repeated way. An inspection shows that this cannot be matched by the top labels of any row of type $\mathcal{W}_3$, which have at most two consecutive 0s. So far, we know that any legal tiling, if one exists at all, must comprise rows of either type. Moreover, rows of type $\mathcal{W}_3$ must always be isolated, while those of type $\mathcal{W}_{\frac{1}{2}}$ must be isolated or come in pairs.

Consider a single row of tiles comprising n tiles from $\mathcal{W}_q$ with labels a_i, $1 \leqslant i \leqslant n$, along the top and c_i, $1 \leqslant i \leqslant n$ along the bottom, and b and d on the left and right. Summing Eq. (5.4) for the n tiles of the row gives

$$(5.5) \qquad b + q \sum_{i=1}^{n} a_i = d + \sum_{i=1}^{n} c_i$$

after proper cancellation of equal terms. Next, consider a rectangular block consisting of m rows of length n. Each row leads to an equation of type (5.5). Combining the m equations via suitable insertion and successive elimination of the labels for the internal horizontal edges, one can show that

$$(5.6) \qquad Q_m \sum_{i=1}^{n} a_i = \sum_{i=1}^{n} c_i + \sum_{j=1}^{m} \frac{Q_m}{Q_j}(d_j - b_j),$$

where $Q_j := \prod_{k=1}^{j} q_k$ denotes the product of the values of the parameter q according to the sequence of rows.

Now, assume that a periodic legal tiling exists. By Proposition 5.9, there also exists a crystallographic tiling, whose lattice of periods, Γ say, is a sublattice of $\mathbb{Z}^2$. If the index is $[\mathbb{Z}^2 : \Gamma] = n$, we know from Lemma 3.1 that $n\mathbb{Z}^2$ is a sublattice of Γ, with the vectors $(n, 0)^T$ and $(0, n)^T$ spanning a fundamental domain. Since Eq. (5.6) applies to this $n \times n$ block, and $d_i = b_i$ and $c_i = a_i$ for $1 \leqslant i \leqslant n$ due to periodicity, one obtains

$$Q_n \sum_{i=1}^{n} a_i = \sum_{i=1}^{n} a_i.$$

Observe that $Q_n \neq 1$, because it is a product of n factors, each being either 3 or $\frac{1}{2}$. This implies $\sum_{i=1}^{n} a_i = 0$, which would require that all $a_i \in \{0, \bar{0}\}$. By inspection of the prototile set, one sees that this is not possible. This shows that there cannot be any legal tiling with a non-trivial period.

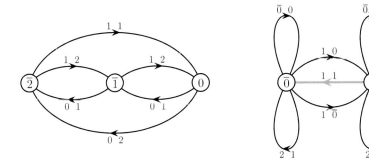

FIGURE 5.5. Labelled adjacency graphs for the Kari–Culik tilings.

It remains to sketch an argument for the existence of at least one legal tiling. This can be achieved by starting from a suitable bi-infinite row of tiles, which is then consistently extended in both vertical directions in a deterministic way to cover the plane. To do so, one employs *Beatty sequences*, which means sequences $([n\alpha])_{n\in\mathbb{Z}}$ with $\alpha \in \mathbb{R}$ and $[x]$ denoting the largest integer $\leqslant x$, and their first difference sequences; see [LP92] and references therein for background. We call the latter *derived* Beatty sequences. Given an arbitrary $\alpha \in \mathbb{R}$, its derived Beatty sequence $(B_n(\alpha))_{n\in\mathbb{Z}}$ is defined by

$$B_n(\alpha) = [n\alpha] - [(n-1)\alpha].$$

Such a sequence is constantly equal to k if $\alpha = k \in \mathbb{Z}$, and takes values in $\{k, k+1\}$ for $k \in \mathbb{Z}$ when $k < \alpha < k+1$. Select the interval $I = [\frac{1}{3}, 2)$ and choose an arbitrary $\alpha \in I$ together with $q\alpha$ where $q \in \{\frac{1}{2}, 3\}$ is the unique multiplier such that $q\alpha \in I$, too. In fact, the prototile set of Figure 5.4 is constructed such that one can form a bi-infinite row of tiles that carries the derived Beatty sequences $B_n(\alpha)$ and $B_n(q\alpha)$ as top and bottom labels in an aligned way. The fitting intermediate sequence of vertical edge labels emerges from the sequence

$$C_n(\alpha, q) = q[n\alpha] - [nq\alpha]$$

with $n \in \mathbb{Z}$, such that the elementary squares have numbers $B_n(\alpha)$ at the top, $C_{n-1}(\alpha, q)$ on the left, $B_n(q\alpha)$ at the bottom and $C_n(\alpha, q)$ on the right. Adding an additional distinction between 0 and $\overline{0}$ (to separate $\mathcal{W}_3$ and $\mathcal{W}_{\frac{1}{2}}$ and to subdivide $\mathcal{W}_{\frac{1}{2}}$ as detailed below), see [ENP07] for the procedure to do so, one obtains the partitioned prototile set of Figure 5.4.

Starting from a pair of derived Beatty sequences for $\alpha, q\alpha \in I$, we thus know that a corresponding bi-infinite row of Wang tiles exists. Such a row requires either tiles from $\mathcal{W}_3$ (for $q = 3$) or from $\mathcal{W}_{\frac{1}{2}}$ (for $q = \frac{1}{2}$). In the latter case, one actually just uses the first four or the last four tiles of the second row in Figure 5.4 (with the shaded tile belonging to both subsets), or the subset consisting of the second, fourth, fifth and seventh tile of that row. The calculation of the lower derived Beatty sequence from the upper one, together with the fitting sequence C_n, can now be encoded as an infinite path in one of the two directed adjacency graphs [Kar96, Cul96] of Figure 5.5, where the identification of state and edge labels proceeds according to

with arrows pointing from the left to the right label of the square. The edge label of the graph is the pair of top and bottom labels of the square, always in that order. Note that, for each row of type $\mathcal{W}_{\frac{1}{2}}$, at most four of the seven edges of the graph on the right are used. The grey edge may occur in all three cases and corresponds to the shaded tile of Figure 5.4.

Let us expand on the mechanism to construct a complete tiling, starting from a single row. Given an initial $\alpha \in I$, the required sequence of multipliers q is fixed, both upwards and downwards. It can be calculated by the mappings $f_\downarrow, f_\uparrow \colon [\frac{1}{3}, 2) \longrightarrow [\frac{1}{3}, 2)$, defined by

$$f_\downarrow \colon x \mapsto \begin{cases} 3x, & x \in [\frac{1}{3}, \frac{2}{3}), \\ \frac{1}{2}x, & x \in [\frac{2}{3}, 2), \end{cases} \quad \text{and} \quad f_\uparrow \colon x \mapsto \begin{cases} \frac{1}{3}x, & x \in [1, 2), \\ 2x, & x \in [\frac{1}{3}, 1), \end{cases}$$

which are the inverses of each other. We can now iterate the previous construction row by row in either direction, possibly from an initial single square, by the appropriate interpretation of the graphs.

We have thus sketched a proof of the following result; see [Kar96, Cul96, ENP07] for further details and examples.

Theorem 5.2. *The prototile set of Figure* 5.4 *is aperiodic. In particular, the edge labels provide a representation of aperiodic local rules.*   □

It is clear by construction that the ensemble of compatible tilings, as defined on the basis of Figure 5.4, comprises uncountably many translation classes (simply because there are uncountably many numbers $\alpha \in I$). What is more, the ensemble necessarily contains distinct LI classes. The local rules are thus aperiodic but not perfect, and the ensemble even has positive entropy.

5.7.5. LOCAL RULES OF INFLATION ORIGIN

As indicated earlier in Section 5.7.1, an LIDS provides a powerful tool to establish the existence of perfect aperiodic local rules, or to construct non-local covers with such rules. Here, we briefly mention some classic examples.

Example 5.12 (*Local rules for Penrose tilings*). Inspired by Wang tiles and Robinson's example, as well as by Kepler's attempts at tiling the plane with pentagons (see [Lüc00] for details and references), Roger Penrose constructed an aperiodic set of tiles with just two prototiles (up to rotations) [Pen74, Pen78]. One common version consists of the two rhombuses used in Figure 1.2 in Chapter 1, where the local rules are encoded by single and double arrows along the edges.

Another, and even simpler, version is given by the kites and darts prototile set

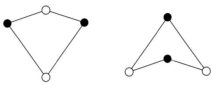

where the dots distinguish two different types of vertices. The local rules now simply consist of a vertex-faithful face to face condition; see also [Gar77]. As we shall see later, these two approaches (and various other ones) lead to equivalent (MLD) tilings; see Figure 6.6 on page 185. For a related version with fractal tile boundaries, we refer to [Ban97].   ◇

More examples of local rules will emerge as we progress; see also [Goo99a] for cases with small prototile sets.

Remark 5.10 (*Ammann's aperiodic prototile sets*). Quite a number of interesting and non-trivial sets of aperiodic prototiles were discovered by Robert Ammann in the 1970s; see [GS87, Sec. 10.4] and [AGS92]. Among them is an eightfold symmetric relative of Penrose's rhombic tiling, which comes in two locally inequivalent versions (only one of which has perfect, aperiodic local rules). This octagonal (or Ammann–Beenker) tiling, compare Figure 1.5, will be one of our guiding examples in Chapters 6, 7 and 9 below. Various other (planar) prototile sets, as noted at the end of Section 5.7.1 above, consist of pairs of 'chair' type tiles with characteristic corner markers. Since these are well described in [GS87, AGS92, Aki12, Sen95], we omit further details.

Less conclusive is the situation for prototile sets in 3-space. Ammann's early contribution to local rules for a rhombohedral tiling are briefly mentioned in [Mac81, p. 521], and a sketch of his local rules is reproduced in [Sen04, Fig. 16]. A detailed (independent and slightly different) treatment is due to Katz [Kat88].   ◇

Remark 5.11 (*Constructing local rules for inflation tilings*). As discussed previously, no aperiodic local rules can exist for tilings in one dimension; compare Lemma 5.7. Examples such as the rhombic Penrose tiling or the (fully decorated) Ammann–Beenker tiling show that aperiodic local rules are possible for inflation tilings in dimensions $d \geqslant 2$ (the underlying inflation rules will be described in detail in Chapter 6). However, a careful inspection leads to the conclusion that a primitive inflation rule alone always defines a unique hull, but not necessarily one with local rules, even if the hull itself is aperiodic. Examples for the latter situation are provided by the undecorated versions of the Ammann–Beenker or the shield tiling discussed below.

On the other hand, with the exception of the Kari–Culik tilings, all known examples with perfect aperiodic local rules possess some underlying inflation structure. This is no accident in the following sense. Given a minimal and aperiodic tiling hull $\mathbb{X}$ (with dimension $d \geqslant 2$) that is defined by a primitive inflation rule, it may or may not have aperiodic local rules. If not, one can introduce non-local information (for instance in the form of symmetry-breaking tile decorations or markers that code the different neighbourhoods) in such a way that one obtains a new (and more complicated) inflation rule for a larger set of prototiles, which defines a new hull $\mathbb{X}'$ that possesses $\mathbb{X}$ as a topological factor. If $\mathbb{X}'$ still fails to possess aperiodic local rules, the procedure can be iterated.

It is an amazing and deep insight [Moz89, Goo98] that, under some mild conditions, there is a systematic way to construct a tower of covers that reach a hull with aperiodic local rules after finitely many steps. This was explored in depth by Goodman-Strauss in [Goo98]; see also [FeS12] for a symbolic dynamics context. The general procedure tends to produce huge prototile sets; see [Goo03] for the example of the sphinx tiling (see Example 6.3). In simple examples (including the Ammann–Beenker and the shield tiling), the level with aperiodic local rules can be reached in one step. Below, we discuss two different half-hex decorations that lead to aperiodic local rules for the half-hex inflation tiling of Example 6.4. Both emerged from direct insight, and not from an application of the general strategy.   ◊

Without starting from an inflation rule, it is a fairly obvious idea to try a systematic search for planar tilings with interesting properties with polyominoes or polyhexes (or similar objects) as prototiles. In particular, this approach has been applied in the search for a monotile; see [Rho05] for a discussion based on an extensive computer search. So far, no monotile has been found this way, though there are some prototiles that tile crystallographically with unexpectedly large fundamental regions.

5.7.6. HEXAGON TILINGS AND PLANAR MONOTILES

While Wang tiles and their generalisations employ the square lattice and then impose aperiodicity via suitable matching conditions at the edges (and possibly also at the vertices) of the square shaped prototiles, an analogous

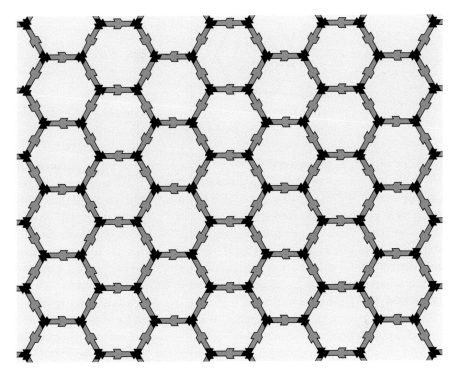

FIGURE 5.6. A patch of Penrose's $(1 + \varepsilon + \varepsilon^2)$-tiling.

attempt is also possible with regular hexagons (and thus on the basis of the root lattice A_2). Two rather interesting examples have been constructed by Roger Penrose [Pen97] and by Joan Taylor [Tay10], which we now discuss, with emphasis on the former. We will return to both examples later in Chapter 6, then with more focus on the latter example and its extension due to Socolar and Taylor [ST11].

Example 5.13 (*Penrose's $(1 + \varepsilon + \varepsilon^2)$-tiling*). This aperiodic tiling due to Penrose [Pen97] is built from the three prototiles

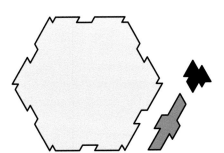

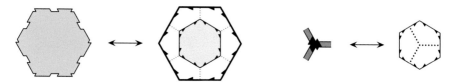

FIGURE 5.7. The mutual local derivation rule between the $(1+\varepsilon+\varepsilon^2)$-tiling and the double hexagon tiling; see text for details.

together with their rotated and reflected versions. The hexagon is viewed as an 'extended' tile in the plane (the '1 tile'), the asymmetric arrow as a 'linear' tile (the 'ε tile', indicating that is can be made arbitrarily thin) and the corner stone as a 'point-like' tile (the 'ε^2 tile', which can be made arbitrarily small). Together, they form an aperiodic set of prototiles, which define tilings of a single LI class. This is proved in [Pen97] with an appropriate inflation rule by the composition-decomposition method described in Section 5.7.1. A variant of the inflation rule will be shown in Example 6.5 below. A legal patch is illustrated in Figure 5.6. Any resulting tiling can be viewed as an aperiodic decoration of the honeycomb packing, where repetitivity is a result of the underlying inflation rule. The legal tilings of the plane form a single LI class, which coincides with the tiling hull of the inflation. As such, the local rules are aperiodic and perfect.   ◊

Inspecting the prototile set of Example 5.13 together with Figure 5.6, one realises that the hexagon has characteristically oriented edges and additional small markers near the corners. As noted in [Pen97], these orientations and markers can be encoded as in the left panel of Figure 5.7, where the corner markings are now translated into orientations of the edges of the inscribed smaller hexagon (shaded). The latter is a rotated and scaled down version of the outer hexagon, with the same arrow pattern. In each hexagon, precisely two parallel edges are oriented in the same direction, while the remaining arrows point towards the two remaining antipodal corners. The scale and orientation of the smaller hexagon is chosen such that the remaining area can be tiled by hexagons of the same size again; compare Figure 5.8. The latter can be derived from the corner stones as shown in the right panel of Figure 5.7. As usual, both rules are to be understood such that rotated or reflected configurations are replaced by the correspondingly rotated or reflected images. Note that, in the $(1 + \varepsilon + \varepsilon^2)$-tiling, the information carried by the ε-tiles is completely contained in the edge direction of the hexagons together with the orientation of the corner stones. This is implicitly covered in our derivation rule.

As one can check explicitly, the local derivation rule of Figure 5.7 maps each legal $(1 + \varepsilon + \varepsilon^2)$-tiling to a new (marked) tiling of oriented hexagons,

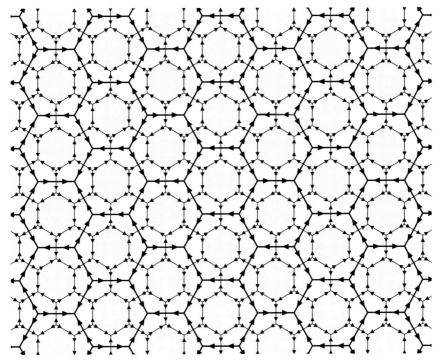

Figure 5.8. Illustration of the patch of Figure 5.6 after the application
of the derivation rule of Figure 5.7; see text for details.

with the property that all hexagons of the smaller scale have the same arrow
pattern type (up to similarity) and that all arrows match; compare Figure 5.8
for a larger patch. We call each image a Penrose *double hexagon pattern*. It
is clear by construction that they form an LI class, as the derivation rule
defines a continuous map on the compact and minimal $(1 + \varepsilon + \varepsilon^2)$-tiling hull.

Conversely, consider any element from the image hull, which is a legal
double hexagon pattern. Then, one can consistently reverse the derivation
rules as also indicated in Figure 5.7 to obtain a legal $(1 + \varepsilon + \varepsilon^2)$-tiling. Note
that the dotted lines indicate the edge directions of the complementary scale
hexagons. In the smaller hexagon, they break the remaining reflection symmetry, and determine the chirality of the corner stone. Note that the arrow
matching on coinciding edges of smaller hexagons that are perpendicular to
edges of the larger hexagons is equivalent to the matching condition of the
two ends of the ε-tiles. In our setting, the orientation of the dotted lines is
determined from the local neighbourhood, but is not needed to formulate the
derivation rules. Combining both directions, and observing that the derivation rules commute with Euclidean motions, the following result is obvious.

Proposition 5.10. *The LI classes of the* $(1 + \varepsilon + \varepsilon^2)$*-tiling of Example* 5.13 *and the Penrose double hexagon pattern are SMLD.* □

The relevance of the reformulation in form of the double hexagon pattern emerges from the observation that the hexagonal building block can be viewed as some kind of a *functional monotile* for the Euclidean plane. Here, we use the attribute 'functional' to indicate that the local rules are given by a finite atlas of local configurations, but not by a geometric nearest neighbour matching condition alone. To substantiate this, let us consider the marked double hexagon from the left panel of Figure 5.7 as a single prototile (up to similarities). The prototile and its reflected version each occur in six different orientations (which means that there are twelve prototiles up to translations). The local rules consist of the following requirements. The prototiles have to be face to face, with matching edge directions. In addition, the edges of the completed small hexagons that surround a vertex of the large hexagons must be directed such that only the allowed arrow pattern occurs and that also all edge directions of the small hexagons match. Strictly speaking, since the remaining orientations produce distinct possibilities, one could argue that we have more than one prototile. We therefore prefer to call this (incomplete) double hexagon a *weak* functional monotile [BGG12b].

It is clear from Proposition 5.10 that there are legal double hexagon patterns of the entire plane. Moreover, the inverse derivation rule can be applied to *any* legal double hexagon pattern, then giving a $(1 + \varepsilon + \varepsilon^2)$-tiling that satisfies the aperiodic local rules of Example 5.13 (which are purely geometric). Consequently, the double hexagon local rule is aperiodic and perfect. Note, however, that the nature of the local rules is effectively a combination of nearest and next-to-nearest neighbour conditions. This is inherited from the $(1 + \varepsilon + \varepsilon^2)$ local rules (via the structure of the ε-tiles).

Corollary 5.5. *The marked double hexagon prototile of Figure* 5.7*, together with the above local rules for edge and corner matching, constitutes a weak functional monotile of the plane. The prototile necessarily occurs in two chiralities, and the local rules comprise both nearest and next-to-nearest neighbour information.* □

Above, we have used the term 'weak' to indicate that the hexagon obtains orientations on the dashed edges only during the tiling process. If the acquired arrow decorations are taken into account from the beginning, one needs several different (completely marked) prototiles. In this sense, the double hexagon is not a true monotile.

A further step towards a 'real' monotile via another decorated hexagon tiling, which is related but subtly different [BGG12b], was independently discovered by Joan Taylor [Tay10]. This gives rise to a functional monotile for

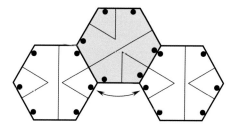

FIGURE 5.9. The functional monotile (left) of Socolar and Taylor [ST11] and a sketch of the corresponding local rules (right).

a non-minimal aperiodic hull [ST11]. One version of the marked prototile is shown in the left panel of Figure 5.9. Once again, it occurs in six different orientations, together with a reflected version (shaded, in another six orientations). The markings consist of lines, which have to continue without bends or discontinuities across tile boundaries, and of dots, which must everywhere satisfy the next-to-nearest neighbour positioning rule indicated in the right panel of Figure 5.9, irrespective of the chiralities.

As was shown in [ST11], these local rules are aperiodic, but not perfect. The corresponding tiling space is non-minimal, and consists of a minimal part that is the LI class of an inflation tiling together with some translation classes of non-repetitive patterns. The local rules can be sharpened to single out the minimal part [Tay10]. More details will be discussed in the inflation context in Example 6.6.

The question whether there is a (simply connected) monotile that enforces aperiodicity in the plane by shape alone is still open.

5.7.7. A CONVEX MONOTILE IN 3-SPACE WITH GEOMETRIC LOCAL RULES

In the plane, it is thus still unknown whether one can enforce aperiodicity with a single prototile and nearest neighbour conditions only, and in particular, whether this is possible with a convex shape and purely geometric local rules. Surprisingly, the answer to this question is affirmative in 3-space. An aperiodic prototile (not yet convex) was discovered by P. Schmitt in 1988, which formed the basis for later work by Conway and Danzer; compare [Dan95, BF05]. The final result is a prototile with several free parameters that, for certain choices of the latter, possesses aperiodic local rules. It should be noted that only the tile and its orientation-preserving isometric copies are permitted, not its mirror image[3]. Such prototiles are called SCD tiles, and the corresponding tilings as SCD tilings. Let us describe their geometric construction and properties.

---

[3]If one allows an SCD tile and its mirror image, periodic tilings become possible.

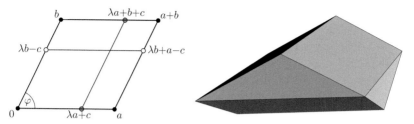

FIGURE 5.10. The construction of the bi-prismatic SCD tile (left) and a
view of one example (right).

The prototile is a bi-prism with rhombic basis, as sketched in Figure 5.10.
Here, $a = (\sqrt{b_1^2 + b_2^2}, 0, 0)$ and $b = (b_1, b_2, 0)$ are vectors in the xy-plane (with
$b_1, b_2 > 0$), while $c = (0, 0, c_3)$ is perpendicular to it (with $c_3 > 0$). We have
$\tan(\varphi) = b_2/b_1$, while $\lambda \in (0, 1)$ is another free parameter. The prototile is
then the convex hull of the 8 vertices

$$\{0, a, b, a + b, \lambda a + c, \lambda a + b + c, \lambda b - c, \lambda b + a - c\}.$$

It has four triangles and four parallelograms as facets, as indicated on the
right of Figure 5.10 for the case $\lambda = \frac{1}{2}$ and $\varphi = \arctan(2)$. In general,
the SCD tile is called *incommensurate* when $\varphi \notin \pi\mathbb{Q}$ and *commensurate*
otherwise. Several variants of this construction are possible, all essentially
leading to the same type of structure; see [Dan95, Sec. 3] for details.

Let us fix a set of parameters and start with one copy of the ensuing
SCD tile T, which contains the vertex point 0. By taking T together with
suitable translates of T along the xy-plane, one can joint them face to face at
the triangular facets. Due to the rhombic basis, this can be done to form a
closed, planar layer L, which is unique and periodic with $\Gamma = \mathbb{Z}a \oplus \mathbb{Z}b$ as its
(planar) lattice of periods. In fact, the layer simply is $L = \{x + T \mid x \in \Gamma\}$.
The top of L shows ridges and valleys, all parallel to each other (and to the
line $0b$), while the bottom shows the corresponding structure 'upside down',
this time parallel to the line $0a$. Starting with a translate of T leads to a
translate of L, wherefore such planar layers (and their rotations around the z-
axis) are the only possibilities to satisfy all triangular face to face conditions.

In order to continue the tiling process, one has to stack the layers. Let
$L' = L - \lambda b$ and take a second layer $L'' = c + RL'$, where R is the rotation
by φ around the z-axis. Then, L'' fits exactly on top of L', which gives
a double layer. Although one has neither gaps nor overlaps this way, the
parallelograms are generally not face to face (this phenomenon is visible in
Figure 5.11). Under certain conditions, and with additional line markings,
one can achieve a face to face matching of the pseudo facets that emerge
from the extra marker lines; see [Dan95] for more. The stacking procedure

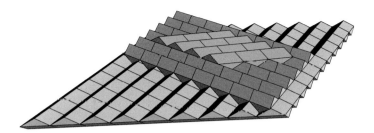

FIGURE 5.11. Illustration of the layer structure of SCD tilings.

can now be iterated as indicated in Figure 5.11. One obtains

$$(5.7) \qquad \mathcal{T} = \bigcup_{m \in \mathbb{Z}} mc + R^m L',$$

which is a tiling of $\mathbb{R}^3$, called an SCD tiling. It may be an FLC tiling or not, which (as repetitivity) depends on $\cos(\varphi)$ being rational or not; see [BF05, Prop. 2.4]. Let us mention in passing that one could also consider arbitrary shifts along the ridges (as is done in [BF05]), though we shall not do that here. An SCD tiling is called *incommensurate*, when it is built from an incommensurate SCD tile, and *commensurate* otherwise.

Lemma 5.8. *An incommensurate SCD tiling $\mathcal{T}$ is non-periodic.*

PROOF. Since $\varphi \notin \pi\mathbb{Q}$, the layers of the tiling have distinct orientations. A translation $x \in \mathbb{R}^3$ with $x + \mathcal{T} = \mathcal{T}$ must thus map each layer onto itself. The translation vectors that fix the mth layer form the lattice plane $R^m \Gamma$ with $\Gamma = \mathbb{Z}a \oplus \mathbb{Z}b$ from above. We thus find

$$\mathrm{per}(\mathcal{T}) = \bigcap_{m \in \mathbb{Z}} R^m \Gamma = \{0\},$$

where the last step is a consequence of the assumed incommensurateness. $\square$

Remark 5.12 (*Remaining periods for finite stacks of layers*). Let us point out that finitely many layers may still possess non-trivial translation symmetries, even in the incommensurate case. This is so because Γ and $R\Gamma$ might still share a sublattice of finite index, the so-called *coincidence site sublattice*; see [Baa97] and references therein for general background. If this happens, any finite stack of layers still has non-trivial translation symmetries in the form of a lattice plane (due to the group structure of the coincidence rotations of a lattice Γ), where the index in Γ grows with the number of layers. In the limit of infinitely many layers, only the trivial translation survives. ◊

Given an incommensurate SCD tiling, we can define its hull in a similar fashion as before. The point to observe here is that, due to the incommensurate rotation involved, we cannot have the FLC property relative to translations, but at most relative to Euclidean motions. Assume that we are in the situation of a repetitive, incommensurate SCD tiling (relative to Euclidean motions), which means that $\cos(\varphi) \in \mathbb{Q}$ by [BF05, Prop. 2.4], with $\varphi \notin \pi\mathbb{Q}$. Replacing the local topology (as induced by a perfect match of finite patches after small translations; see Section 5.4) by the *local rubber topology* (as induced by ε-perfect perfect matches of finite patches after small translations; compare Remark 5.6), the hull is again a compact set, and contains only congruent copies of a single SCD tiling and limits of convergent sequences of congruent copies. Clearly, this construction leaves no freedom to generate non-trivial translation symmetries, and Lemma 5.8 implies the following result.

Theorem 5.3. *A repetitive, incommensurate SCD tiling is aperiodic.*   □

This theorem is perhaps still not entirely satisfactory for the following two reasons. Firstly, if one also admits the reflected tile $-T$, one can build layers with ridges that are rotated against each other in the opposite direction, so that one could then form tilings with non-trivial periods in the z-direction. Secondly, an individual SCD tiling of the restricted type may still possess a symmetry in form of a screw axis. Even though such a tiling admits no non-trivial translation symmetry, its individual symmetry group contains an *infinite* subgroup. This observation motivates the following refined notion.

Definition 5.22. An FLC pattern $\mathcal{T} \sqsubset \mathbb{R}^d$ is called *strongly aperiodic* when it is aperiodic and when the individual symmetry group of each element of $\mathbb{X}(\mathcal{T})$ is a finite group.

In this sense, a repetitive, incommensurate SCD tiling is aperiodic, but not strongly aperiodic. Consequently, the problem of the existence of a strongly aperiodic monotile in 3-space is still open.

CHAPTER 6

Inflation Tilings

Having introduced some general geometric concepts in Chapter 5, we will now discuss inflation tilings in more detail, mainly by way of (planar) examples; see [HFonl] for further cases. Various illustrative colour figures are collected at the end of this chapter. Our setting employs the generalisation of the geometric inflation rules of Chapter 4 to higher dimensions, as developed in Chapter 5. In particular, we again define hulls via fixed points of the geometric inflation and their translation orbit closures. Here, we usually employ the continuous hull as obtained from the translation action of $\mathbb{R}^d$. We explain and further develop the concepts of repetitivity, minimality, ergodicity and aperiodicity as we go along. As intimately related (FLC) tilings may look rather different at first sight, an important point will be the systematic application of the equivalence concept of mutual local derivability (MLD).

6.1. Ammann–Beenker tilings

One of the simplest planar aperiodic tilings is the *octagonal* or *Ammann–Beenker tiling* from Figure 1.5 on page 7. It is built, up to orientation, from two prototiles, a square and a rhombus with opening angle $\pi/4$, both of edge length 1. A simple stone inflation rule can be constructed from the original approach [GS87, AGS92] by cutting the square along one diagonal into two isosceles triangles. An algebraic approach was independently developed in [Bee82], and will later be discussed in the projection context. The corresponding inflation multiplier is the PV unit

$$\lambda_{\mathrm{sm}} = 1 + \sqrt{2} \approx 2.414$$

which we met before in the context of the silver mean chain in Example 4.5. Since the inflation rule breaks the symmetry of the square, the two triangles that form a square inherit an orientation, marked by an arrow. In contrast, the rhombus keeps its symmetry under inflation, so that no extra markings are required at this stage. Despite the arrows at the hypotenuses, we call this tiling *undecorated*, since it is SMLD with a plain square rhombus tiling without any markings. The inflation rule is shown in Figure 6.1, together with a (legal) square-shaped patch of the tiling.

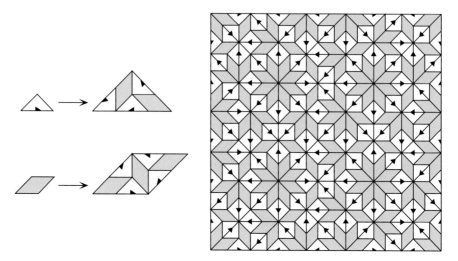

FIGURE 6.1. Inflation rule for the undecorated Ammann–Beenker tiling (left panel), and a patch (right panel) obtained by three inflation steps from a square-shaped patch (consisting of two triangles) in the centre; see text for details. This patch has no rotation symmetry, but is reflection symmetric in the diagonal.

The inflation rule ensures that triangles always come in pairs to form squares, with matching orientations of the hypotenuses (while a mismatch does not give a legal patch). Taking the markings into account, squares occur in eight orientations, as does each of the two chirality types of triangles. In contrast, the rhombus shows up in four orientations only. For convenience, the rhombic tiles have been shaded in Figure 6.1. Rotated or reflected tiles are inflated by the rotated or reflected rules. A square (with its centre as a reference point, say) can be used as a seed for an inflation sequence that converges towards a 2-cycle, in the local topology introduced in Chapter 5. Each member of this cycle is then a fixed point under the square of the inflation rule. As the inflation is primitive, it defines a unique LI class, called the undecorated Ammann–Beenker LI class or tiling (the latter by slight abuse of language, as it is actually a pattern, due to the markings). As a consequence, the two members of the 2-cycle are locally indistinguishable.

The simplest inflation matrix for the Ammann–Beenker inflation rule of Figure 6.1 is

$$M_{\mathrm{AB}} = \begin{pmatrix} 3 & 4 \\ 2 & 3 \end{pmatrix},$$

which only keeps track of the number of triangles and rhombuses, while disregarding their orientations. Contrary to the one-dimensional case, this matrix cannot be interpreted as an Abelianisation of a symbolic substitution on a free group. Note further that the matrix M_{AB} is *not* the square of the silver mean substitution matrix, although the PF eigenvalue of M_{AB} is λ_{sm}^2. If we consider a tiling with rhombus edge length 1, the areas of the two prototiles are $\frac{1}{2}$ (triangle) and $\frac{1}{2}\sqrt{2}$ (rhombus). Indeed, $\frac{1}{2}(1, \sqrt{2})$ is a left eigenvector of M_{AB} for its PF eigenvalue $\lambda_{sm}^2 = 3 + 2\sqrt{2}$. The corresponding right eigenvector, in statistical normalisation, reads

$$\nu_{AB} = \left(2 - \sqrt{2}, \sqrt{2} - 1\right)^T.$$

The two entries are the relative frequencies of the prototiles in an infinite tiling that is a fixed point of (some power of) the inflation. This little calculation has the following interesting consequence.

Proposition 6.1. *Consider a tiling of the plane that is built from the two prototiles (triangle and rhombus) of Figure 6.1 and is a fixed point of the Ammann–Beenker inflation. Then, the triangles taken together cover the same area proportion of the plane as the rhombuses. The same conclusion holds if the tiling is a fixed point under any positive power of the inflation rule, or if it is any tiling in the LI class defined by the inflation.*

When considering the square rhombus version of the tiling, the statements remain true in the sense that squares and rhombuses cover the same area fraction of the plane.

PROOF. Assume that a fixed point under the inflation rule is given.[1] The first claim follows from a simple calculation with the two PF eigenvectors of M_{AB} as introduced above. Observing that their scalar product is $2 - \sqrt{2}$ (which is the average area per tile), it is immediate that the area fraction covered by triangles is $\frac{1}{2}$, as is the area fraction for the rhombuses.

The second claim follows from the fact that the inflation matrix of a power of the inflation map is the corresponding power of the inflation matrix, which keeps the same PF eigenvectors. Also, the ratio of covered areas remains the same for all elements of the LI class. The last claim is clear because two triangles always match to form a square, so that the area covered by triangles equals the one covered by squares.   □

Let us consider the fixed point tiling $\mathcal{T} = \mathcal{T}_{AB}$ of the Ammann–Beenker inflation rule that is illustrated in Figure 6.41 on page 236. It has *exact* individual D_8 symmetry with respect to its centre, which includes an eightfold rotational symmetry. By Corollary 3.2, the vertex point set of $\mathcal{T}$ cannot be crystallographic. Moreover, it cannot have any non-trivial periods, because Proposition 5.7 excludes the existence of more than one rotation centre with perfect eightfold symmetry.

---

[1]Such a fixed point tiling can be constructed by iterated inflation applied to an 8-star built from rhombuses, which is legal; compare Figure 6.41 on page 236.

The Ammann–Beenker inflation is a primitive stone inflation with finitely many prototiles up to translations. The fixed point $\mathcal{T}$ is (linearly) repetitive, as follows by an application of Proposition 5.8. Moreover, the continuous hull of $\mathcal{T}$ is

$$\mathbb{X}(\mathcal{T}) = \overline{\{t + \mathcal{T} \mid t \in \mathbb{R}^2\}},$$

where the closure is taken in the local topology of Chapter 5. Clearly, one has $\mathbb{X}(\mathcal{T}) = \mathrm{LI}(\mathcal{T})$, again by arguments that are analogous to those used before, while minimality of the hull is again equivalent to repetitivity of the tiling by Gottschalk's theorem; compare Proposition 4.3 and [Pet83, Thm. 4.1.2]. Due to the inflation structure and the ensuing linear repetitivity, we know that the frequencies of all finite patches within $\mathcal{T}$ exist uniformly. This is once again equivalent to unique ergodicity by Oxtoby's theorem, compare Proposition 4.4 and its extension in [MR13], so that the dynamical system $(\mathbb{X}(\mathcal{T}), \mathbb{R}^2, \mu)$ is strictly ergodic, where μ is the unique probability measure on $\mathbb{X}(\mathcal{T})$, as defined via the patch frequencies in the usual way; compare Section 5.4. Note that $\mathbb{X}(\mathcal{T})$ is compact due to the FLC property of $\mathcal{T}$. In summary, we have the following result.

Proposition 6.2. *The Ammann–Beenker tiling $\mathcal{T} = \mathcal{T}_{\mathrm{AB}}$ is a linearly repetitive FLC tiling that is aperiodic. It possesses an LIDS with inflation multiplier $1 + \sqrt{2}$. The continuous hull is compact and satisfies $\mathbb{X}(\mathcal{T}) = \mathrm{LI}(\mathcal{T})$. The corresponding dynamical system $(\mathbb{X}(\mathcal{T}), \mathbb{R}^2, \mu)$ is strictly ergodic.* □

As earlier, we have combined the non-periodicity of a fixed point with its repetitivity to infer the (stronger) aperiodicity property. This will work in general for primitive stone inflation tilings with the FLC property. The precise statement can be formulated as follows, which is a consequence of Proposition 5.5 together with a standard MLD argument.

Corollary 6.1. *A repetitive tiling $\mathcal{T}$ in $\mathbb{R}^d$ is FLC. Moreover, if $\mathcal{T}$ is also non-periodic, it is aperiodic.* □

While repetitivity and finite local complexity are often direct consequences of the inflation rule, we need further tools to infer non-periodicity. This is so because non-crystallographic symmetries need not manifest themselves as *exact* symmetries of a given tiling, or even of any member of an LI class; compare the discussion in Section 5.6.

Returning to the Ammann–Beenker tiling, one would expect to see interesting relations to lower-dimensional structures via suitable sections. This is indeed a general phenomenon, though one need not find the standard examples from Chapter 4, but possibly locally equivalent systems.

Remark 6.1 (*Alternative silver mean rules*). The inflation matrix of the Ammann–Beenker tiling satisfies $M_{\mathrm{AB}} = M^2$ with $M = \left(\begin{smallmatrix} 1 & 2 \\ 1 & 1 \end{smallmatrix}\right)$, which is the only possible relation of this kind with a non-negative integer matrix on the right-hand side. The matrix M differs from the substitution matrix M_{sm} of the silver mean chain of Example 4.5, though the two matrices have the same eigenvalues. This observation was also made in [Lüc93, Sec. 3] in the context of Ammann bar grids.

One choice for an underlying substitution rule is $A \mapsto AB$, $B \mapsto AAB$. The corresponding sequences are thus effectively built from the words AB and AAB. One can now check that the locally invertible rule $AB \leftrightarrow a$, $A \leftrightarrow b$ brings us back to the original silver mean substitution ($a \mapsto aba$, $b \mapsto a$) of Example 4.5. These two substitution rules thus define distinct LI classes that are MLD. Three of the remaining five possible (non-negative) substitution rules for the same matrix M are conjugate to our first choice, and thus define the same LI class. The remaining two ($A \mapsto AB$, $B \mapsto BAA$ and its reflected version) define distinct MLD classes.   ◇

One could also distinguish the possible orientations of the prototiles in the tiling, and then deal with a larger inflation matrix. It turns out, however, that the distinct orientations of a prototile are equally frequent (as a consequence of the underlying symmetry), so that this larger matrix does not contain substantial additional information. Whenever we meet a situation of this kind, we will use the simplest possible matrix. Here, it means that we may consider tiles up to similarity as prototiles, not just up to translations. This simplification (with the mentioned consequences on the frequencies) is possible whenever the inflation rule commutes with the relevant rotations and reflections, which is relevant for the analysis of SMLD classes.

A larger patch of the undecorated Ammann–Beenker tiling (in its square rhombus version) is shown in Figure 6.41 on page 236. As implicitly alluded to already, one can also consider a *decorated* version of the Ammann–Beenker tiling. The prototiles with their inflations are displayed in Figure 6.42, with a larger patch shown in Figure 6.43, both on page 237. The benefit of these additional decorations is that they constitute perfect aperiodic local rules in the sense of Definition 5.20, similar to those of the rhombic Penrose tiling in Figure 1.2 on page 3. This property can be proved by the classic composition-decomposition method [Soc89, AGS92, Gäh93]; compare Section 5.7.1.

Remark 6.2 (*Decorated versus undecorated AB tilings*). It is clear that the decorated version of the Ammann–Beenker tiling can be reduced to the undecorated one by simply removing all markings except for the arrows on the hypotenuses of the triangles. Interestingly, the converse is not true (as we discuss later on; see also [AGS92, Gäh93]), because the decorated version of the tiling contains information that cannot be derived locally from the undecorated tiling. This can be seen from the fact that there is no local way to decide upon the position of the symmetry-breaking (house-shaped) vertex marker in the D_8-symmetric tiling of Figure 6.41. This means that the two inflation rules of Figure 6.1 and Figure 6.42 define two LI classes that are *not* MLD. This is distinctively different from the situation for the rhombic Penrose tiling, where the arrow decorations derive locally from the naked tiling; see Theorem 6.1 below.   ◇

6.2. Penrose tilings and their relatives

An interesting pair of prototiles is provided by the two golden triangles shown in the middle column of Figure 6.2. Two distinct edge lengths of ratio $\tau : 1$ occur, while all angles are integer multiples of $\pi/5$. These prototiles permit various inflation rules with inflation multiplier $\lambda_T = \tau$, where the subscript T refers to the triangular nature of the prototiles; two significantly different stone inflation rules are presented in Figure 6.2. They relate to the Penrose–Robinson tiling (PRT, left) and the Tübingen triangle tiling (TTT, right). Let us first discuss the PRT tiling and its relation to various classic Penrose-type tilings, in a way that is complementary to the treatment in [Sen95]; for dynamical systems aspects, see [Rob96].

The patch shown in Figure 6.3 grows into a fixed point under the fourth power of the inflation map ϱ_{PRT}. There are three other fixed points of ϱ_{PRT}^4 in the hull with individual fivefold symmetry. Together, they form a 4-cycle under ϱ_{PRT}, all members of which are LI. The central seeds of the 4-cycle are illustrated in Figure 6.4. For a general analysis of inflation orbits and their behaviour under the group D_{10}, we refer to [BHP97].

Thus, we have a primitive stone inflation ϱ_{PRT} for an FLC tiling, and a fixed point under its fourth power that is (individually) fivefold symmetric.

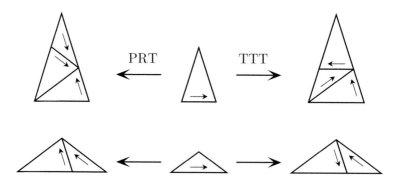

FIGURE 6.2. Stone inflation rules ϱ_{PRT} and ϱ_{TTT} for the Penrose–Robinson tiling (PRT) and the Tübingen triangle tiling (TTT). They define distinct LI and MLD classes; see text for details.

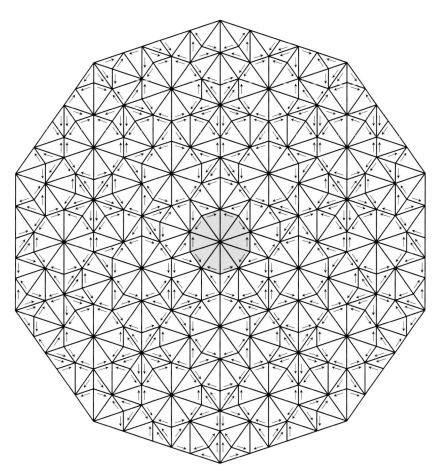

FIGURE 6.3. An exactly D_5-symmetric patch of the PRT, obtained by four inflation steps from the central decagon (shaded).

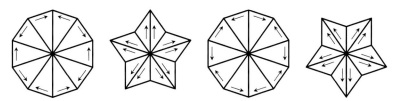

FIGURE 6.4. Central vertex stars of the the four PRTs with individual fivefold symmetry that form a 4-cycle under the inflation map ϱ_{PRT}.

Since this is a non-crystallographic symmetry for $\mathbb{R}^2$, the situation is completely analogous to that of the Ammann–Beenker tiling, wherefore we have the following result.

Proposition 6.3. *The Penrose–Robinson tiling* $\mathcal{T} = \mathcal{T}_{\mathrm{PRT}}$ *is a linearly repetitive FLC tiling of* $\mathbb{R}^2$ *that is aperiodic, with an LIDS with inflation multiplier* τ. *The continuous hull is compact and satisfies* $\mathbb{X}(\mathcal{T}) = \mathrm{LI}(\mathcal{T})$. *The corresponding dynamical system* $\big(\mathbb{X}(\mathcal{T}), \mathbb{R}^2, \mu\big)$ *is strictly ergodic.* □

Let us now employ the full power of the MLD concept to extend this result to an entire collection of related tilings.

Example 6.1 (*Robinson's stone inflation*). A variant of the PRT inflation rule from Figure 6.2, which is due to Robinson [GS87, Fig. 10.3.14], is obtained by using the inflated shape of the obtuse triangle as a prototile, while keeping the acute one (in its previous size). One possible rule is:

This provides a simple stone inflation for the rhombic Penrose tiling of Figure 1.2 on page 3, where the thin (thick) rhombus is dissected along the short (long) diagonal; see below for details.

The precise relation to the PRT tiling of Figure 6.3 is given by the following mutual derivation rule (which is local in both directions):

Note that the thick arrow in the obtuse Robinson triangle points in the opposite direction to the thin arrow of the corresponding PRT level-1 supertile (which is an obtuse triangle of the same size); compare Figure 6.2.

The direction from the Robinson tiling to the PRT is obvious. For the converse direction, one first has to deal with all obtuse triangles of the PRT tiling, by pairing each such triangle with the uniquely matching acute triangle to fill the obtuse Robinson triangle, according to the left part of the rule. All remaining triangles are then acute ones, which directly give the acute triangles of the Robinson tiling, according to the right part of the rule. ◇

Remark 6.3 (*Inflation matrices for golden triangles*). The matrix for both inflation rules of Figure 6.2 is given by

$$M_{\mathrm{T}} = \begin{pmatrix} 2 & 1 \\ 1 & 1 \end{pmatrix} = \begin{pmatrix} 1 & 1 \\ 1 & 0 \end{pmatrix}^2,$$

when taking the acute triangle as the first prototile. Moreover, the same matrix applies to Robinson's rule of Example 6.1 (taking the obtuse triangle

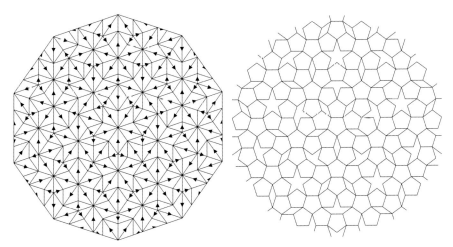

FIGURE 6.5. A patch of a Penrose-type tiling, as obtained via Robinson's stone inflation rule (left), and the corresponding patch of the Penrose pentagon tiling (right).

first this time). The PF eigenvalue is the square of the golden ratio τ, the area ratio of the prototiles is τ, and the frequency vector is

$$\nu_{\mathrm{T}} = \left(\frac{1}{\tau}, \frac{1}{\tau^2}\right)^T = (\tau - 1, 2 - \tau)^T.$$

In all three cases, the larger triangles cover an area fraction $\frac{1}{5}(2 + \tau) \approx 0.7236$ of the plane.   ◊

Robinson's version of the inflation rule results in the tiling shown on the left panel of Figure 6.5. It has the advantage that one can formulate rather simple local derivation rules to obtain other widely used Penrose-type tilings. They include the pentagon tiling displayed on the right panel of Figure 6.5, as well as the rhombic and the kites and darts version; see Figure 6.6. The local derivation rule from the Robinson triangles to the pentagon tiling is

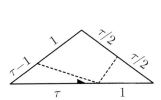   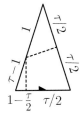

which is formulated for a scale where the short edge of the acute triangle has length 1. The dashed lines will produce the edges of the pentagon tiling, which have length $\sqrt{3 - \tau}$ in this scale. The converse derivation is also local, but requires a finite atlas with larger patches. It can be constructed from [GS87, Fig. 10.3.22] together with the derivation between the kites and darts version and the Robinson tiling (which is given below).

Let us next relate Robinson's stone inflation with the rhombic Penrose tiling. To obtain the rhombuses without matching rule decorations, one simply removes all oriented edges of the Robinson tiling together with their arrows, thus keeping precisely all medium length edges (length τ in our present scale). These constitute the rhombus edges. Moreover, the matching rule decorations of Figure 1.2 on page 3 can be recovered by a local rule as well, namely by

Conversely, starting from the undecorated rhombic Penrose tiling, observe first that each rhombus is part of at least one vertex star of 'simpleton' type, which refers to the vertex stars with coordination number three. Here, there exist two distinct ones up to Euclidean motions. Introducing oriented lines along diagonals according to the rule

  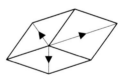

gets one back to the Robinson tiling, thus establishing the mutual local derivability of these three versions.

To derive the kites and darts version of Example 5.12 from the Robinson tiling, one can proceed as follows. From the obtuse triangle, only the long edge is kept, but its arrow is removed. This produces all long edges of the kites and darts tiling. From the acute triangle, one only keeps one of its longer edges, by removing the oriented edge (with its arrow) and the edge it runs into. This produces all short edges of the kites and darts tiling; see Figure 6.6. Conversely, the kites and the darts are dissected and decorated as follows.

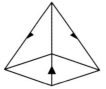 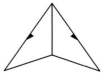

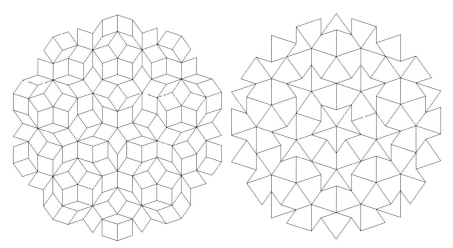

FIGURE 6.6. A patch of the rhombic Penrose tiling (left), obtained from the left patch of Figure 6.5 by a local derivation rule. Also shown is the kites and darts version (right), which is obtained by a similar local derivation rule; see text for details.

One can check that the orientations of the long edges always match consistently under this rule, which transforms the kites and darts into the Robinson tiling. The above discussion can be summarised as follows.

Theorem 6.1. *The LI classes of the following planar tilings (in appropriate scale and relative orientation) belong to the same MLD class:*

(1) *Robinson's triangular version of the Penrose tiling;*
(2) *the Penrose–Robinson tiling (PRT);*
(3) *the rhombic Penrose tiling with edge decorations;*
(4) *the rhombic Penrose tiling without decorations;*
(5) *the Penrose pentagon tiling;*
(6) *the kites and darts version of the Penrose tiling;*
(7) *the vertex point set of the rhombic Penrose tiling.*

SKETCH OF PROOF. The above arguments demonstrate that (1) is MLD to (2)–(6). It is obvious that (7) is locally derivable from (4) by marking the vertex points and removing the edges. Conversely, the edges can be recovered from the vertex point set Λ as follows. As Λ has finite local complexity, $\Lambda - \Lambda$ is locally finite. The non-zero elements of $\Lambda - \Lambda$ of smallest norm correspond to the shortest distance along the short diagonal of the thin rhombus. The next larger norm corresponds to the edges in the tiling, and cannot occur otherwise. Hence, connecting all pairs of vertex points with this distance produces the rhombic Penrose tiling, which we usually take as the representative from now on. Due to finite local complexity, this is a local rule.   □

We are now in the position to extend Proposition 6.3.

Corollary 6.2. *The properties asserted in Proposition 6.3 apply to each of the seven patterns listed in Theorem 6.1, as well as to any other FLC pattern in the MLD class of the rhombic Penrose tiling.*

PROOF. Let $\mathcal{P}$ be any FLC pattern that is MLD with the PRT $\mathcal{T}$. The MLD rule defines a homeomorphism $\psi : \mathbb{X}(\mathcal{T}) \longrightarrow \mathbb{X}(\mathcal{P})$ which commutes with translation. The dynamical systems $(\mathbb{X}(\mathcal{T}), \mathbb{R}^2, \mu)$ and $(\mathbb{X}(\mathcal{P}), \mathbb{R}^2, \nu)$ are thus topologically conjugate via an MLD rule, where $\nu(B) = \mu(\psi^{-1}(B))$ for any measurable set $B \subset \mathbb{X}(\mathcal{P})$. The conjugacy implies non-periodicity of $\mathcal{P}$, compactness and minimality of $\mathbb{X}(\mathcal{P})$, as well as unique ergodicity by standard arguments [DGS76]. It also implies repetitivity (via Gottschalk's theorem) and hence aperiodicity of $\mathcal{P}$. Finally, the FLC property together with linear repetitivity follow from the local nature of the homeomorphism (as an MLD rule).   □

Let us briefly mention that there are different (and perhaps also simpler) derivation rules to and from the kites and darts version of the Penrose tiling. These break the symmetry in an intermediate step (and thus show that MLD need not mean SMLD in general). Since this is quite nicely sketched in [GS87, pp. 539–540], we refer the reader to this source for details.

6.2.1. GUMMELT'S DECAGON COVERING

So far, we considered proper tilings of the plane, without gaps or overlaps. The minimum number of prototiles (up to similarity) for the Penrose family is two. Motivated by the importance of fundamental domains in crystallography, various attempts have been made to construct Penrose-type tilings from a single motif. The most promising approach is based on *coverings*, which are still free of gaps but may contain overlaps. Recall from Chapter 5 that coverings are patterns as well, so that the general notions apply.

For the Penrose family, following a suggestion by Bandt, Gummelt showed that one can construct a covering with a single, decorated decagon cluster [Gum96, Gum99]. It is depicted in Figure 6.7, together with its dissection into an arrow-decorated patch of the PRT. The shaded areas restrict the possible overlaps of neighbouring decagons. The three possible configurations of pairs of overlapping decagons (with two different overlap regions) are shown in Figure 6.8. Within the overlap region, the shaded markings have to agree. Any gapless covering of the entire plane by the decagonal protocluster (together with its rotations under multiples of $\pi/5$), where each pair of overlapping clusters is in accordance with the configurations of Figure 6.8, is called a *Gummelt covering*. A predecessor of a similar type, though with more than one cluster, is described in [Sas86].

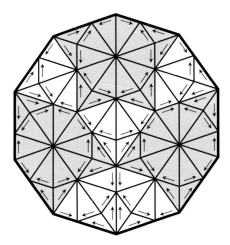

FIGURE 6.7. Gummelt's aperiodic decagon, with vertical reflection symmetry, and the corresponding patch of the PRT in a suitable scale.

A patch of the Gummelt covering is shown in Figure 6.45 on page 239 of this book. Note that more than two decagons may have a joint overlap region, but each pair has to obey the rule. This has striking similarity with the local (or matching) rule approach to the Penrose tiling, which is no accident. Indeed, the dissection of Figure 6.7 can be used to establish a mutual local derivation rule between the family of Gummelt coverings and the LI class of the PRT. As a result, the Gummelt coverings of the plane form a single LI class with perfect aperiodic local rules (as defined by an appropriate finite atlas of local configurations). For the original proof in terms of Penrose rhombuses, we refer to [Gum96].

Corollary 6.3. *The ensemble of all decagon coverings subject to the pairwise overlap rule of Figure 6.8, in a fixed orientation, forms a single LI class that belongs to the MLD class of the rhombic Penrose tiling.* □

Motivated by the Gummelt covering, a systematic search for other examples of this type was initiated. A reformulation of Gummelt's result in

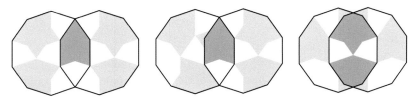

FIGURE 6.8. The three types of permissible pairwise overlaps for the decorated decagon of Gummelt's covering rule.

terms of kites and darts was given in [LRK00], together with some extension to icosahedral coverings. An example with eightfold symmetry was constructed in [B-AG99]. Up to now, no other example with a *single* motif was found, though several examples with two (or more) protoclusters have been constructed by means of the projection method; see [KP03, Ch. 5] for a systematic approach and a general survey.

An alternative strategy was suggested in [JS94], where the maximisation of the density of one specific cluster led to the Penrose rhombus tilings, or to tilings that deviate from them at most in structures of zero density. Rules of this type were later nicknamed *maxing rules* in [Hen98, GGB-A03]. Covering rules of either type, as well as further variants (such as maximal coverings, to capture aspects of both ideas) have since become quite fashionable in materials science; see [RG03, Ste04] and references therein.

6.2.2. THE TÜBINGEN TRIANGLE TILING

Let us now consider the TTT inflation rule ϱ_{TTT}, displayed on the right panel of Figure 6.2 on page 180, which is a primitive stone inflation (with markings). A fixed point under ϱ^2_{TTT} can be obtained by starting from a legal decagon consisting of ten acute triangles. Such a decagon, which is *not* fivefold symmetric, is shaded in Figure 6.9, which also shows a larger patch of the (marked) tiling. The legality can be verified, for instance, by extracting the decagon from the fifth inflation step of a single acute triangle.

The PF analysis of Remark 6.3 also applies to this tiling. To fix the scale, let us assume that the longer edge of the prototiles has length 1 (and the shorter thus length $\tau^{-1} = \tau - 1$). The areas for the acute and the obtuse triangles then follow as

$$A_1 = \frac{\sqrt{\tau + 2}}{4\tau} \quad \text{and} \quad A_2 = \frac{A_1}{\tau}.$$

This implies that the average area per tile in a TTT is

(6.1) $\overline{A} = A_1 \nu_1 + A_2 \nu_2 = \frac{1}{4}(7 - 4\tau)\sqrt{\tau + 2} \approx 0.251,$

with the frequencies $\nu_1 = \tau^{-1}$ and $\nu_2 = \tau^{-2}$ from Remark 6.3.

Lemma 6.1. *When the scale of the TTT is fixed so that the longer edge of the prototiles has length 1, the vertex point set is a Delone set of density* $\frac{2}{5}\tau^2\sqrt{\tau + 2} \approx 1.992.$

PROOF. The Delone property is clear from the explicit construction as an inflation fixed point. Since the density of the vertex point set exists uniformly, we can derive it from the average prototile area $\overline{A}$ by observing that each triangle has angle sum π and thus effectively 'carries' half a point, so that $1/(2\overline{A})$ gives the result via Eq. (6.1).   □

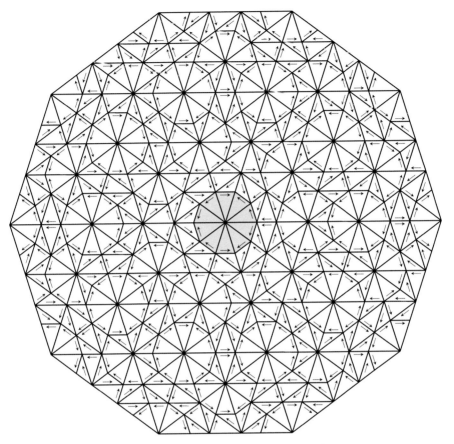

FIGURE 6.9. A patch of the TTT, with horizontal reflection symmetry, via four inflation steps from the legal decagon (shaded) at its centre.

It is clear from the markings in Figure 6.9 that this tiling fixed point does not have individual fivefold rotation symmetry (though it is reflection symmetric in the central horizontal axis). Since a rotation of the entire tiling through $\pi/5$ is another fixed point under ϱ_{TTT} that is in the same LI class, the TTT hull has D_{10}-symmetry in the sense of Definition 5.14. Moreover, it has an LIDS with inflation multiplier τ. One direction is clear from the inflation construction, while the converse direction follows from the observation that the level-1 supertiles can (easily) be recognised locally; compare [KSB93, Fig. 3.4]. In our present terminology,[2] this defines a *deflation* of the tiling. It is interesting to note that the positions of the decagonally shaped 10-stars

---

[2]Note that the terms 'inflation' and 'deflation' are sometimes used with interchanged meaning. This also applies to the treatment in [KSB93].

correspond to the vertices of the triply inflated tiling, which provides an easy way to recognise the level-3 supertiles. Although this is fairly obvious from the inflation structure, a formal argument can most easily be derived within the projection formalism; see [BKSZ90, Figs. 4.2 and 4.3]. We shall explain the corresponding situation for the Ammann–Beenker tiling in Example 7.9.

Remark 6.4 (*Relation between TTT and Penrose tilings*). The hulls of the rhombic Penrose tilings and the TTT are both D_{10}-symmetric. Neither hull contains tilings with individual tenfold symmetry. There are individual D_5-symmetric Penrose tilings, but no such tiling exists in the TTT hull. Consequently, the two hulls cannot be SMLD. What is more, although one can construct a local rule to turn each TTT element into a rhombic Penrose tiling, one can show that the converse is not possible [BSJ91]. The proof employs the model set description of the vertex point sets; see Chapter 7 below. Consequently, the TTT and the Penrose tilings define distinct MLD classes, where the latter is derivable from the former but not vice versa.   ◇

In contrast to the previous examples, the TTT inflation does not admit a fixed point with individual non-crystallographic symmetry. In fact, there is no (arrow decorated) vertex configuration with fivefold symmetry, see [KSB93, Fig. 3.2], which means that there cannot be any tiling with individual fivefold symmetry in the entire LI class, as explained in Remark 6.4.

To infer non-periodicity, we may use the fact that the aperiodic Penrose tiling is locally derivable from the TTT. Alternatively, we may employ the following independent argument, which is based on Definition 5.15.

Theorem 6.2. *Let $\mathcal{T} \subset \mathbb{R}^d$ be an FLC pattern that satisfies $\lambda \mathcal{T} \overset{LD}{\leadsto} \mathcal{T}$ for some $\lambda > 1$. If λ is irrational, the pattern $\mathcal{T}$ is non-periodic.*

PROOF. Assume to the contrary that $t + \mathcal{T} = \mathcal{T}$ for some $0 \neq t \in \mathbb{R}^d$. Then, $\lambda t + \lambda \mathcal{T} = \lambda \mathcal{T}$, and an application of the local derivation rule implies that λt must also be a period of $\mathcal{T}$ itself. If λ is irrational, the group generated by t and λt is dense in the direction $\mathbb{R}t$. Since $\mathcal{T}$ is FLC, this is a contradiction.   □

The TTT obviously satisfies the conditions of Theorem 6.2, wherefore we have the following consequence.

Corollary 6.4. *The TTT inflation rule is a primitive stone inflation that defines a pattern which is a face to face tiling with markings. It is linearly repetitive and aperiodic, and has an LIDS with inflation multiplier τ.*   □

The LI class of the TTT can also be characterised by a set of local rules.

Remark 6.5 (*Local rules for TTT*). Let us mention that the unmarked tiling, which is trivially locally derivable from the TTT, carries the full local

information. In particular, one can restore the markings by a local derivation
rule, as follows from the proof in [KSB93, Sec. 3]. Moreover, one can locally
derive edge and vertex decorations that establish perfect aperiodic local rules;
compare [KSB93, Fig. 3.6]. ◇

Later, in Example 7.10, we will revisit the vertex point set of this tiling
in the context of the projection method.

6.3. Square triangle and shield tilings

From the beginning of quasicrystal theory in materials science, planar
tilings with twelvefold symmetry have been important due to the early discovery of dodecagonal quasicrystals [INF85]. Here, we describe two prominent examples in some detail, one solely based on squares and equilateral
triangles, and the other one with an additional shield-shaped prototile.

6.3.1. SQUARE TRIANGLE TILING

Tilings with squares and equilateral triangles are natural from a crystallographic perspective, and occur in various crystalline structures, as well as in
dodecagonal quasicrystals [CKH98]. Indeed, one can use these prototiles to
construct tilings with twelvefold symmetry in the LI sense. It is an interesting
observation that there is an abundance of possibilities to do so [Gäh88].

However, defining LI classes of such tilings explicitly is considerably
more difficult than in our previous examples. One construction (due to
Schlottmann, though only recorded in the appendix of [HRB97]) is based
on the *pseudo inflation rule* of Figure 6.10. We do not know any stone inflation rule that would produce a locally equivalent tiling, though a non-local
cover with a stone inflation was constructed in [Fre11]. The latter is based
on prototiles obtained as dissections of the squares and triangles, equipped
with suitable (additional) decorations.

As mentioned in Section 5.6, the attribute 'pseudo' refers to the fact that
the tiles on the smaller scale protrude from the inflated tile shape in a consistent way, in the sense that any overlapping tiles in the iterated inflation
procedure are always of the same type and match (which is the case here).
One consequence is that one needs a modified definition of the attached inflation matrix. It is clear that the latter has to be compatible with the irrational
inflation multiplier

$$\lambda_{ST} = 2 + \sqrt{3} \approx 3.732,$$

which is a PV unit with defining polynomial $x^2 - 4x + 1$. A larger patch
of the tiling (with perfect individual D_6 symmetry) is shown in Figure 6.46
on page 240 of this book, in a colour-coded way that distinguishes the five

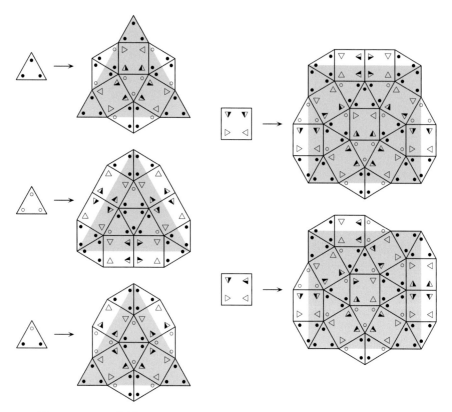

FIGURE 6.10. Schlottmann's pseudo inflation rule for a square triangle tiling, formulated via five prototiles with markings (up to similarity).

prototiles. Strictly speaking, we are again dealing with a decorated tiling (due to the markers), which is an example of a pattern.

The induced inflation matrix has to keep track of the number of the various prototiles in one inflation step. For this matrix, we count the contributions according to the proportion of the area that lies within the rescaled tile (shaded in Figure 6.10). Starting with the three triangles followed by the two squares, one obtains the integer matrix

$$(6.2) \qquad M_{\mathrm{ST}} = \begin{pmatrix} 3 & 4 & 3 & 7 & 7 \\ 1 & 0 & 1 & 2 & 2 \\ 3 & 3 & 3 & 7 & 7 \\ 3 & 3 & 1 & 5 & 4 \\ 0 & 0 & 2 & 2 & 3 \end{pmatrix}.$$

In this approach, the two mirror images of the second square prototile are not distinguished. M_{ST} is a non-negative integer matrix which is primitive,

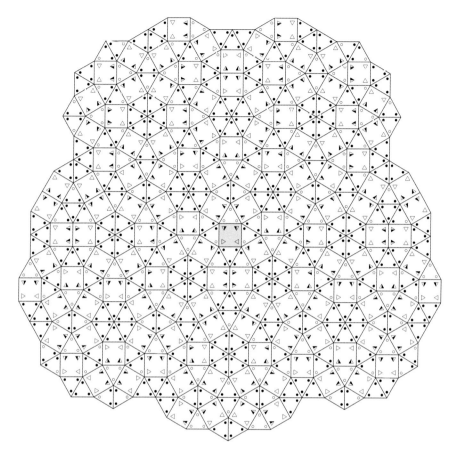

FIGURE 6.11. A patch of the (marked) square triangle tiling, obtained by two inflation steps from the reflection symmetric square prototile (shaded).

so that Perron–Frobenius theory applies, via Theorems 2.2 and 2.3. The PF eigenvalue is $\lambda_{\mathrm{ST}}^2 = (2+\sqrt{3})^2 = 7+4\sqrt{3}$. The corresponding left eigenvector can be chosen as $\frac{1}{4}(\sqrt{3}, \sqrt{3}, \sqrt{3}, 4, 4)$, which reflects the areas of the prototiles when the edge length is chosen to be 1. The statistically normalised right eigenvector, which encodes the prototile frequencies in an infinite tiling, is

$$\nu_{\mathrm{ST}} = \frac{1}{26}\big(34-15\sqrt{3}, 4-\sqrt{3}, 8\sqrt{3}-6, 32\sqrt{3}-50, 44-24\sqrt{3}\big)^T$$
$$\approx (0.3084, 0.0872, 0.3022, 0.2097, 0.0935)^T.$$

Let us now summarise some of the properties of this tiling.

Proposition 6.4. *Consider an infinite square triangle tiling that is a fixed point under some power of the pseudo inflation rule of Figure 6.10. Then, the proportion of the area covered by triangles equals that covered by squares.*

The same conclusion holds for any square triangle tiling that is locally indistinguishable from such a fixed point tiling.

PROOF. In principle, the first claim follows as before from an elementary calculation with the frequency vector v_{ST}. However, we prefer to give an alternative argument here.

Observe that the triangle and square counts for the inflated prototiles are consistent with defining a summatory substitution matrix

$$(6.3) \qquad M = \begin{pmatrix} 7 & 16 \\ 3 & 7 \end{pmatrix},$$

which only keeps track of triangles and squares, irrespective of their types. The matrix M is again primitive, with the same PF eigenvalue λ_{ST}^2 as above. The left PF eigenvector now reads $v = \frac{1}{4}(\sqrt{3}, 4)$, in line with the prototile areas, while the statistically normalised right eigenvector becomes

$$(6.4) \qquad \nu = (\nu_1, \nu_2)^T = \frac{1}{13}\left(16 - 4\sqrt{3}, 4\sqrt{3} - 3\right)^T.$$

The area proportion covered by squares is thus given by

$$\frac{\nu_2 v_2}{\nu v} = \frac{1}{2} = \frac{\nu_1 v_1}{\nu v},$$

which proves the first claim. The second claim follows as in the proof of Proposition 6.1. $\qquad \square$

Remark 6.6 (*Non-existence of integer roots of M*). The summatory substitution matrix M from Eq. (6.3) is the unique non-negative integer matrix with left eigenvector $(\sqrt{3}, 4)$ for the eigenvalue λ_{ST}^2. A simple calculation shows that M has precisely four square roots as a 2×2 matrix, only two of which are non-negative, namely

$$\begin{pmatrix} 2 & 4 \\ \frac{3}{4} & 2 \end{pmatrix} \quad \text{and} \quad \begin{pmatrix} \sqrt{3} & \frac{8}{3}\sqrt{3} \\ \frac{1}{2}\sqrt{3} & \sqrt{3} \end{pmatrix}.$$

Neither of these matrices seems to correspond to an inflation (or pseudo inflation) rule for square triangle tilings, although an inflation multiplier of the form $\sqrt{2 + \sqrt{3}}\, e^{\pi i / 12}$ would be possible in principle, because the latter is a unit in $\mathbb{Z}[\xi_{12}]$; compare Example 2.16. $\qquad \lozenge$

Remark 6.7 (*Area ratio for general square triangle tilings*). There are many other square triangle tilings of the plane. Most of them do not emerge from an inflation rule, such as the tiling constructed in [BKS92] from a projection tiling with squares, triangles and rhombuses by an iterative 'depletion' rule; see also [Fre13]. We will return to this example in Remark 7.9.

The statement about the area proportions of Proposition 6.4 extends to other square triangle tilings with twelvefold symmetry, provided they are quasiperiodic in the sense of the projection method to be described in Chapter 7 below.[3] In fact, the square triangle area relation even applies to almost all square triangle tilings with (statistical) twelvefold symmetry within the much larger class of random square triangle tilings (see Section 11.6.2).   ◊

Let us close this paragraph with a short digression on the mean coordination number of twelvefold symmetric square triangle tilings. The relative tile frequencies from Eq. (6.4) imply that the average number of edges per tile is $\overline{n}_e = \frac{4}{13}(9 + \sqrt{3}) \approx 3.302$. Using Remark 2.4, the average coordination number is obtained as

$$\overline{n}_c = \frac{2\,\overline{n}_e}{\overline{n}_e - 2} = 12 - 4\sqrt{3} \approx 5.072,$$

which, unlike in previous cases, is an irrational number. So, twelvefold symmetry of a square triangle tiling enforces a prototile ratio that is irrational and also implies an irrational mean coordination number. This is incompatible with full periodicity, as expected from the crystallographic restriction.

Despite the fact that we are dealing with a pseudo inflation rule for the square triangle tiling $\mathcal{T}$, we still have $\lambda_{\mathrm{ST}}\mathcal{T} \overset{\mathrm{LD}}{\leadsto} \mathcal{T}$, so that Theorem 6.2 applies to the square-triangle inflation tiling as well.

Corollary 6.5. *The marked square triangle tiling defined by the pseudo inflation rule of Figure* 6.10 *is linearly repetitive and aperiodic. It has an LIDS with inflation multiplier* $2+\sqrt{3}$. *The continuous hull defines a strictly ergodic dynamical system for the translation action of* $\mathbb{R}^2$.   □

6.3.2. SHIELD TILING

If one allows a third prototile, several new opportunities emerge to construct tilings with twelvefold symmetry. One important example is due to Gähler [Gäh88] (in the full version described below), where the third prototile is a 'shield'; compare also [NM87, Fig. 2] for the independently discovered unmarked version. The generalised inflation rule is shown in Figure 6.12, where a decoration is included that plays the same role as the full decoration of the Ammann–Beenker tiling. One can convince oneself that this inflation rule is consistent in the sense that it produces gapless and overlap-free patches under iteration. In particular, one can again define sequences of patches that converge to a proper (marked) tiling of the plane.

Remark 6.8 (*Prototile set for the shield inflation*). A careful inspection of Figure 6.12 reveals the phenomenon that the complete prototile set consists of two sets of equal cardinality that emerge from one another via a rotation

---

[3]A proof for this claim was sketched by A. Katz (unpublished), based on an argument with differential forms for squares and rhombuses.

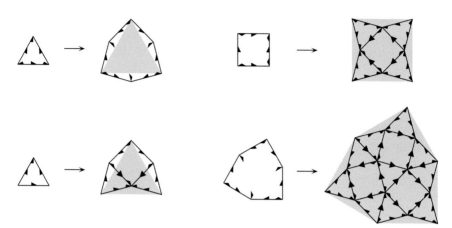

FIGURE 6.12. Inflation rule for Gähler's shield tiling in its fully decorated version. The inflation multiplier is $\sqrt{2+\sqrt{3}}$; see text for details.

by $\pi/12$. Any legal tiling in the inflation process uses one of them only. The reason is that Gähler's shield inflation is not built with a pure homothety, but includes a rotation by $\pi/12$. Consequently, the complete inflation is irreducible, but not primitive. However, the square of the inflation can be reduced to either of the two prototile subsets, and is then primitive. $\Diamond$

Figure 6.13 shows a patch of the decorated tiling, which already contains all possible (legal) vertex configurations (up to rotations). This patch, upon iteration of the square of the inflation rule ϱ, converges to a plane-filling fixed point of ϱ^2. The corresponding hull (under the translation action of $\mathbb{R}^2$) is the LI class of the (marked) shield tiling. By removing the edge and vertex decorations (which clearly is a local operation), one obtains the undecorated shield tiling and its LI class. A larger patch of the undecorated tiling is displayed in Figure 6.48 on page 241. It was obtained by the projection method, to be discussed below in Example 7.12.

Note that it is not possible to reconstruct the complete decorations by a local rule (analogous to the Ammann–Beenker tiling); see [Gäh93] for a proof. More precisely, only part of the vertex decoration can be derived from local information, namely the isosceles triangles of the vertex 'windmill vanes', while this is not possible for the complete information encoded in the remaining vertex and edge markers. In particular, the full decoration of the square and of the two types of triangles requires non-local information. This results from the existence of reflection symmetric elements in the hull of the undecorated tiling, while the complete decoration breaks this symmetry. Consequently, the marked and the unmarked shield tilings define two distinct MLD classes. Only the class with the complete markings possesses perfect

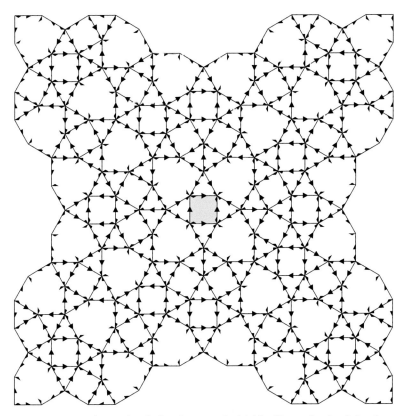

FIGURE 6.13. A patch of the decorated shield tiling, obtained by four inflation steps from the central square (shaded) as seed.

aperiodic local rules [Gäh93]. A locally equivalent (MLD) tiling was independently constructed by Socolar [Soc89, Soc90]; see [GGB-A03, Sec. 3.5] for pictures and a detailed discussion of the relation between the two tilings.

Since Figure 6.12 does not define a stone inflation, the shading indicates the shape of the scaled prototiles. Despite this slightly different appearance, compare the discussion in Section 5.6, we consider the 'inflation' (or incidence) matrix, which reads

$$(6.5) \qquad M = \begin{pmatrix} 0 & 0 & 4 & 7 \\ 0 & 2 & 0 & 3 \\ 0 & 1 & 1 & 3 \\ 1 & 0 & 0 & 0 \end{pmatrix}.$$

It is primitive (with $M^3 \gg 0$), with PF eigenvalue $2 + \sqrt{3}$. The inflation multiplier is then $\lambda = \sqrt{2 + \sqrt{3}}$, which corresponds to the unit $e^{i\pi/12}\lambda$ in $\mathbb{Z}[\xi_{12}]$. The matrix M has a slightly different meaning in comparison with

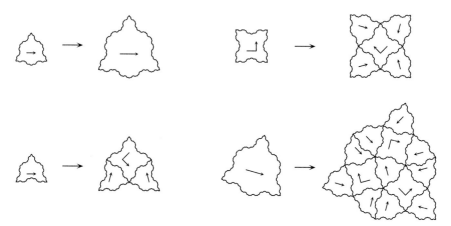

FIGURE 6.14. Reformulation of the original shield tiling inflation rule of Figure 6.12 as a stone inflation for a fractally bounded prototile set.

our previous examples. Its PF eigenvalue is still the square of the inflation multiplier, but the corresponding left eigenvector does not encode the prototile area ratios. Nevertheless, the statistically normalised right eigenvector still gives the relative tile frequencies (the counting is correct due to the consistency of the rule). One obtains the ratios

$$(2 + 2\sqrt{3}) : \sqrt{3} : 1 \qquad \text{for} \quad \text{triangles : squares : shields.}$$

The corresponding ratio for the areas covered by triangles, squares and shields in an infinite tiling is $1 : (\sqrt{3} - 1) : 1$.

Remark 6.9 (*Fractiles for the shield tiling*). The inflation rule of Figure 6.12, together with the outer shape of the patch of Figure 6.13, suggests the alternative use of a different set of prototiles. Indeed, using a 'fractal curve' construction [GK97] of von Koch type

on the edges of the original prototiles, one obtains new prototiles with a fractal boundary of Hausdorff dimension $d_{\mathsf{H}} = \log(2)/\log(2\cos(\frac{\pi}{12})) \approx 1.053$. Such tiles are called *fractiles* [Ban97, BSS+11]. We will explain the underlying reasoning in more detail in a later example. The 'denting' occurs in alternating directions, as indicated by the grey arrowheads. This procedure leads to the stone inflation of Figure 6.14. Indeed, Figure 6.13 gives an impression of how the boundary of the square prototiles evolves under inflation. We have condensed the original edge and vertex decorations into a single arrow inside each tile, breaking all its symmetries. It is thus clear that the two substitution rules of Figures 6.12 and 6.14 define LI classes that are MLD.

FIGURE 6.15. Patch of the fractal version of the shield tiling, as defined by the inflation rule of Figure 6.14. It is obtained by four inflation steps from the shaded fractile in the centre.

The area ratio of the four proto-fractiles is now given by the left PF eigenvector of the inflation matrix M of Eq. (6.5), and reads

$$(2\sqrt{3} - 3) : (6\sqrt{3} - 10) : (18 - 10\sqrt{3}) : \sqrt{3},$$

while their frequencies ratio is $(2 + \sqrt{3}) : \sqrt{3} : \sqrt{3} : 1$. Now, the ratio for the areas covered by the four proto-fractiles is $1 : (6\sqrt{3} - 10) : (18 - 10\sqrt{3}) : 1$. A patch of the fractal version of the shield tiling is shown in Figure 6.15, which also demonstrates the orientations of the fractiles relative to each other. ◊

Applying our previous arguments to the stone inflation version of the fully decorated shield tiling, and using the MLD property, we can formulate the following conclusions.

Corollary 6.6. *The LI class of the fully decorated shield tiling is linearly repetitive and aperiodic, with an LIDS with inflation multiplier $\sqrt{2+\sqrt{3}}$. The LI class possesses perfect aperiodic local rules, and the corresponding translation dynamical system is strictly ergodic. The same properties hold for all FLC patterns in the same MLD class.* □

A close inspection of Figure 6.12 shows that the edge and vertex decorations break all symmetries of the shapes of the prototiles. The same comment applies to the fractalised version of Figure 6.14. Both inflation rules correspond to the unit $z = \sqrt{2+\sqrt{3}}\,\xi_{24}$ of $\mathbb{Z}[\xi_{12}]$ that was discussed in Example 2.16 on page 37. This is reflected in the $\pi/12$ rotation of edge directions in Figures 6.12 and 6.14.

It is also possible to formulate a pseudo inflation rule for the multiplier $2+\sqrt{3}$, starting from the square of the above rule. In this case, a partial symmetry breaking suffices. One version is shown in Figure 6.16, which works with a triangle, a square and a shield (in two chiralities) as prototiles. The symmetry of the square and the shield is broken by the arrow decoration, which is MLD with the original corner decorations according to the rule

(6.6)

One can check that this pseudo inflation rule is consistent, in the sense that the protruding triangles always match. As before, one can now assign the inflation matrix

$$M = \begin{pmatrix} 7 & 8 & 24 \\ 3 & 1 & 6 \\ 0 & 4 & 7 \end{pmatrix}$$

by counting all protruding tile areas proportionally. The PF eigenvalue is then $(2+\sqrt{3})^2$, with left and right eigenvectors now in agreement with the prototile areas and frequencies, the former with ratios $\frac{1}{4}\sqrt{3} : 1 : \frac{1}{2}(3+\sqrt{3})$.

This new rule also defines an aperiodic hull, which consists of a single LI class. It can be derived locally from the fully decorated shield tiling in an obvious way. As mentioned above, the converse is not true [Gäh88, Gäh93]. A proof of this claim employs the description of the shield tiling as a projection structure; see Example 7.12 below. In fact, one can show that the naked (or undecorated) shield tiling and the partially decorated one as defined by the rule of Figure 6.16 are in the same MLD class. In particular, the arrows of Figure 6.16 as well as the corresponding corner decorations according to Eq. (6.6) are locally reconstructible from the undecorated tiling in three steps.

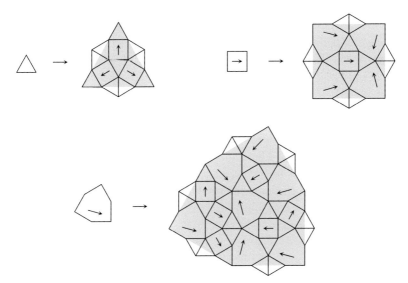

FIGURE 6.16. Pseudo inflation rule for the (undecorated) shield tiling, with irrational inflation multiplier $2 + \sqrt{3}$. The reflected shield is inflated by the reflected rule.

All vertices of the shield tiling have coordination number four or five. The left panel of Figure 6.17 shows how to decorate all five-coordinated vertices (which may correspond to different tile arrangements). This first step produces the corner markings in all three rectangular corners of all shields. In the second step, they are transported to the opposite obtuse corners, according to the middle panel of Figure 6.17. Finally, in all four-coordinated vertices, the missing isosceles triangle is added as in the right panel of Figure 6.17. One can check that this local rule determines all isosceles triangle corner markers. Breaking the remaining reflection symmetry turns out to be the essential step to obtain a class with aperiodic local rules [Gäh93].

Corollary 6.7. *The undecorated and the fully decorated shield tilings represent two distinct MLD classes, both with an LIDS. The undecorated shield tiling hull is aperiodic, but does* not *possess aperiodic local rules.* □

FIGURE 6.17. Reconstruction of the isosceles corner triangle decorations of the shield tiling from its undecorated version; see text for details.

So far, all examples showed an irrational inflation multiplier, which immediately implied aperiodicity by Theorem 6.2. Let us now turn our attention to tilings that are aperiodic for a different reason.

6.4. Planar tilings with integer inflation multiplier

An interesting and widely used planar tiling with a single prototile (up to similarity) and an integral inflation multiplier is the *chair tiling* [GS87, Sol97, BMS98, Rob99]. The L-shaped prototile is scaled by a factor of 2 and then dissected into four congruent copies, as shown in Figure 6.18. The finite patch shown in the same figure converges (under iterated inflation) to a plane-filling tiling with global (individual) D_4 symmetry. A larger patch (with D_4 colour symmetry) is shown in Figure 6.55 on page 246.

This inflation rule is a geometric reformulation of the block substitution that was introduced in Eq. (4.27) on page 121. The identification is usually done [BMS98, Rob99] by means of four oriented squares as follows

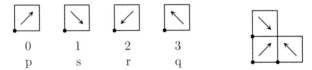

$$
\begin{array}{cccc}
0 & 1 & 2 & 3 \\
\text{p} & \text{s} & \text{r} & \text{q}
\end{array}
$$

where the digits refer to our previous labels in Section 4.9 and the letters are Robinson's labels [Rob99]. The oriented squares form chair shapes as shown above, and are distinguished by four colours (shades of grey) in Figure 4.1 on page 122. The coordinatisation given in Eq. (4.30) refers to the control points, located in the lower left corner of each square as indicated above. The solution has a limit-periodic (or Toeplitz) structure, which is a generalisation of the structure we derived for the period doubling substitution in Section 4.5.1. This structure is incompatible with any non-trivial period, so that the chair fixed point is non-periodic. Since the inflation is primitive, face to face and FLC, this property is inherited by every member of the hull, which consists of a single LI class; compare Proposition 5.4.

Corollary 6.8. *The chair inflation rule of Figure 6.18 is aperiodic. The corresponding hull consists of linearly repetitive FLC tilings and defines a strictly ergodic dynamical system under the translation action of $\mathbb{R}^2$.* $\qquad\square$

For the chair tiling, we extracted its aperiodicity constructively from the explicit coordinatisation discussed in Section 4.9. In general, this is not an efficient strategy. Theorem 6.2 is of no immediate help, since the inflation multiplier is an integer. However, when the tiling has an LIDS, one has the following complementary criterion; compare Example 5.7. We formulate it for homotheties only, as this is sufficient, possibly after taking suitable powers.

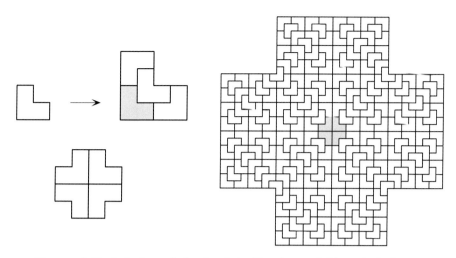

FIGURE 6.18. Inflation rule for the chair tiling (upper left), with the lower left corner being the implicit inflation reference point. Also shown is a D_4-symmetric patch of the tiling (right), obtained by three inflation steps from the legal configuration in the lower left panel (shaded in the patch).

Theorem 6.3. *Let* $\mathcal{T} \subset \mathbb{R}^d$ *be an FLC pattern that satisfies* $\lambda\mathcal{T} \overset{MLD}{\leftrightsquigarrow} \mathcal{T}$ *for some* $\lambda > 1$, *which means that* $\mathcal{T}$ *possesses an LIDS. Then, the pattern* $\mathcal{T}$ *is non-periodic.*

PROOF. In comparison with Theorem 6.2, we have the additional assumption $\mathcal{T} \overset{LD}{\rightsquigarrow} \lambda\mathcal{T}$, which means that $\mathcal{T}$ has a local deflation rule (or is locally recognisable). By similarity, this also implies $\lambda^{-1}\mathcal{T} \overset{LD}{\rightsquigarrow} \mathcal{T}$.

Assume now, contrary to the claim, that $t + \mathcal{T} = \mathcal{T}$ holds for some $0 \neq t \in \mathbb{R}^d$. Since $\mathcal{T}$ is FLC, we may without loss of generality assume that t is a fundamental period in this direction. Dividing by λ leads to $\lambda^{-1}t + \lambda^{-1}\mathcal{T} = \lambda^{-1}\mathcal{T}$. The local derivation $\lambda^{-1}\mathcal{T} \overset{LD}{\rightsquigarrow} \mathcal{T}$ then implies that $\mathcal{T}$ must also possess the period $\lambda^{-1}t$. This contradicts the assumption that t is a fundamental period.   □

This theorem applies to many inflation tilings, some of which will now be described, with special emphasis on hexagon-based examples.

Example 6.2 (*Table tiling*). A close relative of the chair tiling is the table (or domino) tiling [Sol97, Rob99, Fre02] defined by the inflation rule

which is based on a prototile with edge length ratio 2 : 1. The vertical rectangle is dissected accordingly. A fixed point can be obtained, for instance, by starting from two rectangles joined along a long edge to form a square. This tiling is related to the block substitution of Example 4.14 on page 125; see [Rob99] for the original exposition. The horizontal (vertical) rectangle carries the digits 0 and 2 (1 and 3) from left to right (top to bottom). A larger patch (in the form of a level-6 supertile) of the undecorated table tiling is shown in Figure 6.47 on page 240.

By construction, the table tiling is composed of two types of level-1 supertiles, the rectangular patch shown above and its rotated version. It is easy to check (by an inspection of all collared versions of these patches, which are finitely many) that these supertiles are locally recognisable. This means that we have a local deflation rule, and hence an LIDS. Theorem 6.3 then implies non-periodicity. The other standard properties follow, once again, from the fact that the table tiling (which is FLC) is defined by a primitive inflation rule, which can be considered as a face to face stone inflation after the introduction of a pseudo vertex at the centre of each long edge. Consequently, we obtain linear repetitivity and hence also aperiodicity. ◊

Example 6.3 (*Sphinx tiling*). A commonly studied relative of the chair tiling is the *sphinx tiling* [GS87, Fig. 10.1.6]; see also [Gar77, God89] and [Sol97, Ex. 7.2]. Its prototile is constructed from six equilateral triangles in a somewhat similar fashion as the chair is built from squares. Again, it is a tiling with inflation multiplier 2, but the sphinx prototile has no reflection symmetry, and occurs with two chiralities in the tiling. The inflation rule reads

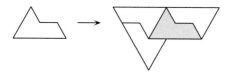

and is again an aperiodic stone inflation. For the version shown here, the inflation step (prior to dissection) consists of multiplication by 2 followed by reflection in a suitable (horizontal) axis, so it is a stone inflation for the expansive matrix $L = \mathrm{diag}(2, -2)$. The advantage of this choice of L is the existence of a fixed point under the direct inflation (rather than its square), as indicated by the shading. A patch of the sphinx tiling is shown in Figure 6.49 on page 241. This is another example where aperiodicity follows either from the underlying limit-periodic structure [God89] or from an application of Theorem 6.3.

Despite the aperiodicity of the inflation rule, the undecorated tiling cannot possess aperiodic local rules. Indeed, the parallelogram formed by two sphinx tiles clearly tiles the plane crystallographically. However, one can follow the general strategy of [Goo98] to add non-local information in order to turn it into a (more complicated) primitive inflation rule with perfect aperiodic local rules; see [Goo03] for an explicit description.   ◊

Example 6.4 (*Half-hex tiling and a decorated variant*). An interesting inflation rule appears in [GS87, Exc. 10.1.3 and Fig. 10.1.7], which was later studied in detail in [Fre02, Sec. 4.1]. The inflation rule (with multiplier 2) is

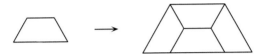

and produces an aperiodic tiling, once again as a consequence of Theorem 6.3, for instance from a hexagonal seed (there are no aperiodic local rules for this tiling, as will be discussed later). The half-hex rule can be augmented by some decoration as

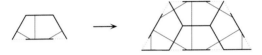

where the base line has been modified to make the underlying hierarchical structure more apparent; see Figure 6.19 for an illustration. Here, the thin lines form a hierarchical arrangement of regular triangles of ever increasing size, which illustrates the limit-periodic structure. This is reminiscent of the analogous structure with squares in the Robinson tiling of Example 5.11. ◊

The orientation of the isolated line segments inside the hexagons is locally derivable from the thin lines. The remaining 'stripped-off' pattern is built from the marked hexagon

which occurs in six different orientations. They are fitted together face to face such that no line terminates or jumps at the boundary of the hexagons. This defines a local packing rule for the decorated hexagons. However, it fails to be aperiodic, because it still permits a periodic tiling of the plane, for instance by periodic repetition of the shaded motif in Figure 6.20.

Remark 6.10 (*Coordinatisation of the undecorated half-hex tiling*). In analogy to the coordinatisation of the block substitution of Section 4.9, one can

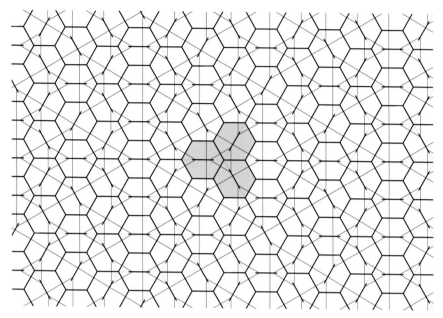

FIGURE 6.19. A patch of the decorated half-hex tiling of Example 6.4 (obtained by inflation from the shaded seed), with indication of the limit-periodic structure via triangles of ever increasing scale (thin lines).

also find a representation of the half-hex tiling via a three-colouring of a lattice, this time a triangular one. Starting from a half-hex with short edges of length 1, in the orientation shown in Example 6.4, one can represent the three possible orientations of the completed hexagons by points of colour (or type) 0 (horizontal), 1 (rotated by $\pi/3$) and 2 (rotated by $-\pi/3$) in the hexagon

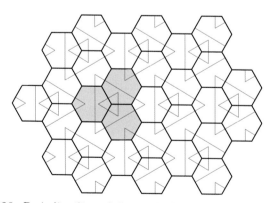

FIGURE 6.20. Periodic tiling of the marked hexagon with minimum possible area for its fundamental domain (shaded).

centres. Let H_i be the set of all points of type i in a fixed point of the inflation (which can start from any of the three types of hexagons in the centre).
Then,

$$H_0 \,\dot{\cup}\, H_1 \,\dot{\cup}\, H_2 \;=\; \Gamma \;=\; \sqrt{3} \left\langle \xi, \xi^3 \right\rangle_{\mathbb{Z}}$$

with $\xi = \mathrm{e}^{\pi \mathrm{i}/6}$, where Γ is a triangular lattice of density $\frac{2}{9}\sqrt{3}$.

In this setting, the fixed point equations for the point sets simplify to

$$H_i \;=\; 2H_i \cup \left(2\Gamma + \sqrt{3}\,\xi^{3+2i} \right)$$

for $0 \leqslant i \leqslant 2$, where the previous identity was used together with the observation that $\sqrt{3}\,\xi^{3+2i} \in \Gamma$. Iteration leads to the solution

$$(6.7) \qquad\qquad H_i \;=\; A_i \,\dot{\cup}\, \bigcup_{n \geqslant 0} 2^n \left(2\Gamma + \sqrt{3}\,\xi^{3+2i} \right),$$

where the A_i are empty sets except for one, which is the singleton set $\{0\}$.
Where the latter occurs depends on the seed of the fixed point, so there are
precisely three possibilities. Note that 0 is the unique limit point of any of
the three unions in the 2-adic topology (the correct space for the analysis of
the set-valued iteration is the 2-adic completion Γ_2 of Γ). $\qquad\qquad \Diamond$

The marked hexagon can occur in six orientations, which can be distinguished by colours. As in Remark 6.10, we choose the centres as reference
points, so that a fixed point tiling with a central hexagon leads to a partition
$\Gamma = \bigcup_{i=0}^{5} \Lambda_i$, where $H_i = \Lambda_i \,\dot{\cup}\, \Lambda_{i+3}$. To derive a reasonable set of fixed point
equations, one better avoids the multiplicities caused by a hexagon pseudo
inflation rule. Since we have fixed point tilings that cover the entire plane, we
can extract proper fixed point equations from the condition in one sector (of
angle $\pi/3$). This can be done by selecting four (of the seven) hexagons, where
we use the version of Figure 6.21. One then finds a neat set of equations,

$$
\begin{aligned}
\Lambda_0 &= \left(2\Lambda_0 + \sqrt{3}\,\{0, \xi^3\} \right) \cup \left(2(\Lambda_2 \cup \Lambda_5) + \sqrt{3}\,\xi^3 \right) \\
\Lambda_1 &= \left(2\Lambda_1 + \sqrt{3}\,\{0, \xi^5\} \right) \cup \left(2(\Lambda_0 \cup \Lambda_3) + \sqrt{3}\,\xi^5 \right) \\
\Lambda_2 &= \left(2\Lambda_2 + \sqrt{3}\,\{0, \xi^7\} \right) \cup \left(2(\Lambda_1 \cup \Lambda_4) + \sqrt{3}\,\xi^7 \right) \\
\Lambda_3 &= \left(2\Lambda_3 + \sqrt{3}\,\{0, \xi^3\} \right) \cup \left(2(\Lambda_1 \cup \Lambda_4) + \sqrt{3}\,\xi^3 \right) \\
\Lambda_4 &= \left(2\Lambda_4 + \sqrt{3}\,\{0, \xi^5\} \right) \cup \left(2(\Lambda_2 \cup \Lambda_5) + \sqrt{3}\,\xi^5 \right) \\
\Lambda_5 &= \left(2\Lambda_5 + \sqrt{3}\,\{0, \xi^7\} \right) \cup \left(2(\Lambda_0 \cup \Lambda_3) + \sqrt{3}\,\xi^7 \right)
\end{aligned}
$$

(6.8)

with $\xi = \mathrm{e}^{\pi \mathrm{i}/6}$ as before. We may now employ the solution of the undecorated
half-hex from Remark 6.10 to see that also these equations are essentially
decoupled. With $H_i' = H_i \setminus A_i$, with the A_i from Eq. (6.7), one obtains via

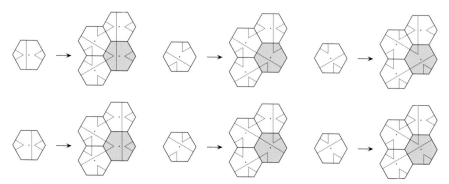

FIGURE 6.21. Consistent core of the pseudo inflation rule for the marked hexagon tiling; see text for details.

iteration

$$\Lambda_0 \;=\; B_0 \mathbin{\dot{\cup}} \bigcup_{n \geqslant 1} \left(2^n H_2' + \sqrt{3}\, \{ \ell \xi^3 \mid 2^{n-1} \leqslant \ell < 2^n \} \right)$$

$$(6.9) \qquad \Lambda_1 \;=\; B_1 \mathbin{\dot{\cup}} \bigcup_{n \geqslant 1} \left(2^n H_0' + \sqrt{3}\, \{ \ell \xi^5 \mid 2^{n-1} \leqslant \ell < 2^n \} \right)$$

$$\Lambda_2 \;=\; B_2 \mathbin{\dot{\cup}} \bigcup_{n \geqslant 1} \left(2^n H_1' + \sqrt{3}\, \{ \ell \xi^7 \mid 2^{n-1} \leqslant \ell < 2^n \} \right)$$

and analogously for the three remaining Λ_i. Here, the sets B_i have to be composed such that the missing points, which are all points along the lines $\mathbb{R}\xi^{3+2i}$ with $i \in \{0,1,2\}$, are added in a way that is consistent with the inflation rule. A close inspection shows that the six possible choices for the central hexagon determine the sets B_i uniquely; see Table 6.1 for the solution. Note that the six choices correspond to the three possibilities to assign the point 0 to one of the sets $H_i = \Lambda_i \mathbin{\dot{\cup}} \Lambda_{i+3}$ from Remark 6.10, refined by the assignment of the point 0 to either Λ_i or Λ_{i+3}.

TABLE 6.1. Possible choices of the sets B_i in Eq. (6.9).

Origin	B_0	B_1	B_2	B_3	B_4	B_5
$0 \in \Lambda_0$	$\sqrt{3}\,\xi^3 \mathbb{Z}$	$\sqrt{3}\,\xi^5 \mathbb{N}$	$-\sqrt{3}\,\xi^7 \mathbb{N}$	$\varnothing$	$-\sqrt{3}\,\xi^5 \mathbb{N}$	$\sqrt{3}\,\xi^7 \mathbb{N}$
$0 \in \Lambda_1$	$-\sqrt{3}\,\xi^3 \mathbb{N}$	$\sqrt{3}\,\xi^5 \mathbb{Z}$	$\sqrt{3}\,\xi^7 \mathbb{N}$	$\sqrt{3}\,\xi^3 \mathbb{N}$	$\varnothing$	$-\sqrt{3}\,\xi^7 \mathbb{N}$
$0 \in \Lambda_2$	$\sqrt{3}\,\xi^3 \mathbb{N}$	$-\sqrt{3}\,\xi^5 \mathbb{N}$	$\sqrt{3}\,\xi^7 \mathbb{Z}$	$-\sqrt{3}\,\xi^3 \mathbb{N}$	$\sqrt{3}\,\xi^5 \mathbb{N}$	$\varnothing$
$0 \in \Lambda_3$	$\varnothing$	$\sqrt{3}\,\xi^5 \mathbb{N}$	$-\sqrt{3}\,\xi^7 \mathbb{N}$	$\sqrt{3}\,\xi^3 \mathbb{Z}$	$-\sqrt{3}\,\xi^5 \mathbb{N}$	$\sqrt{3}\,\xi^7 \mathbb{N}$
$0 \in \Lambda_4$	$-\sqrt{3}\,\xi^3 \mathbb{N}$	$\varnothing$	$\sqrt{3}\,\xi^7 \mathbb{N}$	$\sqrt{3}\,\xi^3 \mathbb{N}$	$\sqrt{3}\,\xi^5 \mathbb{Z}$	$-\sqrt{3}\,\xi^7 \mathbb{N}$
$0 \in \Lambda_5$	$\sqrt{3}\,\xi^3 \mathbb{N}$	$-\sqrt{3}\,\xi^5 \mathbb{N}$	$\varnothing$	$-\sqrt{3}\,\xi^3 \mathbb{N}$	$\sqrt{3}\,\xi^5 \mathbb{N}$	$\sqrt{3}\,\xi^7 \mathbb{Z}$

The natural next step consists in breaking the reflection symmetry of the decorated hexagon. There are different possibilities to do so.

Example 6.5 (*Inflation rule for Penrose's* $(1 + \varepsilon + \varepsilon^2)$-*tiling*). In [Pen97], a pseudo inflation rule is presented for the $(1 + \varepsilon + \varepsilon^2)$-tiling of Example 5.13. A reformulation as a proper inflation (with multiplier 2) leads to the rule

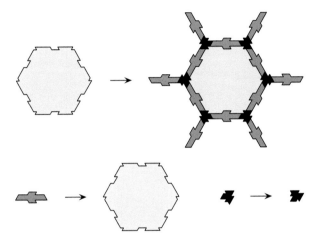

This rule is non-primitive, but still defines a unique tiling LI class in the plane via a fixed point tiling with a hexagon at its centre. By the composition-decomposition method, it is shown in [Pen97] that this LI class is precisely the one specified by the perfect local rules of Example 5.13.

The substitution matrix reads

$$M = \begin{pmatrix} 1 & 1 & 0 \\ 12 & 0 & 0 \\ 6 & 0 & 1 \end{pmatrix},$$

which has spectrum $\sigma(M) = \{4, -3, 1\}$, with right PF eigenvector $\frac{1}{6}(1, 3, 2)^T$. The entries code the relative frequencies of cells, edges and vertices in the (crystallographic) honeycomb packing. This holds because the iterated inflation rule, when starting from a hexagon as a seed, produces a tiling of the plane without any multiple cover of tiles. Figure 6.22 illustrates the parity structure of the hexagons in the tiling, which is MLD with the full tiling [BGG12b]. The non-trivial direction of this claim can be checked by constructing a sliding block map (based on the underlying triangular lattice of hexagon positions) from the possible hexagonal coronae (of sufficient size) to the unique central hexagon with its perimeter tiles.[4]   ◇

---

[4]This is indeed possible, as was checked by Franz Gähler by a computer search. It turns out that hexagonal coronae of order three are sufficient.

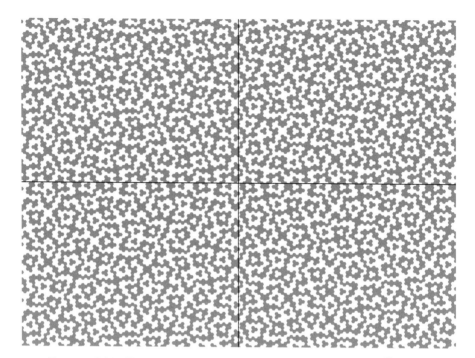

FIGURE 6.22. Hexagon parity pattern of Penrose's $(1 + \varepsilon + \varepsilon^2)$-tiling; compare [Pen97, Fig. 25]. This particular pattern (and its infinite extension) possesses two almost reflection colour symmetries with respect to the black lines. The word 'almost' refers to the property that the pattern is mapped onto itself under reflection and exchange of colours except for those hexagons that are dissected by the reflection lines.

Example 6.6 (*Taylor and Socolar–Taylor tilings*). In principle, it is possible to define a (presumably complicated) cover of the half-hex hull with aperiodic local rules (by adding suitable non-local information), via an application of the ideas of [Moz89, Goo98] mentioned earlier. Beyond the Penrose tiling just explained, there exists another 'shortcut' to this procedure, which was discovered by Joan Taylor [Tay10] through direct insight. Her tiling is based on a primitive inflation rule with 14 distinct copies of the half-hex (and their mirror images). One version of this rule is shown in Figure 6.51 on page 243. Rotated or reflected tiles are inflated by the rotated or reflected rule. In other words, the inflation rule commutes with Euclidean motions.

It is clear that this rule defines a primitive face to face stone inflation. A little less obvious is the underlying limit-periodic structure, which is essentially inherited from the undecorated half-hex inflation. Consequently, the inflation rule of Figure 6.51 is aperiodic. A larger patch of the corresponding tiling (obtained via a fixed point of the inflation) is shown in Figure 6.52.

Observing that the inflation rule commutes with taking mirror images, one can consider the statistics of the tiling via an inflation matrix of dimension 14. Using prototile labels A - G in the order of Figure 6.51, augmented with alternating subscripts u (up) or d (down) to distinguish the two half-hex parts of a hexagon, one obtains the (primitive) matrix

$$M = \begin{pmatrix}
0 & 0 & 0 & 0 & 0 & 0 & 0 & 2 & 0 & 0 & 0 & 1 & 0 & 1 \\
0 & 2 & 0 & 1 & 0 & 0 & 0 & 0 & 0 & 0 & 0 & 0 & 0 & 1 \\
0 & 0 & 1 & 0 & 0 & 0 & 0 & 0 & 2 & 0 & 1 & 0 & 0 & 0 \\
0 & 0 & 0 & 0 & 0 & 0 & 2 & 0 & 0 & 0 & 1 & 0 & 1 & 0 \\
1 & 0 & 1 & 0 & 1 & 0 & 1 & 0 & 1 & 0 & 1 & 0 & 1 & 0 \\
0 & 1 & 0 & 1 & 0 & 1 & 0 & 1 & 0 & 1 & 0 & 1 & 0 & 1 \\
0 & 0 & 0 & 0 & 0 & 2 & 0 & 0 & 0 & 0 & 0 & 0 & 0 & 0 \\
1 & 1 & 0 & 0 & 0 & 0 & 1 & 0 & 0 & 0 & 0 & 0 & 1 & 0 \\
0 & 0 & 0 & 0 & 1 & 1 & 0 & 0 & 0 & 0 & 0 & 0 & 0 & 0 \\
0 & 0 & 0 & 0 & 0 & 0 & 0 & 1 & 1 & 1 & 0 & 1 & 0 & 0 \\
0 & 0 & 0 & 0 & 2 & 0 & 0 & 0 & 0 & 0 & 0 & 0 & 0 & 0 \\
0 & 0 & 1 & 1 & 0 & 0 & 0 & 0 & 0 & 0 & 1 & 0 & 0 & 1 \\
0 & 0 & 0 & 1 & 0 & 0 & 0 & 0 & 0 & 2 & 0 & 1 & 0 & 0 \\
2 & 0 & 1 & 0 & 0 & 0 & 0 & 0 & 0 & 0 & 0 & 0 & 1 & 0
\end{pmatrix}$$

with PF eigenvalue 4 (the square of the inflation multiplier $\lambda = 2$) and corresponding right eigenvector

$$v = \frac{1}{16}(1,1,1,1,2,2,1,1,1,1,1,1,1,1)^T,$$

which encodes the relative frequencies, while the corresponding left eigenvector is proportional to $(1,\dots,1)$ in agreement with all prototiles having the same area. As u and d half-hexes always combine to form single-colour hexagons, the frequencies mean that one out of four hexagons in a fixed point is of type C, while all others are equally frequent (one out of eight each).

Disregarding colours, the tiling is built from one hexagonal prototile and its mirror image, both with two types of markings, lines and dots. They can be used to formulate perfect aperiodic local rules for a class of face to face tilings (the Taylor tilings) as follows. First, adjacent hexagons have to respect the line markings in the sense that the latter must never stop or jump at an edge. Second, the dots in any pair of next-to-nearest neighbours have to follow the rule indicated by the curved arrow (irrespective of the local arrangements of the lines or the chirality)

(6.10)

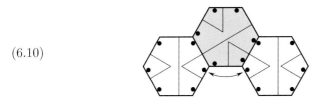

as can also be seen on the lower left panel of Figure 6.51 on page 243. Third, no vertex (where three hexagons meet) is allowed to show a C_3-symmetric arrangement of (nearby) dots; see [BGG12b] and references therein for details. The three conditions constitute local rules which are aperiodic and perfect.

Without the third condition, one still has aperiodic local rules, but they define a larger class of tilings (which we call the Socolar–Taylor tilings). The corresponding hull is not minimal. In addition to the repetitive Taylor tilings, it also contains a mirror image pair of tilings with a special three-fold symmetric defect-type seed (and all its translates). For a proof of these claims by the composition-decomposition method (for the Taylor tilings) or by the forcing of an underlying limit-periodic structure (for the Socolar–Taylor tilings), we refer to [Tay10, ST11, LM13].   ◊

An interesting structure of the Taylor tiling [Tay10] emerges by disregarding all markings and colours, but keeping track of the chirality of the hexagons. We use white and grey for this purpose, as indicated above in the figure of Eq. (6.10). This leads to the pattern shown in Figure 6.23, which is nicknamed the 'llama tiling' (due to the shape of the smallest island); see also Figure 6.54 on page 245. By construction, this pattern is deterministic and repetitive, with equal frequency for grey and white hexagons. Moreover, it has the following interesting property.

Proposition 6.5. *The llama pattern of Figure 6.23 possesses connected components of each colour of unbounded size. Moreover, it is aperiodic.*

PROOF. The patch shown is derived from a unique inflation fixed point with a hexagon (originally of type C) of positive chirality as a seed, which we denote as pattern P1. Another fixed point P2 can be obtained from the hexagon of opposite chirality, which has black and white interchanged relative to P1. Nevertheless, P1 and P2 are LI (in fact, they form a proximal pair, which follows from the almost colour reflection symmetry in Figure 6.23), so that arbitrarily large patches of either pattern occur in the other.

Now, assume that P1 does not contain connected patches of white hexagons of unbounded size, where two hexagons are called connected when they share an edge. If so, there must be a maximal connected component of white hexagons in P1, of diameter r say, which must then be surrounded by a connected 'belt' of grey hexagons (which is possibly part of an even larger connected patch). Then, the same belt exists in P2, this time as a white belt around a grey island, and the belt has diameter $> r$ by construction. Since P1 and P2 are LI, this patch from P2 must also occur somewhere in P1, in contradiction to the assumption, and the first claim is clear.

The aperiodicity claim is a direct consequence of the underlying primitive inflation rule. Let us nevertheless give an independent argument, based on

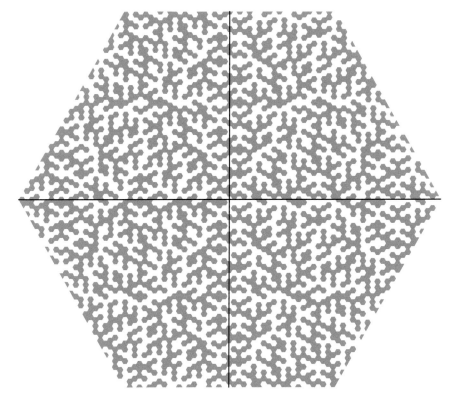

FIGURE 6.23. A patch of the llama pattern, derived from the Taylor tiling of Example 6.6. The horizontal and vertical black lines are almost colour reflection lines in the sense that a reflection in any of these lines, followed by a colour exchange, maps the pattern onto itself, except for the hexagons that are dissected by the line.

proximality. As mentioned above, P1 and P2 form a proximal pair. In fact, one can start from the central hexagon (or its chiral partner) in six different orientations. This leads to twelve distinct fixed points, and hence twelve distinct parity patterns, which can be grouped into a variety of proximal pairs. Some of these can be extracted from Figure 6.23 via its almost reflection symmetries. This structure is incompatible with any non-trivial periodicity by Corollary 5.3. Together with repetitivity, which is inherited from the underlying inflation structure, this implies the aperiodicity of P1 and P2, and hence that of the entire LI class of the llama tiling.   □

Let us add that (two-coloured) hexagonal tilings can be mapped to so-called 'brickwall tilings', via a suitable deformation of the hexagons. The first claim of Proposition 6.5 can then alternatively be proved by the so-called 'brickwall' lemma; compare [LAT04, Fig. 3 and Lemma 2.1]. Unfortunately, neither of these arguments implies the conjectured existence of a sequence of islands of growing size.[5]

The llama pattern thus provides an interesting example of an aperiodic deterministic structure with percolation [BGG12b]. In particular, an element with an infinite connected cluster must exist in the llama LI class. This follows from a compactness argument, because any sequence of llama tilings with connected clusters of increasing diameters around the origin must contain a subsequence that converges in the local topology. Consequently, the limit must contain an infinite cluster. Note that the same arguments apply to Penrose's hexagon parity pattern of Figure 6.22; see [BGG12b] for details.

Let us close with an observation by Joan Taylor, which is analogous to the relation between Penroses's $(1 + \varepsilon + \varepsilon^2)$-tilings and the corresponding parity patterns; compare Example 5.13.

Proposition 6.6. *The Taylor and the llama tilings are MLD.*

SKETCH OF PROOF. The derivation of a llama tiling from a Taylor tiling is performed by only coding the chirality of the hexagons in white or grey while erasing all markers, which is clearly local.

The converse direction proceeds in several steps. First, one identifies the positions of all C-hexagons and all 'hexagon streets' defined by linear arrangements of hexagons of type C and E (light blue and yellow in Figure 6.52), with occasional interruptions by a single hexagon of type F (orange) or type D (dark brown). This is possible via the 'eyes' of the llamas, compare Figure 6.54 on page 245, which define one such 'street' each. Together with their parallels, the streets form a triangular mesh with nine hexagons along each edge and a hexagon of type C at each corner. This fixes the position of *all* C-hexagons, which occupy a triangular lattice. Note that this step is local, because llamas (in all possible orientations) occur repetitively.

The C-hexagon at each 'street crossing' is the centre of a unique level-2 supertile of the form shown in the lower left panel of Figure 6.51, which is locally recognisable from the parity pattern displayed on its right-hand side (in fact, one only needs the 'inner' configuration indicated by the heavy contour). The reflected patch (in the left panel) corresponds to a parity-inverted patch (in the right panel). This step specifies the line markings of the entire mesh of streets, and thus determines the hierarchic triangle structure apart from the smallest scales.

The smallest triangles are already fixed by the C-hexagon sublattice. The missing colours and line markings can now be extracted (locally) from the supertile parity patterns of Figure 6.53. Note that the line decorations and

---

[5]We thank Joan Taylor for sending us a sketch of a possible deflation argument.

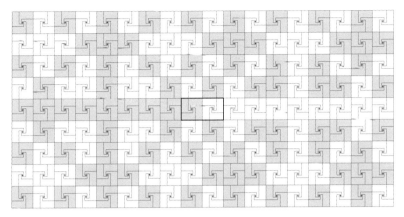

FIGURE 6.24. A rectangular patch of the squiral tiling with a 'black and white' reflection symmetry in the vertical axis. The displayed patch was obtained via two inflation steps from the central rectangular seed (marked) comprising four tiles of each chirality.

the parity of a hexagon determine its point markings (irrespective of colour). Since all operations are clearly local, the claim follows.   □

There are two alternative ways to prove Proposition 6.6. First, as in Example 6.5, a lattice-based sliding block map is also possible, again with hexagonal coronae of order three. Second, based on an idea by Joan Taylor, an explicit derivation is spelt out in [LM13].

Example 6.7 (*Squiral tiling*). An interesting relative of the chair tiling is obtained from a prototile with infinitely many straight edges; see [GS87, Fig. 10.1.4]. The inflation rule (with multiplier 3) is

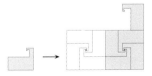

where the colours distinguish the two possible chiralities. This is known as the squiral tiling, a name built from 'square' and 'spiral'. A finite patch is shown in Figure 6.24. Note that this inflation permits several versions that include shifts and rotations through multiples of $\frac{\pi}{2}$, and hence different routes to fixed point tilings. It is clear that they all define the same (minimal) hull.

The geometric shape of the prototile enforces that four tiles of the same chirality combine to fill a square (without gaps or overlaps), sharing the special vertex in the centre of the square. Since this filling is fourfold symmetric, there are thus only *two* different such squares, which can simply be coded by the two colours (for the two chiralities). The squiral inflation now induces

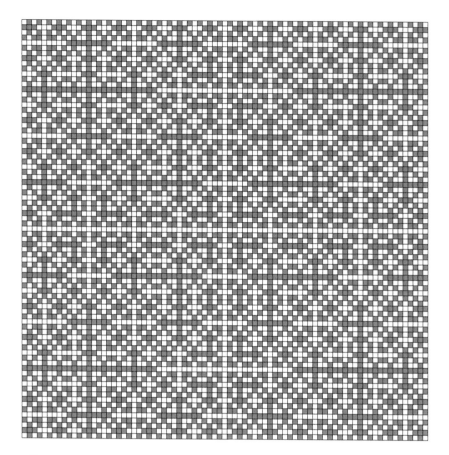

FIGURE 6.25. A patch of the block tiling version with exact D_4 symmetry.

the simpler block inflation

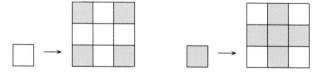

which is primitive; see Figure 6.25 for a larger patch. Moreover, the hulls defined by the squiral and by the derived block inflation are MLD. In both cases, the two tiles are equally frequent. The further analysis of this example is simpler on the basis of the block inflation, which can also be interpreted as a block substitution (on two letters), thus leading to a dynamical system under $\mathbb{Z}^2$-action (via the discrete hull); see [Fra05, BG14] for more information.

The block inflation shows striking analogies to the structure of the Thue–Morse substitution, and can be viewed as a truly planar generalisation. In particular, it has no coincidence, and will thus provide an example that cannot be constructed as a model set; see Chapter 7 for more. The underlying theory, which is an extension of Dekking's result [Dek78] to higherdimensional lattice systems, is developed in [LM01, Fre02, FSi07]. ◇

6.5. Examples of non-Pisot tilings

So far, all our examples in this chapter had PV numbers as inflation multipliers. Most of these examples possess an interpretation in terms of the projection method (to be discussed in Chapter 7). One non-Pisot tiling appeared in Example 5.8. Here, we discuss two examples with a non-PV inflation multiplier, which are significantly different.

6.5.1. CHIRAL TILING WITH FIVEFOLD SYMMETRY

One of the first examples of a tiling with a chiral structure (on the basis of the Penrose rhombus shapes) is the Lançon–Billard (LB) tiling [LB88]; see also [Fra08]. It is chiral in the sense that the rhombuses are oriented as in Figure 6.26, but do not occur in their reflected versions. Note that this property rules out any reflection symmetry of the tiling in a line, both individually and in the sense of the LI class.

Although the inflation rule of Figure 6.26 is not a stone inflation, it is still a valid inflation rule according to our definition, because the second step of the inflation procedure is consistent (a situation analogous to that of the shield tiling above) and preserves the area (it will be modified into a stone inflation in Remark 6.11). Starting with the thick rhombus, the primitive substitution matrix reads

$$M_{\mathrm{LB}} = \begin{pmatrix} 3 & 1 \\ 1 & 2 \end{pmatrix}.$$

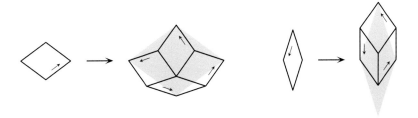

FIGURE 6.26. Chiral inflation rule due to Lançon and Billard [LB88], on the basis of the two Penrose rhombuses. Note that only one chirality of the tiles occurs in the corresponding tiling.

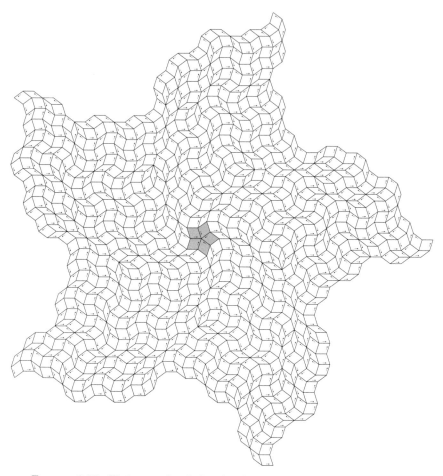

FIGURE 6.27. Finite patch of the chiral Lançon–Billard tiling with C_5 symmetry, obtained by four inflation steps from the shaded vertex star.

The PF eigenvalue is $\frac{1}{2}(5+\sqrt{5})$, which is the square of the inflation multiplier

$$\lambda_{\mathrm{LB}} = 2\cos\left(\frac{\pi}{10}\right) = \sqrt{\frac{1}{2}\left(5+\sqrt{5}\right)} \approx 1.902.$$

This is an algebraic integer of degree 4, with defining polynomial $x^4 - 5x^2 + 5$, which is irreducible over $\mathbb{Q}$. All four roots of this polynomial are real numbers of absolute value > 1, which means that λ_{LB} is not a PV number. In fact, the PF eigenvalue is a Lind number; compare Theorem 2.4. The left and right PF eigenvectors of M_{LB} equal those of the matrix M_{T} introduced in Remark 6.3. In particular, the inflation rule is aperiodic, by an application of Theorem 6.2 together with the primitivity of the inflation rule.

FIGURE 6.28. Finite patch of the chiral Lançon–Billard tiling, obtained by five inflation steps from the thick rhombus.

A patch with C_5 rotational symmetry is shown in Figure 6.27, which also illustrates the chiral nature. One can check that the central five-star is a legal configuration (it occurs in Figure 6.28 on page 219). Upon iterating the fourth power of the inflation rule, one obtains a sequence of patches that converges (in the local topology) towards a globally fivefold symmetric tiling of the plane without any reflection symmetries. Also the hull is C_5-symmetric, but not D_5-symmetric.

Remark 6.11 (*Stone inflation for the LB tiling*). It is possible to define a stone inflation that results in inflation tilings of the same MLD class as the chiral rhombus tiling of Figures 6.27 and 6.28. However, unlike the situation of the rhombic Penrose tiling, this requires an iterated function system approach and leads to two prototiles with fractal boundaries (fractiles). Their construction is a variation of the von Koch curve algorithm for the boundaries of the tiles (similar to the construction for the shield tiling). The crucial observation is that the boundaries of the inflation images in Figure 6.26 comprise alternating sequences of oriented and unoriented edges, a structure that is preserved under further inflation. This can be turned into a recursion for the tile boundaries and thus into an iterative construction of new prototiles with fractal boundaries.

It is desirable that each fractile equals the closure of its interior (regularity). Since the inflation image of the thin rhombus in Figure 6.26 does not contain all four vertices of the rescaled rhombus, the above construction creates a 'tadpole' with an (undesirable) fractal line of dimension < 2. Since this would coincide with boundary lines of other tiles, it can be removed consistently. Altogether, this results in the following stone inflation:

Each new fractile is now a simply connected compact set which is the closure of its interior. The ratio of the prototile areas is unchanged (meaning that it is still τ). The boundary of each fractile is a fractal of Hausdorff dimension

$$d_{\mathsf{H}} = \frac{\log(2)}{\log\left(2\cos\frac{\pi}{10}\right)} = \frac{2\log(2)}{\log\left(\frac{5+\sqrt{5}}{2}\right)} \approx 1.078.$$

A larger patch of the tiling is shown in Figure 6.59 on page 249.   ◇

The non-PV nature of the inflation multiplier λ_{LB} can be used to show that no point set that is MLD to a chiral tiling of this type can be a Meyer set, via an application of Theorem 2.4. Since we will spell out this argument for the following tiling, compare Proposition 6.8, we skip details here.

6.5.2. A SEVENFOLD TILING WITH PERFECT LOCAL RULES

As a planar example with several distinct properties, we describe an inflation tiling with three isosceles triangles as prototiles. It was discovered by Danzer, but only later described in [ND96]. Three different base lengths of the triangles occur, and one common length for all other edges. One variant of the inflation rule is shown in Figure 6.29. Here, the triangles have oriented base lines, and the two larger triangles are equipped with one pseudo vertex each on the base line, marked by a dot. This way, the inflation rule becomes a stone inflation for a face to face tiling. Taking the pseudo vertices into account, there are actually only three distinct edge lengths, with length ratios $\sin(\pi/7) : \sin(2\pi/7) : \sin(3\pi/7)$. All three triangular prototiles occur with both chiralities. A patch of the tiling (with reflection symmetry in the vertical axis) is shown in Figure 6.30. This is a central patch of a fixed point under twofold inflation.

The tiling contains 14-stars of small acute triangles (which can be seen to emerge at the boundary of the patch of Figure 6.30). Figure 6.31 shows the twofold inflation of such a vertex star. Clearly, the decoration completely

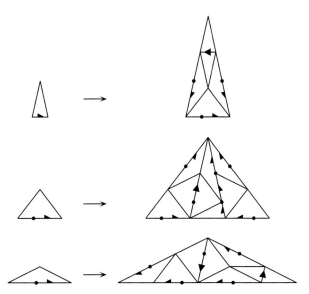

FIGURE 6.29. Inflation rule for the three triangular prototiles of the sevenfold inflation tiling due to Danzer. Two of the triangles are equipped with a pseudo vertex which makes the tiling face to face.

breaks the symmetry. Interestingly, the symmetry is also broken in the undecorated tiling (with all arrows and pseudo vertices removed), as can be seen from the tiles along the boundary of the patch in Figure 6.31. A larger portion of an undecorated tiling (the 'buffalo' tiling with individual reflection symmetry) is shown in Figure 6.60 on page 250, and on the cover of this book. The symmetry of the tiling space is as follows.

Proposition 6.7. *The inflation rule of Figure* 6.29 *is primitive and defines a unique hull, which consists of a single LI class, the latter defined by a suitable fixed point tiling. The hull has D_{14} symmetry, but no member of the hull has individual sevenfold rotational symmetry.*

PROOF. The primitivity of the inflation rule is obvious from Figure 6.30 (or from the matrix M_{D7} below), which implies the statements on the hull and the LI class.

The claim on the D_{14} symmetry of the hull follows from the primitivity together with the observation that the first triangle of Figure 6.29 occurs in all 14 orientations, compare Figure 6.31, while the inflation rule is reflection symmetric by construction.

Since the arrow decoration breaks the sevenfold rotation symmetry of the 14-gonal vertex star, no tiling with individual sevenfold symmetry can exist in the hull; see also [H–Y11]. □

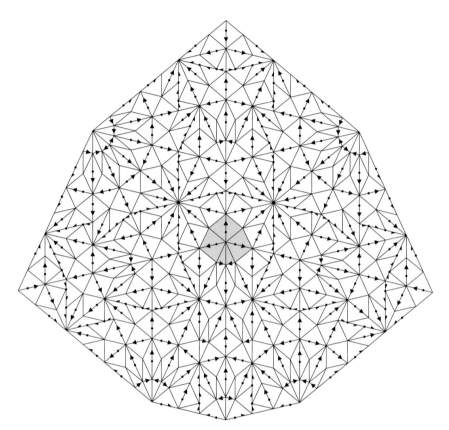

FIGURE 6.30. This reflection-symmetric patch of the sevenfold Danzer tiling, with point and arrow decorations, was obtained by two inflation steps from the central seed (shaded), which is legal.

We encountered this phenomenon earlier for the hull of the TTT of Section 6.2.2, while the situation is different for the rhombic Penrose tiling and for the eightfold Ammann–Beenker tiling. This shows that the relation between the symmetry of a hull, in the sense of Definition 5.14, and that of its individual members, in the spirit of Section 5.6, is non-trivial and needs to be analysed for each example separately.

Danzer's tiling, in the sense of its LI class, has the inflation multiplier

$$\lambda_{\mathrm{D7}} = 1 + \frac{\sin\left(\frac{2\pi}{7}\right)}{\sin\left(\frac{\pi}{7}\right)} = 1 + 2\sin\left(\frac{5\pi}{14}\right),$$

which is an element of $\mathbb{Z}[\xi_7 + \bar{\xi}_7]$. Its defining irreducible monic polynomial is $x^3 - 4x^2 + 3x + 1$, with the roots

$$2\sin\left(\tfrac{5\pi}{14}\right) + 1 \approx 2.802\,, \quad 2\sin\left(\tfrac{3\pi}{14}\right) - 1 \approx 0.247\,, \quad 2\sin\left(\tfrac{\pi}{14}\right) + 1 \approx 1.445\,.$$

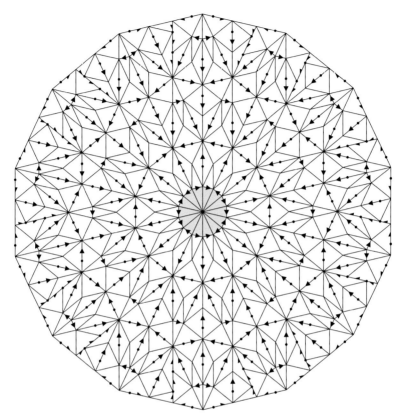

FIGURE 6.31. Another patch of the sevenfold Danzer tiling, again obtained by two inflation steps from the central seed (shaded), which is legal.

The first is the inflation multiplier λ_{D7}, which is a unit but not a PV number. It satisfies $\lambda_{\mathrm{D7}}\mathbb{Z}[\xi_7] = \mathbb{Z}[\xi_7]$ and $\lambda_{\mathrm{D7}}\mathbb{Z}[\xi_7 + \bar{\xi}_7] = \mathbb{Z}[\xi_7 + \bar{\xi}_7]$.

The primitive and symmetric substitution matrix reads

$$M_{\mathrm{D7}} = \begin{pmatrix} 2 & 1 & 2 \\ 1 & 5 & 3 \\ 2 & 3 & 3 \end{pmatrix}.$$

Its characteristic polynomial is $x^3 + 10x^2 - 17x + 1$, whose roots are the squares of the roots of the previous polynomial. In particular, the PF eigenvalue of M_{D7} is λ_{D7}^2. The prototile areas and frequencies have the same ratios, namely $(1 - 2\sin(\pi/14)) : \sin(3\pi/14) : 1$. This gives the following numerical values for the relative tile frequencies, 0.1981, 0.4450 and 0.3569.

Let Λ be the set of vertex and pseudo vertex points of the fixed point (under the square of the inflation) which emerges from the patch of Figure 6.30. It is possible to choose the scale so that $\Lambda \subset \mathbb{Z}[\xi_7]$. Since $0 \in \Lambda$ by construction, we also know that

$$\lambda_{\mathrm{D}7}^2 \Lambda \subset \Lambda,$$

which implies that Λ cannot be a Meyer set, by an application of Theorem 2.4, as $\lambda_{\mathrm{D}7}^2$ is a real Perron number that is neither a PV nor a Salem number.

Proposition 6.8. *If Λ is the point set of all vertex and pseudo vertex positions of any element of the sevenfold Danzer hull, Λ is Delone and FLC, but not Meyer, while the hull is aperiodic.*

PROOF. Since the sevenfold Danzer tiling is face to face (when including the pseudo vertices), the vertex point set Λ is Delone (from the tiling property) and FLC (from the face to face property). However, Λ cannot be Meyer due to the non-Pisot scaling relation mentioned above. Aperiodicity is an immediate consequence of the irrationality of $\lambda_{\mathrm{D}7}$, via Theorem 6.2. $\square$

Moreover, the list of vertex stars in the inflation tiling constitutes an atlas of perfect aperiodic local rules [ND96]; see also [H–Y11].

6.6. Pinwheel tilings

All tilings considered so far, with the exception of the Frank–Robinson tiling of Example 5.8, have the FLC property with respect to translations. That simple but intriguing planar face to face tilings exist which are not FLC in this sense was first shown by Conway and Radin in form of the now classic *pinwheel tilings*; see [Rad99] and references therein. The simplest example is constructed by inflation of a single prototile, a rectangular triangle with adjacent sides of lengths 1 and 2, whence the hypotenuse has length $\sqrt{5}$. After a linear expansion of this triangle by a factor $\sqrt{5}$ (step 1), the image can be dissected into five triangles (step 2), all congruent to the original one. Among those, two triangles occur in the original chirality, while the

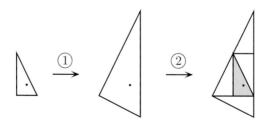

FIGURE 6.32. The inflation rule for the classic pinwheel tiling, which includes a rotation by $\vartheta = -\arctan(\frac{1}{2})$. The dot marks the reference point of the prototile (left), which is also the centre of the inflation process. The fixed point tiling thus covers the entire Euclidean plane.

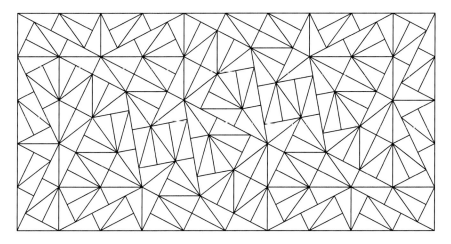

FIGURE 6.33. A finite patch of the pinwheel tiling.

other three are reflected copies, see Figure 6.32. Note that the inflated copy
has also been rotated, such that the original triangle reappears (shaded).
The underlying linear expansion is thus not a (pure) homothety, while the
inflation rule commutes with reflections. This version produces a well-defined
plane-filling fixed point under inflation.

The relevant rotation angle of this inflation is $\vartheta = -\arctan\left(\frac{1}{2}\right)$, so that
$z = \mathrm{e}^{\mathrm{i}\vartheta} = \frac{1}{\sqrt{5}}(2 - \mathrm{i})$. Since $z^n = 1$ is only possible with $n = 0$, the angle
ϑ is irrational modulo 2π. As a result, the inflation fixed point contains
rotated copies of the original prototile in *infinitely* many distinct orientations,
wherefore the tiling cannot be FLC relative to translations, although it still
is FLC with respect to Euclidean motions. As a result of this property, the
tiling looks a lot more irregular than our previous examples, although it is
still constructed from a completely deterministic inflation rule. A rectangular
patch is shown in Figure 6.33, produced by three inflation steps from two
triangular prototiles that fill a rectangle with sides of lengths 1 and 2. A
larger patch with triangles of one chirality being coloured according to their
orientation is shown in Figure 6.57 on page 247.

The formulation of the pinwheel tiling used so far, with a single prototile,
does not give access to interesting statistical properties by Perron–Frobenius
theory. In particular, one cannot extract the relative frequencies of the vertex
stars directly from the original inflation rule. Since also other methods based
on embedding techniques (see Chapter 7) fail for this tiling, an alternative
inflation rule with more than one prototile will be advantageous. One possibility, based on the square of the original inflation, was described in [BFG07]
and is shown in Figure 6.34. It follows from the observation that, in any

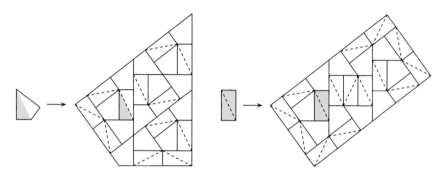

FIGURE 6.34. The kite domino inflation rule for the pinwheel tiling. It is derived from the square of the original inflation rule of Figure 6.32.

legal pinwheel tiling, the triangles always match face to face along their hypotenuses (see Figure 6.33), and that there are only two possible ways how this is realised — one yielding a kite, the other a rectangle.

Lemma 6.2. *Consider an infinite kite domino tiling that is a fixed point of the kite domino inflation rule of Figure* 6.34. *Then, the relative frequencies of kites and dominos are given by* $\frac{5}{11}$ *and* $\frac{6}{11}$, *respectively.*

PROOF. The substitution matrix for the kite domino inflation reads

$$M_{\mathrm{KD}} = \begin{pmatrix} 13 & 10 \\ 12 & 15 \end{pmatrix},$$

which is primitive with PF eigenvalue 25. Kite and domino have the same area, which corresponds to $(1,1)$ being the left PF eigenvector. The corresponding right eigenvector, in statistical normalisation, is $\frac{1}{11}(5,6)^T$, which establishes the claim.   □

Remark 6.12 (*Vertex stars of the pinwheel tiling*). Analysing a tiling fixed point of the pinwheel inflation shows that there are 11 distinct vertex stars up to Euclidean motions (including reflections). Employing the kite domino formulation, one can determine their frequencies with Perron–Frobenius methods. The resulting absolute frequencies are

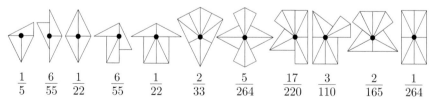

$$\frac{1}{5} \quad \frac{6}{55} \quad \frac{1}{22} \quad \frac{6}{55} \quad \frac{1}{22} \quad \frac{2}{33} \quad \frac{5}{264} \quad \frac{17}{220} \quad \frac{3}{110} \quad \frac{2}{165} \quad \frac{1}{264}$$

As one can easily check, the frequencies sum to $\frac{39}{55}$ and not to $\frac{1}{2}$, as one would naively expect for a tiling with triangular prototiles of unit area. This is due to the existence of a pseudo vertex at the midpoint of the long leg (cathetus) of the triangle prototile, which features in vertex stars 1, 2 and 4 of the list. Taking the pseudo vertex properly into account, there is actually one additional pseudo vertex star, namely $\bigwedge$. This configuration has absolute frequency $\frac{16}{55} = \frac{1}{5} + 2 \cdot \frac{1}{22}$ (derived from the absolute frequencies of vertex stars 1 and 3), which completes the previous fraction to 1. Indeed, the numbers can now be interpreted as relative frequencies (for the vertex and pseudo vertex stars together, and up to Euclidean motions including reflections), because each triangle is now a 'degenerate' quadrangle and thus possesses a total of one vertex. Consequently, the total density of vertex and pseudo vertex points together is 1.

Various related quantities like frequencies of more complicated patches or radial autocorrelation coefficients are accessible by the same methods, see [BFG07] for details. $\diamond$

Let us formulate one result that was proved by direct means in [BFG07] and by methods from K-theory in [Mou10].

Proposition 6.9. *The* (*absolute*) *frequency module of the pinwheel tiling with triangular prototile of unit area is*

$$\mathcal{F} = \frac{1}{264}\, \mathbb{Z}[\tfrac{1}{5}] = \Big\{ \frac{m}{264 \cdot 5^{\ell}} \,\Big|\, m \in \mathbb{Z}, \ell \in \mathbb{N}_0 \Big\},$$

which is a countably generated $\mathbb{Z}$-module.

SKETCH OF PROOF. Each finite patch of the pinwheel tiling occurs in one of the inflated prototiles, or in one of the inflated vertex stars. Using the minimal inflation power in either case, it is clear that one can express the frequency as a linear combination of the frequencies of the inflated tiles and vertex stars, with coefficients in $\mathbb{Z}[\tfrac{1}{5}]$. Since the prototile frequencies (in the kite domino formulation) are in the $\mathbb{Z}$-span of the vertex star frequencies, the frequency module is the $\mathbb{Z}[\tfrac{1}{5}]$-module of the frequencies derived in Remark 6.12. The claim now follows by a straightforward calculation. $\square$

Remark 6.13 (*A pinwheel relative with fractiles*). Although the pinwheel inflation is a stone inflation, so that there is no need to fractalise the prototile, there is an interesting and beautiful tiling in the pinwheel MLD class that is built via a stone inflation with 13 fractiles [FW11]. The origin of the fractiles is rather different from the mechanism discussed in Remark 6.11, and the resulting tiling can be considered as some kind of a 'dual' tiling. It is not a dual in the usual sense, because there is no simple correspondence between the 11 (respectively 12) vertex stars of Remark 6.12 and the 13 fractiles. $\diamond$

The pinwheel inflation can be generalised to various families of tiling spaces with full circular symmetry; we refer to [Sad98, Rad99, Fre08, FR14]

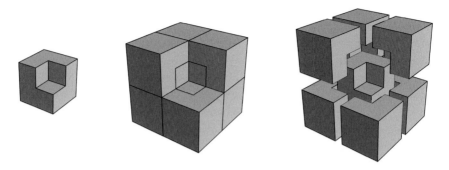

FIGURE 6.35. Illustration of the inflation rule for the chair in 3-space.

and references therein for further details. In contrast, the absence of the pinwheel phenomenon can be proved for some specific prototile sets. For instance, in parallelogram tilings using a finite number of shapes, all tiles occur in finitely many orientations only [FH13].

6.7. Tilings in higher dimensions

The inflation method applies to any dimension d, though matters can quickly become rather complex. One reason is that the number of tiles in a stone inflation grows with the PF eigenvalue of the inflation matrix. The eigenvalue is λ^d, when λ is the inflation multiplier. However, tilings in 3-space are important for physical applications, in particular tilings with icosahedral symmetry. Before we concentrate on this case, we briefly mention a generalisation of the chair and of the pinwheel tiling.

Example 6.8 (*Chair tiling in dimension d*). The construction of the chair tiling can be generalised to d-space [Goo99b, LM01]. One starts from a single prototile of volume $2^d - 1$, which emerges from the cube of edge length 2 by removing a unit cube in one corner. Expanding this prototile by a factor of 2 leads to a region that can be covered, face to face, by 2^d (translated and possibly rotated) copies of the original prototile. The case $d = 2$ was discussed in Sections 4.9 and 6.4, while the case $d = 3$ is sketched in Figure 6.35.

The prescription is formally extendable to $d = 1$, but reduces to an inflation rule for the integer lattice which is periodic. For $d \geqslant 2$, the tiling is recognisable and is thus non-periodic, by Theorem 6.3. As usual, aperiodicity then follows from repetitivity.   ◊

Remark 6.14 (*Quaquaversal tiling*). An interesting extension of the pinwheel construction to 3-space is provided by the *quaquaversal tiling* of Conway and Radin [CR98]; see also [Rad99, Ch. 4.3]. Here, the tiles come in orientations that are uniformly distributed in SO(3), while the convergence of spatial averages is faster than in the pinwheel tiling [DSV00]. This tiling motivated group-theoretic questions about subgroups of SO(3), which are investigated in [RS98, RS99] and have later found applications in the context of similar submodules in 3-space [Gli10]. ◊

Let us turn our attention to the simplest inflation tiling with icosahedral symmetry.

6.7.1. DANZER'S ABCK TILING

One of the simplest inflation generated tilings in 3-space with icosahedral symmetry is the ABCK tiling due to Danzer [Dan89]. It is built from 4 prototiles, which are selected from the family $\mathfrak{F}$ of the altogether 15 similarity classes of tetrahedra with faces parallel to mirror planes of the icosahedron. These classes are systematically labelled $\mathcal{A}, \mathcal{B}, \ldots, \mathcal{H}$ and $\mathcal{J}, \ldots, \mathcal{P}$ (thus leaving out the letter $\mathcal{I}$). All faces of the possible tetrahedra have normal vectors that are elements of the non-crystallographic root system Δ_{H_3}, and all dihedral angles are multiples of $\pi/2$, $\pi/3$ or $\pi/5$, with at least one being a proper multiple of such an angle. Bisecting (trisecting) such a dihedral angle cuts the corresponding tetrahedron in two (three). The resulting tetrahedra are still from $\mathfrak{F}$, so that one obtains a tree of possible successors for each class. The first step towards a useful inflation rule is now to search for closed subtrees with a small (or minimal) number of classes involved. It turns out that this is possible using just the four classes $\mathcal{A}, \mathcal{B}, \mathcal{C}$ and $\mathcal{K}$, which explains the name; see [DSvOD93] for details.

One particular choice of the tetrahedra, called A, B, C and K, is spelt out in Table 6.2. In comparison with the original work, we have doubled all coordinates for ease of presentation and compatibility with the icosahedral modules (and with $\mathcal{M}_\mathsf{F}$ in particular) from Example 2.20. Table 6.2 also contains the solid angle Ω at each vertex corner and the areas of the faces, following the convention that a vertex of a tetrahedron labels its opposite face. The meaning of the vertex type will be explained below in the projection context; see Example 7.13. Further geometric information is collected in Table 6.3, where the 6 edges of each tetrahedron are labelled by their defining vertex pairs, and both the squared length (ℓ^2) and the dihedral angle ($\measuredangle$) of each edge are listed. The volumes of the 4 prototiles A, B, C and K are

$$(6.11) \qquad \frac{1}{3}\tau^5, \quad \frac{1}{3}\tau^4, \quad \frac{1}{3}\tau^4 \quad \text{and} \quad \frac{1}{6}\tau^3.$$

Figure 6.36 gives a first impression of their geometric shape. Note that both possibilities of folding the three outer faces to form tetrahedral tiles occur in the tiling. They result in mirror image pairs of equally frequent tiles.

TABLE 6.2. Vertex data, solid angles Ω, and face areas of ABCK tiles.

	A-vertices			opp.		B-vertices			opp.
no.	coordinate	type	Ω	face	no.	coordinate	type	Ω	face
1	$(0,0,0)$	I	$\frac{\pi}{15}$	τ^3	1	$(0,0,0)$	I	$\frac{\pi}{30}$	τ^2
2	$(\tau^3,0,\tau^2)$	III	$\frac{\pi}{10}$	τ^3	2	$(\tau^3,0,\tau^2)$	III	$\frac{\pi}{15}$	τ^2
3	(τ^2,τ^2,τ^2)	II	$\frac{2\pi}{15}$	τ^3	3	(τ^2,τ^2,τ^2)	II	$\frac{2\pi}{15}$	τ^3
4	$(\tau^2,1,0)$	III	$\frac{13\pi}{30}$	τ^4	4	$(\tau^2,\tau,1)$	I	$\frac{23\pi}{30}$	τ^4

	C-vertices			opp.		K-vertices			opp.
no.	coordinate	type	Ω	face	no.	coordinate	type	Ω	face
1	$(0,0,0)$	I	$\frac{\pi}{5}$	τ^3	1	$(0,0,0)$	I	$\frac{7\pi}{30}$	$\frac{1}{2}\tau^3$
2	$(-\tau,0,1)$	II	$\frac{\pi}{10}$	τ^3	2	$(-1,\tau,0)$	II	$\frac{\pi}{10}$	$\frac{1}{2}\tau^2$
3	(τ^2,τ^2,τ^2)	II	$\frac{\pi}{30}$	τ^2	3	(τ,τ,τ)	III	$\frac{\pi}{30}$	$\frac{1}{2}\tau$
4	$(0,\tau^2,1)$	III	$\frac{4\pi}{15}$	τ^3	4	$\frac{1}{2}(-1,\tau^{-1},\tau)$	IV	$\frac{\pi}{2}$	τ^2

The four prototiles (up to similarity) permit a stone inflation with golden ratio inflation multiplier τ, meaning that the τ-inflated versions can be dissected into copies of the original tiles (up to similarity, of course). The inflation rule is compatible with reflection (in the sense that the reflected tiles are dissected with the reflected rule). The inflation of the four prototiles is indicated in Figures 6.37 and 6.38.

The integer inflation matrix for the four prototiles is primitive and reads

$$M_{\mathrm{ABCK}} = \begin{pmatrix} 0 & 0 & 1 & 0 \\ 3 & 2 & 0 & 1 \\ 2 & 1 & 2 & 0 \\ 6 & 4 & 2 & 1 \end{pmatrix},$$

TABLE 6.3. Squared edge lengths and dihedral angles of ABCK tiles.

	A		B		C		K	
edge	ℓ^2	$\angle$	ℓ^2	$\angle$	ℓ^2	$\angle$	ℓ^2	$\angle$
1–2	$11\tau+7$	$\frac{2\pi}{5}$	$11\tau+7$	$\frac{\pi}{5}$	$\tau+2$	$\frac{\pi}{5}$	$\tau+2$	$\frac{2\pi}{5}$
1–3	$9\tau+6$	$\frac{\pi}{3}$	$9\tau+6$	$\frac{\pi}{3}$	$9\tau+6$	$\frac{\pi}{3}$	$3\tau+3$	$\frac{\pi}{3}$
1–4	$3\tau+3$	$\frac{\pi}{3}$	$4\tau+4$	$\frac{\pi}{2}$	$3\tau+3$	$\frac{2\pi}{3}$	1	$\frac{\pi}{2}$
2–3	$4\tau+3$	$\frac{\pi}{5}$	$4\tau+3$	$\frac{\pi}{5}$	$12\tau+8$	$\frac{2\pi}{5}$	$4\tau+3$	$\frac{\pi}{5}$
2–4	$4\tau+4$	$\frac{\pi}{2}$	$3\tau+3$	$\frac{2\pi}{3}$	$4\tau+3$	$\frac{2\pi}{5}$	$\tau+1$	$\frac{\pi}{2}$
3–4	$4\tau+3$	$\frac{3\pi}{5}$	$\tau+2$	$\frac{3\pi}{5}$	$4\tau+3$	$\frac{\pi}{5}$	$3\tau+2$	$\frac{\pi}{2}$

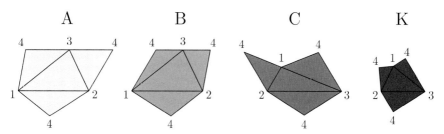

FIGURE 6.36. Net representation of the ABCK prototiles. The vertices are numbered as in Table 6.2.

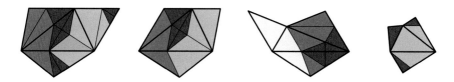

FIGURE 6.37. Net representation of the inflated ABCK prototiles.

with PF eigenvalue $\lambda = \tau^3$. It is the third power of the inflation multiplier because we are in 3-space. The other eigenvalues of M_{ABCK} are τ and the algebraic conjugates of these two, $-1/\tau$ and $-1/\tau^3$. As one can easily check, the four volumes of Eq. (6.11), in the order given there, constitute the left eigenvector for λ, as it must be for a stone inflation. The statistically normalised right eigenvector reads

$$\frac{1}{11}\left(12\tau - 19, 29 - 16\tau, 5 - 2\tau, 6\tau - 4\right)^T \approx \left(0.0379, 0.2829, 0.1604, 0.5189\right)^T$$

and gives the *relative* frequencies of the prototiles A, B, C and K in the inflation tiling. Thus, more than half of the tiles are of type K, while tiles of type A are quite rare. The *absolute* tile frequencies (calculated per unit volume) derive from here as

$$\frac{3-\tau}{15}, \quad \frac{8-\tau}{15}, \quad \frac{2+\tau}{15} \quad \text{and} \quad \frac{2+\tau}{15}.$$

These frequencies can now be used in conjunction with the solid angles Ω from Table 6.2 and the prototile volumes of Eq. (6.11) to calculate the densities of the four vertex types. They turn out to be

$$(6.12) \quad \begin{aligned} \frac{32 - 19\tau}{20} &\approx 0.062868, & \frac{2 - \tau}{20} &\approx 0.019098, \\ \frac{7\tau - 11}{20} &\approx 0.016312, & \frac{3\tau - 4}{20} &\approx 0.042705. \end{aligned}$$

In our setting, the density of all vertices together is $(19 - 10\tau)/20 \approx 0.14098$, while those of types I, II and III add up to $(23 - 13\tau)/20 \approx 0.09828$.

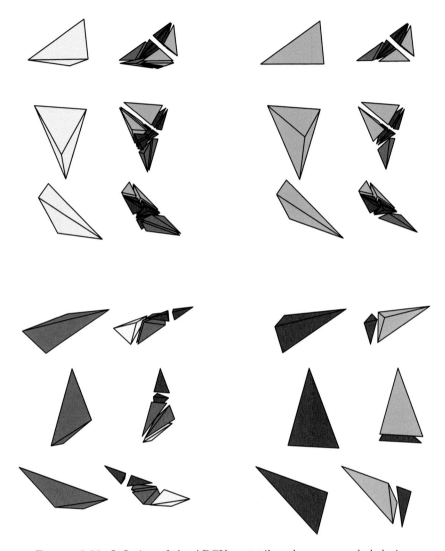

FIGURE 6.38. Inflation of the ABCK prototiles, shown as exploded view drawings from three different angles. Note that the original prototiles are on different (relative) scales in this figure.

The tilings defined by this inflation rule, and all elements of the hull constructed from it, are called ABCK tilings. Due to the primitivity of the inflation rule, the hull coincides with the LI class of any fixed point tiling under some power of the inflation. Linear repetitivity follows from the stone inflation structure, while non-periodicity can be inferred from Theorem 6.2 or from the non-crystallographic symmetry with irreducible representation.

FIGURE 6.39. The four octahedral prototiles of the $\langle$ABCK$\rangle$ tiling, consisting of four or eight tetrahedral tiles; $\langle$A$\rangle$ and $\langle$B$\rangle$ are not convex.

Corollary 6.9. *Danzer's* ABCK *tiling is FLC, linearly repetitive and aperiodic. The corresponding dynamical system under the action of* $\mathbb{R}^3$ *is strictly ergodic. The hull has full icosahedral symmetry, and possesses an LIDS with inflation multiplier* τ. $\qquad\square$

The inflation is constructed so that the tiling is face to face. Moreover, only vertices of the same type can meet. It can be seen from the geometric data in the two tables that the tiles group in clusters of four around a twofold edge. In particular, if a face of a tetrahedron T contains an edge with dihedral angle $\pi/2$, the mirror image of T meets T in precisely this face. The prototiles A, B and C have one edge of this type, while K has three of them, sharing a common vertex. Consequently, tiles of type A, B and C always come in quadruples as mentioned, while tiles of type K actually come in groups of eight around a common vertex. The latter is of type IV, with solid angle $\Omega = \frac{\pi}{2}$. The four, respectively eight, tiles together form the (topological) octahedron shown in Figure 6.39, denoted by $\langle$A$\rangle$, $\langle$B$\rangle$, $\langle$C$\rangle$ and $\langle$K$\rangle$. Note that $\langle$A$\rangle$ and $\langle$B$\rangle$ are not convex. The four octahedra may also be chosen as prototiles, then leading to $\langle$ABCK$\rangle$ tilings. All ABCK vertices of type IV of Table 6.2 become inner vertices of type $\langle$K$\rangle$ octahedra (and hence 'vanish'), while all other vertices remain. Nevertheless, the two tiling versions are SMLD.

The tetrahedral version is preferable for the inflation picture (as it gives a face to face stone inflation), while the octahedral version will result in a simpler formulation for the perfect local rules and for the description via the projection method in Example 7.13.

In the ABCK tiling, there are three legal vertex configurations where 120 tiles of the same type meet in an icosahedrally symmetric vertex star; see Figure 6.40 for an illustration. The 'B-polytope' consists of 120 tiles of type B, which meet in a vertex of type I. This polytope is legal because it occurs in the fifth inflation of a single A tile. Under the inflation action, a central 'B-polytope' leads to a central 'C-polytope' (with central vertex of type II) as shown in the centre of Figure 6.40, and then to a central 'K-polytope' (with central vertex of type III) as shown on the right. In the next step,

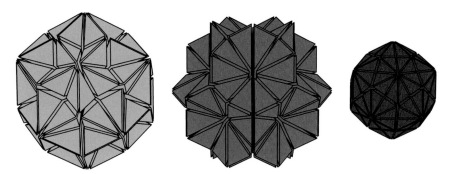

FIGURE 6.40. Icosahedrally symmetric seeds consisting of 120 tiles of types
B (left), 120 tiles of type C (centre) and 120 tiles of type K (right). Under
inflation, they give rise to the only elements of the hull with individual
icosahedral symmetry; see text for details.

one comes back to a 'B-polytope', whence inflation produces a 3-cycle. As
a consequence, these three vertex configurations are legal seeds of globally
icosahedrally symmetric fixed point tilings under threefold inflation. Recall
from Section 5.6 that, due to non-periodicity, any ABCK tiling can at most
have one centre with perfect (individual) icosahedral symmetry. The three
tilings (and their translates) are the only members of the LI class with full
(individual) icosahedral symmetry. A detailed (general) analysis of inflation
orbits and their symmetries is given in [BHP97, Sec. 4].

The ABCK tiling possesses perfect aperiodic local rules, which is stated
(somewhat implicitly, and without proof) in the main theorem of [Dan89].
Interestingly, these rules can be formulated as purely geometric packing rules
on the level of the octahedra. This is one of the rare cases (similar to the
kites and darts tiling of Example 5.12) where the local rules are realisable
via polytopes and their geometric packing rules alone (without geometric
markers or any further distinction of different types). For a proof, recall that
the existence of local rules is an MLD invariant. Observe that the ABCK
tiling is MLD with the Socolar–Steinhardt tiling (SST) from [SS86], which
follows from the detailed (and independent) analysis in [Rot93] and [DPT93];
see also [PHK00] for a related inflation tiling. The perfect aperiodic local rules
of the SST are in turn established in [SS86] by the composition-decomposition
method, which completes the argument.

Proposition 6.10. *The MLD class of Danzer's* ABCK *tiling possesses perfect aperiodic local rules. For the* $\langle$ABCK$\rangle$ *tiling, the local rules are realised
by the geometric packing rules of the four prototiles* $\langle$A$\rangle$, $\langle$B$\rangle$, $\langle$C$\rangle$ *and* $\langle$K$\rangle$ *of
Figure* 6.39.   □

6.7.2. FURTHER EXAMPLES

Historically the first, and arguably the most influential, tiling with icosahedral symmetry is built from two rhombohedra (up to rotations), a thin and a thick one, which are described below in more detail in Section 7.4 around Figure 7.11. These rhombohedra can be packed (face to face) to fill Kepler's triacontahedron of the same edge length [Kow38]. The corresponding tiling is called the *primitive icosahedral tiling* and was discovered independently via the projection method by Kramer and Neri [KN84] and by Ammann via local rules; compare Section 5.7.2. Ammann's work was only published later or became known via other authors; see in particular [GS87, Sen04]. The first (very brief) mentioning of the primitive icosahedral tiling (without details or proofs) can be found at the end of an article by Mackay [Mac81].

This tiling (in proper scaling and orientation) has the limit translation module $\mathcal{M}_{\mathsf{P}}$ from Example 2.20. Since the latter is not invariant under multiplication by τ (but only by τ^3), the primitive tiling cannot possess an inflation with multiplier τ. Though a local inflation with multiplier τ^3 exists in principle, it seems too complex to be pinned down explicitly (and, to our knowledge, has never been published; some preliminary results in this direction are contained in [Lüc88, Sec. 5]). We will return to this example in the context of the projection method.

In dimensions $d \geqslant 4$, apart from the chair tilings of Example 6.8, few tilings have been constructed and analysed explicitly. A notable exception is the E_8 projection tiling due to Elser and Sloane [ES87, BG98], which is a highly symmetric tiling in 4-space; see also Example 7.14 below. It contains the classic τ-inflation tilings in lower-dimensional substructures, in line with the hierarchy of Remark 3.8. However, we are neither aware of any inflation approach to the Elser–Sloane tiling, nor of any results concerning local rules for this tiling.

Let us now return to the planar case and inject some colour to the graphical presentation of inflation tilings.

6.8. Colourful examples

The theory of tilings undeniably also has an aesthetic dimension, compare [Pen74], which has led to the use of various tilings in architecture and the visual arts; compare the Introduction. The pinwheel tiling facades at the Federation Square in Melbourne, Australia, provide one of the exciting examples of recent times; see Figure 6.58 on page 248.

Some aspects of aperiodic order can be highlighted by the use of different colours for the prototiles or for other characteristic features. In this section, we present some of the classic examples in a colourful fashion. All of them are discussed in some detail elsewhere in this book.

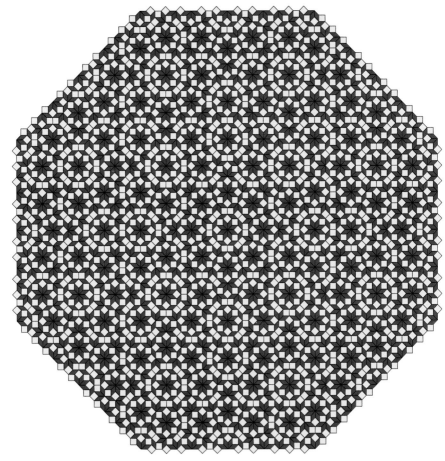

FIGURE 6.41. Undecorated Ammann–Beenker tiling with exact individual D_8 symmetry; compare Section 6.1. The patch shown contains 1912 squares and 2704 rhombuses.

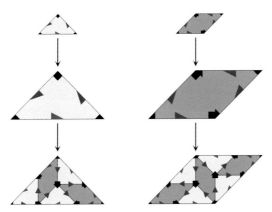

FIGURE 6.42. Inflation rule for decorated Ammann–Beenker tiles. The edge and vertex decorations together provide perfect aperiodic local rules.

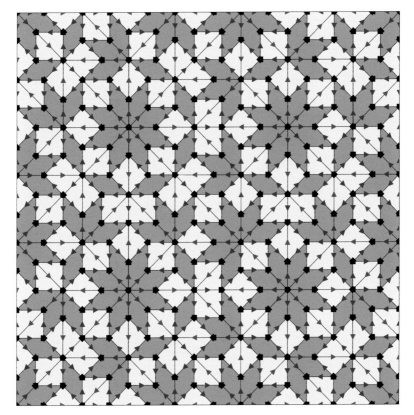

FIGURE 6.43. A patch of the Ammann–Beenker tiling with local rule decorations. The arrows on the edges are locally derivable from the undecorated tiling, while the vertex motif adds local information.

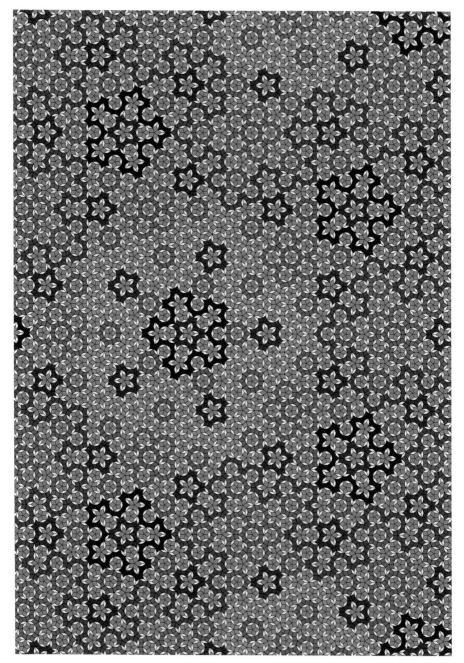

FIGURE 6.44. A patch of the rhombic Penrose tiling. Adjacent large rhombuses form a hierarchy of loops, highlighted by distinct colours.

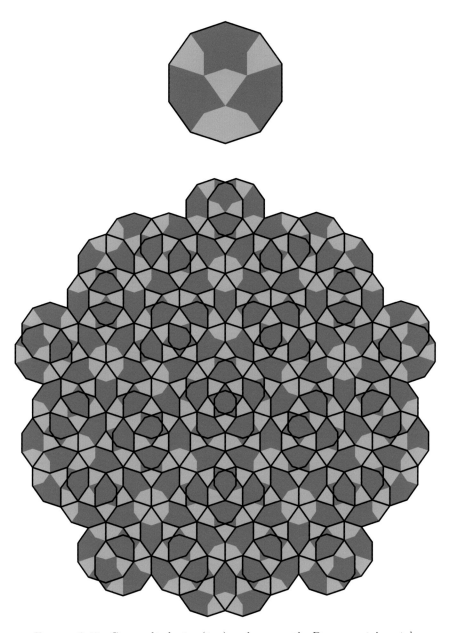

FIGURE 6.45. Gummelt cluster (top) and an exactly D_5-symmetric patch of the corresponding decagon covering of the plane; see Section 6.2.1.

FIGURE 6.46. Square triangle tiling of Section 6.3.1 with exact D_6 symmetry.

FIGURE 6.47. Table tiling of Example 6.2 with two reflection symmetries.

FIGURE 6.48. A patch of the undecorated shield tiling of Section 6.3.2.

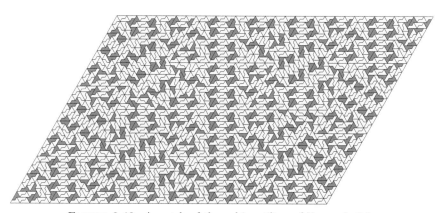

FIGURE 6.49. A patch of the sphinx tiling of Example 6.3.

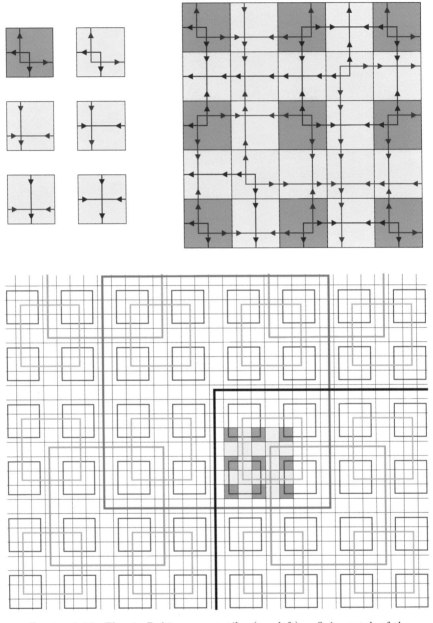

FIGURE 6.50. The six Robinson prototiles (top left), a finite patch of the tiling (top right), and a sketch of the hierarchy of one type of interlocking square loops (bottom, with previous patch highlighted), which demonstrates aperiodicity of the prototile set; see Example 5.11 on page 160.

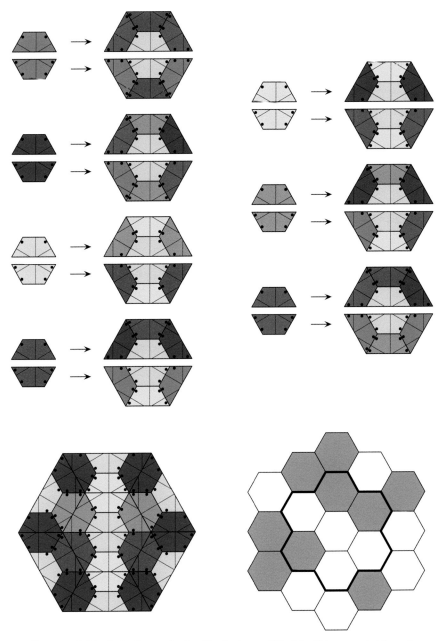

FIGURE 6.51. Inflation rule for Taylor's modified half-hex prototile set (upper part) and the patch produced by two inflation steps from the central hexagon of C type; see Example 6.6 for details. The corresponding parity pattern is shown on the right.

FIGURE 6.52. A patch of Taylor's half-hex inflation tiling (without markings). It is obtained by successive inflation from the central hexagon.

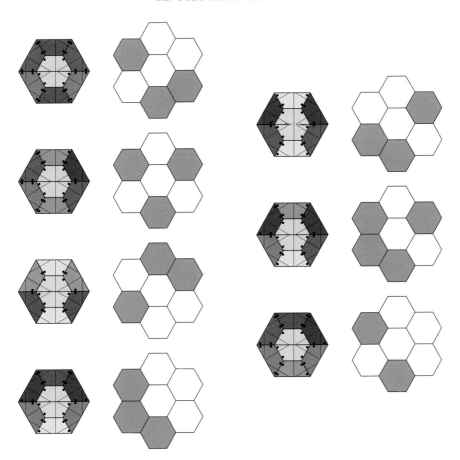

FIGURE 6.53. Parity patterns of level-1 supertiles of the Taylor tiling.

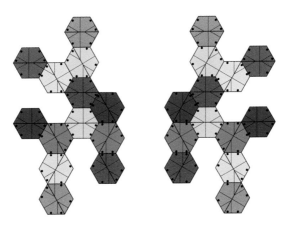

FIGURE 6.54. An opposite pair of Taylor's llamas.

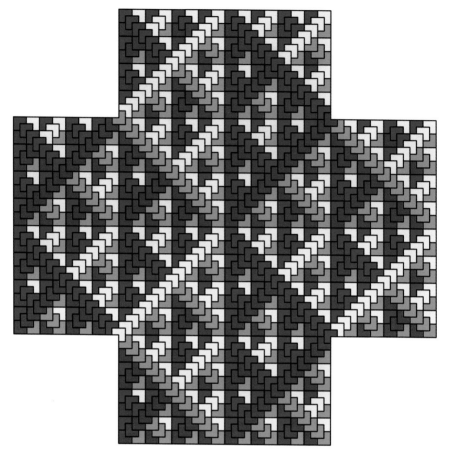

FIGURE 6.55. Chair tiling of Section 6.4 with D_4 colour symmetry.

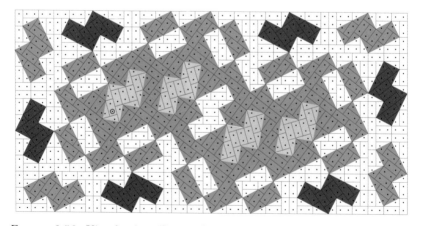

FIGURE 6.56. Kite domino tiling with control points and lattice orientations.

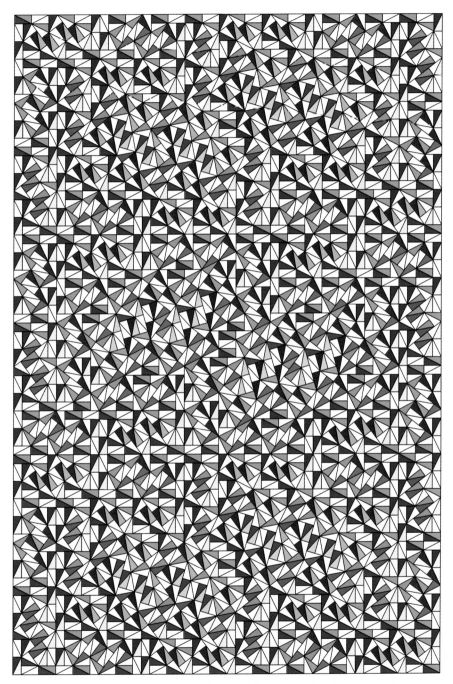

FIGURE 6.57. Pinwheel tiling of Section 6.6 with cyclic orientation colouring for one type of triangle (and none for its mirror image).

FIGURE 6.58. View of one of the buildings at Melbourne's Federation Square (top) and detail of a facade (bottom). Photographs © Uwe Grimm.

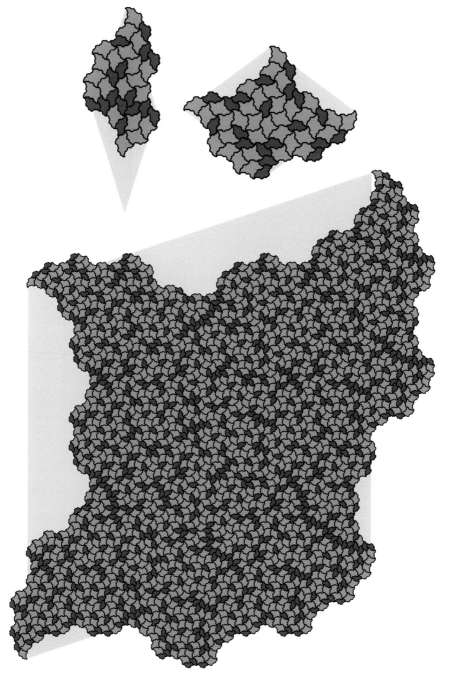

FIGURE 6.59. Chiral tiling obtained from the stone inflation of Remark 6.11 on page 219 with two fractally shaped prototiles.

FIGURE 6.60. Danzer's sevenfold tiling of Section 6.5.2 with three triangular prototiles (shown without decorations).

CHAPTER 7

Projection Method and Model Sets

While substitution and inflation rules provide many examples with aperiodic order, there are alternative methods that give access to other structures and their properties. Arguably the most important and versatile is the projection method, which fortunately has some overlap with the inflation approach. We thus use the opportunity to motivate the projection approach with a simple example from Chapter 4 before embarking on the more general (and formal) setting of the projection method. It may be surprising that we choose the silver mean chain rather than the ubiquitous Fibonacci chain as our guiding example. However, as will become clear at various points, the silver mean chain is slightly simpler and (in a certain sense) more typical.

7.1. Silver mean chain via projection

The silver mean chain emerged from the reflection symmetric bi-infinite fixed point w of the silver mean substitution rule introduced in Example 4.5 on page 88, via the standard geometric interpretation as an aperiodic Delone set in $\mathbb{R}$. Explicitly, a and b are turned into intervals of lengths $1 + \sqrt{2}$ and 1, while the reference point of w is identified with $0 \in \mathbb{R}$. Let now Λ_a and Λ_b denote the sets of left endpoints of the intervals of type a and b, respectively. Both point sets (as well as their union) then are subsets of the $\mathbb{Z}$-module

$$L = \mathbb{Z}[\sqrt{2}] = \{m + n\sqrt{2} \mid m, n \in \mathbb{Z}\},$$

which is a dense subset of $\mathbb{R}$. Arithmetically, it is the ring of integers in the quadratic field $K = \mathbb{Q}(\sqrt{2})$, as discussed in Example 2.14 on page 34. Algebraic conjugation is the field automorphism defined by $\sqrt{2} \mapsto -\sqrt{2}$, with the conjugate of x denoted by x'. This mapping will be crucial for our construction. For reasons that will become clear later, we introduce a mapping $\star\colon K \longrightarrow K$ by $x \mapsto x^\star := x'$, called the $\star$-map (star map).

Recall that the diagonal embedding of $L = \mathbb{Z}[\sqrt{2}]$ in $\mathbb{R}^2$,

$$\mathcal{L} = \{(x, x^\star) \mid x \in \mathbb{Z}[\sqrt{2}]\},$$

is the Minkowski embedding of L, and thus a lattice in $\mathbb{R}^2$, in complete analogy to the discussion of Section 3.4.1. Explicitly, $\mathcal{L}$ is the rectangular lattice spanned by the two basis vectors $(\sqrt{2}, -\sqrt{2})$ and $(1,1)$; see Figure 7.1. Viewed as a point set in $\mathbb{R}^2$, it thus has density $\mathrm{dens}(\mathcal{L}) = \sqrt{2}/4$.

By construction, the canonical projection $\pi : \mathbb{R}^2 \longrightarrow \mathbb{R}$ onto the first coordinate satisfies $\pi(\mathcal{L}) = L$, and its restriction to $\mathcal{L}$ is bijective (note that $x = y$ holds in K if and only if $x' = y'$). One implication is that $\left(\pi|_{\mathcal{L}}\right)^{-1}(\Lambda_a)$ and $\left(\pi|_{\mathcal{L}}\right)^{-1}(\Lambda_b)$ are well-defined subsets of the lattice $\mathcal{L}$. Similarly, the projection π_{int} of $\mathbb{R}^2$ onto the complementary direction shows the same properties, this time involving $L^\star = \pi_{\mathrm{int}}(\mathcal{L})$, which is again $\mathbb{Z}[\sqrt{2}\,]$.

The fixed point property of the word w (under the silver mean substitution $b \mapsto a \mapsto aba$) and its consequence for the inflation structure of the geometric realisation imply that the point sets Λ_a and Λ_b satisfy the equations

$$\Lambda_a = s\Lambda_a \,\dot{\cup}\, \left(s\Lambda_a + (1+s)\right) \dot{\cup}\, s\Lambda_b$$
$$\Lambda_b = s\Lambda_a + s$$

with $s = \lambda_{\mathrm{sm}} = 1 + \sqrt{2}$, and $\dot{\cup}$ denoting the disjoint union of sets. Under algebraic conjugation followed by taking the closure, one obtains a new set of equations for the closed sets $W_a := \overline{\Lambda_a^\star}$ and $W_b := \overline{\Lambda_b^\star}$,

$$(7.1) \qquad \begin{aligned} W_a &= s^\star W_a \cup \left(s^\star W_a + (1+s^\star)\right) \cup s^\star W_b \\ W_b &= s^\star W_a + s^\star \end{aligned}$$

where $s^\star = s' = 1 - \sqrt{2}$ is less than 1 in absolute value, which is the PV property of s. This new set of equations constitutes a coupled iterated function system (IFS) that is a contraction in the Hausdorff metric, with contraction constant $s^\star$; see [Hut81, BM00a, Wic91] for general background. In this setting, one needs to work with compact sets, which was the reason for taking closures when deriving Eq. (7.1). Note that, due to taking closures, the unions in Eq. (7.1) no longer need to be disjoint.

Proposition 7.1. *The coupled system of equations in Eq. (7.1) has a unique solution within the set of pairs of compact subsets of $\mathbb{R}$. It is given by the closed intervals*

$$W_a = \left[\tfrac{\sqrt{2}-2}{2}, \tfrac{\sqrt{2}}{2} \right] \quad and \quad W_b = \left[-\tfrac{\sqrt{2}}{2}, \tfrac{\sqrt{2}-2}{2} \right].$$

PROOF. The existence and uniqueness of the solution with compact subsets of $\mathbb{R}$ follow from standard Hutchinson theory, which is an application of Banach's contraction mapping principle to this setting; compare [Hut81] and [BM00a, Thm. 1.1 and Sec. 4] for details. Verifying that the specified intervals W_a and W_b solve Eq. (7.1) is a straightforward exercise. □

The above construction (with algebraic conjugation followed by taking the closure) now has an immediate consequence.

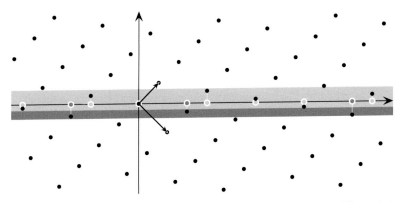

FIGURE 7.1. Cut and project setting for the silver mean chain. The point sets Λ_a (light grey dots) and Λ_b (dark grey dots) on the horizontal axis emerge from the lattice $\mathcal{L}$ (black dots) by the projection of points within the strips defined by the windows W_a (light grey) and W_b (dark grey).

Corollary 7.1. *The point sets Λ_a and Λ_b of the silver mean chain as constructed above satisfy the inclusions*

$$\Lambda_a \subset \{x \in L \mid x^\star \in W_a\} \quad and \quad \Lambda_b \subset \{x \in L \mid x^\star \in W_b\}.$$

With $\Lambda = \Lambda_a \,\dot\cup\, \Lambda_b$, this also implies the inclusion $\Lambda \subset \{x \in L \mid x^\star \in W\}$, where $W = W_a \cup W_b = \left[-\frac{\sqrt{2}}{2}, \frac{\sqrt{2}}{2}\right]$. □

The sets W_a and W_b thus have a selection property for the point sets Λ_a and Λ_b, or for certain supersets of them. Since the boundary points of W_a and W_b are not in L, we also have the slightly stronger property

$$(7.2) \qquad \Lambda_a \subset \{x \in L \mid x^\star \in W_a^\circ\} \quad and \quad \Lambda_b \subset \{x \in L \mid x^\star \in W_b^\circ\},$$

where A° denotes the interior of A; see Figure 7.1 for an illustration.

So far, this amounts to the introduction of a *cut and project scheme* (or CPS for short) for the silver mean chain, as summarised in the diagram

$$
(7.3) \qquad
\begin{array}{ccccc}
\mathbb{R} & \xleftarrow{\;\pi\;} & \mathbb{R} \times \mathbb{R} & \xrightarrow{\;\pi_{\mathrm{int}}\;} & \mathbb{R} \\
{\scriptstyle\text{dense }} \cup & & \cup & & \cup {\scriptstyle\text{ dense}} \\
\mathbb{Z}[\sqrt{2}] & \xleftarrow{\;1\text{--}1\;} & \mathcal{L} & \xrightarrow{\;1\text{--}1\;} & \mathbb{Z}[\sqrt{2}] \\
\| & & & & \| \\
L & & \xrightarrow{\qquad\star\qquad} & & L^\star
\end{array}
$$

for this case. Let us now introduce the notation

$$(7.4) \qquad \curlywedge(A) := \{x \in L \mid x^\star \in A\}$$

for a *projection set* (or cut and project set) within this CPS, on the basis of an arbitrary (but fixed) set $A \subset \mathbb{R}$, which is called the corresponding *window*.

Proposition 7.2. *Consider the projection set $\curlywedge(A)$ for the CPS (7.3). If the window A is bounded, $\curlywedge(A)$ is uniformly discrete; if $A^\circ \neq \varnothing$, $\curlywedge(A)$ is relatively dense. Hence, if A is relatively compact with non-empty interior, $\curlywedge(A)$ is Delone. Moreover, $\curlywedge(A)$ is both FLC and Meyer in this case.*

PROOF. If A is bounded, let J be a compact interval that contains A. Since $\mathcal{L}$ is a lattice, the (convex) rectangle $(t + I) \times J$, with I another compact interval and $t \in \mathbb{R}$, contains finitely many points of $\mathcal{L}$. Since $\mathcal{L}$ is FLC, these rectangles can only select finitely many distinct patches (up to translations) from $\mathcal{L}$. In turn, the projection can only produce finitely many distinct patches of $\curlywedge(J)$ of a given length, again up to translations, so that $\curlywedge(J)$ is FLC as well. This implies in particular that $\curlywedge(J)$, and thus also $\curlywedge(A)$, are uniformly discrete.

If $A^\circ \neq \varnothing$, there exists a non-empty open interval $U \subset A^\circ$ of length $2\varepsilon > 0$, and one has $\curlywedge(U) \subset \curlywedge(A)$. Since $L^\star$ is dense in $\mathbb{R}$, $U \cap L^\star \neq \varnothing$, which implies that $\curlywedge(U)$ contains at least one point, z say. Since also the set $\{x^\star \mid x \in L, \, x > 0\}$ is dense in $\mathbb{R}$, we can choose two lattice points $(x, x^\star)$ and $(y, y^\star)$ in $\mathcal{L}$ with $x, y > 0$, $0 < x^\star < \varepsilon$ and $-\varepsilon < y^\star < 0$. For any $t \in \curlywedge(U)$, consider the four points $t \pm x$ and $t \pm y$. By construction, at least two of these points are again in $\curlywedge(U)$, one on either side of t. Starting with the point z, one sees by induction that $\curlywedge(U)$ continues in both directions with gaps bounded by $\max(x, y)$. This implies $\curlywedge(U)$, and hence also $\curlywedge(A)$, to be relatively dense.

The claim about the Delone property is now obvious. Observe next that

$$\curlywedge(A) - \curlywedge(A) = \{x - y \mid x, y \in L \text{ and } x^\star, y^\star \in A\}$$
$$\subset \{z \in L \mid z^\star \in A - A\} = \curlywedge(A - A).$$

Since $A - A$ is still relatively compact, $\curlywedge(A - A)$ and hence $\curlywedge(A) - \curlywedge(A)$ are uniformly discrete. This implies $\curlywedge(A)$ to be Meyer and FLC. $\square$

Returning to the silver mean point set $\Lambda = \Lambda_a \,\dot\cup\, \Lambda_b$, our previous findings can be summarised as

$$\Lambda_a \subset \curlywedge(W_a^\circ) = \curlywedge(W_a) \quad \text{and} \quad \Lambda_b \subset \curlywedge(W_b^\circ) = \curlywedge(W_b)$$

and similarly for the entire set Λ, where all windows are sets that satisfy $\overline{A^\circ} = A$ and have boundaries of measure 0. We now aim to show $\Lambda = \curlywedge(W^\circ)$, which will then imply the corresponding identities for Λ_a and Λ_b.

The construction of the point set Λ from a primitive substitution rule implies that Λ is linearly repetitive; compare Remark 4.4. It thus has a well-defined density which can be calculated from the PF eigenvectors of the substitution matrix as $\text{dens}(\Lambda) = \frac{1}{2}$. On the other hand, one can derive from the explicit geometry of the CPS (7.3) that also $\curlywedge(W^\circ)$ has a well-defined density. Noting its bijection to the set $\mathcal{L} \cap (\mathbb{R} \times W)$, this density can be calculated as

$$(7.5) \qquad \mathrm{dens}\big(\lambda(W^\circ)\big) = \mathrm{dens}(\mathcal{L})\,\mathrm{vol}(W) = \frac{1}{2} = \mathrm{dens}(\Lambda),$$

where the first equality needs some justification; compare [Els86, Schl98]. The (deeper) reason is the uniform distribution of the points $\pi_{\mathrm{int}}\big(\mathcal{L}\cap(\mathbb{R}\times W)\big)$ in W, for instance when exhausting the discrete set $\mathcal{L}\cap(\mathbb{R}\times W)$ by the points of $\mathcal{L} \cap \big([-n,n] \times W\big)$ with $n \in \mathbb{N}$. Recall that a sequence $(x_i)_{i\in\mathbb{N}}$ of points in a compact interval I of length $|I|$ is *uniformly distributed* in I if

$$(7.6) \qquad \frac{1}{N}\sum_{i=1}^{N} f(x_i) \xrightarrow{N\to\infty} \frac{1}{|I|}\int_I f(x)\,\mathrm{d}x$$

holds for all continuous functions f on I. The above density formula then follows by means of the characteristic function 1_W for f. Since uniform distribution will turn out to be a crucial property for the understanding of cut and project sets, let us discuss this point in more detail.

Example 7.1 (*Weyl sequences*). If $\alpha \in \mathbb{R}$ is irrational, the sequence $(a_m)_{m\in\mathbb{N}}$ with $a_m = (m\alpha) \bmod 1$ is uniformly distributed in the unit interval $[0,1]$. This follows from Weyl's criterion [KN74, Ch. 1, Thm. 2.1], because

$$\lim_{N\to\infty} \frac{1}{N}\sum_{m=1}^{N} \mathrm{e}^{2\pi i k a_m} = 0$$

holds in this case for all $0 \neq k \in \mathbb{Z}$. The proof uses an inequality that readily derives from the sum formula for the geometric series; see [KN74, Ch. 1, Ex. 2.1]. More generally, by an application of the same criterion, one sees that also the sequence defined by $a_m = \big((m\alpha + \beta) \bmod 1\big)$ is uniformly distributed in the unit interval, for any irrational α and any $\beta \in \mathbb{R}$. $\lozenge$

Remark 7.1 (*Powers of PV and Salem numbers*). Although uniform distribution is a rather common phenomenon, there are important sequences in our context which do *not* possess this property. A striking example is the sequence $(a^m \bmod 1)_{m\in\mathbb{N}}$ with $a = 1+\sqrt{2}$, which is a PV unit. This sequence tends to 0 (mod 1) as $m \to \infty$, as do all other sequences of this form where a is a PV number [Sal63, Thm. 1, p. 3]; see also [Dub06].

If a is a Salem number, the corresponding sequence is everywhere dense in $(0,1)$, but nevertheless fails to be uniformly distributed [Sal63, Thm. V, p. 33]. This is even more striking because any sequence that is everywhere dense in the unit interval can be rearranged to a sequence that is uniformly distributed there [KN74, Ch. 2, Cor. 4.2]. $\lozenge$

Let us return to the uniform distribution property of the $\star$-image of the projection set $\lambda(W^\circ)$, and begin with a technical result about its geometry.

Lemma 7.1. *If $x \in \curlywedge(W^\circ)$ is an arbitrary element of the silver mean point set, its successor to the right is either $x + 1$ or $x + 1 + \sqrt{2}$.*

PROOF. Observe first that neither any boundary point of the window W nor the point $(\sqrt{2} - 2)/2$ lies in $L = \mathbb{Z}[\sqrt{2}]$. Now, $x \in \curlywedge(W^\circ)$ means $x \in L$ with $x^\star \in W^\circ$. If $x^\star \in \left(-\frac{\sqrt{2}}{2}, \frac{\sqrt{2}-2}{2}\right) = W_b^\circ$, we have $x^\star + 1 \in W^\circ$ and hence $x + 1 \in \curlywedge(W^\circ)$. Similarly, $x^\star \in \left(\frac{\sqrt{2}-2}{2}, \frac{\sqrt{2}}{2}\right) = W_a^\circ$ implies $x^\star + 1 - \sqrt{2} \in W^\circ$, so that $x + 1 + \sqrt{2} \in \curlywedge(W^\circ)$. The maximal distance between neighbouring points is thus $1 + \sqrt{2}$. Our claim now follows by showing that only the distances 1 and $1 + \sqrt{2}$ occur between neighbouring points of $\curlywedge(W^\circ)$.

The distance z between neighbouring points of $\curlywedge(W^\circ)$ must be a positive element of L, hence $z = m + n\sqrt{2}$ with $m, n \in \mathbb{Z}$ and $0 < z \leqslant 1 + \sqrt{2}$. In addition, we get the restriction $-\sqrt{2} < z^\star = m - n\sqrt{2} < \sqrt{2}$ from the window condition. For $m, n \geqslant 0$, the only solutions are $z = 1$ and $z = 1 + \sqrt{2}$, as $z = \sqrt{2}$ is excluded by the second inequality. The only possible cases left to consider are $m, n \neq 0$ with opposite signs. However, none of them satisfies the window condition, which completes the argument.   □

Consider an ordered coordinatisation of the projection set $\curlywedge(W^\circ)$ in the form $\curlywedge(W^\circ) = \{x_m \mid m \in \mathbb{Z}\}$, where we can choose $x_0 = 0$ so that $x_{-m} = -x_m$ for $m \in \mathbb{Z}$ due to the reflection symmetry of the window. The ordering results in the recursion $x_{m+1} = \min\{x \in \curlywedge(W^\circ) \mid x > x_m\}$ for $m \geqslant 0$. By Lemma 7.1, the distance between neighbouring points of $\curlywedge(W^\circ)$ is either 1 or $s = 1 + \sqrt{2}$, meaning that $x_{m+1} = x_m + 1$ or $x_{m+1} = x_m + s$. By induction, with $x_0 = 0$, this implies

$$(7.7) \qquad\qquad x_m = m + c_m \sqrt{2},$$

where $(c_m)_{m \in \mathbb{Z}}$ is an integer-valued sequence with $c_0 = 0$ and $c_{-m} = -c_m$ due to the symmetry of the sequence. Let us now determine the c_m (and hence the coordinates) explicitly, using the projection setting.

Our above analysis implies that $x_m^\star \in W = W_a \cup W_b$ for all $m \in \mathbb{Z}$, but never an element of $\partial W_a \cup \partial W_b$. Taking the $\star$-image, the successor of x_m is then determined by the recursion

$$(7.8) \qquad\qquad x_{m+1}^\star = x_m^\star + \begin{cases} 1, & \text{if } x_m^\star \in W_b, \\ s^\star, & \text{if } x_m^\star \in W_a. \end{cases}$$

This follows from the structure of the underlying fixed point word w and the fact that, for any $x^\star \in W^\circ$, precisely one of the two choices is again in W°. Below, we need the Gauss bracket $[x] := \max\{n \in \mathbb{Z} \mid n \leqslant x\}$ and the fractional part $\{x\}$ of a number $x \in \mathbb{R}$, where we define the latter as $\{x\} = x - [x]$. This implies the relation $\{-x\} = 1 - \{x\}$.

Lemma 7.2. *The $\star$-image of the above coordinatisation of $\curlywedge(W^\circ)$ satisfies*

$$x_m^\star = y_m := z_m - \frac{1}{2}\sqrt{2} \quad with \quad z_m := \left(\frac{1}{2}\sqrt{2} + m\right) \bmod \sqrt{2},$$

which is valid for all $m \in \mathbb{Z}$. The corresponding direct space coordinates are

$$x_m = m + c_m\sqrt{2} \quad with \quad c_m = \left[\frac{1 + m\sqrt{2}}{2}\right],$$

again for all $m \in \mathbb{Z}$.

PROOF. One clearly has $y_0 = 0$, and a careful consideration of the modulo condition shows that $y_{-m} = -y_m$ for all $m \in \mathbb{N}$. It thus suffices to prove the claim for $m \in \mathbb{N}_0$ by induction. Indeed, one has $z_0 = \frac{1}{2}\sqrt{2}$, and then obtains z_{m+1} from z_m by adding 1 when $z_m \in [0, \sqrt{2} - 1)$, or by adding $s^\star = 1 - \sqrt{2}$ when $z_m \in [\sqrt{2} - 1, \sqrt{2})$, where the boundary points of the intervals are never met. This recursion implies $z_m = (z_0 + m) \bmod \sqrt{2}$, where $z_m \in [0, \sqrt{2})$ by construction. This means

$$\frac{z_m}{\sqrt{2}} = \frac{1 + m\sqrt{2}}{2} - \left[\frac{1 + m\sqrt{2}}{2}\right] = \left\{\frac{1 + m\sqrt{2}}{2}\right\},$$

with $[.]$ and $\{.\}$ as defined above.

As mentioned before, x_m for $m \geqslant 0$ is of the form given in Eq. (7.7). Since c_m is an integer, one has $x_m^\star = m - c_m\sqrt{2} = z_m - \frac{1}{2}\sqrt{2}$. The claimed formula emerges by solving this equation for c_m. The coordinates for negative m follow by the symmetry of the silver mean chain, while one checks that the Gauss bracket $[.]$ indeed implies $c_{-m} = -c_m$.   □

An interesting consequence of Eq. (7.7) and Lemma 7.2 is the relation

$$(7.9) \qquad\qquad x_m + x_m^\star = 2m,$$

which holds for all $m \in \mathbb{Z}$. Since $|x_m^\star| < \frac{1}{2}\sqrt{2}$ by Eq. (7.2), we may conclude that $2\mathbb{Z}$ is an *average lattice* for $\varLambda$ in the sense that we have a bijection between $2\mathbb{Z}$ and $\varLambda$ with bounded distance between a point and its image; compare the concept of wobbling equivalence in Remark 5.2. This implies $\mathrm{dens}(\varLambda) = \frac{1}{2}$, as we already know from the substitution property. In fact, the projection approach results in a stronger property of $\varLambda^\star$.

Proposition 7.3. *The sequence $(x_m^\star)_{m\in\mathbb{N}}$ derived from the projection set $\curlywedge(W^\circ)$ is uniformly distributed in the interval W. The same statement also holds for the two-sided sequence $(x_m^\star)_{m\in\mathbb{Z}}$.*

PROOF. Use Lemma 7.2 and observe that one can rewrite z_m as

$$z_m = \left(\frac{1}{2}\sqrt{2} + m\right) \bmod \sqrt{2} = \sqrt{2} \cdot \left(\left(\frac{1}{2} + \frac{m}{2}\sqrt{2}\right) \bmod 1\right).$$

Example 7.1 with $\alpha = \frac{\sqrt{2}}{2}$, which is irrational, and $\beta = \frac{1}{2}$ shows that $(z_m/\sqrt{2})_{m \in \mathbb{N}}$ is uniformly distributed in the unit interval, so that $(x_m^\star)_{m \in \mathbb{N}}$ is then uniformly distributed in the interval $\left[-\frac{\sqrt{2}}{2}, \frac{\sqrt{2}}{2} \right] = W$.

The second claim is clear from the reflection symmetry of $\lambda(W^\circ)$. □

Remark 7.2 (*Well-distributed sequences*). The sequences $(a_m)_{m \in \mathbb{N}}$ of Example 7.1 satisfy an even stronger property than uniform distribution. They are *well-distributed* modulo 1 in the sense that

$$\lim_{N \to \infty} \frac{1}{N} \sum_{m=k+1}^{k+N} g(a_m) = \int_0^1 g(x)\, \mathrm{d}x$$

holds uniformly in $k \in \mathbb{N}_0$, for all continuous functions g on $[0, 1]$; compare [KN74, Ch. 1.5]. As a consequence, the sequence $(x_m^\star)_{m \in \mathbb{N}}$ from Proposition 7.3 is well-distributed in the interval W. However, this stronger property is not needed for our further discussion. ◇

At this point, we already know that $\Lambda \subset \lambda(W)$, and the previous argument is a strong indication that we actually have equality here.

Theorem 7.1. *The silver mean point set Λ satisfies $\Lambda = \lambda(W^\circ) = \lambda(W)$ within the CPS of Eq. (7.3), with window $W = \left[-\frac{\sqrt{2}}{2}, \frac{\sqrt{2}}{2} \right]$. The corresponding identities hold for Λ_a and Λ_b as well, with windows W_a and W_b.*

PROOF. The identity $\lambda(W^\circ) = \lambda(W)$ follows from $\mathcal{L} \cap (\mathbb{R} \times \partial W) = \varnothing$, as mentioned before. Moreover, we have the inclusion $\Lambda \subset \lambda(W^\circ)$, where both sets have density $\frac{1}{2}$ and Λ is repetitive by Lemma 4.4. The claim now follows by showing that also $\lambda(W^\circ)$ is repetitive, for the following reason. If the set $\lambda(W^\circ) \setminus \Lambda \neq \varnothing$, there must be at least one extra point in $\lambda(W^\circ)$. Since the distance between neighbouring points in Λ can only be 1 or $1 + \sqrt{2}$ by construction, the addition of any single point creates a new distance. This means that $\lambda(W^\circ)$ would then contain a patch that Λ does not, which has to repeat with bounded gaps. Consequently, $\lambda(W^\circ) \setminus \Lambda$ must have positive density, which is a contradiction.

It remains to prove the repetitivity of the set $\lambda(W^\circ)$. Let $I \subset \mathbb{R}$ be a closed interval and consider the patch $P = \lambda(W^\circ) \cap I$, assuming $P \neq \varnothing$. By construction, $P^\star \subset W^\circ$, as well as $P^\star + (-\varepsilon, \varepsilon) \subset W^\circ$ for some $\varepsilon > 0$, because $P^\star$ is a finite set while W° is open. The set $t + P$ is now a patch of $\lambda(W^\circ)$ for all $t \in \lambda\big((-\varepsilon, \varepsilon)\big)$ because $x \in P$ implies

$$(t + x)^\star = t^\star + x^\star \in (-\varepsilon, \varepsilon) + P^\star \subset W^\circ.$$

Due to the properties of the CPS (7.3), the projection set $\lambda\big((-\varepsilon, \varepsilon)\big)$ is relatively dense by Proposition 7.2, wherefore the patch P repeats with bounded gaps in $\lambda(W^\circ)$. Since this argument applies to any finite patch of $\lambda(W^\circ)$, and $\lambda(W^\circ)$ is clearly FLC, repetitivity follows. The proof for the subsets Λ_a and Λ_b follows from $\Lambda = \Lambda_a \dot\cup \Lambda_b$ together with the bijectivity of $\pi|_\mathcal{L}$. □

The attentive reader may have noticed that the previous result can be obtained also without using the density and repetitivity arguments. Indeed, we knew $\Lambda \subset \lambda(W)$ from Corollary 7.1, with both sets sharing the property that the distance between neighbouring points is either 1 or $1 + \sqrt{2}$. For Λ, this follows from the substitution structure, while Lemma 7.1 shows this for $\lambda(W)$. A moment's reflection reveals that this is not possible unless both sets are equal. The more elaborate argument used above, however, is applicable in many similar or more general situations. To this end, we need a different (and less special) approach to prove uniform distribution later on.

A projection set for the CPS (7.3) with a relatively compact window A is called a *model set*[1]. It is called *regular* when the boundary ∂A of the window has measure 0, and *generic* (or *non-singular*) when $L^\star \cap \partial A = \varnothing$. Note that two properties of the CPS from Eq. (7.3) have not been used so far, namely the denseness of L in $\mathbb{R}$ and the bijectivity of $\pi_{\mathrm{int}}|_\mathcal{L}$. These properties will later be waived to reach a formulation of greater generality.

Corollary 7.2. *The silver mean point set Λ is a regular, generic model set for the CPS of Eq. (7.3) with the window W of Theorem 7.1.* □

It is clear that any translate $t + \Lambda = t + \lambda(W)$ with $t \in \mathbb{R}$ should also be considered as a model set. A little less obvious is what happens with $\lambda(h+W)$, again with $h \in \mathbb{R}$. Two situations are possible for the intersection $L^\star \cap (h + \partial W)$, which can either be empty (*generic* or *non-singular* case) or not (*singular* case).

Lemma 7.3. *If $h \in \mathbb{R}$ is a generic translation of the window W of Theorem 7.1, the point set $t + \lambda(h + W)$ is an element of $\mathrm{LI}(\Lambda)$, for all $t \in \mathbb{R}$.*

PROOF. Since it is clear that a point set is LI from any translate of itself, it suffices to show $\lambda(h + W) \in \mathrm{LI}(\Lambda)$, where $\Lambda = \lambda(W)$ by Theorem 7.1. Let $P = \Lambda \cap I$ be any non-empty finite patch of Λ (with suitably chosen compact interval I). Since Λ is generic, $P^\star \subset W^\circ$. For all $x \in L$, the shifted set $x^\star + W$ is again in a generic position. Since $L^\star \subset \mathbb{R}$ is dense, there is an $x^\star \in L^\star$ such that $x^\star + P^\star \subset h + W$. This shows that $x + P$ is a patch of $\lambda(h + W)$. Since the same type of argument also works backwards, and for arbitrary patches, Λ and $\lambda(h + W)$ are LI. □

This result does *not* extend to $\lambda(h + W)$ for singular h, because one picks up an extra point from the boundary. This is caused by the fact that the length of the window W is $\sqrt{2} \in \mathbb{Z}[\sqrt{2}] = L^\star$, so that both boundary

---

[1]The formal definitions will be repeated in a more general setting in Section 7.2.

points then contribute to the projection setting. To remain in the LI class of Λ, one must select one of the two points arising from the boundary, which gives two locally indistinguishable point sets that differ just in these two points. This is the *singular* situation we met before in Example 4.6 and the discussion following it. The two choices are obtained as limits of sequences of the form $\bigl(\lambda(h_i + W)\bigr)_{i \in \mathbb{N}}$ in the local topology, with all shifts h_i generic and approaching the singular parameter $h \in \mathbb{R}$ from above or from below.

Recall that the geometric (or continuous) hull of Λ is

$$\mathbb{X}(\Lambda) \;=\; \overline{\{t + \Lambda \mid t \in \mathbb{R}\}},$$

where the closure is taken in the local topology; compare Section 4.2.

Proposition 7.4. *The geometric hull $\mathbb{X}(\Lambda)$ of the silver mean point set Λ coincides with* $\mathrm{LI}(\Lambda)$.

PROOF. The inclusion $\mathrm{LI}(\Lambda) \subset \mathbb{X}(\Lambda)$ follows from the fact that each element of $\mathrm{LI}(\Lambda)$ is the limit (in the local topology, since Λ is FLC) of a converging sequence $\bigl(t_n + \Lambda\bigr)_{n \in \mathbb{N}}$ and hence an element of $\mathbb{X}(\Lambda)$.

For the converse direction, let $\Lambda' \in \mathbb{X}(\Lambda)$ be arbitrary. There is always a small translation t (bounded by $(1+\sqrt{2}\,)/2$) such that $0 \in t + \Lambda'$, where Λ' and $t + \Lambda'$ are clearly LI. Observe that there is a natural (continuous) bijection between the transversal $\mathbb{X}_0(\Lambda) := \{\Lambda' \in \mathbb{X}(\Lambda) \mid 0 \in \Lambda'\}$ and the discrete hull of the silver mean chain. Our claim now follows from the corresponding statement for the discrete hull in Proposition 4.1. □

Let us also mention at this point that the somewhat delicate distinction between generic and singular cases, and how to treat the latter, is intimately connected with the Cantor structure of the transversal $\mathbb{X}_0(\Lambda)$. From the projection perspective, the latter emerges by replacing the naive singular projection set (which is not in the hull) by the one-sided limits of generic (non-singular) sets, which are then distinct members of the hull.

Despite the intrinsic aperiodicity of the silver mean chain, it is sometimes useful to consider this chain as a limit of a sequence of periodic systems.

Example 7.2 (*Periodic approximants of the silver mean chain Λ*). One systematic approach to periodic approximants employs the finite continued fractions for $\sqrt{2}$, which are the rational numbers $1, \frac{3}{2}, \frac{7}{5}, \frac{17}{12}, \ldots$; see Example 8.1 below for details. These fractions approximate $\sqrt{2}$ in an optimal way. Let $r_i = \frac{p_i}{q_i}$ be one of these numbers (with $r_0 = 1$ say), and rotate the strip of Figure 7.1 around the origin such that it is parallel to the lattice direction $(t_i, t_i^\star) \in \mathcal{L}$ with $t_i = p_i + q_i \sqrt{2}$. Projecting all lattice points in this rotated strip orthogonally to the horizontal axis produces a point set Λ_i that is crystallographic with period t_i. Note that $t_i \xrightarrow{i \to \infty} \infty$ while $t_i^\star \xrightarrow{i \to \infty} 0$. By construction, these approximants, as $i \to \infty$, agree with Λ on patches around the origin of increasing size, so that $\Lambda_i \xrightarrow{i \to \infty} \Lambda$ in the local topology.

Note that Λ_i consists of the periodic repetition of intervals, coded by a finite word of length p_i. One choice to write down the fundamental word leads to a, aba, $abaaaba$, $abaaabaabaabaaaba$ and so on. By a comparison with Example 4.5 on page 88, this is easily recognised as the sequence $\varrho^i(a)$ with $i \in \mathbb{N}_0$. This way, the approximants on the basis of the projection formalism can be connected with the underlying substitution rule ϱ. $\Diamond$

This approach works for many other cases in complete analogy, and can also be generalised to structures in higher dimensions; see [GKQ95] for a systematic exposition.

7.1.1. CLUSTER FREQUENCIES AND INVARIANT MEASURE

The model set description of the silver mean point set Λ permits the calculation of absolute and relative frequencies of arbitrary *finite* subsets $P \subset \Lambda$. Indeed, for any such $P \neq \varnothing$, there is a compact set $K \subset \mathbb{R}$ (even with non-empty interior and $K = \overline{K^\circ}$, as Λ is locally finite) such that $P = \Lambda \cap K$. If $t + P \subset \Lambda$ for some $t \in \mathbb{R}$, we must have $t \in L$. Since $\Lambda = \curlywedge(W^\circ)$ by Theorem 7.1 and thus $P^\star \subset W^\circ$, one finds that the *repetition set* of P is

$$\mathrm{rep}(P) := \big\{ t \in L \mid t + P \subset \Lambda \big\} = \big\{ t \in L \mid t^\star + P^\star \subset W^\circ \big\} = \curlywedge\big((-\varepsilon_-, \varepsilon_+)\big),$$

where $\varepsilon_\pm = \inf\big\{ |\pm\frac{\sqrt{2}}{2} - y| \ \big| \ y \in P^\star \big\}$, and hence a regular model set for the CPS (7.3). If $t \in L$, the inclusion $t^\star + P^\star \subset W^\circ$ is equivalent to $x^\star \in W^\circ - t^\star$ for all $x \in P$, so that

$$(-\varepsilon_-, \varepsilon_+) = \bigcap_{x \in P} (W^\circ - x^\star) \neq \varnothing.$$

Since the repetition set of P is a regular model set, it has uniform density $\mathrm{dens}\big(\mathrm{rep}(P)\big) = \mathrm{dens}(\mathcal{L}) \, \mathrm{vol}\big(\bigcap_{x \in P} (W - x^\star)\big)$, which is the *absolute frequency* (per unit length) $\mathrm{abs\,freq}_\Lambda(P)$. Consequently, the *relative frequency* (per point of Λ) of the cluster P exists uniformly and is given by

$$(7.10) \qquad \mathrm{rel\,freq}_\Lambda(P) = \frac{\mathrm{abs\,freq}_\Lambda(P)}{\mathrm{dens}(\Lambda)} = \frac{\mathrm{vol}\big(\bigcap_{x \in P} (W - x^\star)\big)}{\mathrm{vol}(W)}.$$

Since the cluster P was arbitrary, this also shows that the point set Λ possesses the UCF property.

The relative cluster frequencies are always numbers in the unit interval $[0, 1]$ and can be used to define an invariant probability measure on the set $\mathbb{X}_0(\Lambda) = \{ \Lambda' \in \mathbb{X}(\Lambda) \mid 0 \in \Lambda' \}$ (which contains Λ) as follows. Let P be a non-empty finite cluster of Λ. To the corresponding cylinder set

$$Z_P := \big\{ \Lambda' \in \mathbb{X}_0(\Lambda) \mid P \subset \Lambda' \big\},$$

we assign the measure

$$\mu_0(Z_P) \,=\, \operatorname{rel\,freq}_\Lambda(P).$$

This extends to a positive measure on $\mathbb{X}_0(\Lambda)$ by standard arguments, where one uses the σ-algebra generated by the cylinder sets; see [Bil95, Sec. 3]. In particular, $\mu_0\big(\mathbb{X}_0(\Lambda)\big) = 1$.

The matching measure μ on the full hull $\mathbb{X}(\Lambda)$ is obtained by a similar procedure. If P is again a finite cluster (with defining compact set K), and if $\varepsilon > 0$ is smaller than the packing radius of Λ, we use the cylinder sets

$$Z_{P,\varepsilon} \,:=\, \big\{\Lambda' \in \mathbb{X}(\Lambda) \mid (\Lambda' - t) \cap K = P \text{ for some } t \in B_\varepsilon(0)\big\}$$

and define their measures by

$$\mu(Z_{P,\varepsilon}) \,=\, \operatorname{abs\,freq}_\Lambda(P)\,\operatorname{vol}(B_\varepsilon).$$

This extends to a probability measure on the compact hull $\mathbb{X}(\Lambda)$, as can be shown by the methods from [Par67, Ch. V] around the Kolmogorov extension theorem. The measure μ is translation invariant because $\mu(Z_{P,\varepsilon}) = \mu(Z_{t+P,\varepsilon})$ for all $t \in \mathbb{R}$ and arbitrary clusters P. The connection between the two measures μ and μ_0 emerges from the filtration property

$$\lim_{\varepsilon \searrow 0} \frac{\mu(Z_{P,\varepsilon})}{\operatorname{dens}(\Lambda)\,\operatorname{vol}(B_\varepsilon)} \,=\, \mu_0(Z_P),$$

which reflects the fact that $\mathbb{X}_0(\Lambda)$ is a subset of $\mathbb{X}(\Lambda)$ with $\mu\big(\mathbb{X}_0(\Lambda)\big) = 0$.

For FLC point sets, the frequencies of finite patches are relevant with respect to questions of ergodicity. Due to Oxtoby's theorem, we know that such a system is uniquely ergodic if and only if all patch frequencies exist uniformly; compare [Oxt52] and [Que10, Thm. 4.3] as well as Proposition 4.4 and its extension in [MR13]. The difference to Chapter 4 lies in the group acting on the system, which is now $\mathbb{R}$ (general translations) rather than $\mathbb{Z}$ (discrete shifts); see [FR14] for a generalisation to the corresponding situation (beyond FLC) with the local rubber topology.

Example 7.3 (*Fibonacci chain as model set*)**.** Let us contrast the above derivation with the corresponding situation for the golden mean τ. To this end, consider the Minkowski embedding of $\mathbb{Z}[\tau]$ of Figure 3.3 on page 60 as a CPS, with lattice $\mathcal{L} = \big\{(x, x^\star) \mid x \in \mathbb{Z}[\tau]\big\}$. As above, $x^\star = x'$ is algebraic conjugation, this time within the quadratic field $\mathbb{Q}(\sqrt{5}\,)$. Consider the regular model set $\Lambda = \lambda\big((-1, \tau - 1]\big)$ and write it as the disjoint union $\Lambda = \Lambda_a \dot\cup \Lambda_b$ with point sets $\Lambda_a = \lambda\big((\tau - 2, \tau - 1]\big)$ and $\Lambda_b = \lambda\big((-1, \tau - 2]\big)$. Let us define the half-open intervals $V_a = (\tau - 2, \tau - 1]$ and $V_b = (-1, \tau - 2]$, of lengths 1 and $1/\tau$. It is an exercise to check that they satisfy the equations

$$\tau^2 V_a \,=\, V_a \,\dot\cup\, V_b \,\dot\cup\, (V_a + 1) \quad \text{and} \quad \tau^2 V_b \,=\, (V_a - \tau) \,\dot\cup\, (V_b - \tau),$$

where the disjointness of the unions is a consequence of the special choice made for V_a and V_b. Algebraic conjugation turns this into the equations

$$\Lambda_a = \tau^2 \Lambda_a \,\dot{\cup}\, \tau^2 \Lambda_b \,\dot{\cup}\, (\tau^2 \Lambda_a + \tau^2) \quad \text{and} \quad \Lambda_b = (\tau^2 \Lambda_a + \tau) \,\dot{\cup}\, (\tau^2 \Lambda_b + \tau)$$

for the point sets Λ_a and Λ_b. This is precisely the set of equations induced by the substitution $a \mapsto aba$, $b \mapsto ab$, when a and b are represented by intervals of length τ and 1. These intervals define point sets by their left endpoints. This substitution is the square of the Fibonacci rule ϱ_F of Example 4.6, thus with inflation multiplier τ^2. The regular model set Λ corresponds to the fixed point of ϱ_F^2 with seed $a|a$, which can be proved as above. In particular, Λ is a repetitive point set of density $(\tau + 2)/5$ with nearest neighbour distances either 1 or τ, which must thus be one of the two fixed points constructed in Example 4.6. The observation that $-\tau$ is in Λ while -1 is not determines the fixed point. The other fixed point corresponds to choosing the windows as $V_a' = [\tau - 2, \tau - 1)$ and $V_b' = [-1, \tau - 2)$, now including -1 but not $-\tau$. This example once more demonstrates the subtle role of the boundary of the windows for the description of singular elements of the hull by projection. Here, the difficulty can be resolved by using half-open intervals as windows, which corresponds to the different choices of limits. $\Diamond$

We hope that the examples are sufficiently transparent to motivate the general setting which comes next. In particular, the internal space will be a more general LCAG than $\mathbb{R}^m$, the necessity of which will be justified later by further examples, in particular of Toeplitz type.

7.2. Cut and project schemes and model sets

Let us begin by generalising the CPS of Eq. (7.3).

Definition 7.1. A *cut and project scheme* (CPS) is a triple $(\mathbb{R}^d, H, \mathcal{L})$ with a (compactly generated) LCAG H, a lattice $\mathcal{L}$ in $\mathbb{R}^d \times H$ and the two natural projections $\pi : \mathbb{R}^d \times H \longrightarrow \mathbb{R}^d$ and $\pi_{\text{int}} : \mathbb{R}^d \times H \longrightarrow H$, subject to the conditions that $\pi|_{\mathcal{L}}$ is injective and that $\pi_{\text{int}}(\mathcal{L})$ is dense in H.

As before, we write $L = \pi(\mathcal{L})$. Since, for a given CPS, π is then a bijection between $\mathcal{L}$ and L, there is a well-defined mapping $\star : L \longrightarrow H$ specified by

$$x \mapsto x^\star := \pi_{\text{int}}\big((\pi|_{\mathcal{L}})^{-1}(x)\big),$$

where $(\pi|_{\mathcal{L}})^{-1}(x)$ is the unique point in the set $\mathcal{L} \cap \pi^{-1}(x)$. This mapping is called the *star map* of the CPS. The $\star$-image of L is denoted by $L^\star$ as before. Furthermore, $\mathcal{L}$ can again be viewed as a diagonal embedding of L,

$$\mathcal{L} = \big\{ (x, x^\star) \mid x \in L \big\}.$$

In analogy to Eq. (7.3), the setting of a general CPS is conveniently summarised in the following diagram.

$$
\begin{array}{ccccc}
\mathbb{R}^d & \xleftarrow{\ \pi\ } & \mathbb{R}^d \times H & \xrightarrow{\ \pi_{\mathrm{int}}\ } & H \\
\cup & & \cup & & \cup\ \text{dense} \\
\pi(\mathcal{L}) & \xleftarrow{\ 1-1\ } & \mathcal{L} & \longrightarrow & \pi_{\mathrm{int}}(\mathcal{L}) \\
\| & & & & \| \\
L & & \xrightarrow{\qquad\quad \star \qquad\quad} & & L^{\star}
\end{array}
$$

(7.11)

Let us briefly mention that there is a matching CPS that involves the dual lattice $\mathcal{L}^*$, and the dual group $\widehat{\mathbb{R}^d \times H} = \widehat{\mathbb{R}^d} \times \widehat{H}$. This is known as the *dual* CPS; see [Moo97a, Sec. 5]. Since $\mathbb{R}^d$ is self-dual, this is straightforward for *Euclidean model sets*, where $H = \mathbb{R}^m$ for some $m \in \mathbb{N}$. Here, the notion of the dual lattice is precisely the one introduced in Eq. (3.4).

Remark 7.3 (*General CPS*). The restriction of H to compactly generated groups is not necessary, but sufficient in practice. A further generalisation is achieved by replacing $\mathbb{R}^d$ with a σ-compact LCAG G, thus leading to a CPS of the form $(G, H, \mathcal{L})$, where the lattice $\mathcal{L}$ is a co-compact discrete subgroup of the LCAG $G \times H$. Most of the theory can be formulated in this generality, see [Mey72, Moo97a, Schl98, Schl00, Moo00] for details. Since our exposition deals mainly with the case $G = \mathbb{R}^d$, we stick to $\mathbb{R}^d$ for simplicity. $\lozenge$

For a given CPS $(\mathbb{R}^d, H, \mathcal{L})$ and a (general) set $A \subset H$,

(7.12) $\curlywedge(A) := \{x \in L \mid x^{\star} \in A\}$

denotes a projection set within the CPS. The set A is called its acceptance set, coding set or *window*. Let us now expand on the most important situation.

Definition 7.2. Let $(\mathbb{R}^d, H, \mathcal{L})$ be a CPS according to Definition 7.1. If $W \subset H$ is a relatively compact set with non-empty interior, the projection set $\curlywedge(W)$, or any translate $t + \curlywedge(W)$ with $t \in \mathbb{R}^d$, is called a *model set*. A model set is termed *regular* when $\mu_H(\partial W) = 0$, where μ_H is the Haar measure of H. If $L^{\star} \cap \partial W = \varnothing$, the model set is called *generic*.

Remark 7.4 (*Various notions*). The term 'regular' refers to Riemann integrability of the characteristic function 1_W. Such functions can be approximated from above and from below by continuous functions of compact support; compare [BM00a, Lemma A.6] and the discussion following it. If the window W is not in a generic position (meaning that $L^{\star} \cap \partial W \neq \varnothing$), the corresponding model set is called *singular*. A standard Baire category argument shows that the set of translations $\{y \in H \mid L^{\star} \cap (y + \partial W) = \varnothing\}$ is a set of second category in this case (which explains the term 'generic'); see [Schl98, Moo00, BM04, BLM07] for details on this type of argument in the present context. If W is relatively compact, possibly also with empty interior, the projection set $\curlywedge(W)$ is referred to as a *weak model set*. We will meet examples of this kind in Section 10.4. $\Diamond$

For the further development of the theory, we need the following covering result for the embedding space [Moo97a, Lemma 2.5].

Lemma 7.4. *Let $(\mathbb{R}^d, H, \mathcal{L})$ be a CPS as in Eq. (7.11), and let $U \subset H$ be a non-empty open set. Then, there exists a compact set $K \subset \mathbb{R}^d$ such that $\mathbb{R}^d \times H = \mathcal{L} + (K \times U)$.*

PROOF. Since $\mathcal{L} \subset \mathbb{R}^d \times H$ is a lattice, there is a compact set $C \subset \mathbb{R}^d \times H$ with $\mathcal{L} + C = \mathbb{R}^d \times H$. The natural projections of C define compact sets $K_1 = \pi(C) \subset \mathbb{R}^d$ and $K_2 = \pi_{\mathrm{int}}(C) \subset H$ so that

$$\mathbb{R}^d \times H \;=\; \mathcal{L} + (K_1 \times K_2).$$

The denseness of $\pi_{\mathrm{int}}(\mathcal{L})$ in H implies that $\bigcup_{p \in \mathcal{L}} \big(\pi_{\mathrm{int}}(p) + U\big) = H \supset K_2$. Since K_2 is compact, there is a finite subcover, meaning that

$$\bigcup_{q \in F} \big(\pi_{\mathrm{int}}(q) + U\big) \;\supset\; K_2$$

for a finite set $F \subset \mathcal{L}$. Let $z \in \mathbb{R}^d \times H$ be arbitrary but fixed. By construction, there is some $p \in \mathcal{L}$ so that $z - p \in K_1 \times K_2$, and thus some $q \in F$ with $\pi_{\mathrm{int}}(z - p - q) \in U$. Consequently, $\pi(z - p - q) \in K_1 - \pi(F) =: K \subset \mathbb{R}^d$, which is compact, and

$$z \;=\; p + q + (z - p - q) \;\in\; \mathcal{L} + (K \times U).$$

Since z was arbitrary, this establishes the claim. $\square$

Proposition 7.5. *Let $(\mathbb{R}^d, H, \mathcal{L})$ be the CPS of Eq. (7.11) and consider a projection set of the form $\Lambda = t + \curlywedge(W)$ with $t \in \mathbb{R}^d$ and window $W \subset H$. If W is relatively compact, Λ is FLC and thus also uniformly discrete; if $W^\circ \neq \varnothing$, Λ is relatively dense. If Λ is a model set, it is also a Meyer set.*

PROOF. Since $\mathcal{L}$ is a lattice in $\mathbb{R}^d \times H$, it is FLC and the first claim follows by an argument analogous to that used in Proposition 7.2. If W is relatively compact and K is a compact subset of $\mathbb{R}^d$, the 'rectangles' $(s + K) \times \overline{W}$ with $s \in \mathbb{R}^d$ contain only finitely many distinct patches of $\mathcal{L}$ up to translation. The projection leads to finitely many distinct K-patches up to translation, which implies the FLC property of Λ and hence uniform discreteness.

If $W^\circ \neq \varnothing$, there is a non-empty open set $U \subset (-W)$. By Lemma 7.4, there is a compact set $K \subset \mathbb{R}^d$ so that $\mathbb{R}^d \times H = \mathcal{L} + (K \times U)$. For any $x \in \mathbb{R}^d$, there are some $s \in L$, $k \in K$ and $u \in U$ so that

$$(x, 0) \;=\; (s, s^\star) + (k, u).$$

This implies $s^\star = -u \in W$, hence $s \in \lambda(W)$ and $x = s + k \in \lambda(W) + K$. Consequently, $\mathbb{R}^d = \lambda(W) + K$, whence $\lambda(W)$ is relatively dense, as is any translate of it.

If Λ is a model set, the window W is relatively compact with non-empty interior by definition. From the previous two properties, Λ is Delone. The observation $\lambda(W) - \lambda(W) \subset \lambda(W - W)$ (as in the proof of Proposition 7.2), where $W - W$ is still relatively compact, shows that $\Lambda - \Lambda$ is again uniformly discrete. This implies the Meyer property by Lemma 2.1. □

Example 7.4 (*Period doubling chain as model set*)**.** In Section 4.5.1, we discussed the period doubling chain and derived an explicit coordinatisation for the geometric counterpart Λ of the bi-infinite fixed point (under ϱ^2 of Eq. (4.7)) with seed $a|a$; see Eq. (4.9) on page 96. If we compare this with the situation of the silver mean chain described in Section 7.1, we now need a $\star$-map that turns the image of 4^i into a small number for large i. This cannot work via algebraic conjugation, but 4^i for large i becomes a small number in the compact Abelian group $\mathbb{Z}_2$ of the 2-adic integers; see Example 2.10. Indeed, we may choose $H = \mathbb{Z}_2$ as the 2-adic completion of $\mathbb{Z}$ to obtain a CPS $(\mathbb{R}, \mathbb{Z}_2, \mathcal{L})$ with the 'diagonal' lattice

$$\mathcal{L} = \big\{ (x, \iota(x)) \mid x \in \mathbb{Z} \big\} \subset \mathbb{R} \times \mathbb{Z}_2,$$

where $\iota \colon \mathbb{Z} \hookrightarrow \mathbb{Z}_2$ is the canonical embedding; compare Example 2.11. It also is the $\star$-map in this case. All requirements of a CPS are satisfied.

To turn Λ_a and Λ_b into model sets, we need two windows, W_a and W_b. Observe that

$$\overline{\Lambda_a}^2 \cup \overline{\Lambda_b}^2 = \mathbb{Z}_2 \quad \text{and} \quad \overline{\Lambda_a}^2 \cap \overline{\Lambda_b}^2 = \{-1\},$$

where $\overline{.}^2$ denotes closure in the 2-adic topology. Consequently, one obtains

$$\Lambda_a = \lambda(W_a) \quad \text{and} \quad \Lambda_b = \lambda(W_b)$$

with $W_a = \overline{\Lambda_a}^2$ and $W_b = \overline{\Lambda_b}^2 \setminus \{-1\}$. If we had chosen the other fixed point of ϱ^2, $\{-1\}$ would have gone to W_b instead of W_a. Although this construction looks a bit artificial, it actually is not. The deeper reason for this is that the autocorrelation coefficients (to be discussed in Section 9.4.4 below) give rise to a pseudo-metric on $\mathbb{Z}$ that defines the 2-adic topology, and the internal group H. The latter is thus intrinsically coded in the distance structure of the point set Λ; see [BM04] for a general derivation. ◇

Let us note in passing that the paper folding sequence of Section 4.5.2, as well as the maximal pure point factor of the Rudin–Shapiro sequence, can be described as model sets as well, in complete analogy to the previous example, and even with the same CPS.

A key to the understanding of a model set Λ is the uniform distribution of $\Lambda^\star$ (properly put into a sequence) inside the window. In fact, this is perhaps the crucial property that was overlooked in the physics community at the beginning, in the sense that it is plausible but really needs a proof. This was pointed out in [Els86, App. 1], with an explicit argument given for the icosahedral case. The general case was studied and proved in [Schl98].

Theorem 7.2. *Let $\Lambda = \curlywedge(W)$ be a regular model set for the CPS $(\mathbb{R}^d, H, \mathcal{L})$ of Eq. (7.11), with a compact window $W = \overline{W^\circ}$. Order the points of Λ according to their distance from 0, and collect them in an exhaustive sequence $(x_i)_{i \in \mathbb{N}}$ (such that $\|x_{i+1}\| \geqslant \|x_i\|$ for all $i \in \mathbb{N}$, for some norm $\|.\|$ in $\mathbb{R}^d$). Then, the sequence $(x_i^\star)_{i \in \mathbb{N}}$ is uniformly distributed in W.*

SKETCH OF PROOF. Given any (measurable) subset U of the window with $\mu_H(\partial U) = 0$ (where μ_H is the Haar measure of H as in Definition 7.2), one has to show that the fraction of points from the sequence $(x_i^\star)_{i \leqslant N}$ that lie in U becomes proportional to $\mu_H(U)$ as $N \to \infty$; compare [KN74, Ch. 1.6] for the general criteria. Of course, it suffices to show this for a family of sets that generate the topology. This follows from [Schl98, Prop. 2.1 and Thm. 1]; see also [Schl93b, Lemma 2.7].

The result is then proved by showing that

$$\lim_{r \to \infty} \frac{\operatorname{card}\bigl(\mathcal{L} \cap (t + B_r + U)\bigr)}{\operatorname{vol}(B_r)} = \operatorname{dens}(\mathcal{L})\,\mu_H(U)$$

holds uniformly in $t \in \mathbb{R}^d$, where B_r denotes the centred open ball of radius r in $\mathbb{R}^d$. Note that, if $\mathcal{A} = (A_n)_{n \in \mathbb{N}}$ is an averaging sequence for the internal group H, the density of $\mathcal{L}$ is given by

$$\operatorname{dens}(\mathcal{L}) = \lim_{r,n \to \infty} \frac{\operatorname{card}\bigl(\mathcal{L} \cap (B_r \times A_n)\bigr)}{\operatorname{vol}(B_r)\,\mu_H(A_n)},$$

where r and n are assumed to grow simultaneously (for instance, proportionally). The detailed (technical) argument is spelt out in [Schl98, Sec. 3]. The analogous argument for Euclidean internal space and polytopal windows is also sketched in [Els86, App. 1].

For a slightly more general approach, we refer to [Moo02]. $\quad\square$

Note that the precise ordering of the points is not relevant as long as this only amounts to local changes in the sequence. The uniform distribution property enables us once more to determine relative frequencies of clusters by means of the window, as in Eq. (7.10) for the silver mean chain.

Corollary 7.3. *Let Λ be a regular model set for the general CPS $(\mathbb{R}^d, H, \mathcal{L})$, with a compact window $W = \overline{W^\circ}$. If $P \subset \Lambda$ is a finite cluster, its relative* *frequency (per point of Λ) is given by*

$$\operatorname{rel\,freq}_\Lambda(P) \;=\; \frac{\operatorname{vol}\!\left(\bigcap_{x\in P}(W-x^\star)\right)}{\operatorname{vol}(W)},$$

which is related to the absolute frequency of P by

$$\operatorname{abs\,freq}_\Lambda(P) \;=\; \operatorname{dens}(\Lambda)\,\operatorname{rel\,freq}_\Lambda(P).$$

PROOF. Let $P \subset \Lambda$ be a non-empty finite cluster, where we may work with $\Lambda \subset L = \pi(\mathcal{L})$, as the other model sets for the CPS with the same window are translates. Now, $P \subset \Lambda$ means that $x^\star \in W$ for all $x \in P$, which is equivalent with $0 \in W_P := \bigcap_{x\in P}(W - x^\star) \neq \varnothing$, where $W_P = \overline{W_P^\circ}$ is compact. Now, $t + P \subset \Lambda$ if and only if $t \in L$ together with $t^\star \in W_P$. So, we have

$$\operatorname{rep}(P) \;=\; \{t \in L \mid t + P \subset \Lambda\} \;=\; \{t \in L \mid t^\star \in W_P\} \;=\; \curlywedge(W_P),$$

where the $\star$-map need not be injective, but $y \mapsto \operatorname{card}\{t \in L \mid t^\star = y\}$ is constant on $L^\star$. Theorem 7.2 now implies that the relative frequency of the cluster P per point of Λ is given by $\operatorname{vol}(W_P)/\operatorname{vol}(W)$ as claimed. The formula for the absolute frequency (counted per unit volume) is then clear. $\square$

The set $W_P \subset W$ is often referred to as the *window* or the *coding region* of the cluster $P \subset \Lambda$. Notice that a symmetry of the window manifests itself in the cluster structure of the corresponding model set, both in the occurrence of symmetry-related clusters and in the equality of their frequencies. This nicely matches the structure of LI classes and their symmetry in the sense of Definition 5.14. Corollary 7.3 can often be used to calculate the frequency module of a model set, as long as the window is sufficiently simple.

Example 7.5 (*Model set description of Kolakoski* $(3,1)$). The substitution rule of Example 4.8 on page 110, in its natural geometric realisation, gives rise to a cut and project scheme with a lattice $\Gamma \subset \mathbb{R}^3$ that is the Minkowski embedding of $\mathbb{Z}[\lambda]$, with λ the real root of $x^3 - 2x^2 - 1$. Let ϑ be the algebraic conjugate of λ with positive imaginary part. One convenient choice of the embedding uses the generators 1, λ and $\lambda^2 - \lambda$ of $\mathbb{Z}[\lambda]$, together with an embedding as explained in Example 3.5; compare [BS04a].

The coupled iterated function system (IFS) for the three windows in internal space $\mathbb{R}^2$ reads

$$W_a \;=\; \vartheta W_a \cup \vartheta W_b,$$
$$W_b \;=\; (\vartheta W_a + \vartheta^2 - \vartheta) \cup (\vartheta W_b + \vartheta^2 - \vartheta) \cup \vartheta W_c,$$
$$W_c \;=\; \vartheta W_a + \vartheta^2.$$

Setting $W_{ab} := W_a \cup W_b$, this system decouples to the simple IFS given by $W_{ab} = \bigcup_{i=1}^{3} f_i(W_{ab})$ with the similitudes

$$f_1(z) \;=\; \vartheta z, \quad f_2(z) \;=\; \vartheta^3(z+1) \quad \text{and} \quad f_3(z) \;=\; \vartheta z + \vartheta^2 - \vartheta.$$

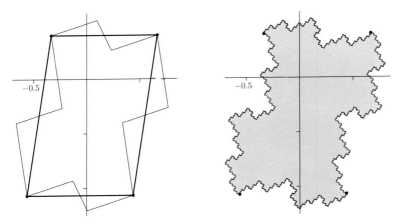

FIGURE 7.2. Graphical representation (right) of the window W_{ab} for Example 7.5. The boundary is obtained by a von Koch curve type construction as indicated (left). The four corner points remain fixed in this process.

Since $|\vartheta| < 1$, the IFS for W_{ab} is contractive, whence Hutchinson's theorem [Hut81, Sec. 3.1(3)] guarantees a unique compact set as a solution, called the *attractor* of the IFS. The individual windows can be calculated from it as

$$W_a = f_1(W_{ab}), \quad W_b = f_2(W_{ab}) \cup f_3(W_{ab}) \quad \text{and} \quad W_c = \vartheta^2 W_{ab} + \vartheta^2.$$

This defines three regular compact sets with fractal boundaries of dimension $d_{\mathrm{H}} = -\log(\tau)/\log(|\vartheta|) \approx 1.2167$, where τ is the golden ratio. The intersection of the interior of distinct windows is empty; compare [BS04a] for the details. We illustrate the result in Figure 7.2, which shows an image of W_{ab}.

The CPS $(\mathbb{R}, \mathbb{R}^2, \Gamma)$ with $\mathbb{Z}[\lambda] = \pi(\Gamma)$ is equipped with the $\star$-map that is the unique field extension of the mapping $\lambda \mapsto \vartheta$, thus using $\mathbb{R}^2 \simeq \mathbb{C}$. The model set description (of the left end points of the intervals) now reads

$$\Lambda_s = \lambda(W_s) \quad \text{and} \quad \Lambda = \lambda(W),$$

where $s \in \{a, b, c\}$, $W = W_a \cup W_b \cup W_c$ and $\lambda(A) = \{x \in \mathbb{Z}[\lambda] \mid x^\star \in A\}$. Although the windows have fractal boundaries (and are thus examples of Rauzy fractals [PF02, Sec. 7.4]), the model set described here is both regular and generic; see [BS04a] for a proof. Note that the three windows can also be used as disk-like fractiles for the Euclidean plane [BW01]. For more on planar tilings by fractiles, we refer to [Vin00] and references therein. ◇

Remark 7.5 (*Related tilings of the plane*). The windows of Example 7.5 provide interesting prototiles for a number of tilings. First of all, W_{ab} tiles the plane periodically, as does the total window $W = W_{ab} \cup W_c$. Moreover, the three windows W_a, W_b and W_c can be used as prototiles for an aperiodic

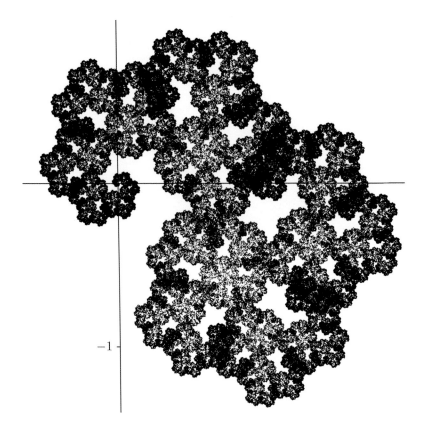

FIGURE 7.3. Graphical representation of the window W_b for Example 7.6.

tiling with a stone inflation, which is dual to the original substitution rule of Example 4.8; see [Sin02, BS04a, BFS12] and [Sin06, Ex. 6.114] for details. ◊

Example 7.6 (*Windows for the minimal PV substitution of Example 4.4*). The geometric realisation (with intervals of lengths 1, β and β^2) of the three-letter substitution rule $\varrho\colon a \mapsto b \mapsto c \mapsto ab$ of Example 4.4 from page 79 is another model set with $\mathbb{R}^2$ as internal space. Consider the complex algebraic conjugate $\alpha \approx -0.66236 + 0.56228\,\mathrm{i}$ of the minimal PV number β. Starting from a two-sided fixed point for ϱ^6, the three windows W_a, W_b and W_c are the unique compact sets that solve the equations (a coupled, contractive IFS)

$$W_a = \alpha^6 W_a \cup \alpha^6 W_b + (\alpha + 2\alpha^2) \cup \alpha^6 W_c + \{\alpha^2, 1 + \alpha + \alpha^2\},$$

$$W_b = \alpha^6 W_a + \{1, 1+\alpha\} \cup \alpha^6 W_b + \{0, 1+\alpha+2\alpha^2\}$$
$$\cup\ \alpha^6 W_c + \{1+\alpha^2, 2+\alpha+\alpha^2, 2+2\alpha+\alpha^2\},$$

$$W_c = \alpha^6 W_a + (1+2\alpha) \cup \alpha^6 W_b + \{\alpha, \alpha+\alpha^2\} \cup \alpha^6 W_c + \{0, 2+3\alpha+\alpha^2\}.$$

Due to the action of ϱ, the windows are related by $W_a = \alpha W_c = \alpha^2 W_b$, whence it is sufficient to show one of them. Figure 7.3 gives a representation of W_b, obtained by the usual random approximation to a fractal set; see [BV11] and references therein. Despite its more complicated appearance, this is a proper window, in the sense that it is the closure of its interior. This demonstrates one of the many possibilities with fractal sets. $\Diamond$

The above example, and Figure 7.3 in particular, is also discussed in [ST09, Fig. 2.3 and Ch. 2.4]. This reference also gives a general account of the theory of Rauzy fractals, which provides a natural frame for the discussion of windows of cut and project sets; see also [PF02, SW02].

Remark 7.6 (*Local derivability for model sets*)**.** The projection formalism for model sets provides additional tools to investigate (mutual) local derivability and its symmetry-preserving counterpart. Since two patterns that are MLD must share the same limit translation module (LTM) by Corollary 5.1, it is most natural to compare two model sets that are obtained from the same CPS, but with different windows, W_1 and W_2 say. For simplicity, we assume each window to be compact and to be the closure of its interior, and consider only model sets with internal space $\mathbb{R}^m$ (Euclidean model sets).

The following criterion is explained informally in [Baa02a] and proved in detail in [BSJ91]. The two model sets satisfy $\curlywedge(W_1) \overset{\text{LD}}{\leadsto} \curlywedge(W_2)$ if and only if W_2 can be expressed as a finite union of sets each of which is a finite intersection of translates of W_1 (or its complement), with translations from $L^\star$. Obviously, the model sets are MLD when both $\curlywedge(W_1) \overset{\text{LD}}{\leadsto} \curlywedge(W_2)$ and $\curlywedge(W_2) \overset{\text{LD}}{\leadsto} \curlywedge(W_1)$. Each window W, under the translations by elements from $L^\star$ and the set operations $\cap$, $\cup$ together with taking the complement in internal space, generates a Boolean algebra $\mathcal{B}\mathcal{A}(W)$. The derivability relation $\curlywedge(W_1) \overset{\text{LD}}{\leadsto} \curlywedge(W_2)$ is then equivalent to $\mathcal{B}\mathcal{A}(W_2)$ being a subalgebra of $\mathcal{B}\mathcal{A}(W_1)$, while $\curlywedge(W_1) \overset{\text{MLD}}{\leadsto} \curlywedge(W_2)$ if and only if $\mathcal{B}\mathcal{A}(W_1) = \mathcal{B}\mathcal{A}(W_2)$.

Note that only translations are admitted in the criterion. In many examples, the corresponding derivation rules will be the same for patches that are rotated copies of each other. In this sense, the criterion can be extended to symmetry-preserving local derivability, which is needed for the discussion of SMLD classes. In practice, the criterion may be easy to verify, provided the windows are sufficiently simple. In particular, this works well for polytopes, including standard cases of multi-component model sets such as the vertex set of the rhombic Penrose tiling; compare Remark 7.8 below. In contrast, it is often difficult to apply the criterion to windows with fractal boundaries, such as those of the previous case; compare [Gäh10] for some more favourable examples. $\Diamond$

7.3. Cyclotomic model sets

An efficient and systematic approach to planar model sets with n-fold symmetry (with $n \geqslant 3$) can be formulated on the basis of the $\mathbb{Z}$-modules $\mathbb{Z}[\xi_n]$ of Section 3.4. Recall that, since ξ_n is a primitive nth root of unity, the point set $\mathbb{Z}[\xi_n] \subset \mathbb{C} \simeq \mathbb{R}^2$ has N-fold rotational symmetry with $N = \mathrm{lcm}(2, n)$. Because $\mathbb{Z}[\xi_3]$ and $\mathbb{Z}[\xi_4]$ are planar lattices, we restrict to $n \geqslant 5$ for the remainder of this section (and also to $n \not\equiv 2 \bmod 4$ as before). The CPS is then given by the diagram

$$(7.13) \quad
\begin{array}{ccccc}
\mathbb{R}^2 & \xleftarrow{\;\pi\;} & \mathbb{R}^2 \times \mathbb{R}^{\phi(n)-2} & \xrightarrow{\;\pi_{\mathrm{int}}\;} & \mathbb{R}^{\phi(n)-2} \\
\cup & & \cup & & \cup \;\; \text{dense} \\
\pi(\mathcal{L}_n) & \xleftarrow{\;1\text{–}1\;} & \mathcal{L}_n & \longrightarrow & \pi_{\mathrm{int}}(\mathcal{L}_n) \\
\| & & & & \| \\
\mathbb{Z}[\xi_n] & & \xrightarrow{\qquad\star\qquad} & & \mathbb{Z}[\xi_n]^\star
\end{array}$$

where we always consider the Euclidean spaces grouped in appropriate pairs to correspond to copies of $\mathbb{C}$, according to the Minkowski embedding of Eq. (3.12) on page 61. The $\star$-map is defined by $x \mapsto (\sigma_2(x), \ldots, \sigma_{\frac{1}{2}\phi(n)}(x))$, with the automorphisms σ_i as described in Section 3.4. It is particularly simple when $\phi(n) = 4$, which is discussed in Example 7.7 below.

A model set that emerges from such a CPS with a window $W \subset \mathbb{R}^{\phi(n)-2}$ is called a *cyclotomic model set*.

Example 7.7 (*The usual suspects*). Let us recall the Minkowski embedding $\mathcal{L}_n$ of $\mathbb{Z}[\xi_n]$ for $n \in \{5, 8, 12\}$, so that $\phi(n) = 4$; compare Example 2.16. In each case, we only need to choose one non-trivial automorphism σ, which then takes the role of the $\star$-map. One possible choice for the automorphisms and the resulting lattice data are summarised in Table 7.1.

Of particular relevance is the *dual module*, which will show up later as the Fourier module of a model set. It is obtained as $\pi(\mathcal{L}_n^*)$, where $*$ refers to the dual lattice as introduced in Eq. (3.4). For $n = 8$ and $n = 12$, the explicit calculation follows from Example 3.6. In the case $n = 5$, one may use

TABLE 7.1. Minkowski embedding of $\mathbb{Z}[\xi_n]$ for $n \in \{5, 8, 12\}$.

ring of integers	$\mathbb{Z}[\xi_5]$	$\mathbb{Z}[\xi_8]$	$\mathbb{Z}[\xi_{12}]$
choice of $\star$-map	$\xi_5 \mapsto \xi_5^2$	$\xi_8 \mapsto \xi_8^3$	$\xi_{12} \mapsto \xi_{12}^5$
lattice	$\mathcal{L}_5$	$\mathcal{L}_8$	$\mathcal{L}_{12}$
volume of FD	$\frac{5}{4}\sqrt{5}$	4	3
dual module	$\frac{2}{5}(1-\xi_5)\mathbb{Z}[\xi_5]$	$\frac{1}{2}\mathbb{Z}[\xi_8]$	$\frac{1}{\sqrt{3}}\mathbb{Z}[\xi_{12}]$

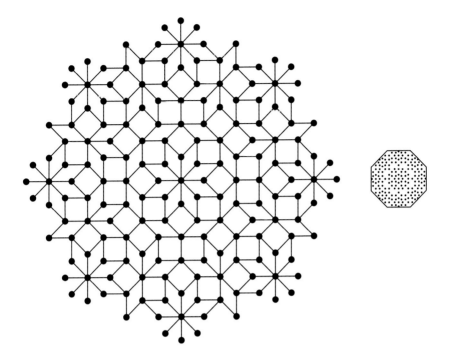

FIGURE 7.4. The octagonal tiling via projection as a cyclotomic model set.

Example 3.7 to derive that the Minkowski embedding of the four complex numbers $\{\frac{2}{5}(\xi_5^\ell - \xi_5^4) \mid 0 \leqslant \ell \leqslant 3\}$ into $\mathbb{C}^2 \simeq \mathbb{R}^4$ gives the correct dual basis. A simple calculation then shows that $\pi(\mathcal{L}_5^*) = \frac{2}{5}(1 - \xi_5)\mathbb{Z}[\xi_5]$.

As an alternative to our algebraic approach, one can start from certain root lattices in 4-space [BJKS90, BKSZ90, BJS90]; see Examples 3.6 and 3.7 and Remarks 3.5 and 3.6 for the connection. For a comparison of concrete results, one has to carefully check the relative scale and orientation used in the corresponding CPS. ◇

Example 7.8 (*Ammann–Beenker point set*). Consider the cyclotomic model set $\Lambda_{\mathrm{AB}} = \lambda(W_{\mathrm{AB}})$ for the module $\mathbb{Z}[\xi_8]$, with the star map of Example 7.7 and a centred regular octagon of edge length 1 as its window W_{AB}. It then has inradius $\frac{1}{2}(1 + \sqrt{2})$, circumradius $\sqrt{(2 + \sqrt{2})/2}$ and area $2(1 + \sqrt{2})$. The octagon is oriented so that all edges are perpendicular to some 8th root of unity; see Figure 7.4 for a finite central patch of Λ_{AB} and its lift into the window. The edges are obtained by connecting all pairs of points at unit distance, which only occur in the directions of the 8th roots of unity.

The model set Λ_{AB} is regular and generic, and hence repetitive. Regularity is clear because the window is a polygon. Genericity, which means that

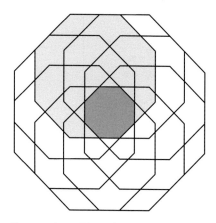
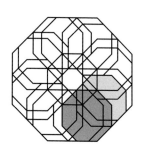

FIGURE 7.5. The left panel shows the rescaled window $\lambda^{\star} W_{AB}$ (dark grey) as the intersection of eight copies of the original window W_{AB} (light grey); the right panel shows one possibility to cover W_{AB} by translates of $\lambda^{\star} W_{AB}$. All translations used are elements of $\mathbb{Z}[\xi_8]$.

$\mathbb{Z}[\xi_8] \cap \partial W_{AB} = \varnothing$, follows from the observation that the window intersects the positive real axis in $\frac{1}{2}\lambda_{sm} = \frac{1}{2}(1+\sqrt{2}\,)$ and that the line through this point parallel to the imaginary axis does not meet any point of $\mathbb{Z}[\xi_8] = \langle 1, \xi_8 \rangle_{\mathbb{Z}[\sqrt{2}]}$. The claim then follows from the eightfold symmetry of ∂W_{AB} and $\mathbb{Z}[\xi_8]$.

Since the Minkowski embedding $\mathcal{L}_8$ of $\mathbb{Z}[\xi_8]$ is a lattice in 4-space with $\operatorname{dens}(\mathcal{L}_8) = \frac{1}{4}$, compare Example 3.6, and since the window W_{AB} has area $2\lambda_{sm}$, the model set Λ_{AB} is a point set of density

$$\operatorname{dens}(\Lambda_{AB}) = \operatorname{dens}(\mathcal{L}_8)\,\operatorname{vol}(W_{AB}) = \frac{1}{2}\lambda_{sm} \approx 1.207,$$

which coincides with the value derived from the substitution approach. To show that the model set Λ_{AB} is the set of vertex points of the Ammann–Beenker tiling constructed in Section 6.1, one can proceed as follows. Observe that the centres of the regular 8-stars form another point set of the same kind, with the scale multiplied by λ_{sm}^2. This can be proved by determining the window of the 8-star as a patch, which is again a centred regular octagon in the same orientation as W_{AB}, rescaled by a factor $(\lambda_{sm}^2)' = \lambda_{sm}^{-2}$; see Example 7.9 below for details. This shows that Λ_{AB} is a fixed point under twofold geometric inflation, wherefore the claim follows from the usual combination of density and repetitivity arguments; compare Theorem 7.1. $\Diamond$

Remark 7.7 (*LIDS of Ammann–Beenker point set*). The natural inflation multiplier (in direct space) is $\lambda = \lambda_{sm}$. The corresponding action on the window is multiplication (scaling) by $\lambda^{\star} = -1/\lambda$. The rescaled octagon can be expressed as the intersection of eight translated copies of the original window, with translations that are elements of $\mathbb{Z}[\xi_8]$. Explicitly, one has

$$\lambda^\star W_{\mathrm{AD}} = \bigcap_{\ell=0}^{7} \left(W_{\mathrm{AD}} - \mathrm{e}^{\pi\mathrm{i}\ell/4}(1-\xi_8) \right),$$

as illustrated in Figure 7.5. In view of our results on local derivability, this implies the existence of a *local* inflation rule for the Ammann–Beenker model set. In fact, using the formulation with undecorated prototiles, this is the inflation rule of Figure 6.1 on page 176.

Likewise, W_{AB} can be written as a union of translated copies of the rescaled window $\lambda^\star W_{\mathrm{AB}}$, for instance as

$$W_{\mathrm{AB}} = \lambda^\star W_{\mathrm{AB}} \cup \bigcup_{\ell=0}^{7} \left(\lambda^\star W_{\mathrm{AB}} + \mathrm{e}^{\pi\mathrm{i}\ell/4}(1-\xi_8) \right)$$
$$\cup \bigcup_{\ell=0}^{7} \left(\lambda^\star W_{\mathrm{AB}} + \mathrm{e}^{\pi\mathrm{i}\ell/4}(1-\xi_8-\lambda^\star\xi_8) \right)$$

which is slightly more involved because $\lambda > 2$, so that one needs an extra copy in the centre and along each edge; see Figure 7.5 for an illustration. Note that $(1-\xi_8-\lambda^\star\xi_8) \in \mathbb{Z}[\xi_8]$ because $\mathbb{Z}[\xi_8] = \langle 1, \xi_8 \rangle_{\mathbb{Z}[\sqrt{2}]}$ and $\lambda^\star$ is a unit in $\mathbb{Z}[\sqrt{2}]$. This construction corresponds to the deflation rule that emerges as the inverse of the above mentioned inflation rule, by identifying (or recognising) the level-1 supertiles (which is clearly local here). Together, this establishes an LIDS in the sense of Definition 5.16, without reference to the local recognisability criterion of [Sol98a].   ◊

Example 7.9 (*Ammann–Beenker vertex stars*). With the two prototiles of the octagonal tiling, the square and the $\pi/4$ rhombus, one can form 16 distinct complete vertex stars (up to rotations and reflections). Only six of these show up in the Ammann–Beenker tiling of Figure 6.41, namely those

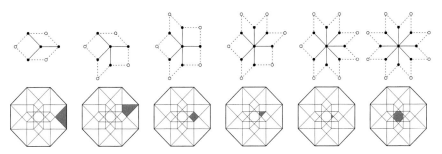

FIGURE 7.6. The six vertex stars of the Ammann–Beenker tiling, ordered with increasing coordination number. Below each vertex configuration, its corresponding window (inside the general window) is displayed.

TABLE 7.2. Relative frequencies of the six Ammann–Beenker vertex stars of Figure 7.6, summed over their possible orientations.

vertex	coordination	orbit length	relative frequency		
1	3	8	$-1 + \sqrt{2}$	$= \lambda^{-1}$	≈ 0.41421
2	4	8	$6 - 4\sqrt{2}$	$= 2\lambda^{-2}$	≈ 0.34315
3	5	8	$-14 + 10\sqrt{2}$	$= 2\lambda^{-3}$	≈ 0.14214
4	6	8	$34 - 24\sqrt{2}$	$= 2\lambda^{-4}$	≈ 0.05887
5	7	8	$-41 + 29\sqrt{2}$	$= \lambda^{-5}$	≈ 0.01219
6	8	1	$17 - 12\sqrt{2}$	$= \lambda^{-4}$	≈ 0.02944

of Figure 7.6. Note that the solid dots in Figure 7.6 alone do not define the vertex configurations uniquely, because they define stars that can still be subsets of others (and of the 8-star in particular). One thus needs to include the open circles as well. The relative frequencies of the vertex configurations now follow by an application of Corollary 7.3. The corresponding windows are also shown in Figure 7.6. Apart from the regular 8-star, all other vertex stars occur in 8 different orientations, each with its own window.

The window and its subdivision give access to the relative frequencies of the vertex stars, essentially via Corollary 7.3. The result is summarised in Table 7.2, which also lists the coordination numbers (meaning the number of neighbours at edge distance 1). The relative frequencies add to 1, and the average coordination number is 4, as it must be for all rhombus tilings; compare Remark 2.4. An alternative approach via the dualisation method is described in [BJ90], which starts from the lattice $\mathbb{Z}^4$ and constructs the cut and project setup by arguments from group representation theory. This approach leads to a natural subdivision of the window (as shown in Figure 7.6).

The $\mathbb{Z}$-span of the six relative frequencies from Table 7.2 is $\mathbb{Z}[\sqrt{2}]$. Recalling from Example 7.8 that Λ_{AB} has density $\frac{1}{2}\lambda$, and that λ is a unit in $\mathbb{Z}[\sqrt{2}]$, the (absolute) frequency module is $\frac{1}{2}\mathbb{Z}[\sqrt{2}]$ in our setting.

Note that $\lambda^2 \Lambda_{AB}$ is a homothetic subset of Λ_{AB}. It is precisely the set of locations of the regular 8-star (vertex 6) in the rhombus tiling that emerges from Λ_{AB}. This follows from the observation that the window of the 8-star is $\lambda^{-2} W_{AB}$, as is obvious from Figure 7.6 and the last entry of Table 7.2. $\Diamond$

Example 7.10 (*TTT point set*). Here, we start from the lattice $\mathcal{L}_5$ according to Table 7.1, which has $\mathrm{dens}(\mathcal{L}_5) = 4/(5\sqrt{5})$. We select a regular decagon W_{10} of edge length $\sqrt{(\tau + 2)/5}$ as window for the CPS (7.13), whose area then is $\frac{1}{2}(3\tau + 1)\sqrt{\tau + 2}$. The orientation is such that each edge is perpendicular

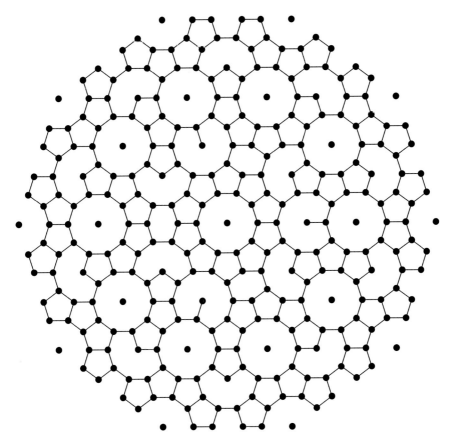

FIGURE 7.7. A patch of a cyclotomic model set based on $\mathbb{Z}[\xi_5]$ with a decagonal window of edge length $\sqrt{(\tau+2)/5}$; compare Figure 6.9.

to a 10th root of unity. It turns out that centring this decagon at the origin leads to a singular model set. Figure 7.7 shows a patch of $\curlywedge(W_{10}+\varepsilon)$ with a tiny generic translation ε along the positive real axis of internal space. It is turned into a tiling patch (with some isolated pseudo vertices) by connecting all pairs of points of distance τ^{-1}.

Any model set of the form $t + \curlywedge(W_{10}+y)$ is regular and Delone, with

$$\mathrm{dens}\big(\curlywedge(W_{10})\big) = \mathrm{dens}(\mathcal{L}_5)\,\mathrm{vol}(W_{10}) = \frac{2}{5}\tau^2\sqrt{\tau+2}.$$

This equals the density of the vertex point set of the TTT calculated in Lemma 6.1. In fact, the patch of Figure 7.7 is chosen to precisely match the vertex point set of the inflation-generated TTT of Figure 6.9 on page 189. The triangles of the TTT can locally be reconstructed from the projection point set [KSB93].

The vertex star of maximal symmetry is a 10-star, whose window turns out to be $\tau^{-3}W_{10}$. Their centres thus occupy a τ^3-scaled version of the model set. More generally, the TTT of Figure 6.9 has nine distinct vertex stars up to similarity. A detailed analysis including the determination of their relative frequencies was given in [BKSZ90, Fig. 4.3]. Note that the orientation of the tiles and of the window in [BKSZ90] is rotated by $\pi/10$ relative to our formulation as cyclotomic model sets, because an approach via the dualisation method on the basis of the root lattice A_4 is used. $\lozenge$

Example 7.11 (*Model set description of rhombic Penrose tilings*). The vertex points of the rhombic Penrose tiling (RPT) can be described as a four-component model set. Consider $L = \mathbb{Z}[\xi]$ with $\xi = \xi_5$ (compare Section 2.5.2) and its Minkowski embedding $\mathcal{L}_5$ as described in Examples 3.7 and 7.7, which means that we work with a lattice in 4-space. A crucial extra ingredient is the homomorphism $\kappa\colon L \longrightarrow \mathbb{Z}/5\mathbb{Z}$ defined by

$$\kappa\big(\textstyle\sum_j m_j \xi^j\big) \,=\, \textstyle\sum_j m_j \bmod 5,$$

which is consistent and does not depend on the way an arbitrary element $x \in L$ is expressed as an integer linear combination of powers of ξ. Observe that the golden mean satisfies $\tau = 1 + \xi + \bar{\xi} = 1 + \xi + \xi^4$, hence $\kappa(\tau) = 3$. More generally, $\kappa(\tau^n) = 3^n \bmod 5$ for all $n \in \mathbb{Z}$. Consequently, the inflation mapping induces the permutation (1342) on the cosets. These properties are consistent with the observation that $\mathbb{F}_5 = \mathbb{Z}/5\mathbb{Z}$ is a (finite) field and that κ corresponds to a field automorphism of $\mathbb{F}_5$.

Defining the cosets $L^{(i)} = \{x \in L \mid \kappa(x) = i\}$ for $0 \leqslant i \leqslant 4$, one obtains the partition $L = \bigcup_{i=0}^4 L^{(i)}$. In particular, one has $L^{(0)} = (1-\xi)\mathbb{Z}[\xi]$, which is an ideal of $\mathbb{Z}[\xi]$ of index 5. Let P be the convex hull of $\{1, \xi, \xi^2, \xi^3, \xi^4\}$ which is a closed pentagon, and define the four windows

$$W^{(1)} = P, \quad W^{(4)} = -P, \quad W^{(3)} = \tau P, \quad W^{(2)} = -\tau P$$

together with the corresponding regular model sets

$$\Lambda^{(i)} \,=\, \big\{x \in L \mid \kappa(x) = i \text{ and } x^\star \in W^{(i)}\big\} \,=\, \big\{x \in L^{(i)} \mid x^\star \in W^{(i)}\big\}.$$

Here, the $\star$-map is the Galois automorphism of $L = \mathbb{Z}[\xi]$ that sends ξ to ξ^2; compare Table 7.1. Note that the LTM in this approach is not L, but the index 5 submodule $L^{(0)} = (1-\xi)\mathbb{Z}[\xi]$. The Minkowski embedding of the latter is the lattice $\mathcal{L}'$, with $[\mathcal{L}_5 : \mathcal{L}'] = 5$. Since $\mathbb{Z}/5\mathbb{Z} \simeq C_5$ as an Abelian group, we are thus using a CPS $(\mathbb{R}^2, \mathbb{R}^2 \times C_5, \mathcal{L})$ with embedding lattice $\mathcal{L} = \mathcal{L}' \times C_5$.

The disjoint union $\Lambda := \bigcup_{i=1}^4 \Lambda^{(i)}$ is a four-component model set. It is singular because $L^{(i)} \cap \partial W^{(i)} \neq \varnothing$. A generic case is obtained by a suitable shift of all four windows by the same complex number. This gives the vertices of (proper) rhombic Penrose tilings with edge length 1. Examples are shown in Section 7.5.2 below via the equivalent formulation due to de Bruijn [dBr81]. An alternative interpretation of the four windows will be mentioned in the context of Danzer's icosahedral tiling in Remark 7.11.   ◇

As explained in Example 3.7, $\mathcal{L}_5$ is the weight lattice A_4^*, while a suitable change of the relative scale between direct and internal space permits to use the root lattice A_4 instead; compare Remark 3.6. The latter choice also explains the presence of $\mathbb{Z}/5\mathbb{Z}$, which is isomorphic with A_4^*/A_4 as a group. This setting is widely used in the literature, and gives perhaps the most natural minimal embedding description of the RPT. Rather than starting from the algebraic setting of cyclotomic fields, it is then advantageous to employ the cell structure of the mutually dual Voronoi and Delone complexes of Section 2.2. The Penrose rhombuses are then the projection of selected 2-facets of the Voronoi complex. This amounts to a modification of the projection scheme that is known as Kramer's klotz construction [Kra87a, Kra88]. It was later reformulated and proved in sufficient generality in [KS89], which also contains a more complete account of the history of this concept. We will explain this variant in a simpler one-dimensional example later, and refer to [BKSZ90] for the completely and step by step elaborated case of the rhombic Penrose tiling.

As briefly mentioned in Example 7.10, the same A_4-based approach also produces the TTT, this time via projections of selected 2-facets of the Delone complex. In this sense, the tiling families RPT and TTT are dual to each other. For the same reason, a classification of their vertex configurations, together with all relative frequencies, can be obtained in a systematic way; see [BKSZ90, Figs. 4.3 and 5.2].

Remark 7.8 (*Non-minimal embedding in 5-space*). It is possible to embed the 'lattice' $\mathcal{L}_5 \times C_5$ from Example 7.11 into 5-space in such a way that it is (after proper scaling) a subset of the integer lattice $\mathbb{Z}^5$. The four pentagonal windows reoccur as sections of the projection of the five-dimensional hypercube into 3-space, which is the rhombic icosahedron illustrated in Figure 7.8. From this description, it is apparent that the additional embedding dimension carries no relevant information on the tiling, though it can profitably be used to describe the family of generalised Penrose tilings; compare Section 7.5.2.

Since non-minimal embeddings result in various pathologies and difficulties in connection with equivalence concepts and diffraction formulas, we prefer to avoid them altogether from now on.   ◇

Example 7.12 (*Gähler's shield tiling* [Gäh88, Ch. 5]). Here, one uses $\mathbb{Z}[\xi_{12}]$ for the CPS (7.13) with a regular dodecagon of edge length 1 as window W_{12}, with the right-most edge oriented as the corresponding one of the octagon in Example 7.8. The dodecagon has inradius $(2+\sqrt{3})/2$, circumradius $\sqrt{2+\sqrt{3}}$

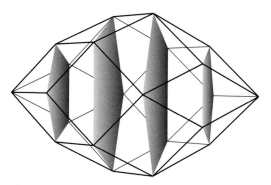

FIGURE 7.8. Aspect of the rhombic icosahedron with the four pentagonal windows of the RPT as sections. The fivefold symmetry axis (along the horizontal) corresponds to the space diagonal in 5-space, and the shaded intersections represent P, $-\tau P$, τP and $-P$ (from left to right).

and area $3(2+\sqrt{3}\,)$. The resulting model set is regular, but not generic if the window is centred (its vertex points then belonging to $\mathbb{Z}[\xi_{12}]$). A patch of a generic shield tiling is shown in Figure 6.48 on page 241; it was obtained by shifting the window (by some translation s) into a nearby generic position (so that $\partial(s + W_{12}) \cap \mathbb{Z}[\xi_{12}] = \varnothing$). The tiling is obtained from the (vertex) point set by drawing all edges between points at distance $\sqrt{2 - \sqrt{3}}$. This produces a pattern with three prototiles, an equilateral triangle (which occurs in four orientations), a square (in three orientations) and a 'shield' (in four orientations). Their area ratio is $\frac{1}{4}\sqrt{3} : 1 : \frac{1}{2}(3 + \sqrt{3}\,)$; compare Section 6.3.2.

As mentioned above, the centred window leads to a singular case. Figure 7.9 shows a patch of a model set with the (closed) window W_{12}. The grey points are the points $x \in \mathbb{Z}[\xi_{12}]$ of the patch with $x^\star \in \partial W_{12}$. Infinitely more instances occur along the twelve directions, but all of them together are still of zero density. These points lead to local configurations that do not occur in the hull of the shield tiling. Indeed, the hull (as defined by the pseudo inflation rule of Figure 6.16 on page 201) contains all limit points obtained from sequences $s_i + W_{12}$ with $s_i \xrightarrow{i \to \infty} 0$ such that $\partial(s_i + W_{12}) \cap \mathbb{Z}[\xi_{12}] = \varnothing$ holds for all i. This limiting procedure always selects the configurations

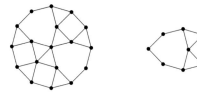

or rotated versions thereof. The grey points come in pairs (on opposite edges in internal space), precisely one of which gets selected. Combinatorially, two

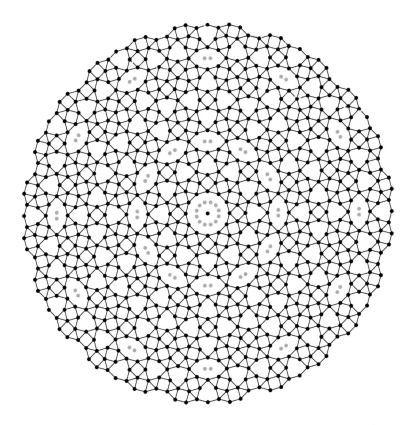

FIGURE 7.9. Singular patch of the shield tiling; see text for details.

other dissections of the central dodecagon are possible by the three prototiles, but the one shown above is the only configuration that occurs in the shield tiling. One consequence is that the shield tiling hull contains no element with perfect (individual) twelvefold symmetry, while the hull as a whole is D_{12}-symmetric in the sense of Definition 5.14.

There is an LIDS with inflation multiplier $\lambda = 2 + \sqrt{3}$, which satisfies $\lambda^\star = 1/\lambda$. Its existence follows from $1 + i \in \mathbb{Z}[\xi_{12}]$ together with

$$\frac{W_{12}}{2 + \sqrt{3}} = \bigcap_{\ell=0}^{11} \left(W_{12} - e^{\pi i \ell/6}(1 + i)\right).$$

One possible realisation of the underlying inflation mapping as a pseudo inflation rule was illustrated in Figure 6.16.

Note that the inflation rule (for the fully decorated shield tiling) described earlier in Figure 6.12 on page 196, which has inflation multiplier $\sqrt{\lambda}$, is still local, despite the rotation of the prototile orientations by $\pi/12$. This

FIGURE 7.10. Fractally shaped window of the square triangle tiling.

manifests itself in a corresponding rotation of the window as well. The local information encoded in the markers corresponds to further subdivisions of the window [Gäh93]. The latter ensure that the tiling and its image under the inflation mapping are still MLD, even though we are now dealing with two LI classes with distinct orientations of the prototiles.   ◊

Remark 7.9 (*Vertex point set and window of the square triangle tiling*). The vertex point set of the square triangle tiling of Section 6.3 can again be described as a cyclotomic model set with $\mathbb{Z}[\xi_{12}]$. This time, the window has a fractal boundary, as shown in Figure 7.10; see [HRB97, Appendix] for details. It is a closed set that is the closure of its interior. The actual image was obtained by lifting the vertex points of a large inflation-generated patch.

Alternatively, one can employ the tiling cover constructed in [Fre11], which leads to the same set of vertex points.

Following a suggestion by Elser, it has been conjectured [Fre11] that this window (or a closely related one) can serve as the closure of a fundamental domain of the triangular lattice A_2. This is remarkable as the latter has only sixfold rotational symmetry. A similar phenomenon is expected for the square lattice, with a domain with eightfold rotational symmetry and again a fractal boundary. Another fractally bounded window (for a different square triangle tiling), arises from the depletion construction of [BKS92], which was mentioned in Remark 6.7. It was further analysed in [Fre13], and can be used as (the closure of) a fundamental domain of the lattice A_2. ◊

Despite the complicated structure of the window of the square triangle tiling, one knows from the underlying inflation that the patch complexity function is quadratic in the radius. At present, a realistic estimate of the complexity of projection tilings is only possible under rather stringent conditions on the window; compare [Jul10]. In particular, windows with fractal boundary are presently out of reach. The latter, however, are rather typical for inflation tilings that permit a projection description, while they are pretty special from a fractal point of view. The understanding of patch complexity is thus still fairly limited.

Let us now turn our attention to higher dimensions, and to the important case of model sets in 3-space with icosahedral symmetry.

7.4. Icosahedral model sets and beyond

From the projection perspective, the simplest tiling with icosahedral symmetry is obtained from the CPS (7.11) with $d = 3$, $H = \mathbb{R}^3$ and the lattice $\mathcal{L} = \mathbb{Z}^6$. The orientation of the lattice relative to direct and internal space follows from the group theoretic arguments explained in Appendix A. This approach was pioneered by Kramer and Neri [KN84] and (with an appropriate window) leads to what is known as the P-type or *primitive icosahedral tiling*. A more detailed analysis can be found in [Kra88]. As mentioned in Section 6.7.2, the tiling is built from two rhombohedral prototiles, a thick (or prolate, denoted by T_p) and a thin (or oblate, T_o) one. Using the vectors from Example 2.20 on page 43, the prototiles can be defined as the convex hulls of their vertices by

$$T_p = \mathrm{conv}\{0, v_1, v_2, v_3, v_1 + v_2, v_1 + v_3, v_2 + v_3, v_1 + v_2 + v_3\},$$
$$T_o = \mathrm{conv}\{0, v_1, v_2, v_5, v_1 + v_2, v_1 + v_5, v_2 + v_5, v_1 + v_2 + v_5\}.$$

Up to Euclidean motions, these two tiles are rhombohedra with solid angles $\pi/5$, $3\pi/5$ and $7\pi/5$ as illustrated in Figure 7.11. More precisely, the solid

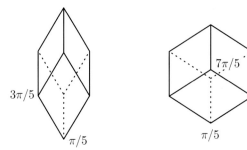

FIGURE 7.11. The two Kepler rhombohedra of the primitive icosahedral tiling due to Kramer and Neri [KN84] and Ammann [Mac81, Sen95].

angle at the bottom and top vertices of T_p (left) is $\pi/5$, while it is $3\pi/5$ at the other six vertices. Likewise, T_o (right) has two (opposite) vertices of solid angle $7\pi/5$ and six of $\pi/5$. The solid angles in both cases add up to 4π.

These prototiles possess only one type of facet, which is a rhombus of edge length $\sqrt{2+\tau}$ with angles $\arccos(1/\sqrt{5}\,) \approx 0.3524\pi$ and $\arccos(-1/\sqrt{5}\,) \approx 0.6476\pi$, and area 2τ. In particular, this is *not* one of the Penrose rhombuses. The latter arise as suitable sections of the rhombohedra. Ten rhombohedra of each type can be assembled to fill Kepler's triacontahedron of Figure 2.6, as is discussed in detail in [Kow38]; see also [Mac81, Fig. 8] and the surrounding discussion. The prototiles T_p and T_o have volumes $2\tau^2$ and 2τ, so that the triacontahedron of edge length $\sqrt{2+\tau}$ has volume $20\tau^3$ and surface area 60τ.

Any face to face tiling of 3-space by the two rhombohedral prototiles is MLD with the Delone set of its vertex points. While the derivation of the point set from the tiling is obvious, the converse direction follows from the observation that the distance between vertices along an edge does not occur otherwise in the tiling. Consequently, a complete description is possible via the vertex point set. We now describe a version that is equivalent to the construction in [KN84]. In line with our number-theoretic approach to planar model sets, we prefer to work with the icosahedral modules of Example 2.20. Here, we need the primitive module $L = \mathcal{M}_\mathsf{p}$ together with its (modified) Minkowski embedding explained in Example 3.8. The $\star$-map acts as $(a, b, c) \mapsto \tau(a', b', c')$ on $\mathcal{M}_\mathsf{p}$, where $'$ denotes algebraic conjugation in $\mathbb{Q}(\sqrt{5}\,)$. In this formulation, the embedding lattice $\mathcal{L} = \{(x, x^\star) \mid x \in L\}$ is similar to $\mathbb{Z}^6$, and explicitly generated by the $\mathbb{Z}$-basis $\{(v_i, v_i^\star) \mid 1 \leqslant i \leqslant 6\}$ that emerges from Example 2.20. Consequently, the fundamental cell of $\mathcal{L}$ has volume $40(4\tau + 3)$, so that $\mathrm{dens}(\mathcal{L}) = (7 - 4\tau)/200$.

If we choose Kepler's triacontahedron with edge length $\sqrt{2+\tau}$ as the window, centred at the origin and oriented as in Figure 7.12, the model set

(7.14) $\Lambda = \curlywedge(W) = \{x \in L \mid x^\star \in t + W\}$

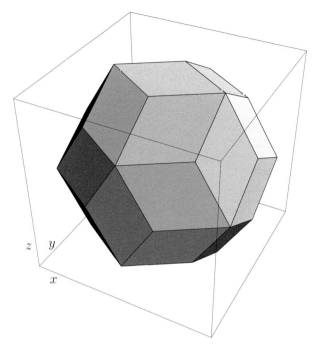

FIGURE 7.12. Kepler's triacontahedron as the window W of the primitive icosahedral model set. Up to translation, it is the convex set obtained as $\bigoplus_{i=1}^{6}[0,1]v_i^{\star}$, with the $\star$-map as described in the text.

produces the vertex set of a rhombohedral tiling for all translations $t \in \mathbb{R}^3$ that satisfy $(t + \partial W) \cap L^{\star} = \varnothing$, which is the generic case. Note that $t = 0$ results in a singular point set with additional points from the boundary. Such singular point sets are Delone and FLC, but not repetitive. No generic case results in an individual icosahedral symmetry, while the (minimal) hull defined by them is icosahedrally symmetric in the sense of Definition 5.14.

In either case, the density of the resulting model set is

$$\text{dens}(\Lambda) = \text{vol}(W)\,\text{dens}(\mathcal{L}) = \frac{\sqrt{5}}{10} \approx 0.22361.$$

This value can alternatively be obtained from the relative frequencies of the two prototiles. These are $1/\tau$ and $1/\tau^2$ for T_{p} and T_{o}, as follows from a window intersection calculation similar to that explained in Example 7.9. Using the volumes of the prototiles, the corresponding *absolute* frequencies (per unit volume) turn out to be $(2 + \tau)/5$ and $(3 - \tau)/5$. Since each rhombohedron contributes precisely one vertex, the density of vertex points is

$$\frac{2 + \tau}{5}\,\frac{1}{2\tau^2} + \frac{3 - \tau}{5}\,\frac{1}{2\tau} = \frac{\sqrt{5}}{10},$$

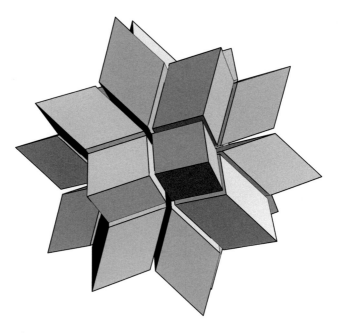

FIGURE 7.13. Exploded view of the icosahedrally symmetric vertex star of the P-type icosahedral tiling. It is built from 20 thick rhombohedra.

which highlights once more the importance of distinguishing relative and absolute frequencies.

Remark 7.10 (*Vertex stars and frequency module*). Combinatorially, the two rhombohedra admit 5450 distinct vertex configurations (up to Euclidean motions, including reflections). Of these, only 24 occur in a generic primitive icosahedral tiling. Their relative frequencies can be calculated via suitable dissections of W into the subwindows for the 24 vertex configurations. All (relative) vertex frequencies are elements of $\mathbb{Z}[\tau]$ and have been estimated numerically in [Hen86]; later, they were calculated exactly in [BB-A+94]. One can check that the $\mathbb{Z}$-span of the vertex frequencies is $\mathbb{Z}[\tau]$. Observing that the density of vertex points in our formulation is $\sqrt{5}/10$, the (absolute) frequency module of the primitive icosahedral tiling (on our scale) is

$$\mathcal{F} = \frac{\sqrt{5}}{10}\, \mathbb{Z}[\tau].$$

The vertex density was calculated from the volume of the window $(20\tau^3)$ and the density of the embedding lattice (which is $(7 - 4\tau)/200$). Note that the (absolute) frequency module depends on the scale, but is an MLD invariant, so that it is the same also for inflated versions of the tiling. $\Diamond$

Only one of the 24 possible vertex stars of the primitive icosahedral tiling has full icosahedral symmetry. It is illustrated in Figure 7.13. In any tiling obtained from a generic model set of Eq. (7.14), this vertex type occupies a subset that itself is a model set with a triacontahedron as its window. The latter turns out to be $\tau^{-3}W$, so that the centres of the vertex stars occupy a model set that is similar to the original one (or, more precisely, similar to one in the hull of the tiling); compare the analogous situation for the eightfold symmetric Ammann–Beenker tiling in Example 7.9. This subset property corresponds to the invariance of the module $\mathcal{M}_{\mathsf{P}}$ under multiplication by τ^3 and reflects the inflation symmetry of the tiling. Unfortunately, the corresponding (local) inflation rule is rather complicated, and has not been spelt out in complete detail.

Let us briefly mention that the positions of the maximally symmetric vertex star (and hence also those of all vertex stars) do not lead to sphere packings of high density. However, choosing a suitable subset of vertices, one can essentially achieve the density of randomly close-packed spheres; see [Hen86, Sec. III D] for details.

Example 7.13 (*Projection method for Danzer's ABCK tiling*). The ABCK tiling of Section 6.7.1 possesses a model set structure as follows. The coordinatisation chosen is based on the F-type icosahedral module $\mathcal{M}_{\mathsf{F}}$ of Example 2.20, which is also the LTM of this tiling. The vertex points are of four different types, as indicated in Table 6.2. Disregarding type IV, the distances between vertex points of the same type are elements of $\mathcal{M}_{\mathsf{F}}$. This gives rise to a description of the vertex set of the $\langle\text{ABCK}\rangle$ tiling as a three-component model set, which is based on the CPS (7.11) with $d = 3$, $H = \mathbb{R}^3$ and an embedding lattice $\mathcal{L}$ that is similar to the root lattice D_6; compare Example 3.8, again with the modified $\star$-map as used above. Note that the $\langle\text{ABCK}\rangle$ tiling is SMLD with the ABCK tiling. This follows from the observation that all type IV vertices are located in the centres of the $\langle K\rangle$ octahedra shown in Figure 6.39. The projection approach (in slightly different, but equivalent formulation) appeared first in [KPSZ94]. The key results were also derived by Danzer and his coworkers, but we are not aware of a publication.

To expand on the model set structure, one observes that the factor group $G := D_6^*/D_6$ is an Abelian group of order 4 (one can check that it is Klein's 4-group $C_2 \times C_2$). In analogy to the Penrose model set of Example 7.11, one can work with an internal space of the form $\mathbb{R}^3 \times G$. This description corresponds to expressing the body-centred module $\mathcal{M}_{\mathsf{B}}$ as a union of $\mathcal{M}_{\mathsf{F}}$ with three shifted cosets; see Eq. (2.15). This approach requires three windows, one for each class of vertex points. If we use the coordinatisation of Table 6.2, the vertices of type I are elements of $\mathcal{M}_{\mathsf{F}}$ which lift to elements of $\mathcal{L}$, whereas vertices of type II and III lie in the cosets $\mathcal{M}_{\mathsf{F}} + (\tau, 0, 1) = \mathcal{M}_{\mathsf{F}} + \tau^2 u$ and

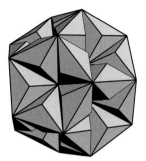 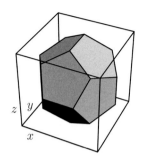 

FIGURE 7.14. The windows for the vertex types I (left), II (centre) and III (right) of the $\langle$ABCK$\rangle$ tiling. They are shown in the same orientation (as indicated for type II) and with the correct relative size.

$\mathcal{M}_{\mathsf{F}} + (\tau, \tau, \tau) = \mathcal{M}_{\mathsf{F}} + \tau u$, with $u = \frac{1}{2}(v_1 - v_2 + v_3 - v_4 + v_5 - v_6) = (1, 1, 1)$ as in Example 2.20. These cosets lift to $\mathcal{L} + (\tau, 0, 1, -1, 0, \tau)$ and $\mathcal{L} + (\tau, \tau, \tau, -1, -1, -1)$, respectively.

A more natural description of the model sets is obtained by shifting all vertices of the $\langle$ABCK$\rangle$ tiling by $(1, 1, 1)$, so that all vertices are elements of the three shifted cosets of $\mathcal{M}_{\mathsf{F}}$ in $\mathcal{M}_{\mathsf{B}}$, which are lifted to deep and shallow holes[2] in the lattice $\mathcal{L}$. The vertices of type I are then deep holes from $\mathcal{L} + (1, 1, 1, \tau, \tau, \tau)$, those of type II deep holes from $\mathcal{L} + (\tau, \tau, \tau, -1, -1, -1)$ and vertices of type III shallow holes from $\mathcal{L} + (\tau, 0, 1, -1, 0, \tau)$. The corresponding three windows have icosahedral symmetry and (as a result of the shift) are centred at (τ, τ, τ). They are shown in Figure 7.14. The window W_{I} of vertex type I is a dodecahedral extension of an icosahedron, with pentagonal edge length 2, the window W_{II} is a dodecahedron of edge length $2/\tau$, and the third window W_{III} is a great dodecahedron (a Kepler–Poinsot polyhedron, see [Cox73, Ch. VI]), with pentagonal edge length 2. The pentagonal cross-sections at the surface of W_{I} and W_{III} have equal size. In fact, the protruding star of W_{III} matches the indentation on the surface of W_{I}.

The volumes of the windows are calculated as

$$\mathrm{vol}(W_{\mathrm{I}}) = 20(4 - \tau), \quad \mathrm{vol}(W_{\mathrm{II}}) = 4(\tau + 2) \quad \text{and} \quad \mathrm{vol}(W_{\mathrm{III}}) = 20(\tau - 1).$$

The modified Minkowski embedding of $\mathcal{M}_{\mathsf{F}}$, according to Example 3.8, leads to a lattice that is similar to the root lattice D_6. It has a fundamental cell of volume $80(4\tau + 3)$, which can be calculated as the determinant of the basis matrix. Its reciprocal is the density $\mathrm{dens}(\mathcal{L})$, which gives $\mathrm{vol}(W_\iota)\,\mathrm{dens}(\mathcal{L})$ with $\iota \in \{\mathrm{I}, \mathrm{II}, \mathrm{III}\}$ for the (absolute) densities of the three vertex types.

---

[2]Holes in a lattice are vertices of the Voronoi cells whose distance from points of the lattice is a local maximum. If the distance is an absolute maximum, the hole is called *deep*, otherwise *shallow*; compare [CS99, Secs. 1.2 and 7.1].

A simple calculation confirms the values previously calculated in Eq. (6.12) from the inflation structure.   ◊

Remark 7.11 (*Alternative interpretation of windows*). For multi-component model sets such as the ⟨ABCK⟩ tiling or the rhombic Penrose tiling, there is a natural alternative description as a section (rather than a projection), with the advantage that the windows are attached to the proper preimages in the embedding space. Here, the windows are interpreted as 'target' polytopes, which also works for ordinary (one-component) model sets. We will describe this approach in Section 7.5.1 for the Fibonacci chain. In the case of the ⟨ABCK⟩ tiling, this amounts to attaching centred space-inverted copies of the three windows of Figure 7.14 at the points of the corresponding lattice cosets. These targets are parallel to internal space, so that their intersections with direct space are either empty or a single point. This remark also applies to the four windows of the rhombic Penrose tiling of Figure 7.8.   ◊

In complete analogy, one can also construct model sets with icosahedral symmetry on the basis of the module $\mathcal{M}_\mathsf{B}$. Since they have rarely shown up in practice so far, we do not go into detail; see [PKK97] for explicit examples.

Example 7.14 (*An E_8 quasicrystal in 4-space*). In [ES87], Elser and Sloane constructed a highly symmetric model set in 4-space with a number of interesting properties. It is based on the CPS (7.11) with $d = 4$, $H = \mathbb{R}^4$ and the root lattice $\mathcal{L} = E_8$. The projection $L = \pi(\mathcal{L}) = \mathbb{I}$ is the icosian ring of Example 2.19, with (modified) Minkowski embedding as described in Example 3.9. The window is the four-dimensional analogue of the triacontahedron, which is a semi-regular polytope with H_4 symmetry known as the 600-cell [Cox73]. It emerges as the projection of the Voronoi cell of E_8 into internal space, as described in [ES87]; see also [BG98] for various properties and the connection with four-dimensional crystallographic groups. Within the resulting tiling, one can find a hierarchy of lower-dimensional tilings related to τ, which complements the observation of Remark 3.8.   ◊

7.5. Alternative constructions

Cut and project methods appear in various different disguises in the literature, each of which has its own merits. For systematic reasons, we have mainly followed the model set approach. In this section, we briefly outline a few common variants in an informal manner, using the Fibonacci chain as a guiding example. For further approaches and aspects, we refer to [GR86, MP96, HKPM97]. We hope that the generalisation to higher dimensions will be intuitively clear. After the general description, we recollect de Bruijn's approach to the Penrose tiling in more detail.

7.5.1. Alternative routes to the Fibonacci chain

In Example 7.3, we gave a description of the Fibonacci chain based on a cut and project scheme with the Minkowski embedding of $\mathbb{Z}[\tau]$ as the lattice. As already mentioned in Remark 3.4 on page 59, one can turn this lattice into a square lattice via a linear shear. The corresponding projection construction is shown in the top panel of Figure 7.15. Unlike before, it is oriented along the lattice directions, which is a version that is often used in the literature. The points represent the scaled square lattice $\sqrt{\tau+2}\,\mathbb{Z}^2$, and direct (or tiling) space is the line with slope $1/\tau$. The window is the cross-section of the strip with the orthogonal direction, and is an interval of length $1+\tau$. This emerges from the original window of Example 7.3 by an application of the shear.

For simplicity, we suppress the non-generic possibility that a lattice point lies on the boundary of the strip in the following discussion. Since precisely the points inside the (shaded) strip are projected, this variant is usually referred to as the *strip projection method*, particularly in the physics literature. The prototiles have lengths 1 and τ as before. For convenience, the lattice points inside the strip are connected by edges such that their projections give the tiles. This works because the width of the strip equals that of the projection of a fundamental square to the perpendicular direction. This setting corresponds to what is often called the 'canonical' projection method.

An obvious alternative construction is shown in the middle panel of Figure 7.15. Here, each lattice point carries a centred 'target' line of length $1+\tau$, which is a translated copy of the (inverted) window. The points of the Fibonacci chain are now obtained as the intersection points of the line in the centre of the original strip (lightly shaded and included for comparison) with the 'targets' (once again restricting to the generic situation of never hitting the end points of the 'targets'). One can indeed check that this produces the same point set as the strip projection method. For a number of reasons connected with the inverse problem of structure determination of quasicrystals, this approach is perhaps the most popular construction in crystallography and materials science. There, the 'targets' are usually called *atomic hypersurfaces*.

A further approach is shown in the bottom panel of Figure 7.15, which is known as Kramer's *klotz construction*; compare [KS89] and references therein. It is a variant of the *dualisation method* (see below). Its main advantage is to immediately produce tilings by an intersection, rather than just point sets (which becomes relevant in higher dimensions). This is achieved by using a fundamental domain with boundaries either parallel or perpendicular to tiling space. In the example at hand, one possible choice consists of two squares of edge lengths 1 and τ, as indicated in the bottom panel of Figure 7.15 (strongly shaded). Cutting through the two types of squares along a line

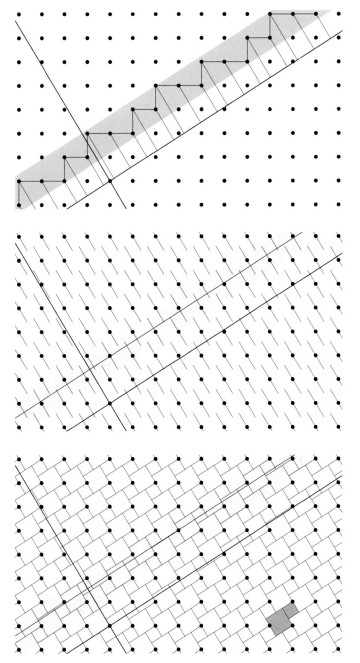

FIGURE 7.15. Alternative constructions of the Fibonacci chain from the square lattice $\sqrt{\tau+2}\,\mathbb{Z}^2$: strip projection (top), atomic hypersurfaces (middle), and klotz construction (bottom); see text for details.

parallel to the tiling space produces the two types of intervals, disregarding the non-generic case of cutting along a boundary. The position of the cut line in this figure is chosen to match the previous two constructions; it dissects the strip into two sub-strips of width ratio τ. This precisely corresponds to the two windows used previously in Example 7.3.

Finally, let us briefly describe the *dualisation method*, for the same example. Recall from Example 2.7 on page 19 that the Voronoi complex of the square lattice is built from a fundamental square (centred at the origin, say) and its lattice translates, while the corresponding Delone complex is a congruent copy, which is shifted such that its vertex points are the lattice points. Both complexes are shown in Figure 7.16. As before, we choose a cut line parallel to the tiling space in a generic position (avoiding the vertices of the Voronoi complex). Whenever this line intersects a k-boundary of the Voronoi complex, the unique dual $(2 - k)$-boundary of the Delone complex is orthogonally projected into tiling space. In particular, intersecting a Voronoi 1-boundary (grey lines) produces a tile via the projection of the dual Delone 1-boundary (heavy lines), while intersecting a Voronoi cell (shaded) produces a vertex point as indicated in the figure.

The possible projections of the Delone 1-boundaries to tiling space and its orthogonal complement also provide the boundaries of the two cells used in the klotz construction described above. Note that the role of Voronoi and Delone complex can be interchanged. This does not produce anything new in this case (because the two complexes are congruent), but it does in general [KS89], which is intuitively clear for the triangular lattice from Figure 2.2. For an explicit derivation of the rhombic Penrose tiling in this setting, we refer to [BKSZ90]. This paper also contains a step by step derivation of the dualisation method for the TTT.

7.5.2. DE BRUIJN'S GRID METHOD

In 1981, de Bruijn [dBr81] devised an algebraic description of the rhombic Penrose tilings, based on the dualisation of a pentagrid. This method immediately generalises to other symmetries, and can easily be adapted to produce planar rhombic tilings with n-fold rotational symmetry for arbitrary values of n. It is essentially equivalent to the cut and project approach of Section 7.3 with a suitable choice of the window; see [GR86] for a detailed discussion. Here, we first describe de Bruijn's approach for the Penrose tiling, and afterwards show some examples of other tilings obtained by this method. One advantage of de Bruijn's method is its easy applicability for large n, without the need to choose a window in internal space, which has dimension $\phi(n) - 2$. It has also been used in non-trivial applications in statistical physics; see [AP07] and references therein.

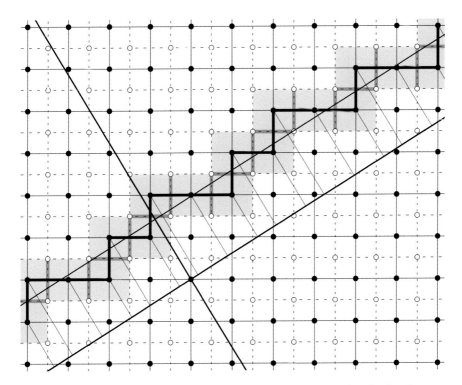

FIGURE 7.16. Construction of the Fibonacci chain by the dualisation method. Full circles mark the lattice points, while open circles are the vertices of the Voronoi cells. The 1-boundaries of the Voronoi (Delone) complex are indicated by dashed (solid) lines; see text for further details.

Consider the complex plane $\mathbb{C}$, and let $\xi \in \mathbb{S}^1$ and $\gamma \in \mathbb{R}$ be arbitrary. For any value of these two parameters, we define a *grid* in $\mathbb{C}$ by

$$\mathcal{G}(\xi, \gamma) = \{z \in \mathbb{C} \mid \mathrm{Re}(z\xi^{-1}) + \gamma \in \mathbb{Z}\}.$$

This defines an infinite set of equidistant parallel lines in the complex plane, with ξ determining the slope, and γ the position in space (modulo 1). The grid divides the complex plane into infinitely many strips, and the function $K \colon \mathbb{C} \to \mathbb{Z}$ defined by

$$K(z) = \lceil \mathrm{Re}(z\xi^{-1}) + \gamma \rceil$$

associates any point $z \in \mathbb{C}$ with the corresponding strip, enumerated by an integer. Here, $\lceil x \rceil = \min\{y \geqslant x \mid y \in \mathbb{Z}\}$ denotes the least integer $\geqslant x$. The non-integral part is captured by the difference function (which differs from the splitting $x = \lfloor x \rfloor + \{x\}$ for $x \in \mathbb{R}$ used previously)

$$\lambda(z) = K(z) - \mathrm{Re}(z\xi^{-1}) - \gamma, \quad 0 \leqslant \lambda(z) < 1.$$

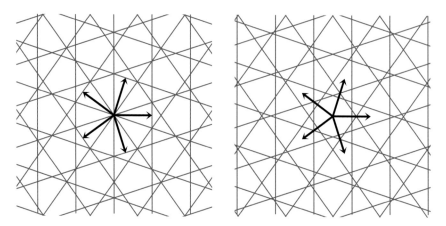

FIGURE 7.17. Part of the singular pentagrid $\mathcal{G}_5(0,0,0,0,0)$ (left) and the regular pentagrid $\mathcal{G}_5(\frac{1}{2},\frac{1}{2},\frac{1}{2},\frac{1}{2},\frac{1}{2})$ (right). The five unit vectors $\{1,\xi,\xi^2,\xi^3,\xi^4\}$ are also shown.

We are interested in a setting with fivefold (resp. tenfold) symmetry. Choosing ξ as a primitive 5th root of unity, say $\xi = \exp(2\pi i/5)$ for definiteness, we obtain a *pentagrid* as the union of five rotated grids

$$\mathcal{G}_5(\gamma_0, \gamma_1, \gamma_2, \gamma_3, \gamma_4) = \bigcup_{j=0}^{4} \mathcal{G}(\xi^j, \gamma_j),$$

which is parametrised by five real parameters γ_j, with $j \in \{0, 1, 2, 3, 4\}$. The pentagrid is called *regular* (or generic) if no point in the complex plane belongs to more than two of the five grids (which means no more than two lines intersect in any single point), and it is called *singular* otherwise. Two examples are shown in Figure 7.17. Using the corresponding coordinate functions

$$K_j(z) = \lceil \operatorname{Re}(z\xi^{-j}) + \gamma_j \rceil, \quad 0 \leqslant j \leqslant 4,$$

we associate to any $z \in \mathbb{C}$ a point $(K_0(z), K_1(z), K_2(z), K_3(z), K_4(z)) \in \mathbb{Z}^5$ of the hypercubic lattice. The function

$$f(z) = \sum_{j=0}^{4} K_j(z)\xi^j,$$

which is constant on every mesh of the pentagrid, associates one point in $\mathbb{C}$ to every mesh, and obviously corresponds to a projection from the lattice $\mathbb{Z}^5$ to the complex plane. Noting that $\sum_{j=0}^{4} \operatorname{Re}(z\xi^{-j})\xi^j = \frac{5}{2}z$, the function f can also be written as

$$f(z) = \frac{5}{2}z + \sum_{j=0}^{4} (\lambda_j(z) + \gamma_j)\xi^j,$$

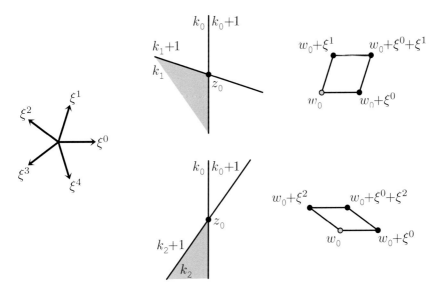

FIGURE 7.18. The dualisation of the grid in the vicinity of an intersection point z_0 of two grid lines yields a Penrose rhombus. The shaded area corresponds to the vertex $w_0 = f(z_0)$ of the rhombus tiling.

which shows that $f(z) - \frac{5}{2}z$ is bounded. Provided that the pentagrid is regular, the set $\{f(z) \mid z \in \mathbb{C}\}$ is the vertex set of a rhombus tiling of the plane with the two Penrose rhombuses, also known as a *generalised* Penrose rhombus tiling [PK87]. The tiling property can be seen by the following argument.

Consider a point z_0 which is the intersection of two grid lines of a regular grid, so it satisfies the equations

$$(7.15) \qquad \operatorname{Re}(z_0\xi^{-r}) + \gamma_r = k_r \quad \text{and} \quad \operatorname{Re}(z_0\xi^{-s}) + \gamma_s = k_s$$

for some $0 \leqslant r < s \leqslant 4$ and $k_r, k_s \in \mathbb{Z}$. Because the grid is regular, no further grid line passes through z_0. Hence, in a sufficiently small neighbourhood of z_0, the function f takes precisely four values, which are

$$(7.16) \qquad f(z_0) + \{0, \xi^r, \xi^s, \xi^r + \xi^s\}.$$

For any choice of r and s, these four points form the vertices of one of the two Penrose rhombuses; more precisely, they form a thick rhombus (with angles $2\pi/5$ and $3\pi/5$) for $s - r \in \{1, 4\}$ and a thin rhombus (which angles $\pi/5$ and $4\pi/5$) for $s - r \in \{2, 3\}$; see Figure 7.18 for examples. Note that the point $f(z_0)$ is always located at a $2\pi/5$ angle of the thick rhombus or at a $4\pi/5$ angle of the thin rhombus.

FIGURE 7.19. Rhombus tiling with exact (individual) D_{10} symmetry, as obtained by dualisation of the grid $\mathcal{G}_5(\frac{1}{2}, \frac{1}{2}, \frac{1}{2}, \frac{1}{2}, \frac{1}{2})$.

The resulting rhombic tiling is dual to the pentagrid in the sense that every intersection point of the grid corresponds to a unique rhombus of the tiling, and, conversely, every vertex of the rhombic tiling corresponds to a particular mesh of the grid. Furthermore, two vertices of the tiling are connected by an edge of a rhombus if and only if the corresponding two meshes share a common edge, and the edge of the rhombus is orthogonal to the corresponding edge of the mesh. Note that the correspondence via the function f, however, does not respect the position in the sense that the vertices of the rhombic tiling are not, in general, located within the corresponding mesh.

Since each mesh of the grid is a polygon with at least 3 and at most 10 edges, between 3 and 10 rhombuses meet at each vertex of the dual tiling. By construction, the rhombuses do not overlap, and completely cover the angle around a vertex, so the corresponding tiling has neither gaps nor overlaps [dBr81]. An example based on the grid $\mathcal{G}_5(\frac{1}{2}, \frac{1}{2}, \frac{1}{2}, \frac{1}{2}, \frac{1}{2})$ is shown in Figure 7.19. Due to the symmetry of the grid, this tiling has an individual

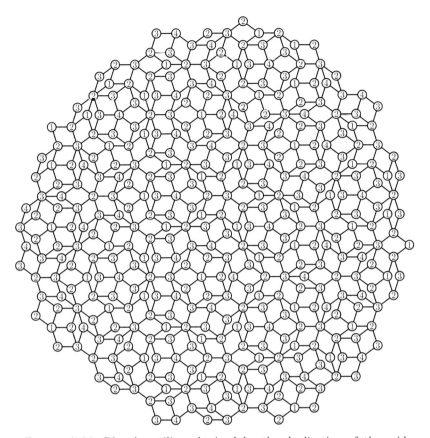

FIGURE 7.20. Rhombus tiling obtained by the dualisation of the grid $\mathcal{G}_5(\frac{1}{7}, \frac{2}{7}, \frac{3}{7}, -\frac{2}{7}, -\frac{4}{7})$. The four types of vertices are distinguished by numbers according the distinct cosets; see text for details.

tenfold rotational symmetry with respect to the origin. It is sometimes called the *anti*-Penrose tiling.

In the generic case, the rhombus tilings obtained in this way are not proper Penrose tilings (RPT), as they will contain local configurations of rhombuses that are not present in rhombic Penrose tilings (such as the ten thin rhombuses meeting in one vertex in the centre of the patch of Figure 7.19). There is, however, a simple condition on the parameters γ_j to ensure that the resulting tiling is indeed a proper rhombic Penrose tiling, namely $\sum_{j=0}^{4} \gamma_j = 0$ (or, more generally, the same condition mod 1). As a consequence of this constraint, we get

$$\sum_{j=0}^{4} \lambda_j(z) = \sum_{j=0}^{4} \big(K_j(z) - \mathrm{Re}(z\xi^{-j}) - \gamma_j \big) = \sum_{j=0}^{4} K_j(z),$$

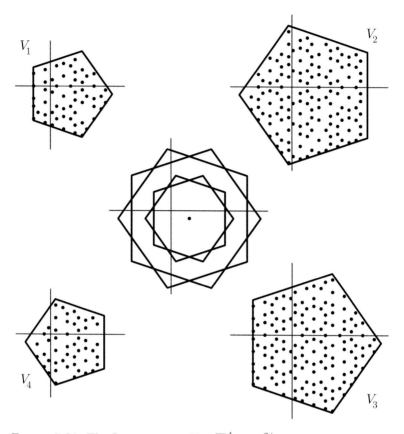

FIGURE 7.21. The five pentagons $V_r + \sum_{j=0}^{4} \gamma_j \xi^{2j}$ with the positions and indices of x^* for vertices x of the tiling of Figure 7.20. The dot in the central part denotes the common centre $\sum_{j=0}^{4} \gamma_j \xi^{2j}$ of the four pentagons.

since $\sum_{j=0}^{4} \xi^{-j} = 0$. By definition, we have $0 < \sum_{j=0}^{4} \lambda_j(z) < 5$, whence it follows that $\sum_{j=0}^{4} K_j(z) \in \{1, 2, 3, 4\}$. This value is called the *index* of the corresponding vertex $f(z)$. Moving along the edge of a rhombus, the index increases by 1 if the edge is along one of the directions ξ^j, and decreases by 1 if it is in one of the opposite directions $-\xi^j$, where $0 \leqslant j \leqslant 4$. Hence, for the rhombus of Eq. (7.16) corresponding to the grid point z_0 of Eq. (7.15), the possible indices of the four vertices are either $1, 2, 2, 3$ or $2, 3, 3, 4$; in particular, the vertex $f(z_0)$ is always the one with the smallest index. An example with vertex indices is shown in Figure 7.20.

The edge decorations of the Penrose tiles are locally derivable from the vertex indices as follows. Edges connecting vertices with indices 1 and 2 or vertices with indices 3 and 4 carry double arrows, pointing from index 2 to index 1 or from index 3 to 4. Edges connecting vertices with index 2 and 3 carry single arrows; their orientations are determined by the orientations of the double arrows on the same rhombus. Here, de Bruijn [dBr81] proved that this assignment of arrows is consistent along edges that are shared by two adjacent rhombuses.

For the case of the RPT, the connection to the cut and project approach is provided by [dBr81, Thm. 8.1]. It states that $x = \sum_{j=0}^{4} k_j \xi^j$, with $(k_0, k_1, k_2, k_3, k_4) \in \mathbb{Z}^5$, is a vertex of the tiling if and only if

$$x^\star = \sum_{j=0}^{4} k_j \xi^{2j} \in V_r + \sum_{j=0}^{4} \gamma_j \xi^{2j},$$

where $r = \sum_{j=0}^{4} k_j \in \{1, 2, 3, 4\}$ is the index of x and V_r are four pentagon-shaped regions defined by

$$V_r = \Big\{ \sum_{j=0}^{4} \lambda_j \xi^{2j} \ \Big|\ 0 < \lambda_j < 1, \sum_{j=0}^{4} \lambda_j = r \Big\}.$$

Hence, V_1 is the interior of the pentagon with vertices $\{1, \xi, \xi^2, \xi^3, \xi^4\}$, while $V_2 = (1 + \xi)V_1$, $V_3 = -V_2$ and $V_4 = -V_1$. The four pentagons thus play the role of the window system for the projection from the hypercubic lattice $\mathbb{Z}^5$. Note that $\xi \mapsto \xi^2$ is the $\star$-map for $\mathbb{Z}[\xi]$ from Table 7.1. For the example of Figure 7.20, the shifted pentagons and the positions of the images of the vertices x under the $\star$-map are shown in Figure 7.21; compare Remark 7.8. It is clear that this approach is essentially equivalent to our number-theoretic model set description of Example 7.11.

Singular grids are grids which contain points that belong to more than two grid lines. Such grids can be seen as limits of sequences of regular grids approaching the singular case. In this way, each singular grid corresponds to several tilings which reflect the different ways that the singular case can be approached. In particular, the exceptionally singular grid $\mathcal{G}_5(0, 0, 0, 0, 0)$, where all five grid lines meet in a single point, corresponds to ten different tilings; the central patch of one of these tilings is shown on the left of Figure 7.22, and the other nine tilings are obtained by rotation through multiples of $2\pi/10$ about the origin. Any two such tilings agree almost everywhere, while they only differ along one or more of the grid directions, as indicated on the right plate of Figure 7.22. This gives proximal pairs in such a way that Theorem 5.1 applies, which is one way to infer aperiodicity. In a singular tiling, images of vertices under the $\star$-map populate part of the boundary of the pentagonal regions $V_r + \sum_{j=0}^{4} \gamma_j \xi^{2j}$, with different parts corresponding to taking different limits. A detailed discussion of singular pentagrids and the corresponding tilings can be found in [dBr81].

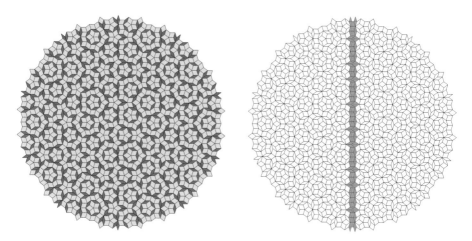

FIGURE 7.22. Central patch of one of the ten Penrose tilings corresponding to the singular grid $\mathcal{G}_5(0,0,0,0,0)$ (left). Any two of the ten different tilings corresponding to this grid differ along one or more of the grid directions. In the example shown on the right, mismatches between two rotated copies occur along the grey 'worm' only.

If we set $\gamma = \sum_{i=0}^{4} \gamma_i$, its fractional part $\{\gamma\}$ labels the LI classes, where $\{\gamma\} = 0$ is the rhombic Penrose tiling and $\{\gamma\} = 1/2$ the anti-Penrose tiling. These are the classes with D_{10} symmetry, where only the anti-Penrose class contains an element with individual tenfold symmetry; see Figure 7.19. For the remaining classes, we have D_5 symmetry, and the classes with $\{\gamma\}$ and $1 - \{\gamma\}$ are related by a π-rotation. The geometry behind these claims can be understood via Figure 7.8, where γ acts as a shift of the cut planes in the rhombic icosahedron; compare [Ing99, Fig. 7]. For aspects around local rules and MLD relations, we refer to [Ing99] and references therein.

This approach readily generalises to other symmetries. One can define n-grids $\mathcal{G}_n$ for any value of n, using a primitive nth root of unity, and taking the union of n (for n odd) or $n/2$ (for n even) rotated and shifted grids. Figure 7.23 shows a patch of the tiling obtained by dualising the grid $\mathcal{G}_7(\frac{1}{2}, \frac{1}{2}, \frac{1}{2}, \frac{1}{2}, \frac{1}{2}, \frac{1}{2}, \frac{1}{2})$, which contains three different rhombic prototiles. Note that the central 14-star appears only once in the patch of Figure 7.23, although it has to appear repetitively in the full tiling.

Figure 7.24 shows part of a tiling for $n = 11$ with a generic choice of the shifts γ_j, which comprises five different rhombic prototiles. The patch shown does *not* contain a high-symmetry vertex star, although such vertex stars do exist in the tiling — they are dual to 22-gonal meshes in the 11-grid.

Remark 7.12 (*Symmetric configurations are rare*). It is an obvious observation that vertex configurations with n-fold symmetry become increasingly

FIGURE 7.23. A central patch of a fourteenfold symmetric tiling with 3 prototiles.

rare as n grows. This can be quantified, and then shows a clear dependence on the codimension of the tiling in the minimal embedding, which means $\phi(n) - 2$ by Theorem 3.2. In a model set formulation of the vertex points, the window for the maximally symmetric star is a small region in the centre of the model set window. The volume thus scales with the length ratio to the power of the codimension, which explains the observation qualitatively. The scaling behaviour as an explanation for the rarity of symmetric configurations was observed in the context of laser beam interference patterns; see [MSR+10, Fig. 4]. ◇

As mentioned before, the grid approach is equivalent to the projection method for cyclotomic model sets with a specific choice of the window (or windows). When one starts with a non-minimal embedding into $\mathbb{R}^n$, one may

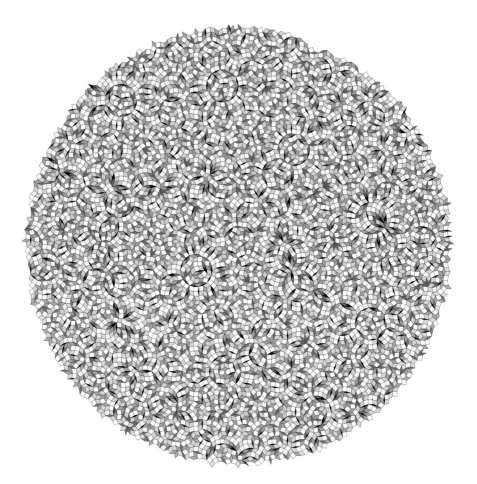

FIGURE 7.24. A generic patch of a 22-fold tiling with 5 prototiles.

employ the hypercubic lattice $\mathbb{Z}^n$ and the perpendicular projection of the hypercube as a window. Here, one also has to select a suitable representative of each lattice direction perpendicular to direct space, which exists due to the non-minimal embedding. Effectively, one ends up with certain slices of the hypercube, similar to the situation we saw in Figure 7.8 for the rhombic Penrose tiling, but with an increasing number of subwindows. The latter can become rather complicated with growing n, and have not been described in general.

This concludes our discussion of tilings and patterns generated by local rules (Chapter 5), inflation rules (Chapter 6) and the projection method (Chapter 7). In Chapter 9, we will analyse the diffraction spectra of projection tilings, which now requires some additional tools from analysis.

CHAPTER 8

Fourier Analysis and Measures

In this chapter, we recall some basic facts from Fourier analysis which we need for our approach to diffraction theory in Chapter 9. Since Fourier series and their generalisations to almost periodic functions occur only occasionally below, we keep their exposition brief and informal (mostly without proofs, but with proper references). Subsequently, Fourier transforms of functions and measures are covered in more detail, including a brief introduction to volume-averaged (or Eberlein) convolutions.

8.1. Fourier series

A (possibly complex-valued) function f of one real variable is called *periodic*, if $f(x + T) = f(x)$ holds for some $T \neq 0$ and all $x \in \mathbb{R}$. Clearly, one then has $f(x + nT) = f(x)$ for all $n \in \mathbb{Z}$. Assuming that the periodic function f is continuous and not constant, the smallest $T > 0$ with this property is the *fundamental period* of f. It is simply called the period of f when misunderstandings are unlikely. Examples are provided by trigonometric functions, such as $\sin(x)$, $\cos(x)$ or $\mathrm{e}^{\mathrm{i}x}$, all three with fundamental period $T = 2\pi$. A natural way to look at such functions in a more general setting is to consider T-periodic functions that are locally integrable (over any compact set $K \subset \mathbb{R}$ say), so that one can also view them as elements of the Banach space $L^1([0, T])$; see [DMcK72, Kat04, Pin02] for general background. Note that, depending on the context, different notions of convergence will show up, such as convergence in the mean (i.e., in the norm $\|.\|_1$) as well as pointwise and uniform convergence (in the supremum norm $\|.\|_\infty$) for continuous functions.

Recall that a real-valued (but not necessarily continuous) function f has *bounded variation* on an interval $[a, b]$ if a constant C exists such that

$$\sum_{i=1}^{n} \left| f(x_i) - f(x_{i-1}) \right| \leqslant C$$

holds for every partition defined by $a = x_0 < x_1 < \cdots < x_n = b$. Note that $n \in \mathbb{N}$ is arbitrary and that C does not depend on n. A function of bounded variation on $[a, b]$ is bounded on $[a, b]$. A real-valued function f is said to be of bounded variation if it has bounded variation on any compact interval. Examples include non-increasing or non-decreasing functions (not necessarily continuous, though all one-sided limits in $[a, b]$ exist) as well as Lipschitz functions. In fact, by Jordan's theorem, f is of bounded variation on $I = [a, b]$ if and only if it can be expressed (on I) as the difference of two non-decreasing functions; compare [Lan93, Thm. X.1.2].

The relevance of bounded variation is evident from the following convergence result [Cha87, Thm. 15.5]; see also [Pin02, Sec. 1.2.5].

Proposition 8.1. *If g is a T-periodic continuous function of bounded variation on $[0, T]$, it has a uniformly converging Fourier series of the form*

$$g(x) = \sum_{m \in \mathbb{Z}} c_m \, \mathrm{e}^{2\pi \mathrm{i} \frac{m}{T} x}$$

with Fourier coefficients

$$c_m = \frac{1}{T} \int_0^T \mathrm{e}^{-2\pi \mathrm{i} \frac{m}{T} x} g(x) \, \mathrm{d}x.$$

When g is not a constant function, this holds in particular if T is the fundamental period of g.   □

The Fourier series also exists and converges pointwise everywhere for any locally integrable, periodic function g of bounded variation, which is then differentiable (and thus continuous) almost everywhere. At points of discontinuity, convergence is towards the mean $\frac{1}{2}\big(g(x+0)+g(x-0)\big)$; compare [Pin02, Sec. 1.2.5]. Beyond this class, by theorems of Carleson and Hunt, one still has pointwise convergence almost everywhere for periodic functions that are locally L^p with $p > 1$, though it is difficult to determine the exceptional set. In fact, any subset of $\mathbb{R}$ of measure 0 occurs as the pointwise divergence set for some continuous function [Kat04, Thm. II.3.4].

Remark 8.1 (*Hilbert space approach to Fourier series*). For many purposes, it is more natural to consider T-periodic functions that are locally square integrable, hence employing a formulation with the Hilbert space $L^2([0, T])$ and convergence in the L^2-norm. Each locally square integrable periodic function then possesses a Fourier series of the above form, with L^2-convergence; see [Pin02, Sec. 1.3] for details.   ◇

We also need a generalisation of Proposition 8.1 to Fourier series of lattice periodic functions. We state it for $\mathbb{Z}^d$-periodic functions and refer to [DMcK72, Rud62, Lan93] for further generalisations. Here, we continue to write kx (rather than $\langle k|x \rangle$) for the Euclidean scalar product of k and x. We expand on the uniform convergence later when we need the result.

Theorem 8.1. *Let $g \in C(\mathbb{R}^d)$ be a $\mathbb{Z}^d$-periodic function that is also $(d+1)$-times continuously differentiable. Then, it has a uniformly converging Fourier* *series of the form*

$$g(x) = \sum_{k \in \mathbb{Z}^d} c_k e^{2\pi i k x},$$

with Fourier coefficients $c_k = \int_{[0,1]^d} e^{-2\pi i k x} g(x) \, \mathrm{d}x.$

PROOF. By assumption, all partial derivatives up to order $d+1$ are continuous, and they are also $\mathbb{Z}^d$-periodic. Via suitable integration by parts, in the direction of the maximal component of k, one can derive the estimate

$$|c_k| \leqslant \frac{C}{1 + \|k\|_\infty^{d+1}},$$

where the constant C does not depend on k. As $\sum_{k \in \mathbb{Z}^d} \left(1 + \|k\|_\infty^{d+1}\right)^{-1} < \infty$, the formal Fourier series is absolutely convergent, which also implies the claimed uniform convergence. The integral expression for the coefficients is standard; compare [DMcK72] or [AAP92]. □

There are important cases of uniform convergence without absolute convergence, but they subtly depend on the summation method for multiple Fourier series. Since we do not need more than the result of Theorem 8.1, we refer to [AAP92] for a systematic and general exposition.

A different extension (needed later) concerns the Fourier–Stieltjes series of measures on the unit circle $\mathbb{S}^1$. We will return to this case in Section 8.7.

8.2. Almost periodic functions

Periodic functions are important in many areas, but they form a special class that often is too restrictive. Indeed, even if a function fails to be periodic, it need not be irregular. Examples are provided by *trigonometric polynomials* on the real line, which are functions of the form

$$(8.1) \qquad f(x) = \sum_{j=1}^n a_j \, e^{2\pi i k_j x}$$

with $n \in \mathbb{N}$ and $a_j \in \mathbb{C}$. The numbers $k_j \in \mathbb{R}$ are called the *frequencies* of f. Note that a trigonometric polynomial need not be periodic.

Example 8.1 (*Quasiperiodic functions*)**.** Consider the uniformly continuous function f defined by

$$(8.2) \qquad f(x) = \cos(2\pi x) + \cos(2\pi \alpha x),$$

with $\alpha \in \mathbb{R} \backslash \mathbb{Q}$. This function cannot have any non-trivial period because α is irrational (observe that $f(x) = 2$ holds for $x = 0$ only). Nevertheless, it is the sum of two trigonometric functions, and hence a trigonometric polynomial. Moreover, it is related to a $\mathbb{Z}^2$-periodic function of two variables via

$$f(x) = \cos(2\pi x) + \cos(2\pi y)\big|_{y=\alpha x}.$$

Functions of this type are simple examples of *quasiperiodic* functions, first studied by Bohl [Boh93] and Esclangon [Esc04]. They are characterised by Fourier series with *finitely* many fundamental frequencies and their harmonics (see Remark 8.2 below for more). An analogous mechanism is at the heart of the theory of model sets, as we already noticed, to some extent, in the context of the cut and project method in Chapter 7. We shall see a consequence later in more detail when we derive the diffraction formula for model sets.

When f is a non-periodic function, it can still show some 'almost repetition' under certain translations. Let us get some insight into the existence of such ε-almost periods for the function f from Eq. (8.2). Let $n \in \mathbb{Z}$ and observe that $\cos(2\pi(x+n)) = \cos(2\pi x)$. This gives

$$
\begin{aligned}
\left| f(x+n) - f(x) \right| &= \left| \cos(2\pi\alpha(x+n)) - \cos(2\pi\alpha x) \right| \\
&= \left| -2\sin(\pi\alpha n)\sin(\pi\alpha(n+2x)) \right| \leqslant 2\left| \sin(\pi\alpha n) \right|,
\end{aligned}
$$

where the second step follows by standard trigonometric identities. Note that the derived bound is sharp, which need not be the case for alternative estimates involving the triangle inequality. Because α is irrational, the sequence $(\alpha n \bmod 1)_{n\in\mathbb{Z}}$ is dense in the interval $[0,1)$; see also Example 7.1. It is thus clear that, for any $\varepsilon > 0$, there are infinitely many integers $n \in \mathbb{Z}$ such that $\left| f(x+n) - f(x) \right| < \varepsilon$ holds for all $x \in \mathbb{R}$. Observing that $|\sin(x)| \leqslant |x|$ on $\mathbb{R}$, one derives that a sufficient condition for this estimate reads $\alpha n \bmod 1 < \varepsilon/(2\pi)$. One can now see by an elementary geometric argument (involving the corresponding irrational rotation on the unit circle) that such integers n occur with bounded gaps.

Concretely, consider $\alpha = \sqrt{2}$, which is well approximated by the rational numbers obtained from truncating the continued fraction expansion $\sqrt{2} = [1; 2, 2, 2, \ldots]$; see [Per54] for background. The first few of them read

$$
1, \frac{3}{2}, \frac{7}{5}, \frac{17}{12}, \frac{41}{29}, \frac{99}{70}, \frac{239}{169}, \cdots
$$

which are of the form a_i/b_i with $a_{i+1} = a_i + 2b_i$ and $b_{i+1} = a_i + b_i$ for $i \geqslant 0$, together with the initial condition $a_0 = b_0 = 1$. By construction, a_i and b_i are coprime for all $i \geqslant 0$. The denominators b_i satisfy the second order recurrence $b_{i+1} = 2b_i + b_{i-1}$ for $i \geqslant 1$, with $b_0 = 1$ and $b_1 = 2$. In view of our above estimate, b_i is an ε_i-almost period with $\varepsilon_i = 2\left| \sin(\pi\sqrt{2}\, b_i) \right|$. Here, one has $b_i \sim (1+\sqrt{2})^{i+1}/(2\sqrt{2})$ and thus $\varepsilon_i \sim 2\pi/(1+\sqrt{2})^{i+1}$ as $i \to \infty$.

Some cases are illustrated in Figure 8.1. Reducing ε leads to a closer proximity of the function and its translates, however with sparser ε-almost periods. Nevertheless, the latter are always relatively dense, for the reason mentioned before.   ◊

Let us put this in a more systematic perspective. Consider the space $C(\mathbb{R})$ of continuous functions in one variable, equipped with the topology of

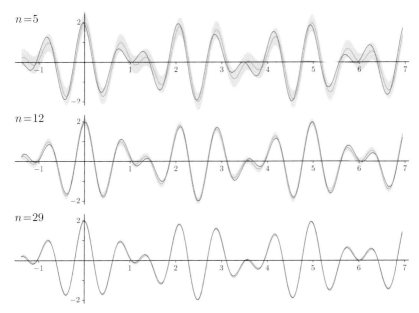

FIGURE 8.1. A plot of $f(x+n)$ (black line) within the ε-tube (shaded) around the function $f(x)$ (grey line) of Eq. (8.2), for $\alpha = \sqrt{2}$ and three different translations n. Their values derive from the denominators of the rational approximants to the continued fraction expansion of $\sqrt{2}$.

uniform convergence, as induced by the supremum norm

$$\|g\|_\infty := \sup_{x \in \mathbb{R}} |g(x)|$$

on the subspace $C_{\mathsf{b}}(\mathbb{R})$ of bounded continuous functions and then suitably extended to $C(\mathbb{R})$. For $\varepsilon > 0$, a number $t \in \mathbb{R}$ is called an ε-almost period of a function $g \in C(\mathbb{R})$ if $\|g - T_t g\|_\infty < \varepsilon$, where $T_t g$ denotes the translated function, defined by $T_t g(x) = g(x - t)$. For a periodic function g and for any $\varepsilon > 0$, each period of g is also an ε-almost period of g. If g is uniformly contin-uous, every sufficiently small translation is an ε-almost period. Quasiperiodic functions possess ε-almost periods for all $\varepsilon > 0$, and this even with bounded gaps, though they need not have any non-trivial periods, as we saw in Ex-ample 8.1. This motivates the following more general concept, which was introduced by H. Bohr in the 1920s [Boh47]; see also [Bes54, Cor89], and [Kat04, Sec. VI.5] for a brief modern account.

Definition 8.1. A continuous function $g \in C(\mathbb{R})$ is called *almost periodic* if, for every $\varepsilon > 0$, the set of its ε-almost periods is relatively dense in $\mathbb{R}$.

Clearly, all periodic functions are almost periodic, as are all trigonometric polynomials. An almost periodic function is bounded [Kat04, Lemma VI.5.3] and uniformly continuous [Kat04, Lemma VI.5.4]. Almost periodic functions form an algebra, so that (finite) sums, differences, and products of almost periodic functions are again almost periodic, as is the absolute value; see [Kat04, Thm. VI.5.7]. Moreover, one has the following fundamental result, which contains an important extension of Bohr's work due to Bochner and von Neumann; see [Cor89, Kat04] for details and references.

Proposition 8.2. *For $g \in C(\mathbb{R})$, the following properties are equivalent.*

(1) *The function g is almost periodic;*

(2) *g is the limit of a sequence of trigonometric polynomials, with uniform convergence;*

(3) *the translation orbit $\{T_t g \mid t \in \mathbb{R}\}$ is precompact in the topology of uniform convergence.*

SKETCH OF PROOF. The equivalence of (1) and (3) follows from [Kat04, Thm. VI.5.5]. Next, since all trigonometric polynomials are almost periodic, the implication (2) $\Longrightarrow$ (1) follows from a 2ε-argument. The converse direction, (1) $\Longrightarrow$ (2), is proved in [Kat04, Thm. VI.5.18]. $\qquad\square$

This result says that the $\|.\|_\infty$-closure of the set of trigonometric polynomials equals the set of almost periodic functions. It implies that this class of functions provides a natural generalisation of periodic functions, and points a way towards more general Fourier series, which are often called *Fourier–Bohr series*. Since almost periodic functions possess no analogue of a fundamental period, their description as a series involves coefficients that are defined by an average over $\mathbb{R}$. If $g \in C(\mathbb{R})$ is bounded, one fixes some $a \in \mathbb{R}$ and sets

$$(8.3) \qquad M(g) := \lim_{T \to \infty} \frac{1}{2T} \int_{a-T}^{a+T} g(x)\,\mathrm{d}x,$$

provided the limit exists. For almost periodic functions, and hence for periodic functions in particular, the limit exists for all a, and is independent of the parameter a. In fact, the limit is uniform in a [Cor89, Thm. I.1.12].

Theorem 8.2. *Let $g \in C(\mathbb{R})$ be an almost periodic function, and define, for arbitrary $k \in \mathbb{R}$, the corresponding Fourier–Bohr coefficient*

$$a(k) = M\big(\mathrm{e}^{-2\pi \mathrm{i} k x} g(x)\big),$$

which exists. Then, the set $I := \{k \in \mathbb{R} \mid a(k) \neq 0\}$ is at most a countable subset of $\mathbb{R}$. The (formal) Fourier–Bohr series attached to g is defined as

$$\sum_{k \in I} a(k)\,\mathrm{e}^{2\pi \mathrm{i} k x}.$$

If this series is uniformly convergent, its limit is the function g.

SKETCH OF PROOF. This theorem can be traced back to several results from [Cor89] as follows. If g is almost periodic, then so is the product function defined by $x \mapsto e^{-2\pi i k x} g(x)$ (Thm. I.1.5). Consequently, the Fourier–Bohr coefficient $a(k)$ exists for all $k \in \mathbb{R}$ (Thm. I.1.12). It differs from 0 for at most countably many $k \in \mathbb{R}$ (Thm. I.1.15). The convergence claim is Thm. I.1.20. □

Note that the uniform approximability of an almost periodic function f by trigonometric polynomials does *not* imply the (uniform) convergence of its Fourier–Bohr series. In general, the convergence of the Fourier–Bohr series is a difficult problem; see [Cor89, Sec. I.4] for some sufficient criteria. Nevertheless, distinct almost periodic functions have distinct Fourier–Bohr series, and the coefficients always satisfy Parseval's identity [Cor89, Thm. I.1.18]

$$(8.4) \qquad \sum_{k \in I} |a(k)|^2 = M(|g|^2).$$

The inner product $\langle g | h \rangle_M := M(\bar{g}h)$ turns the space of almost periodic function on $\mathbb{R}$ into a Hilbert space; see [Kat04, Sec. VI.5.15]. It is thus evident that the Fourier–Bohr coefficients carry the essential information on g, even if the series does not converge everywhere. However, the Parseval identity (8.4) implies that the Fourier–Bohr series always converges in the mean; see also [Ebe49, Ebe55] for further details.

Remark 8.2 (*Subclasses of almost periodic functions*). The structure of the set I from Theorem 8.2 is used to distinguish important subclasses of almost periodic functions. When $g \in C(\mathbb{R})$ is *periodic* with fundamental period T, the coefficient $a(k)$ reduces to the ordinary Fourier coefficient introduced earlier, with $I \subset \frac{1}{T}\mathbb{Z}$. For a general almost periodic function, there are (at most) countably many pairwise incommensurate fundamental frequencies $k_j \in I$, and I may also contain their integer multiples. When the number of fundamental frequencies is finite, the series represents a *quasiperiodic* function [Boh93, Esc04]. An interesting case emerges when the set I is not contained in a finitely generated $\mathbb{Z}$-module, but is still a subset of the module $\mathbb{Z}[q]$ for some $q \in \mathbb{Q}$ (which implies that all elements of I are linearly dependent over $\mathbb{Q}$). Then, the corresponding function is called *limit periodic*. A prominent example that we shall meet later, in the context of quasiperiodic measures and systems of Toeplitz type such as the period doubling chain, is $I = \mathbb{Z}[\frac{1}{2}]$; see [GK97] for related concepts and examples, and [Bes54, Ch. I.6] for a slightly wider definition of limit periodicity. ◇

There are various generalisations of almost periodic functions, both with respect to the topology used and the number of variables; compare [GLA90] for a general exposition in the setting of measures.

8.3. Fourier transform of functions

Let $\mathcal{S}(\mathbb{R}^d)$ be the space of rapidly decreasing C^∞-functions on $\mathbb{R}^d$, also known as Schwartz functions [Schw98, Ch. VII.3]. This space is equipped with its usual metric topology; see [RS80, Thm. V.9] for details. Besides all C^∞-functions with compact support, it also contains functions such as $P(x)\exp(-|x|^2)$, where $P(x)$ is an arbitrary polynomial in $x = (x_1, \ldots, x_d)$. By the *Fourier transform* of a Schwartz function $\phi \in \mathcal{S}(\mathbb{R}^d)$, we mean

$$(8.5) \qquad (\mathcal{F}\phi)(k) = \widehat{\phi}(k) := \int_{\mathbb{R}^d} e^{-2\pi i k x}\, \phi(x)\, \mathrm{d}x,$$

which is well-defined and again a Schwartz function, with $k \in \mathbb{R}^d$. The mapping $\mathcal{F}\colon \mathcal{S}(\mathbb{R}^d) \longrightarrow \mathcal{S}(\mathbb{R}^d)$ is a homeomorphism, with inverse

$$(8.6) \qquad (\mathcal{F}^{-1}\psi)(x) = \widecheck{\psi}(x) = \int_{\mathbb{R}^d} e^{2\pi i k x}\, \psi(k)\, \mathrm{d}k;$$

compare [RS80, Thm. IX.1]. With this definition (with the factor 2π in the exponent), one has

$$\widecheck{\widehat{\phi}} = \phi \quad \text{and} \quad \widehat{\widecheck{\psi}} = \psi.$$

Note that $\mathcal{F}$ has a unique extension to the Hilbert space $L^2(\mathbb{R}^d)$, often called the Fourier–Plancherel transform [Rud87], which is a unitary operator of fourth order, $\mathcal{F}^4 = \mathrm{Id}$. This follows from $(\mathcal{F}^2\phi)(x) = \phi(-x)$; see [RS80] for details. Moreover, Eq. (8.5) is also well-defined for functions from the Banach space $L^1(\mathbb{R}^d)$, though the existence of an inverse then becomes a delicate matter; see [Rud62, ReSt00] for a detailed discussion.

Example 8.2 (*Gaussian profiles*). Consider the normalised Gaussian density function on $\mathbb{R}$, as defined by $x \mapsto \psi(x) = \exp(-\pi x^2)$. It is a well-known fact that ψ is an eigenfunction of $\mathcal{F}$ with eigenvalue 1, meaning that $\widehat{\psi} = \psi$; see [Pin02, Ex. 2.2.7] for more. Let $\varepsilon > 0$ and define

$$\psi_\varepsilon(x) := \frac{1}{\varepsilon}\exp\left(-\frac{\pi x^2}{\varepsilon^2}\right),$$

which gives a family of positive Schwartz functions with $\int_{\mathbb{R}} \psi_\varepsilon(x)\, \mathrm{d}x = 1$. In particular, one has $\psi_1 = \psi$. If g is any integrable function that is continuous at 0, one obtains the representation

$$g(0) = \lim_{\varepsilon \searrow 0} \int_{\mathbb{R}} g(x)\, \psi_\varepsilon(x)\, \mathrm{d}x$$

by standard arguments from calculus.

Moreover, the Fourier transform of ψ_ε is

$$\widehat{\psi_\varepsilon}(x) = \int_{\mathbb{R}} e^{-2\pi i x y}\, e^{-\pi(y/\varepsilon)^2}\, \mathrm{d}(y/\varepsilon) = e^{-\pi \varepsilon^2 x^2}.$$

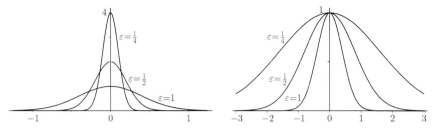

FIGURE 8.2. Illustration of the functions ψ_ε (left) of Example 8.2 and their Fourier transforms $\widehat{\psi_\varepsilon}$ (right), for the parameters $\varepsilon \in \{1, \frac{1}{2}, \frac{1}{4}\}$.

As $\varepsilon \searrow 0$, $\widehat{\psi_\varepsilon}$ is a monotonically increasing family of functions that pointwise converges to the constant function 1; see Figure 8.2 for an illustration. ◇

Let us mention another interesting property of the functions ψ_ε, namely

$$\text{(8.7)} \qquad \lim_{\varepsilon \searrow 0} (\psi_\varepsilon * f) = f,$$

which holds for every integrable f. Here, $*$ denotes the usual convolution of L^1-functions (compare Eq. (8.8) below). Recalling that the Banach algebra $L^1(\mathbb{R})$ contains no unit for the multiplication $*$, this limiting property is the closest equivalent. The family $(\psi_\varepsilon)_{\varepsilon > 0}$ is thus called an *approximate unit*; see [Rud62] for details. The generalisation to $\mathbb{R}^d$ is obvious.

Remark 8.3 (*Eigenfunctions and unitarity of $\mathcal{F}$*). As mentioned previously, the Fourier transform has a unique extension to a unitary operator on the Hilbert space $L^2(\mathbb{R})$. Clearly, the function ψ of Example 8.2 is an L^2-function, as is any function of the form $P(x)\psi(x)$ with P a polynomial. Define (one version of) the *Hermite polynomials* H_n by the recursion

$$H_{n+1}(x) = 2x\, H_n(x) - 2n\, H_{n-1}(x),$$

for $n \in \mathbb{N}$, together with the initial condition $H_0(x) = 1$ and $H_1(x) = 2x$. Note that H_n is a polynomial of degree n. The *Hermite functions*

$$h_n(x) := \left(\frac{\sqrt{2}}{2^n\, n!}\right)^{\frac{1}{2}} H_n(\sqrt{2\pi}\, x)\, \psi(x),$$

form an orthonormal basis of $L^2(\mathbb{R})$, relative to the inner product defined by $\langle \varphi \,|\, \psi \rangle = \int_{\mathbb{R}} \overline{\varphi(x)}\, \psi(x)\, \mathrm{d}x$; see [DMcK72, Sec. 2.5] for more and Figure 8.3 for an illustration.

This basis is special in that the Hermite functions h_n are eigenfunctions of $\mathcal{F}$ with eigenvalues $(-\mathrm{i})^n$, so that Fourier transform is a unitary diagonal matrix in this basis. All of this can be proved by induction, observing that the function $f(x) := x\, g(x)$ satisfies $-2\pi\mathrm{i}\, \widehat{f}(k) = \frac{\mathrm{d}}{\mathrm{d}k}\, \widehat{g}(k)$, together with the fact that $\frac{\mathrm{d}}{\mathrm{d}x} H_n(x) = 2n\, H_{n-1}(x)$. ◇

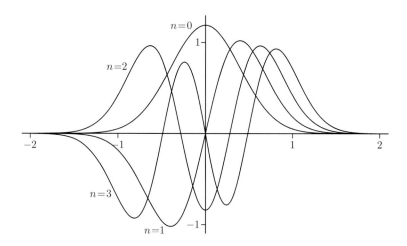

FIGURE 8.3. The Hermite functions h_n of Remark 8.3 for $n \in \{0, 1, 2, 3\}$.

The ordinary *convolution* of two Schwartz functions reads

$$(8.8) \qquad \bigl(f * g\bigr)(x) := \int_{\mathbb{R}^d} f(x - y)\, g(y)\, \mathrm{d}y \; = \; \int_{\mathbb{R}^d} f(y)\, g(x - y)\, \mathrm{d}y,$$

which is again a Schwartz function. The same definition applies to functions $f, g \in L^1(\mathbb{R}^d)$, resulting in $f * g \in L^1(\mathbb{R}^d)$; see [Rud62, Sec. 1.1.6]. Convolution is both commutative and associative. In fact, $L^1(\mathbb{R}^d)$ together with pointwise addition and the above convolution forms a Banach algebra. An important property of the Fourier transform is the *convolution theorem*.

Lemma 8.1 (Fourier convolution theorem [RS80, Thm. IX.3]). *For two arbitrary functions* $\phi_1, \phi_2 \in \mathcal{S}(\mathbb{R}^d)$, *one has* $\bigl(\phi_1 * \phi_2\bigr)^{\widehat{}} = \widehat{\phi_1} \cdot \widehat{\phi_2}$. *The corresponding result also holds for* $\phi_1, \phi_2 \in L^1(\mathbb{R}^d)$. □

Example 8.3 (*Convolution and FT of characteristic functions*). Consider the characteristic function $1_{[a,b]}$ of a compact interval in $\mathbb{R}$ with $a < b$. Let g be a function on $\mathbb{R}$ and define $\tilde{g}$ by $\tilde{g}(x) = \overline{g(-x)}$. Then, the convolution $f := 1_{[a,b]} * \widetilde{1_{[a,b]}}$ is the continuous, symmetric, tent-shaped function

$$f(x) = \begin{cases} (b - a) - |x|, & |x| \leqslant b - a, \\ 0, & \text{otherwise.} \end{cases}$$

It is compactly supported on the interval $[a - b, b - a] = [a, b] - [a, b]$. The Fourier transform of the characteristic function $1_{[a,b]}$ reads

$$\widehat{1_{[a,b]}}(k) = \int_a^b \mathrm{e}^{-2\pi \mathrm{i} k x}\, \mathrm{d}x \; = \; (b - a)\, \mathrm{e}^{-\pi \mathrm{i} k(a+b)}\, \mathrm{sinc}\bigl(\pi k(b - a)\bigr),$$

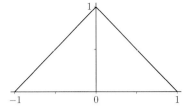 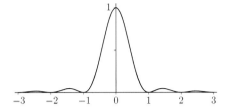

FIGURE 8.4. The function $1_{[0,1]} * \widetilde{1_{[0,1]}}$ (left) and its Fourier transform (right).

where $\mathrm{sinc}(x) = \sin(x)/x$ as before. This is the canonical way to write the result as a product of three terms, one for the length, a phase factor that keeps track of the position, and the essential function of the transform. If the interval is symmetric (so that $a = -b < 0$), the right-hand side simplifies to $2b\,\mathrm{sinc}(2\pi k b)$. Observing that $\widehat{\widetilde{g}} = \overline{\widehat{g}}$ holds for any integrable function g, an application of the convolution theorem results in the continuous function

$$\widehat{f}(k) = \big|\widehat{1_{[a,b]}}(k)\big|^2 = \left(\frac{\sin\big(\pi k(b-a)\big)}{\pi k}\right)^2,$$

which has the maximum value $\widehat{f}(0) = (b-a)^2$; compare Figure 8.4.

In d-space, one often considers the radially symmetric characteristic function $1_{\overline{B_r(0)}}$ of the centred closed ball of radius $r > 0$. By a standard calculation with (d-dimensional) spherical coordinates, one obtains

$$\widehat{1_{\overline{B_r(0)}}}(k) = \widehat{1_{B_r(0)}}(k) = \left(\frac{r}{|k|}\right)^{d/2} J_{d/2}\big(2\pi|k|r\big),$$

where J_α is the Bessel function of the first kind [AAR99, Sec. 4.5]. ◇

A function of the form $g * \widetilde{g}$ is an example of a positive definite function. Recall that a continuous function $f\colon \mathbb{R}^d \longrightarrow \mathbb{C}$ is *positive definite* if, for all $n \in \mathbb{N}$ and for all n-tuples $(x_1, \dots, x_n)$ of elements from $\mathbb{R}^d$, the associated $n \times n$-matrix $\big(f(x_i - x_j)\big)_{1 \leqslant i,j \leqslant n}$ is positive Hermitian [BF75, Def. 3.3]. By Bochner's theorem, a continuous f is positive definite if and only if there is a finite positive measure μ on $\mathbb{R}^d$ such that

$$f(x) = \int_{\mathbb{R}^d} e^{2\pi i k x}\, \mathrm{d}\mu(k)$$

holds for every $x \in \mathbb{R}^d$; see [Rud62] for a proof. For instance, if $f = g * \widetilde{g}$ with $g \in \mathcal{S}(\mathbb{R}^d)$ say, one has the representation

$$f(x) = \int_{\mathbb{R}^d} e^{2\pi i k x}\, |\widehat{g}(k)|^2\, \mathrm{d}k,$$

so that $\mu = |\widehat{g}|^2 \lambda$, with λ the Lebesgue measure, is the finite positive measure in this case. Some of the concepts used here are discussed in more detail later.

8.4. Fourier transform of distributions

Next, we need to extend the Fourier transform beyond functions. A *tempered distribution* is a continuous linear functional on the Schwartz space $\mathcal{S}(\mathbb{R}^d)$. The tempered distributions $T : \mathcal{S}(\mathbb{R}^d) \longrightarrow \mathbb{C}$ thus form the corresponding dual space, denoted by $\mathcal{S}'(\mathbb{R}^d)$. We also use the common notation $(T, \phi) := T(\phi)$ for the evaluation of $T \in \mathcal{S}'(\mathbb{R}^d)$ with a function $\phi \in \mathcal{S}(\mathbb{R}^d)$, called a *test function* in this context. Each continuous function g of at most polynomial growth defines a tempered distribution T_g via

$$T_g(\phi) := \int_{\mathbb{R}^d} \phi(x)\, g(x)\, \mathrm{d}x\,.$$

Such distributions are also called *regular*. In particular, $\mathcal{S}(\mathbb{R}^d)$ is naturally embedded into $\mathcal{S}'(\mathbb{R}^d)$ this way.

The Fourier transform of a tempered distribution $T \in \mathcal{S}'(\mathbb{R}^d)$ is induced by that of the regular distributions and reads

$$(8.9) \qquad\qquad \widehat{T}(\phi) := T\big(\widehat{\phi}\big)$$

for all test functions $\phi \in \mathcal{S}(\mathbb{R}^d)$. The Fourier transform is then a linear bijection of $\mathcal{S}'(\mathbb{R}^d)$ onto itself which is the unique weakly continuous extension of the Fourier transform on $\mathcal{S}(\mathbb{R}^d)$; see [RS80, Thm. IX.2]. This is important, since weak-$*$ convergence of a sequence of tempered distributions, $T_n \longrightarrow T$ as $n \to \infty$ (which means that $T_n(\phi) \xrightarrow{n\to\infty} T(\phi)$ for all $\phi \in \mathcal{S}(\mathbb{R}^d)$), then implies weak-$*$ convergence of their Fourier transforms, so that $\widehat{T_n} \xrightarrow{n\to\infty} \widehat{T}$.

Positive definite Schwartz functions have a counterpart on the level of tempered distributions. An element $T \in \mathcal{S}'(\mathbb{R}^d)$ is called *positive definite* (or of positive type) if $T(\phi * \widetilde{\phi}) \geqslant 0$ for all $\phi \in \mathcal{S}(\mathbb{R}^d)$. This is an extension in the sense that the regular distribution T_g defined by a positive definite function g is positive definite as a distribution. The Bochner–Schwartz theorem [RS80, Thm. IX.10] characterises the positive definite distributions in terms of the Fourier transform of positive measures with at most polynomial growth.

Remark 8.4 (*Compact support and analyticity*). Recall that a continuous function has compact support when it vanishes outside some compact set. A tempered distribution T has support in a closed set K if $T(\phi) = 0$ for every test function ϕ with support in the complement of K. When K is compact, T is said to have *compact support*. In this case, the Fourier transform $\widehat{T}$ is always a regular distribution. Its representing function has an analytic continuation to an entire function on $\mathbb{C}^d$ with certain growth restrictions; see [RS80, Thm. IX.12] for details. Finer results of this type, both for functions and for distributions, are known as Paley–Wiener theorems.   ◇

Example 8.4 (*Dirac distribution*). Consider the continuous linear mapping $\delta_x \colon \mathcal{S}(\mathbb{R}^d) \longrightarrow \mathbb{C}$ defined by $\phi \mapsto \delta_x(\phi) := \phi(x)$, which is also well-defined on all test functions that are merely continuous at x. Clearly, δ_x is an element of the dual space $\mathcal{S}'(\mathbb{R}^d)$, but it has no representation as a regular distribution. It is thus called *singular*.

The Dirac distribution $\delta_x \in \mathcal{S}'(\mathbb{R}^d)$ is the weak-$*$ limit of a sequence $(T_{g_n})_{n \in \mathbb{N}}$ of regular distributions with

$$g_n(y) = n^d g\big(n(y - x)\big),$$

where g is any positive Schwartz function with $g(0) > 0$ and the normalisation condition $\int_{\mathbb{R}^d} g(x)\,\mathrm{d}x = 1$. A concrete choice (for $d = 1$) is provided by the functions $\psi_{1/n}$ of Example 8.2. As mentioned before, such sequences are called *approximate units*, due to their role in the convolution algebra $L^1(\mathbb{R}^d)$.

The Fourier transform of δ_x is easily calculated as

$$\widehat{\delta_x}(\phi) = \delta_x(\widehat{\phi}) = \widehat{\phi}(x) = \int_{\mathbb{R}^d} \mathrm{e}^{-2\pi \mathrm{i} xy}\, \phi(y)\,\mathrm{d}y = T_{h_x}(\phi),$$

where $h_x(y) = \exp(-2\pi \mathrm{i} xy)$ is a bounded function. This makes Remark 8.4 explicit for the Dirac distribution δ_x, whose support is the singleton set $\{x\}$. The previous calculation explains the widely used shorthand

(8.10) $$\widehat{\delta_x} = \mathrm{e}^{-2\pi \mathrm{i} xy}.$$

In particular, $\widehat{\delta_0} = 1$, which matches nicely with the behaviour of the functions $\widehat{\psi_\varepsilon}$ from Example 8.2 in the limit as $\varepsilon \searrow 0$.

Note that δ_x can also be interpreted as a probability measure on $\mathbb{R}^d$, called the *Dirac* or *point measure* at x. Its Fourier transform (compare Section 8.6) is then an analytic function. Interpreting the latter as the Radon–Nikodym density of a measure relative to Lebesgue measure λ, the distribution $\widehat{\delta_x}$ is embedded into the class of complex measures. $\diamond$

8.5. Measures and their decomposition

For simplicity, we introduce (possibly unbounded) measures as linear functionals, and connect them to the standard approach via σ-algebras of measurable sets by means of the Riesz–Markov representation theorem; see [Die70] for a general exposition. This is possible because we only need regular Borel measures. As motivated in Chapter 1, measures provide a natural platform for our formulation. In particular, they are general enough to cover almost periodic functions as well as rigid tiling structures.

In Chapter 4, we used (finite) regular Borel measures on a compact space $\mathbb{X}$. They can also be introduced as linear functionals on $C(\mathbb{X})$, the space of continuous functions on $\mathbb{X}$. Below, our focus is more on the situation that $\mathbb{X}$ is replaced by a non-compact space that, at the same time, is a locally compact Abelian group (LCAG). Since the latter possesses a Haar measure, this will give us an important tool for the decomposition of a measure into distinctive parts. We will formulate this mainly for the group $\mathbb{R}^d$ and mention extensions occasionally when we need them.

8.5.1. LINEAR FUNCTIONALS AND MEASURES

Let $C_c(\mathbb{R}^d)$ be the space of complex-valued continuous functions on $\mathbb{R}^d$ with compact support. A (complex) *measure* μ on $\mathbb{R}^d$ is a linear functional on $C_c(\mathbb{R}^d)$ with the extra condition that, for every compact set $K \subset \mathbb{R}^d$, there is a constant a_K such that

$$|\mu(g)| \leqslant a_K \|g\|_\infty$$

holds for all $g \in C_c(\mathbb{R}^d)$ with support in K. Here, $\|g\|_\infty := \sup_{x \in \mathbb{R}^d} |g(x)| = \sup_{x \in K} |g(x)|$ is the supremum norm of g. If μ is a measure, the *conjugate* of μ is defined by the mapping $g \mapsto \overline{\mu(\bar{g})}$. It is again a measure and denoted by $\bar{\mu}$. A measure μ is called *real* (or signed) if $\bar{\mu} = \mu$, or (equivalently) if $\mu(g)$ is real for all real-valued $g \in C_c(\mathbb{R}^d)$. A real measure μ is called *positive* if $\mu(g) \geqslant 0$ for all $g \geqslant 0$. For every measure μ, there is a smallest positive measure, denoted by $|\mu|$, such that $|\mu(g)| \leqslant |\mu|(g)$ holds for all non-negative $g \in C_c(\mathbb{R}^d)$, and $|\mu|$ is called the (*total*) *variation* (or absolute value) of μ.

A measure μ is *finite* or *bounded* if $|\mu|(\mathbb{R}^d)$ is finite (with obvious meaning, see below), otherwise it is called unbounded. Note that a measure μ is continuous on $C_c(\mathbb{R}^d)$ with respect to the topology induced by the norm $\|.\|_\infty$ if and only if it is finite [Die70, Ch. XIII.20]. In view of this, the vector space $\mathcal{M}(\mathbb{R}^d)$ of measures on $\mathbb{R}^d$ is given the *vague topology*, meaning that a sequence of measures $(\mu_n)_{n \in \mathbb{N}}$ converges vaguely to μ if $\lim_{n \to \infty} \mu_n(f) = \mu(f)$ in $\mathbb{C}$ for all $f \in C_c(\mathbb{R}^d)$. This is the weak-$*$ topology on $\mathcal{M}(\mathbb{R}^d)$, in which all the 'standard' linear operations on measures are continuous; compare [RS80, p. 114] for some consequences of this. The measures defined this way are, by proper decomposition [Die70, Ch. XIII.2 and Ch. XIII.3] and an application of the Riesz–Markov representation theorem, see [RS80, Thm. IV.18] or [Ber65, Thm. 69.1], in one-to-one correspondence with the regular Borel measures on $\mathbb{R}^d$, wherefore we identify them.

For the 'traditional' point of view, the σ-algebra of measurable sets is formed by the Borel sets of $\mathbb{R}^d$, the smallest σ-algebra that contains all open (and hence also all closed) subsets of $\mathbb{R}^d$, in its standard topology. For simplicity, we thus write $\mu(A)$ (measure of a set) and $\mu(f)$ (measure of a function) in parallel. The connection is quite obvious for the characteristic function 1_A of a (relatively compact) Borel set A via $\mu(1_A) = \mu(A)$. Note that 1_A is not an element of $C_c(\mathbb{R}^d)$, but the regularity of μ means that we can find a sequence $(f_n)_{n \in \mathbb{N}}$ of non-negative continuous functions with support in A such that $f_n \leqslant 1_A$ and $\mu(f_n) \xrightarrow{n \to \infty} \mu(1_A)$, and another sequence $(g_n)_{n \in \mathbb{N}}$ with $A \subset \mathrm{supp}(g_n)$ for all $n \in \mathbb{N}$, $1_A \leqslant g_n$ and $\mu(g_n) \xrightarrow{n \to \infty} \mu(1_A)$.

Example 8.5 (*Dirac versus Lebesgue measure*). The Dirac distribution δ_x also defines a measure, where one has

$$\delta_x(A) = \begin{cases} 1, & \text{if } x \in A, \\ 0, & \text{otherwise,} \end{cases}$$

for an arbitrary Borel set $A \subset \mathbb{R}^d$. The Lebesgue measure of an integrable function g is $\lambda(g) = \int_{\mathbb{R}^d} g \, \mathrm{d}\lambda = \int_{\mathbb{R}^d} g(x) \, \mathrm{d}x$, while the measure of a Borel set A is $\lambda(A) = \int_{\mathbb{R}^d} 1_A \, \mathrm{d}\lambda$. The measures δ_0 and λ are related by

$$\widehat{\delta_0} = \lambda \quad \text{and} \quad \widehat{\lambda} = \delta_0,$$

which is a special case of Eq. (8.10), the latter now read as an identity between measures. Indeed, these relations between λ and δ_0 can simultaneously be understood as identities between distributions (acting on $\mathcal{S}(\mathbb{R}^d)$) and between measures (acting on $C_{\mathsf{c}}(\mathbb{R}^d)$). $\diamond$

The space of complex measures is often too general for our purposes, whence we restrict to a natural class of objects as follows; see [BF75] for background material, and [Schl00, Sec. 1] for a systematic approach in the context of general LCAGs.

Definition 8.2. A measure $\mu \in \mathcal{M}(\mathbb{R}^d)$ is called *translation bounded* (or shift bounded) if $\sup_{x \in \mathbb{R}^d} |\mu|(x + K) < \infty$ holds for every compact $K \subset \mathbb{R}^d$.

In other words, translation boundedness means that, for every compact set $K \subset \mathbb{R}^d$, there is a constant b_K such that $|\mu|(x + K) \leqslant b_K$ is satisfied for all translations $x \in \mathbb{R}^d$.

Example 8.6 (*Dirac combs*). Let Λ be a locally finite point set. Consider the *weighted Dirac comb*

$$\omega_\Lambda := \sum_{x \in \Lambda} w(x) \, \delta_x,$$

with δ_x as above and a weight function $w : \Lambda \longrightarrow \mathbb{C}$. Clearly, ω_Λ defines a complex measure. When Λ is uniformly discrete and w is bounded, meaning that $\sup_{x \in \Lambda} |w(x)| < \infty$, the measure ω_Λ is translation bounded. When $w \equiv 1$, the corresponding measure ω_Λ is simply called the (uniform) *Dirac comb* of Λ, written as δ_Λ. In particular, $\delta_\mathbb{Z} = \sum_{x \in \mathbb{Z}} \delta_x$, while $\omega_\Lambda = w \, \delta_\Lambda$ is the corresponding shorthand for the weighted Dirac comb from above. $\diamond$

Though measures and (tempered) distributions are both defined as linear functionals, there is an important difference between them. While the former are defined on continuous functions of compact support, the latter need Schwartz functions as arguments. In particular, measures need not be tempered distributions, for instance if they grow too fast as $|x| \to \infty$. If a measure μ also defines a tempered distribution T_μ, via $T_\mu(\phi) = \mu(\phi)$ for all $\phi \in \mathcal{S}(\mathbb{R}^d)$, it is called a *tempered measure*. A sufficient condition for a measure to be tempered [Schw98, Thm. VII.VII] is that it increases only slowly, in the sense that $\int_{\mathbb{R}^d}(1 + |x|)^{-\ell} \, \mathrm{d}|\mu|(x) < \infty$ for some $\ell \in \mathbb{N}$. Consequently, every translation bounded measure is tempered. Such measures form a good class for most of our purposes.

Conversely, a tempered distribution need not define a measure, which can sometimes cause problems. As a simple example, consider the distribution δ_x' defined by $\delta_x'(\phi) := -\phi'(x)$, where the symbol $'$ denotes the derivative with respect to x. This is unambiguous because ϕ is C^∞, and tempered distributions (also called generalised functions) provide a minimal scheme where all 'functions' are automatically infinitely differentiable (in this distributional sense). However, δ_x' does *not* define a measure, because one cannot give $\delta_x'(g)$ a clear meaning for continuous functions g of compact support — just think of some function g that is not differentiable at x. Nevertheless, if we start from a tempered measure, we usually do not distinguish between the measure and the corresponding distribution, in which case we write $\widehat{\mu}$ for $\widehat{T_\mu}$.

8.5.2. Decomposition of measures

To discuss the decomposition of measures into components of different type, we begin with positive measures and extend to general measures later. Consider a given positive measure $\mu \in \mathcal{M}(\mathbb{R}^d)$, and define the set

$$P_\mu := \{x \mid \mu(\{x\}) \neq 0\},$$

which is called the set of *pure points* (or *atoms*) of μ. Since $0 \leqslant \mu(K) < \infty$ for any compact set $K \subset \mathbb{R}^d$, we know that P_μ is (at most) a countable set [RS80, Lan93]. This follows from the observation that, for all $n \in \mathbb{N}$, we have $\mathrm{card}\{x \in K \mid \mu(\{x\}) > \frac{1}{n}\} < \infty$. Indeed, any pure point $x \in K$ of μ must satisfy $\mu(\{x\}) > \frac{1}{n}$ for some $n \in \mathbb{N}$, while $\mathbb{R}^d$ can be covered by countably many compact sets.

Let A be an arbitrary Borel set, and define

$$\mu_{\mathsf{pp}}(A) := \sum_{x \in A \cap P_\mu} \mu(\{x\}) = \mu(A \cap P_\mu).$$

With this definition, both μ_{pp} and $\mu_{\mathsf{cont}} := \mu - \mu_{\mathsf{pp}}$ are positive measures. By construction, μ_{cont} has no pure points (which means that $\mu_{\mathsf{cont}}(\{x\}) = 0$ for all $x \in \mathbb{R}^d$), while μ_{pp} has nothing but (in the sense that $\mu_{\mathsf{pp}}(A) = \sum_{x \in A} \mu_{\mathsf{pp}}(\{x\})$, which automatically reduces to a countable sum and converges absolutely for all bounded sets A). More generally, a measure is called *pure point* if it has pure points only, and *continuous* if it has none. It is clear from the constructive argument that the decomposition [RS80, Thm. I.13]

$$(8.11) \qquad \mu = \mu_{\mathsf{pp}} + \mu_{\mathsf{cont}}$$

is unique. Moreover, μ_{pp} and μ_{cont} are mutually singular ($\mu_{\mathsf{pp}} \perp \mu_{\mathsf{cont}}$); see Proposition 8.4 below for more. The term 'continuous' for a measure ν refers to the set-theoretic intermediate value property that, if there are Borel sets $A \subset C$ with $\nu(A) < b < \nu(C)$, then there is a Borel set B with $A \subset B \subset C$ and $\nu(B) = b$.

The natural reference measure in $\mathbb{R}^d$ is Lebesgue measure λ, because it is the unique normalised Haar measure of $\mathbb{R}^d$, meaning that λ is the only translation invariant positive measure on $\mathbb{R}^d$ that assigns volume 1 to the unit cube. Following common practice, we write $\mathrm{d}x$ instead of $\mathrm{d}\lambda(x)$.

A measure $\mu \in \mathcal{M}(\mathbb{R}^d)$ is said to be *absolutely continuous* relative to λ if $\mu = f\lambda$ for some function f that is locally integrable (meaning that $\int_A |f(x)|\,\mathrm{d}x < \infty$ for all relatively compact Borel sets A). Such an f is called the *Radon–Nikodym density* of μ relative to λ. This implies

$$(8.12) \qquad \mu(g) = \int g\,\mathrm{d}\mu = \int gf\,\mathrm{d}\lambda = \lambda(gf)$$

for $g \in C_{\mathsf{c}}(\mathbb{R}^d)$, which also explains the notation $\mathrm{d}\mu = f\,\mathrm{d}\lambda$ and $\mathrm{d}\mu(x) = f(x)\,\mathrm{d}x$. More generally, if $\mu, \nu \in \mathcal{M}(\mathbb{R}^d)$, μ is called *absolutely continuous* relative to ν when $\mu = g\nu$ for some function g that is locally L^1 for the measure ν. By the Radon–Nikodym theorem, see [Die70, Thm. (13.15.1)] together with the comments in [Die70, Ch. XIII.16], this happens if and only if $\mu(A) = 0$ for all Borel sets with $\nu(A) = 0$.

A positive measure $\mu \in \mathcal{M}(\mathbb{R}^d)$ is called *singular* relative to λ if there is a Borel subset $S \subset \mathbb{R}^d$ with $\lambda(S) = 0$ so that $\mu(\mathbb{R}^d \setminus S) = 0$. In this situation, $\mu = \mu|_S$ (the restriction of μ to S). As a counterpart to the previous decomposition of Eq. (8.11), there is a unique way to write

$$(8.13) \qquad \mu = \mu_{\mathsf{ac}} + \mu_{\mathsf{sing}},$$

with μ_{ac} being absolutely continuous and μ_{sing} being singular relative to λ [RS80, Thm. I.14]. Putting these properties together, one has the following fundamental result; compare [RS80] for details.

Theorem 8.3 (Lebesgue decomposition theorem). *Any positive, regular Borel measure $\mu \in \mathcal{M}(\mathbb{R}^d)$ has a unique decomposition*

$$\mu = \mu_{\mathsf{pp}} + \mu_{\mathsf{sc}} + \mu_{\mathsf{ac}}$$

relative to Lebesgue measure λ, where μ_{sc} is the unique part of μ that is both continuous and singular relative to λ. □

Below, we will often refer to the attribute $\alpha \in \{\mathsf{pp}, \mathsf{sc}, \mathsf{ac}\}$ of a measure as its *spectral type*. The measure μ_{sc} is the *singular continuous* part of μ, and μ is called purely singular continuous when $\mu = \mu_{\mathsf{sc}}$. There are obvious examples of this spectral type, such as the uniform probability measure on the unit circle, embedded in the plane. However, such a situation is 'decomposable' in the sense that the unit circle is a manifold and the measure is absolutely continuous tangentially and singular radially (then referring to Lebesgue measure in one dimension in both cases). A fundamentally different example is the following.

Example 8.7 (*Cantor measure*). Consider the classic middle thirds Cantor set, constructed from the unit interval $I_0 = [0,1]$ by means of the following iterated function system (compare [Car00, Ch. 2] for a more 'traditional' point of view). Define the two functions (or similitudes) $f_1(x) = \frac{1}{3}x$ and $f_2(x) = \frac{1}{3}x + \frac{2}{3}$, and consider the set-valued map

$$f(A) := f_1(A) \cup f_2(A),$$

which is a contraction on the space of compact subsets of $\mathbb{R}$, with respect to the Hausdorff distance; see [Hut81, BM00a] for details. The Cantor iteration is then given by $I_{n+1} = f(I_n)$ for $n \geqslant 0$, which converges to a compact set $\mathcal{C}$ that is the unique fixed point of the contraction, so that $\mathcal{C} = f(\mathcal{C})$. This is the classic *Cantor set*, which is perfect, totally disconnected, uncountable and of Lebesgue measure $\lambda(\mathcal{C}) = 0$; compare Definition 2.8. It has Hausdorff dimension $d_{\mathrm{H}} = \log(2)/\log(3) \approx 0.631$, and is thus a fractal.

The map f also induces a mapping on $\mathcal{M}(\mathbb{R})$, $\mu \mapsto f.\mu$, via

$$(8.14) \qquad (f.\mu)(A) := \frac{1}{2}\Big(\mu\big(f_1^{-1}(A)\big) + \mu\big(f_2^{-1}(A)\big)\Big)$$

for measurable sets A, where $f_i^{-1}(A)$ is the preimage of A under f_i. This induced map is again a contraction, this time on the space of probability measures on $\mathbb{R}$, equipped with the Hutchinson metric; see [Hut81, BM00a]. Starting with the probability measure $\mu_0 = 1_{I_0}\lambda$ and iterating according to $\mu_{n+1} = f.\mu_n$, one obtains a vaguely converging sequence of probability measures. The limit $\mu_{\mathcal{C}} = \lim_{n\to\infty}\mu_n$ is the classic *Cantor measure*. This measure has no pure points, as the assumption of a point measure at $x \in \mathcal{C}$ leads to a contradiction with μ being a measure via an iterative application of Eq. (8.14) with $A = \{x\}$. Moreover, it is concentrated to $\mathcal{C}$, an uncountable set of Lebesgue measure 0. Consequently, $\mu_{\mathcal{C}}$ is purely singular continuous relative to λ.

Since $\mu_{\mathcal{C}}$ is a probability measure on $\mathbb{R}$ (which vanishes outside $[0,1]$), an illustration is possible via the distribution function

$$F(x) := \mu_{\mathcal{C}}\big((-\infty, x)\big).$$

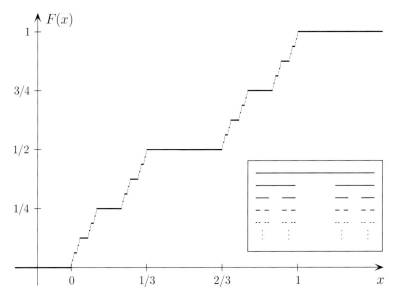

FIGURE 8.5. The continuous distribution function $F(x)$ of the classic Cantor measure, known as the Devil's staircase. The iterative construction of the underlying middle thirds Cantor set is sketched in the inset.

For $x > 0$, this simplifies to $F(x) = \mu_{\mathcal{C}}\big([0,x)\big)$, while one has $F(x) \equiv 0$ for $x \leqslant 0$ and $F(x) \equiv 1$ for $x \geqslant 1$. Since $\mu_{\mathcal{C}}$ has no pure points, $F(x)$ is a nondecreasing continuous function. Moreover, it is constant almost everywhere. In contrast, it fails to be differentiable at the uncountably many points of $\mathcal{C}$. Due to this property, it is called the *Devil's staircase*, shown in Figure 8.5.

The Cantor set is a paradigm of a fractal embedded in $\mathbb{R}$. It is a surprising (though well-known) observation that, despite $\lambda(\mathcal{C}) = 0$, its Minkowski sum satisfies $\mathcal{C} + \mathcal{C} = [0,2]$. This follows from the relation $\mathcal{C} - \mathcal{C} = [-1,1]$ by the reflection symmetry of $\mathcal{C}$ (so that $-\mathcal{C} = \mathcal{C} - 1$), where the difference property was first observed by Steinhaus [Ste17]. A nice illustration of the underlying geometric proof is given in [Car00, Fig. 2.9]. ◇

Let us now extend the above concepts to signed and complex measures. Any complex measure ω can uniquely be written as $\omega = \mu + \mathrm{i}\nu$ with real (or signed) measures μ and ν, while any signed measure μ has a unique Hahn decomposition $\mu = \mu_+ - \mu_-$ with positive measures μ_+ and μ_- such that $\mu_+ + \mu_- = |\mu|$; see [RS80, Thm. IV.16]. Clearly, this also implies

$$\mu_\pm = \frac{1}{2}\big(|\mu| \pm \mu\big).$$

Together, we have the unique decomposition

(8.15) $$\omega = \mu + \mathrm{i}\nu = \mu_+ - \mu_- + \mathrm{i}\nu_+ - \mathrm{i}\nu_-$$

with positive measures $\mu_\pm$ and $\nu_\pm$ such that $|\mu| = \mu_+ + \mu_-$ and $|\nu| = \nu_+ + \nu_-$.

To extend the earlier definition, we recall that two regular Borel measures μ, ν on $\mathbb{R}^d$ are called *mutually singular* (or mutually orthogonal in the measure-theoretic sense), denoted as $\mu \perp \nu$, when a Borel set A exists such that $|\mu|(A) = 0$ and $|\nu|(\mathbb{R}^d \setminus A) = 0$. In particular, this means that each measure is concentrated on a null set with respect to the other. The set A in this definition is essentially unique in the sense that any other set A' with this property leads to $A \triangle A'$ being a null set both for $|\mu|$ and for $|\nu|$. Applying this terminology to the Hahn decomposition of a signed measure $\mu = \mu_+ - \mu_-$ implies that $\mu_+ \perp \mu_-$.

A nice way to summarise (and extend) these properties is the following *polar decomposition* of a complex measure [Die70, Thm. 13.16.3].

Proposition 8.3. *If ω is a complex measure on $\mathbb{R}^d$, with total variation measure $|\omega|$, one has $\omega = h|\omega|$ for some complex-valued, locally $|\omega|$-integrable function h that $|\omega|$-almost everywhere satisfies $|h(x)| = 1$.*   □

Splitting the function h first into its real and imaginary part, and then each part into its positive and negative one, corresponds to splitting the complex measure as in Eq. (8.15), where measurability of functions as well as null sets are defined with respect to $|\omega|$; see [Die70, Sec. XIII.16] for details. The polar decomposition also implies that $|g\omega| = |g|\,|\omega|$, for any $|\omega|$-measurable function g.

Lemma 8.2. *If ω and ω' are two regular Borel measures, one has $\omega \perp \omega'$ if and only if $|\omega| \perp |\omega'|$. Moreover, $\omega \perp \omega'$ implies $|\omega + \omega'| = |\omega| + |\omega'|$.*

PROOF. The first claim is obvious from the definition. For the second claim, let $\omega \perp \omega'$, so that we have a Borel set $A \subset \mathbb{R}^d$ with $|\omega|(A) = |\omega'|(A^c) = 0$, where $A^c = \mathbb{R}^d \setminus A$. If B is an arbitrary Borel set, we thus get

$$
\begin{aligned}
|\omega + \omega'|(B) &= |\omega + \omega'|\big((B \cap A) \,\dot{\cup}\, (B \cap A^c)\big) \\
&= |\omega + \omega'|(B \cap A) + |\omega + \omega'|(B \cap A^c) \\
&= |\omega'|(B \cap A) + |\omega|(B \cap A^c) = |\omega|(B) + |\omega'|(B),
\end{aligned}
$$

which completes the argument.   □

This result has the following important consequence for the Lebesgue decomposition of general complex measures, which extends Theorem 8.3.

Proposition 8.4. *Any regular Borel measure ω on $\mathbb{R}^d$ has a unique decomposition $\omega = \omega_{\mathsf{pp}} + \omega_{\mathsf{sc}} + \omega_{\mathsf{ac}}$ with respect to λ. The components satisfy*

$$
\omega_{\mathsf{pp}} \perp \omega_{\mathsf{sc}} \perp \omega_{\mathsf{ac}} \perp \omega_{\mathsf{pp}},
$$

and one has $|\omega| = |\omega_{\mathsf{pp}}| + |\omega_{\mathsf{sc}}| + |\omega_{\mathsf{ac}}|$.

PROOF. The definition of the pure points of ω works as in the case of positive measures, so that we have a unique decomposition $\omega = \omega_{\mathsf{pp}} + \omega_{\mathsf{cont}}$. From Eq. (8.15), the singular part of each of the four measures $\mu_\pm$, $\nu_\pm$ is a λ-null set, hence also their union, A say. This gives the decomposition $\omega = \omega_{\mathsf{sing}} + \omega_{\mathsf{ac}}$, where the singular continuous part of ω is the one common to ω_{sing} and ω_{cont}, which can be written as $\omega_{\mathsf{sc}} = \omega_{\mathsf{cont}}\big|_A$.

The orthogonality statement will follow from $\omega_{\mathsf{pp}} \perp \omega_{\mathsf{cont}}$ together with $\omega_{\mathsf{sing}} \perp \omega_{\mathsf{ac}}$, where $\omega_{\mathsf{cont}} = \omega_{\mathsf{sc}} + \omega_{\mathsf{ac}}$ and $\omega_{\mathsf{sing}} = \omega_{\mathsf{pp}} + \omega_{\mathsf{sc}}$. We already know that $\omega_{\mathsf{pp}} = \sum_{x \in P_\omega} \omega(\{x\}) \delta_x$, where $P_\omega \subset A \subset \mathbb{R}^d$ is a countable set, wherefore $|\omega_{\mathsf{cont}}|(P_\omega) = |\omega_{\mathsf{pp}}|(P_\omega^{\mathsf{c}}) = 0$. Now, Lemma 8.2 implies that $\omega_{\mathsf{pp}} \perp \omega_{\mathsf{sc}}$ and $\omega_{\mathsf{pp}} \perp \omega_{\mathsf{ac}}$. By definition, ω_{sing} is concentrated to a Borel set A with $|\omega_{\mathsf{ac}}|(A) = 0$ by construction. Since the concentration also means $|\omega_{\mathsf{sing}}|(A^{\mathsf{c}}) = 0$, we may conclude the final relation $\omega_{\mathsf{sc}} \perp \omega_{\mathsf{ac}}$. The formula for the total variation of ω follows from Lemma 8.2. □

Though we have formulated the results for measures on $\mathbb{R}^d$, the same approach works for measures on $\mathbb{S}^1$, again with Lebesgue measure as reference. More on this case will follow in Section 8.7.

8.5.3. NORM CONVERGENCE OF MEASURES

The concept of vague convergence is not always sufficient for the analysis of measures. In particular, the spectral type of a measure (according to its Lebesgue decomposition) need not be preserved under vague convergence.

Example 8.8 (*Change of spectral type*). Fix $n \in \mathbb{N}$ and consider the points $x_{n,\ell} = \frac{2\ell-1}{2^{n+1}}$ for $1 \leqslant \ell \leqslant 2^n$, which all lie in $(0,1)$. It is easy to show that

$$\nu_n := \frac{1}{2^n} \sum_{\ell=1}^{2^n} \delta_{x_{n,\ell}} \xrightarrow{\ n\to\infty\ } \lambda\big|_{[0,1]},$$

which demonstrates that the vague limit of a sequence of pure point measures can be absolutely continuous.

Conversely, in the spirit of Example 8.2, define the function $h_n \in C_{\mathsf{c}}(\mathbb{R})$ by $h_n(x) = n(1 - n|x|)$ for $|x| \leqslant 1/n$ and $h_n(x) = 0$ otherwise. This way, $\lambda(h_n) = 1$ for all $n \in \mathbb{N}$, and $h_n\lambda$ is absolutely continuous. Nevertheless, in the vague topology,

$$h_n\lambda \xrightarrow{\ n\to\infty\ } \delta_0,$$

which means that the vague limit is a pure point measure. ◇

This example is also meant as a warning that the use of periodic approximants has strong limitations in the context of spectral theory. This is well-known from the theory of Schrödinger operators, but equally applies to dynamical and diffraction spectra. To overcome this limitation, one introduces stronger notions of convergence. One natural possibility can be formulated via the total variation measure. Let us first do this for the group $\mathbb{R}^d$, and comment on the extension to other LCAGs in Remark 8.5 below.

Let $\varnothing \neq K \subset \mathbb{R}^d$ be compact with $\overline{K^\circ} = K$ (where we mainly think of a closed ball or a similar type of body), and define the K-*norm*

$$(8.16) \qquad \|\mu\|_K := \sup_{t \in \mathbb{R}^d} |\mu|(t + K),$$

where $|\mu|$ is the total variation of μ as introduced earlier.

Definition 8.3. Let $(\mu_n)_{n \in \mathbb{N}}$ be a sequence of measures on $\mathbb{R}^d$ and consider the K-norm for K a (fixed) closed ball of radius $R > 0$. Then, the sequence is called *norm converging* to a measure μ when $\|\mu_n - \mu\|_K \xrightarrow{n \to \infty} 0$.

Due to the σ-compactness of $\mathbb{R}^d$, the property of convergence does not depend on the actual choice of the radius in Definition 8.3.

Theorem 8.4. *Let $(\mu_n)_{n \in \mathbb{N}}$ be a sequence of regular Borel measures on $\mathbb{R}^d$ that norm converge to μ. If all μ_n are (with respect to λ) of the same spectral type $\alpha \in \{\mathsf{pp}, \mathsf{sc}, \mathsf{ac}, \mathsf{sing}, \mathsf{cont}\}$, the limit measure μ is of type α, too.*

PROOF. Recall Lemma 8.2 for $\mu = \mu_{\mathsf{pp}} + \mu_{\mathsf{cont}}$. If μ_n is pure point, one has

$$\left| (\mu_n - \mu)|_{t+K} \right| = \left| (\mu_n - \mu_{\mathsf{pp}})|_{t+K} \right| + \left| \mu_{\mathsf{cont}}|_{t+K} \right|$$

for any $t \in \mathbb{R}^d$. Then, $\|\mu_n - \mu\|_K \xrightarrow{n \to \infty} 0$ is only possible if $\mu_{\mathsf{cont}}|_{t+K} = 0$ for all $t \in \mathbb{R}^d$, hence $\mu_{\mathsf{cont}} = 0$ and $\mu = \mu_{\mathsf{pp}}$. The analogous argument works for any type $\alpha \in \{\mathsf{pp}, \mathsf{ac}, \mathsf{sing}, \mathsf{cont}\}$.

Finally, if all μ_n are sc, they define both a sequence of singular measures and one of continuous measures. The limit must then be singular and continuous, hence sc. $\square$

This theorem tells us that the spectral type is preserved under norm convergence. In fact, if the μ_n norm converge to μ with $\mu_n \perp \nu$, one also has $\mu \perp \nu$. This provides a powerful constructive tool to establish the spectral type of a measure, by identifying it as a suitable limit in the $\|.\|_K$-topology.

Remark 8.5 (*Norm topology for compactly generated LCAGs*). By Theorem 2.1, we know that any compactly generated LCAG G is isomorphic with $\mathbb{R}^d \times \mathbb{Z}^n \times \mathbb{K}$ for some $d, n \in \mathbb{N}_0$ and some compact Abelian group $\mathbb{K}$. To define the norm topology for G, we need to define it for each of the factors. Above, we described it for $\mathbb{R}^d$, while one simply uses $\|\mu\|_{\mathbb{K}} := |\mu|(\mathbb{K})$ for any compact group $\mathbb{K}$. Finally, for $\mathbb{Z}^n$, one employs the supremum norm $\|\mu\|_\infty := \sup_{\ell \in \mathbb{Z}^n} |\mu(\{\ell\})|$. Note that vague and norm convergence on discrete groups are pointwise and uniform convergence, respectively. $\Diamond$

Note that the result of Theorem 8.4 remains literally true for the (compact) group $\mathbb{S}^1$. In general, one has to start with the analogue of the Lebesgue decomposition (relative to the Haar measure of the group) to formulate the corresponding result.

8.6. Fourier transform of measures

Let us now summarise a direct approach to the Fourier transform of measures. This is needed for the treatment of measures on general LCAGs, where we cannot rely on tempered distributions. Nevertheless, we formulate most results for the group $\mathbb{R}^d$, with occasional reference to the general case.

If $\mu \in \mathcal{M}(\mathbb{R}^d)$ is a *finite* measure, its Fourier transform (or Fourier–Stieltjes transform) can directly be defined as

$$(8.17) \qquad \widehat{\mu}(k) = \int_{\mathbb{R}^d} e^{-2\pi i k x}\, d\mu(x),$$

which is a bounded and uniformly continuous function on $\mathbb{R}^d$; see [Rud62, Thm. 1.3.3(a)]. Seen as the Radon–Nikodym density (relative to Lebesgue measure), $\widehat{\mu}$ coincides with the Fourier transform of μ in the distribution sense introduced in Eq. (8.9).

If μ and ν are finite measures on $\mathbb{R}^d$, their *convolution* $\mu * \nu$ is defined by

$$(8.18) \qquad (\mu * \nu)(g) = \int_{\mathbb{R}^d \times \mathbb{R}^d} g(x+y)\, d\mu(x)\, d\nu(y),$$

with $g \in C_c(\mathbb{R}^d)$. This is again a finite measure, and thus certainly Fourier transformable. Moreover, one has the following variant of the convolution theorem [Rud62, Thm. 1.3.3(b)].

Proposition 8.5. *The convolution $\mu * \nu$ of two finite measures $\mu, \nu \in \mathcal{M}(\mathbb{R}^d)$ satisfies $\widehat{\mu * \nu} = \widehat{\mu}\,\widehat{\nu}$.*   □

Remark 8.6 (*Convolution of measures and functions*). Consider the convolution $f * \nu$ of an L^1-function f with a finite measure ν, which is well-defined because f represents the finite measure $\mu = f\lambda$. The convolution $f * \nu = \mu * \nu$ is then an absolutely continuous measure (relative to λ), which is represented by the Radon–Nikodym density defined by $x \mapsto \int_{\mathbb{R}^d} f(x-y)\, d\nu(y)$.

If both μ and ν are absolutely continuous, with Radon–Nikodym densities f and g say, the convolution $\mu * \nu$ of measures reduces to the convolution $f * g$ of L^1-functions as defined in Eq. (8.8).   ◇

The convolution $\mu * \nu$ can be extended to the situation where one measure (μ say) is still finite while the other is translation bounded; see for instance [BF75, Prop. 1.13]. If μ is represented by a density $f \in C_c(\mathbb{R}^d)$, the convolution $f * \nu$ is uniformly continuous. In general, the measure $\mu * \nu$ is again translation bounded, and $\widehat{\mu * \nu}$ is a tempered distribution, though it need not be a measure. If $\hat{\nu}$ is a measure, then so is $\widehat{\mu * \nu}$, and the convolution theorem applies to this case as well. We may now summarise our various convolution results as follows; see [BF75] for further details.

Theorem 8.5. *Let* $\mu, \nu \in \mathcal{M}(\mathbb{R}^d)$ *with* μ *finite and* ν *translation bounded. Then, the convolution* $\mu * \nu$ *exists and is a translation bounded measure.*

If $\hat{\nu}$ *is not only a tempered distribution, but also a measure, one has the convolution identity* $\widehat{\mu * \nu} = \hat{\mu} \hat{\nu}$. *The latter is again a measure, which is absolutely continuous relative to* $\hat{\nu}$. $\qquad\square$

The condition that $\hat{\nu}$ be a measure is crucial and non-trivial. In general, this is not the case, wherefore the verification of the measure property of $\hat{\nu}$ will require an explicit argument. In fact, as we shall see in our later chapter on diffraction theory, it is this detail that will ultimately pave the way for a simple, constructive proof of the diffraction formula for model sets. Further details on the convolution will be useful for this purpose as well.

Example 8.9 (*Convolution with* λ). If $\nu \in \mathcal{M}(\mathbb{R}^d)$ is a finite measure and λ denotes Lebesgue measure as before, one has

$$\widehat{\nu * \lambda} = c\, \delta_0 \quad \text{with} \quad c = \hat{\nu}(0) = \nu(\mathbb{R}^d).$$

This follows either from the previous convolution theorem or from the observation that $\nu * \lambda$ is translation invariant and hence must be a multiple of λ (by the uniqueness of the Haar measure), so that $\nu * \lambda = c\,\lambda$. A direct calculation gives $c = \nu(\mathbb{R}^d)$, and one has $\widehat{c\lambda} = c\,\delta_0$, by Example 8.5. The relation $c = \hat{\nu}(0)$ follows from the definition of $\hat{\nu}$ via Eq. (8.17). $\qquad\Diamond$

With basically the same argument, one can show the following more general result, which we formulate for a general LCAG G.

Lemma 8.3. *Let* G *be an LCAG with Haar measure* μ_G *and let* ν *be a finite Borel measure on* G. *Then, one has* $\nu * \mu_G = c\,\mu_G$ *with* $c = \nu(G)$.

PROOF. Let g be a compactly supported continuous function on G and observe that, for all $x \in G$, $\mu_G(T_{-x}g) = (T_x \mu_G)(g) = \mu_G(g)$ due to the translation invariance of the Haar measure.

Since the convolution $\nu * \mu_G$ is well-defined, we can now calculate

$$(\nu * \mu_G)(g) = \int_{G \times G} g(x+y)\, \mathrm{d}\mu_G(y)\, \mathrm{d}\nu(x) = \int_G \mu_G(T_{-x}g)\, \mathrm{d}\nu(x)$$

$$= \int_G \mu_G(g)\, \mathrm{d}\nu(x) = \nu(G)\, \mu_G(g).$$

Since g was arbitrary, the claim follows. $\qquad\square$

Remark 8.7 (*Positive definite distributions*). The Fourier transform of a tempered measure on $\mathbb{R}^d$ is always well-defined as a tempered distribution. As mentioned before, the latter need not be a measure. However, if μ is *positive definite* (or of positive type) in the sense that $\mu(\phi * \widetilde{\phi}) \geqslant 0$ for all $\phi \in \mathcal{S}(\mathbb{R}^d)$, where $\widetilde{\phi}(x) = \overline{\phi(-x)}$, then $\widehat{\mu}$ is a positive *measure* by the Bochner–Schwartz theorem [RS80, Thm. IX.10], and $\widehat{\mu}$ is translation bounded.   ◇

Since this property will become important in diffraction theory, we expand a little on this point.

Definition 8.4. A complex measure $\mu \in \mathcal{M}(\mathbb{R}^d)$ is called *positive definite* if $\mu(g * \widetilde{g}) \geqslant 0$ holds for all $g \in C_{\mathsf{c}}(\mathbb{R}^d)$.

The set of positive definite measures on $\mathbb{R}^d$ forms a vaguely closed convex cone inside $\mathcal{M}(\mathbb{R}^d)$. When μ is positive definite, then so are $\bar{\mu}$ and $\widetilde{\mu}$, where the latter is defined by $\widetilde{\mu}(g) = \overline{\mu(\widetilde{g})}$. If μ is positive and positive definite, it is automatically translation bounded [BF75, Prop. 4.4]. Moreover, using the general results of [BF75, Ch. I.4], one obtains the following statement.

Proposition 8.6. *If $\mu \in \mathcal{M}(\mathbb{R}^d)$ is positive definite, its Fourier transform $\widehat{\mu}$ exists, and is a translation bounded positive measure on $\mathbb{R}^d$. The Fourier transform as a mapping from positive definite to positive measures is injective. When, in addition, μ is absolutely continuous relative to λ, one has*

$$\widehat{\mu}(t + K) \xrightarrow{\ |t| \to \infty\ } 0$$

for all compact sets $K \subset \mathbb{R}^d$.   □

Note that the second part of Proposition 8.6 is a formulation of the Riemann–Lebesgue lemma in this case; compare [Pin02, Sec. 2.2.1].

Quite frequently, we shall see positive definite measures μ on $\mathbb{R}^d$ with support contained in the lattice $\mathbb{Z}^d$. Here, one has the following simplifying connection between measures (or Dirac combs) and functions. For a related result, in multiplicative notation, we also refer to [AGL74, Thm. 6.2].

Lemma 8.4. *Let $\eta \colon \mathbb{Z}^d \longrightarrow \mathbb{C}$ be a function on the integer lattice and consider the measure $\mu = \eta \delta_{\mathbb{Z}^d}$ on $\mathbb{R}^d$. Then, μ is positive definite as a measure if and only if η is positive definite as a function on $\mathbb{Z}^d$.*

PROOF. We prove the claim for $d = 1$ and leave the obvious extension to $\mathbb{Z}^d$ to the reader. Recall from [BF75, Ch. I.3] that η is a positive definite function on $\mathbb{Z}$ if $\sum_{i,j\in\mathbb{Z}} \eta(i - j)\, c_i \overline{c_j} \geqslant 0$ holds for all sequences $c = (c_i)_{i\in\mathbb{Z}} \in \mathbb{C}^{\mathbb{Z}}$ that only have finitely many non-zero entries. Another way to write the condition reads $\sum_{m\in\mathbb{Z}} \eta(m) \sum_{i\in\mathbb{Z}} c_i \overline{c_{i-m}} \geqslant 0$.

Let $\mu = \eta \delta_{\mathbb{Z}}$ be positive definite and choose a parameter-dependent family of hat functions ψ_ε (with $\frac{1}{2} > \varepsilon > 0$, say) such that $\psi_\varepsilon \geqslant 0$ together with

$\mathrm{supp}(\psi_\varepsilon) = [-\varepsilon, \varepsilon]$ and $\|\psi_\varepsilon\|_2^2 = 1$ holds for all ε; see Eq. (10.20) below and the text following it for one possibility to do so.

Now, fix an arbitrary sequence $c \in \mathbb{C}^{\mathbb{Z}}$ with only finitely many non-zero entries, and set $h_\varepsilon(x) = \sum_{i \in \mathbb{Z}} c_i \psi_\varepsilon(x - i)$, which clearly defines an element $h_\varepsilon \in C_{\mathsf{c}}(\mathbb{R})$. For $m \in \mathbb{Z}$, one obtains

$$\big(h_\varepsilon * \widetilde{h_\varepsilon}\big)(m) = \sum_{i,j \in \mathbb{Z}} c_i \overline{c_j} \int_{\mathbb{R}} \psi_\varepsilon(x - i) \, \overline{\psi_\varepsilon(x - j - m)} \, \mathrm{d}x.$$

As $\varepsilon \searrow 0$, this converges to $\sum_{i \in \mathbb{Z}} c_i \overline{c_{i-m}}$, due to the properties of the hat functions ψ_ε, so $\mu(h_\varepsilon * \widetilde{h_\varepsilon}) \longrightarrow \sum_{m \in \mathbb{Z}} \eta(m) \sum_{i \in \mathbb{Z}} c_i \overline{c_{i-m}}$. Since $\mu(h_\varepsilon * \widetilde{h_\varepsilon}) \geqslant 0$ for all $\varepsilon > 0$, with c arbitrary but fixed, η is positive definite.

For the converse direction, assume η to be positive definite. By the Herglotz–Bochner theorem, see [BF75, Thm. 3.12] or Theorem 8.6 below, there is a finite positive measure ϱ on $[0,1)$ so that $\eta(m) = \int_0^1 \mathrm{e}^{2\pi i m x} \, \mathrm{d}\varrho(x)$ for all $m \in \mathbb{Z}$. Consider the positive measure $\sigma = \varrho * \delta_{\mathbb{Z}}$, which is well-defined by Theorem 8.5. It is clearly Fourier transformable, and one calculates

$$\widecheck{\sigma} = \big(\varrho * \delta_{\mathbb{Z}}\big)^{\widecheck{}} = \widecheck{\varrho} \cdot \widetilde{\delta_{\mathbb{Z}}} = \widecheck{\varrho} \cdot \delta_{\mathbb{Z}} = \eta \, \delta_{\mathbb{Z}}$$

via the convolution theorem, Poisson's summation formula (Proposition 9.4 below), and the representation of η. Consequently, $\mu = \eta \, \delta_{\mathbb{Z}}$ is transformable, with positive Fourier transform σ, so μ is positive definite by the Bochner–Schwartz theorem; see [RS80, Thm. IX.10] or [BF75, Thm. 4.7].   □

8.7. Fourier–Stieltjes coefficients of measures on $\mathbb{S}^1$

A natural setting of Fourier analysis employs an LCAG G and its dual group $\widehat{G}$; see Section 2.3.3 and [Kat04, Ch. VIII]. Important cases are $G = \mathbb{R}^d$ (which is self-dual) and the dual pair $G = \mathbb{Z}$, $\widehat{G} = \mathbb{S}^1$. Here, we recall some results for the Fourier theory of measures on $\mathbb{S}^1$, which we identify with the unit interval (or 1-torus) $\mathbb{T} = [0,1)$ with addition modulo 1. We may thus use Lebesgue measure as the normalised Haar measure on $\mathbb{T}$.

Let μ be a regular Borel measure on the unit interval (without a point at 1). Its Fourier transform is

$$\widehat{\mu}(k) = \int_0^1 \mathrm{e}^{-2\pi i k x} \, \mathrm{d}\mu(x),$$

which is a bounded and uniformly continuous function on $\mathbb{Z}$ by an application of [Rud62, Thm. 1.3.3]. The numbers $\widehat{\mu}(k)$ are called the *Fourier–Stieltjes coefficients* of the measure μ; compare also [Kat04, Ch. I.7]. When μ is absolutely continuous, with Radon–Nikodym density f, the Fourier–Stieltjes transform reduces to the ordinary Fourier coefficients of a function $f \in L^1(\mathbb{T})$.

The following result is a direct consequence of [Rud62, Thm. 1.3.6].

Proposition 8.7 (Uniqueness). *Let μ and ν be two regular Borel measures on $\mathbb{T}$. When $\widehat{\mu}(k) = \widehat{\nu}(k)$ for all $k \in \mathbb{Z}$, one has $\mu = \nu$.* $\qquad\Box$

The asymptotic behaviour of the Fourier–Stieltjes coefficients is related to the nature of the measure μ. As this will become important in Chapter 10, we now spell out some of the details.

Proposition 8.8. *Let μ be a regular Borel measure on $\mathbb{T}$ that is absolutely continuous relative to Lebesgue measure. Then, $\lim_{|k|\to\infty} \widehat{\mu}(k) = 0$.*

SKETCH OF PROOF. Under the assumption stated, we have $\mu = f\lambda$ with $f \in L^1(\mathbb{T})$. The claim is now a direct consequence of the classic Riemann–Lebesgue lemma [Kat04, Thm. I.2.8], which can be proved by an approximation argument involving trigonometric polynomials. $\qquad\Box$

The Riemann–Lebesgue lemma has various generalisations, one of which we already saw in the second part of Proposition 8.6. The generalisation of Proposition 8.8 to $\mathbb{T}^d$, with Fourier–Stieltjes coefficients in $\mathbb{Z}^d$, is obvious. For a formulation in the context of general LCAGs, see [ReSt00, Sec. 4.4].

Next, we need a characterisation of the pure point part of μ in terms of its Fourier–Stieltjes coefficients, which is possible by Wiener's criterion; see [Wie27, Mah27] for the original source.

Proposition 8.9. *If μ is a regular Borel measure on $\mathbb{T}$, one has*

$$\mu(\{x\}) = \lim_{N\to\infty} \frac{1}{2N+1} \sum_{n=-N}^{N} \widehat{\mu}(n)\, e^{2\pi i n x}$$

for any $x \in \mathbb{T}$. As a consequence, a necessary and sufficient condition for μ to be a continuous measure (which means $\mu_{\mathsf{pp}} = 0$) reads

$$\lim_{N\to\infty} \frac{1}{2N+1} \sum_{n=-N}^{N} \left|\widehat{\mu}(n)\right|^2 = 0.$$

PROOF. Fix an element $x \in \mathbb{T}$ and define the trigonometric polynomial

$$g_N(t) := \frac{1}{2N+1} \sum_{n=-N}^{N} e^{-2\pi i n(t-x)} = \frac{e^{2\pi i N(t-x)}}{2N+1} \frac{1 - e^{-2\pi i(2N+1)(t-x)}}{1 - e^{-2\pi i(t-x)}},$$

where the right hand side obviously converges to 1 as $t \to x$. Observing that $|g_N(t)| \leqslant \left((N + \frac{1}{2})|1 - e^{-2\pi i(t-x)}|\right)^{-1}$, it is clear that the continuous function g_N (which is bounded by 1) tends to 0 uniformly (as $N \to \infty$) on the complement of any (open) neighbourhood of x. Moreover, the sequence

$(g_N)_{N\in\mathbb{N}}$ converges pointwise to the function $1_{\{x\}}$. By construction, we have

$$\mu(g_N) = \int_0^1 g_N \, \mathrm{d}\mu = \frac{1}{2N+1} \sum_{n=-N}^{N} \widehat{\mu}(n) \, \mathrm{e}^{2\pi \mathrm{i} n x},$$

while our above concentration argument implies that

$$\lim_{N\to\infty} \mu(g_N) = \int_0^1 1_{\{x\}} \, \mathrm{d}\mu = \mu(\{x\})$$

by the dominated convergence theorem; see [Ped95, Thm. 6.1.15] for a formulation in the generality needed here. Together, this proves the first claim.

If we decompose μ as $\mu = \mu_{\mathsf{pp}} + \mu_{\mathsf{cont}}$ as in Eq. (8.11), which is also possible for measures on $\mathbb{T}$, we know that $\mu_{\mathsf{pp}} = \sum_{\xi\in P_\mu} \mu(\{\xi\}) \delta_\xi$ with $P_\mu \subset \mathbb{T}$ being (at most) a countable set. As a consequence, we have $\mu * \widetilde{\mu} = \mu_{\mathsf{pp}} * \widetilde{\mu_{\mathsf{pp}}} + \nu$, where ν is a continuous measure. This implies

$$(\mu * \widetilde{\mu})(\{0\}) = (\mu_{\mathsf{pp}} * \widetilde{\mu_{\mathsf{pp}}})(\{0\}) = \sum_{\xi\in P_\mu} |\mu(\{\xi\})|^2,$$

where we have used that $\widetilde{\mu_{\mathsf{pp}}} = \sum_{\xi\in P_\mu} \overline{\mu(\{\xi\})} \, \delta_{-\xi}$. Applying our first claim to the measure $\mu * \widetilde{\mu}$ together with the identity $(\mu * \widetilde{\mu})\widehat{}(n) = |\widehat{\mu}(n)|^2$, we get

$$\sum_{\xi\in P_\mu} |\mu(\{\xi\})|^2 = \lim_{N\to\infty} \frac{1}{2N+1} \sum_{n=-N}^{N} |\widehat{\mu}(n)|^2,$$

which proves the second claim. $\qquad\qquad\qquad\qquad\qquad\qquad\qquad\square$

The last two propositions are significant for the detection of singular continuous measures. Indeed, if μ is a measure on $\mathbb{T}$ that satisfies Wiener's criterion (so that $\mu_{\mathsf{pp}} = 0$) but violates the asymptotic condition of the Riemann–Lebesgue lemma, it has to comprise a singular continuous component. This situation will be met in Section 10.1.

Let us formulate a generalisation of the previous result to higher dimensions.[1] Recall that $C_n = \{x \in \mathbb{R}^d \mid \|x\|_\infty \leqslant n\}$ denotes the closed cube of side length $2n$, centred at the origin. One has $\mathrm{card}(\mathbb{Z}^d \cap C_n) = (2n+1)^d$.

Theorem 8.6. *If $\eta\colon \mathbb{Z}^d \longrightarrow \mathbb{C}$ is positive definite, there is a unique positive and finite measure μ on $\mathbb{T}^d \simeq \mathbb{R}^d/\mathbb{Z}^d$ such that, for all $k \in \mathbb{Z}^d$, one has*

$$\eta(k) = \int_{\mathbb{T}^d} \mathrm{e}^{2\pi \mathrm{i} k x} \, \mathrm{d}\mu(x).$$

Moreover, one obtains the relation

$$\lim_{n\to\infty} \frac{1}{(2n+1)^d} \sum_{k\in\mathbb{Z}^d \cap C_n} |\eta(k)|^2 = \sum_{t\in\mathbb{T}^d} (\mu(\{t\}))^2,$$

---

[1]We thank Daniel Lenz for useful hints; compare [Nad95, Sec. 4.16] for $d = 1$.

where the last sum runs over at most countably many points.

PROOF. Since $\mathbb{Z}^d$ and $\mathbb{T}^d$ form a mutually dual pair of LCAGs, the first claim is just Bochner's theorem [Rud62] for this situation.

Now, define $g_n(z) = \frac{1}{(2n+1)^d} \sum_{k \in \mathbb{Z}^d \cap C_n} e^{2\pi i k z}$, which is bounded by 1. It is not difficult to see that $g_n(z) \xrightarrow{n \to \infty} 1_{\{0\}}(z)$ holds for any $z \in \mathbb{T}^d = [0,1)^d$. Using Fubini's theorem and recalling that μ is a positive measure, one obtains

$$\frac{1}{(2n+1)^d} \sum_{k \in \mathbb{Z}^d \cap C_n} \left| \eta(k) \right|^2 = \frac{1}{(2n+1)^d} \sum_{k \in \mathbb{Z}^d \cap C_n} \int_{\mathbb{T}^d} e^{2\pi i k x} \, d\mu(x) \overline{\int_{\mathbb{T}^d} e^{2\pi i k y} \, d\mu(y)}$$

$$= \int_{\mathbb{T}^d \times \mathbb{T}^d} g_n(x-y) \, d\mu(x) \, d\mu(y) \xrightarrow{n \to \infty} \int_{\mathbb{T}^d \times \mathbb{T}^d} 1_{\{0\}}(x-y) \, d\mu(x) \, d\mu(y)$$

where the last step follows from dominated convergence again. The integral over the diagonal evaluates via Fubini's theorem as

$$\int_{\mathbb{T}^d} \int_{\mathbb{T}^d} 1_{\{0\}}(x-y) \, d\mu(x) \, d\mu(y) = \int_{\mathbb{T}^d} \mu(\{y\}) \, d\mu(y) = \sum_{t \in \mathbb{T}^d} \left(\mu(\{t\}) \right)^2,$$

where the last sum runs over all points $t \in \mathbb{T}^d$ with $\mu(\{t\}) \neq 0$, which are (at most) countably many. $\square$

This result can be extended to the context of more general LCAGs. Since the above version suffices for our context, we omit further details.

8.8. Volume averaged convolutions

The convolution of two finite measures on $\mathbb{R}^d$, or of a finite with a translation bounded measure, is well-defined; see Eq. (8.18) and Theorem 8.5. This is no longer the case if both measures are unbounded. In the latter case, we can attempt to define a *volume averaged convolution* as

$$(8.19) \qquad \mu \circledast \nu := \lim_{R \to \infty} \frac{\mu_R * \nu_R}{\mathrm{vol}(B_R)},$$

where μ_R and ν_R are the restrictions of μ and ν to the (open) ball $B_R(0)$. In general, it is not clear whether this limit exists, although there is always at least one accumulation point if the two measures are translation bounded; compare Proposition 9.1 below. More generally, we call two measures μ and ν *mutually amenable* when the limit in (8.19) exists for balls as well as for arbitrary nested van Hove sequences. The resulting measure is called the *Eberlein convolution* [GLA90] of μ and ν. When both measures are absolutely continuous with respect to Lebesgue measure, the definition reduces to the volume averaged convolution of their densities f and g, given by

$$\left(f \circledast g \right)(x) = \lim_{R \to \infty} \frac{1}{\mathrm{vol}(B_R)} \int_{B_R(0)} f(y) \, g(x-y) \, dy.$$

Using Remark 8.6, the function mean of Eq. (8.3) can now be rewritten as $M(g) = (g \circledast \tilde{\lambda})(0)$, and a similar expression applies to the Fourier–Bohr coefficients of Theorem 8.2.

When μ and ν are mutually amenable and translation bounded, one also has, as a consequence of [Schl00, Lemma 1.2], that

$$(8.20) \qquad \mu \circledast \nu = \lim_{R \to \infty} \frac{\mu_R * \nu_R}{\mathrm{vol}(B_R)} = \lim_{R \to \infty} \frac{\mu * \nu_R}{\mathrm{vol}(B_R)} = \lim_{R \to \infty} \frac{\mu_R * \nu}{\mathrm{vol}(B_R)}.$$

This relation also holds for arbitrary (nested) van Hove sequences.

Example 8.10 (*Eberlein convolutions with λ and $\delta_{\mathbb{Z}}$*). First, observe the relation $\frac{1}{2a}\lambda\big|_{[-a,a]} * \lambda\big|_{[-a,a]} = g_a\,\lambda$ with $g_a(x) = 1 - \frac{|x|}{2a}$ for $|x| \leqslant 2a$ and $g_a(x) = 0$ otherwise; compare Example 8.3. Now, one can see that, for any fixed $x \in \mathbb{R}$, one has $g_a(x) \xrightarrow{a \to \infty} 1$, so that

$$\lambda \circledast \lambda = \lambda.$$

Similarly, by the corresponding argument via discrete convolution, one finds

$$\delta_{\mathbb{Z}} \circledast \delta_{\mathbb{Z}} = \delta_{\mathbb{Z}}$$

as well as $\delta_{\mathbb{Z}} \circledast \lambda = \lambda$. The last identity follows easily from the translation invariance of $\delta_{\mathbb{Z}} \circledast \lambda$, which must thus be a multiple of λ. It is straightforward to see that the corresponding factor is 1; compare Example 8.9 and Lemma 8.3. More generally, when $\Gamma \subset \mathbb{R}^d$ is a lattice, one finds

$$(8.21) \qquad \delta_{\Gamma} \circledast \delta_{\Gamma} = \mathrm{dens}(\Gamma)\,\delta_{\Gamma} \quad \text{and} \quad \delta_{\Gamma} \circledast \lambda = \mathrm{dens}(\Gamma)\,\lambda$$

by a simple density calculation. Also, with $\alpha > 0$, one obtains

$$(8.22) \qquad \delta_{\mathbb{Z}} \circledast \delta_{\alpha\mathbb{Z}} = \begin{cases} \frac{1}{\alpha}\lambda, & \text{if } \alpha \notin \mathbb{Q}, \\ \frac{1}{p}\delta_{\mathbb{Z}/q}, & \text{if } \alpha = \frac{p}{q} \text{ with } p, q \text{ coprime.} \end{cases}$$

Finally, if $\omega = w\,\delta_{\mathbb{Z}} := \sum_{n \in \mathbb{Z}} w(n)\,\delta_n$ with bounded weights $w(n)$ such that $\lim_{N \to \infty} \frac{1}{2N+1} \sum_{n=-N}^{N} w(n) = 0$, one finds

$$\omega \circledast \delta_{\mathbb{Z}} = 0 \quad \text{and} \quad \omega \circledast \lambda = 0.$$

Later on, we will refer to any ω of this kind as a Dirac comb (compare Example 8.6) with *balanced weights*. $\diamond$

With the methods summarised in this chapter, we are now ready to embark on the structure of periodic and aperiodic order from the point of view of mathematical diffraction theory.

CHAPTER 9

Diffraction

Mathematical diffraction theory is concerned with the precise measure-theoretic formulation of kinematic diffraction in the classic Fraunhofer picture [Cow95, Hof95a]. In this setting, the diffraction measure is given by the Fourier transform of the autocorrelation of the initial structure. The corresponding inverse problem deals with the reconstruction of the initial structure from the diffraction measure. The solution is generally not unique, which makes this a hard problem. Here, we introduce the necessary tools for the mathematical theory of kinematic diffraction, followed by the explicit treatment of crystallographic systems and regular model sets. Our approach also permits the explicit construction of homometric model sets.

9.1. Mathematical diffraction theory

A diffraction image emerges via the interference of waves that are scattered by an 'obstacle', or by a collection of obstacles. Diffraction occurs in many situations of wave propagation, and is fully described by the solutions of the corresponding wave equation. Fortunately, standard situations from optics and crystallography allow a considerable simplification via Fourier analysis in Fraunhofer's far field limit. We omit the underlying derivation, and refer to [HZ74, Ch. 10] and [Höf01] for very readable introductions. The essential connection that we need below is the description of the diffraction image as the absolute square of the Fourier transform of the obstacle. The latter is traditionally modelled by a function, which will be extended to a measure in our setting.

Generalising classical predecessors from the treatment of periodic structures, the approach via the autocorrelation, as used below, was introduced by Hof in [Hof95a, Hof97a]. It was inspired by the observation that Wiener's diagram for integrable functions (which we will discuss in Section 9.1.2 below) generally fails to be commutative in the setting of translation bounded measures. Fortunately, the autocorrelation is always Fourier transformable, so that the powerful tools of Fourier analysis are available. Below, we use a formulation for $\mathbb{R}^d$, though quite a bit of what we describe can be extended to general LCAGs; see [Schl00] for a systematic extension.

9.1.1. Autocorrelation

Recall that, for any (continuous) function g, we defined $\widetilde{g}$ as the mapping $x \mapsto \widetilde{g}(x) := \overline{g(-x)}$. The proper extension to measures is $\widetilde{\mu}(g) := \overline{\mu(\widetilde{g})}$, with $g \in C_c(\mathbb{R}^d)$. Of particular interest is the convolution $g * \widetilde{g}$, with $g \in L^1(\mathbb{R}^d)$, which can be written as

$$(9.1) \qquad \big(g * \widetilde{g}\big)(x) = \int_{\mathbb{R}^d} g(x+z)\,\overline{g(z)}\,\mathrm{d}z = \int_{\mathbb{R}^d} g(w)\,\overline{g(w-x)}\,\mathrm{d}w.$$

The function $g * \widetilde{g}$ is positive definite, so that $(g * \widetilde{g}\,)\widehat{} = |\widehat{g}|^2 \geqslant 0$, due to $\widehat{\widetilde{g}} = \overline{\widehat{g}}$.

Recall the definition for the convolution $\mu * \nu$ of two measures μ and ν from Eq. (8.18), which is well-defined if at least one measure is finite while the other is translation bounded. Since we are dealing with regular Borel measures, the evaluation of $(\mu * \nu)(A)$ for a measurable set A is conveniently rewritten via the characteristic function 1_A as

$$(9.2) \qquad \big(\mu * \nu\big)(A) = \int_{\mathbb{R}^d \times \mathbb{R}^d} 1_A(x+y)\,\mathrm{d}\mu(x)\,\mathrm{d}\nu(y).$$

Moreover, the convolution of any measure with the Dirac measure δ_t is well-defined and describes a translation by t, so that $(\delta_t * \mu)(A) = \mu(A - t)$.

Let μ_R denote the restriction of a measure $\mu \in \mathcal{M}(\mathbb{R}^d)$ to the closure of the (open) ball $B_R = B_R(0)$. Since μ_R has compact support (and is a finite measure), the volume-averaged convolution

$$(9.3) \qquad \gamma_\mu^{(R)} := \frac{\mu_R * \widetilde{\mu_R}}{\mathrm{vol}(B_R)}$$

is well-defined, and positive definite by construction. Every accumulation point of the family $\{\gamma_\mu^{(R)} \mid R > 0\}$ in the vague topology, as $R \to \infty$, is called an *autocorrelation measure* of μ, and as such it is a positive definite measure by construction. If only one point of accumulation exists, the autocorrelation measure

$$\gamma_\mu := \lim_{R \to \infty} \gamma_\mu^{(R)}$$

is well-defined. One way to establish the existence of the limit is through the pointwise ergodic theorem, if such methods apply. If not, explicit convergence proofs will be needed, as is apparent from known examples [BMP00] and counterexamples [LP03].

Definition 9.1. When the family of approximating measures in Eq. (9.3) converges to a limit as $R \to \infty$, γ_μ say, the latter is called the *natural autocorrelation measure* (or simply the natural autocorrelation) of μ. The term 'natural' refers to the use of balls for the volume averaging. By the natural autocorrelation of a point set $\Lambda \subset \mathbb{R}^d$, we mean the natural autocorrelation of its Dirac comb δ_Λ.

If the natural autocorrelation of μ exists, we have $\gamma_\mu = \mu \circledast \widetilde{\mu}$, which is one reason why we introduced the Eberlein convolution $\circledast$ in Section 8.8. If the limit (as $R \to \infty$) fails to exist, the structure of the approximating autocorrelations is rich enough to continue their analysis via converging subsequences. The following result is a minor variation of [Hof95a, Prop. 2.2].

Proposition 9.1. *If $\mu \in \mathcal{M}(\mathbb{R}^d)$ is translation bounded, the corresponding family $\{\gamma_\mu^{(R)} \mid R > 0\}$ of approximating autocorrelations is precompact in the vague topology. Any accumulation point of this family, of which there is at least one, is translation bounded.*

PROOF. The claim follows from showing that the measures $\gamma_\mu^{(R)}$ are uniformly translation bounded. Recall from Definition 8.2 that translation bounded-ness of μ means $\sup_{x \in \mathbb{R}^d} |\mu|(x + K) = b_K < \infty$ for any compact set $K \subset \mathbb{R}^d$. Clearly, also $\widetilde{\mu}$ is translation bounded, with $\sup_{x \in \mathbb{R}^d} |\widetilde{\mu}|(x + K) = b_{-K}$. Observe the inequality $0 \leqslant |\mu_R * \widetilde{\mu_R}| \leqslant |\mu_R| * |\widetilde{\mu_R}|$ of positive measures, which is a consequence of the convolution of two finite measures.

For $K \subset \mathbb{R}^d$ compact and $x \in \mathbb{R}^d$, one obtains

$$\left(|\mu_R| * |\widetilde{\mu_R}|\right)(x + K) = \int_{\mathbb{R}^d \times \mathbb{R}^d} 1_{x+K}(u + v)\, \mathrm{d}|\mu_R|(u)\, \mathrm{d}|\widetilde{\mu_R}|(v)$$

$$= \int_{\mathbb{R}^d} |\mu_R|(x - v + K)\, \mathrm{d}|\widetilde{\mu_R}|(v) \leqslant b_K\, |\widetilde{\mu_R}|(\mathbb{R}^d),$$

where $|\widetilde{\mu_R}|(\mathbb{R}^d) = |\widetilde{\mu}|(B_R(0))$. Note that a ball of radius R can be covered by n_R translates of the unit ball $B_1(0)$, where $n_R \leqslant c\, R^d$ with some constant $c > 0$. This gives $|\widetilde{\mu}|(B_R(0)) \leqslant c\, b_{B_1(0)}\, R^d$. Putting everything together leads to the inequality

$$|\mu_R * \widetilde{\mu_R}|(x + K) \leqslant c\, b_K\, b_{B_1(0)}\, R^d.$$

Since the right-hand side is independent of x, the approximating autocorrelation measures $\gamma_\mu^{(R)} = \frac{1}{\mathrm{vol}(B_R)}\, \mu_R * \widetilde{\mu_R}$ are translation bounded, uniformly in the radius R. Consequently, for any given function $g \in C_c(\mathbb{R}^d)$, the mapping $R \mapsto \gamma_\mu^{(R)}(g)$ is bounded which implies the precompactness by standard countability arguments. The final claim is then clear. $\square$

Recall that, when μ is a translation bounded measure on $\mathbb{R}^d$, the continuous hull $\mathbb{X}(\mu) = \overline{\{\delta_t * \mu \mid t \in \mathbb{R}^d\}}$, with closure in the vague topology, is vaguely compact [BL04]. In this context, Proposition 9.1 is an important extension, upon which a fair bit of our later analysis rests. If the limit fails to exist, there is still at least one converging subsequence. Each such subsequence converges to *an* autocorrelation, which is then automatically translation bounded and positive definite.

Remark 9.1 (*Homometric measures*). Different measures can lead to the same natural autocorrelation, for instance if a sufficiently 'meagre' measure ν is added to a given measure μ; see [Hof95a, Prop. 2.3] for details. In particular, adding or removing finitely many points (or even a set of points of density 0) from a point set $\Lambda \subset \mathbb{R}^d$ does not change the autocorrelation of the Dirac comb δ_Λ. Two measures with the same (natural) autocorrelation are called *homometric*; see Section 9.6 below for more.   ◇

Remark 9.2 (*Alternative averaging sets*). Hof [Hof95a] uses cubes rather than balls in his definition of $\gamma^{(R)}$. This simplifies some of his proofs technically, but they also work for balls, which are more natural objects in a physics context. One should keep in mind that the autocorrelation, in general, depends on the shape of the volume over which the average is taken — with obvious meaning for the experimental situation where the shape corresponds to the aperture. To avoid this problem, one often restricts the class of models to be considered, and defines the limits along arbitrary van Hove sequences (recall Definition 2.9), thus also demanding a slightly stricter version of uniqueness; compare [Schl00, p. 145].   ◇

Example 9.1 (*Autocorrelation of weighted FLC sets*). Let $\Lambda \subset \mathbb{R}^d$ be a point set of finite local complexity, and consider the Dirac comb $\omega = \omega_\Lambda$ with a bounded weight function w, as introduced in Example 8.6 on page 317. Assume that its natural autocorrelation γ_ω exists. A short calculation shows that $\widetilde{\omega} = \sum_{x \in \Lambda} \overline{w(x)}\, \delta_{-x}$. Since $\delta_x * \delta_y = \delta_{x+y}$, we get

$$\gamma_\omega \;=\; \sum_{z \in \Lambda - \Lambda} \eta(z)\, \delta_z \,,$$

where $\Lambda - \Lambda$ is locally finite by Proposition 2.1, and where the *autocorrelation coefficient* $\eta(z)$, for $z \in \Lambda - \Lambda$, is given by the limit

(9.4) $$\eta(z) \;=\; \lim_{R \to \infty} \frac{1}{\mathrm{vol}(B_R)} \sum_{\substack{x \in \Lambda_R \\ x - z \in \Lambda}} w(x)\, \overline{w(x - z)} \,,$$

with $\Lambda_R := \Lambda \cap \overline{B_R}$. Eq. (9.4) actually uses the fact that $\{B_R \mid R > 0\}$ has the van Hove property. Here, it simply means that the surface to volume ratio of a sphere of radius R goes to 0 as $R \to \infty$.

Conversely, if the limits in Eq. (9.4) exist for all $z \in \Lambda - \Lambda$, the natural autocorrelation exists as well, because $\Lambda - \Lambda$ is locally finite by assumption, and the measure γ_ω is positive definite and translation bounded. This is one of the benefits of using FLC sets.   ◇

Note that the autocorrelation coefficients in Eq. (9.4) have the form of an orbit average, wherefore the usefulness of ergodicity to establish their existence (and hence that of the measure γ_ω) becomes evident.

9.1.2. DIFFRACTION MEASURE

Let g be an integrable function, and consider the Wiener diagram

(9.5)
$$
\begin{array}{ccc}
g & \xrightarrow{\;\;*\;\;} & g * \tilde{g} \\
\mathcal{F} \downarrow & & \downarrow \mathcal{F} \\
\widehat{g} & \xrightarrow{\;|\cdot|^2\;} & |\widehat{g}|^2
\end{array}
$$

which is commutative, as a result of the convolution theorem and the relation $\widetilde{\widehat{g}} = \overline{\widehat{g}}$. Kinematic diffraction [Cow95] is the diagonal map $g \mapsto |\widehat{g}|^2$.

For mathematical diffraction theory, we have to formulate the analogue of this diagram for translation bounded measures. The first difficulty here is that the Fourier transform $\widehat{\mu}$ of a tempered measure μ is a tempered distribution, but not necessarily a measure. But even if it is, there is then no natural and sufficiently general analogue (for measures) of taking the absolute square, or at least none that would make the diagram consistent. This renders the 'lower path' of Eq. (9.5) useless in general, and suggests to use the approach via the autocorrelation and the 'upper path' in the diagram, thus referring to

(9.6)
$$
\mu \xrightarrow{\;\circledast\;} \gamma_{\mu} = \mu \circledast \tilde{\mu} \xrightarrow{\;\mathcal{F}\;} \widehat{\gamma_{\mu}}
$$

as the chain of mappings from the initial structure μ to $\widehat{\gamma_{\mu}}$.

Definition 9.2. Let μ be a translation bounded complex measure whose natural autocorrelation γ_{μ} exists. The Fourier transform $\widehat{\gamma_{\mu}}$ is then called the *diffraction measure* of μ.

The consistency of this definition follows from the fact that the natural autocorrelation (which is assumed to exist) is a positive definite measure by construction, and thus always possesses a Fourier transform, which is then a positive measure by Proposition 8.6; see also [BF75, Thm. 4.5].

Remark 9.3 (*Interpretation of $\widehat{\gamma_{\mu}}$ and its Lebesgue decomposition*). In terms of the physical process of (kinematic) diffraction, the measure $\widehat{\gamma_{\mu}}$ quantifies how much intensity is scattered into a given volume. This must clearly be a non-negative number, which the above approach guarantees. Moreover, since volume is quantified by Lebesgue measure λ, it is natural to decompose the diffraction measure relative to λ according to Theorem 8.3, thus giving

$$
\widehat{\gamma_{\mu}} = \left(\widehat{\gamma_{\mu}}\right)_{\mathsf{pp}} + \left(\widehat{\gamma_{\mu}}\right)_{\mathsf{sc}} + \left(\widehat{\gamma_{\mu}}\right)_{\mathsf{ac}}.
$$

This offers a rigorous approach to decompose $\widehat{\gamma_{\mu}}$ into its pure point or 'Bragg' part and its continuous or 'diffuse' part. The latter comprises two contributions, with the singular continuous one being rarely observed in the crystallographic practice; compare [Wit05] for interesting exceptions. $\Diamond$

Note that any singular continuous measure ν on $\mathbb{R}^d$ has two (seemingly conflicting) properties. On the one hand, ν is concentrated to a set $S \subset \mathbb{R}^d$ of zero Lebesgue measure, so that $\nu = \nu|_S$ with $\lambda(S) = 0$. On the other hand, since ν gives no weight to single points, the set S must be uncountable, as the Cantor set in Example 8.7. In particular, any isolated point of S may be removed from S, and S can be chosen such that, for any open $U \subset \mathbb{R}^d$, the set $S \cap U$ is either empty or uncountable.

In general, the pure point part of $\widehat{\gamma_\mu}$ contains a Dirac measure at 0 with positive intensity. To explain this, let us simplify our notation and write γ and $\hat{\gamma}$ for the autocorrelation and the diffraction measure of μ. We will make use of the relation $\gamma = \mu \circledast \tilde{\mu}$ together with the continuity of the Fourier transform on $\mathcal{S}'(\mathbb{R}^d)$. To formulate the result, we introduce the function

$$(9.7) \qquad\qquad h_r := \frac{1_{\overline{B_r(0)}}}{\operatorname{vol}(B_r)}$$

for arbitrary $r > 0$.

Proposition 9.2. *Let μ be an unbounded measure with existing natural autocorrelation $\gamma = \mu \circledast \tilde{\mu}$. Then, $\hat{\gamma}$ has a Dirac measure at 0 with intensity*

$$\hat{\gamma}(\{0\}) = \lim_{r \to \infty} \lim_{R \to \infty} \frac{1}{\operatorname{vol}(B_R)} \int_{B_R(0)} \left| (\delta_z * \mu)(h_r) \right|^2 \mathrm{d}z \geqslant 0.$$

PROOF. Let ν be a finite measure on $\mathbb{R}^d$ (so that the measure $\nu * \tilde{\nu}$ is well-defined and positive definite) and let g be a Lebesgue measurable function. Assuming that g and all its translates $T_z g$, where $T_z g(x) := g(x - z)$, are ν-integrable functions, one has

$$\left(\nu * \tilde{\nu} \right)(g * \tilde{g}) = \int_{\mathbb{R}^d} \int_{\mathbb{R}^d} \int_{\mathbb{R}^d} g(x + z)\, \tilde{g}(y - z)\, \mathrm{d}z\, \mathrm{d}\nu(x)\, \mathrm{d}\tilde{\nu}(y)$$

$$= \int_{\mathbb{R}^d} \nu(T_{-z} g)\, \tilde{\nu}(T_z \tilde{g})\, \mathrm{d}z = \int_{\mathbb{R}^d} \left| (\delta_z * \nu)(g) \right|^2 \mathrm{d}z \geqslant 0,$$

where we have used $\nu(T_{-z} g) = (\delta_z * \nu)(g)$ and $\delta_{-z} * \tilde{\nu} = \widetilde{\delta_z * \nu}$.

Observing that $\gamma = \mu \circledast \tilde{\mu}$, one obtains

$$\gamma(h_r * \tilde{h}_r) = \lim_{R \to \infty} \frac{1}{\operatorname{vol}(B_R)} \int_{\mathbb{R}^d} \left| (\delta_z * \mu_R)(h_r) \right|^2 \mathrm{d}z$$

with h_r as defined in Eq. (9.7). The limit exists by assumption, since $h_r * \tilde{h}_r$ is a continuous function, with support $\overline{B_{2r}(0)}$. Taking $r \to \infty$ gives the expression on the right-hand side of our claim, where the existence of the limit $C := \lim_{r \to \infty} \gamma(h_r * \tilde{h}_r)$ will follow from our next argument.

Using Example 8.3 together with $\mathrm{vol}(B_r) = \pi^{d/2} r^d / \Gamma(1 + \frac{d}{2})$ and the convolution theorem, one calculates the Fourier transform of $\phi_r := h_r * \tilde{h}_r$ as

$$\widehat{\phi_r}(k) = \widecheck{\phi_r}(k) = \frac{\left(\Gamma(1 + \frac{d}{2}) J_{\frac{d}{2}}(2\pi|k|r)\right)^2}{(\pi|k|r)^d} \geqslant 0,$$

which is bounded by 1, converges to 1 as $|k| \to 0$, and is asymptotically of order $\mathcal{O}\left((|k|r)^{-d-1}\right)$ when $|k|r \to \infty$. In fact, one has pointwise convergence $\widehat{\phi_r} \xrightarrow{r \to \infty} 1_{\{0\}}$. This means that the positive function $\widehat{\phi_r}$, as r increases, concentrates (symmetrically) around 0 with maximum value 1 at $k = 0$.

Observe next that $\hat{\gamma}$ is a positive, tempered measure, so that

$$\gamma(\phi_r) = \widecheck{\hat{\gamma}}(\phi_r) = \hat{\gamma}\left(\widecheck{\phi_r}\right) = \hat{\gamma}\left(\widehat{\phi_r}\right).$$

By the dominated convergence theorem [Ped95, Thm. 6.1.15], we get

$$\lim_{r \to \infty} \hat{\gamma}\left(\widehat{\phi_r}\right) = \hat{\gamma}(1_{\{0\}}) = \hat{\gamma}(\{0\}),$$

which also implies the convergence of $\gamma(\phi_r)$ as $r \to \infty$. This establishes $C = \hat{\gamma}(\{0\})$ and thus proves the claim. $\qquad\square$

The central intensity vanishes when $C = 0$. This means that the average scattering strength (in physics terminology) is 0, which we will meet repeatedly below. We refer to this situation as the *balanced weight* case. An interesting special case of Proposition 9.2 occurs for point sets.

Corollary 9.1. *Let Λ be a locally finite point set with natural autocorrelation measure γ. Its diffraction measure $\hat{\gamma}$ comprises a Dirac measure at 0, with $\hat{\gamma}(\{0\}) = \left(\mathrm{dens}(\Lambda)\right)^2$, where $\mathrm{dens}(\Lambda)$ is the natural density of Λ.*

PROOF. Since the natural autocorrelation exists, so does the (natural) density. We may now apply Proposition 9.2 with $\mu = \delta_\Lambda$. With the function h_r as in Eq. (9.7) and the previous proof, we have

$$\gamma(h_r * \tilde{h}_r) = \lim_{R \to \infty} \frac{1}{\mathrm{vol}(B_R)} \int_{B_R(0)} \left(\frac{\mathrm{card}\left(\Lambda \cap \overline{B_r(z)}\right)}{\mathrm{vol}(B_r)}\right)^2 \, \mathrm{d}z,$$

where the existence of the limit on the right hand side is once again a consequence of the assumed existence of γ.

By Proposition 9.2, we know that $\hat{\gamma}(\{0\}) = \lim_{r \to \infty} \gamma(h_r * \tilde{h}_r)$. Due to the definition of γ, this is a double limit. The inner limit is the space average of the squared local density of Λ in a closed ball of radius r. Since we know that the (natural) density of Λ exists, this average clearly converges to $C = \left(\mathrm{dens}(\Lambda)\right)^2$ as $r \to \infty$. $\qquad\square$

9.2. Poisson's summation formula and perfect crystals

There is a simple, yet rigorous, approach to the diffraction of crystallographic measures, based on the *Poisson summation formula* (PSF). Since this explains several formal (and correct) calculations used in physics texts, we discuss the PSF in some detail. We start with a derivation of the classic PSF for functions, followed by an extension to tempered measures. This effort pays off in the form of a short derivation for the pure point diffraction measure of crystals. In fact, it also holds the key for the diffraction formula of model sets, as we shall see in Sections 9.3 and 9.4.

9.2.1. PSF FOR FUNCTIONS

Recall that a function $g \colon \mathbb{R}^d \longrightarrow \mathbb{C}$ is $\mathbb{Z}^d$-periodic, if $g(x + k) = g(x)$ for all $x \in \mathbb{R}^d$ and all $k \in \mathbb{Z}^d$. Let $\mathbb{T}^d := \mathbb{R}^d/\mathbb{Z}^d$ denote the d-dimensional torus (identified with a fundamental domain of the lattice $\mathbb{Z}^d$, which we choose as $[0,1)^d$ for convenience). Recall Theorem 8.1 and define the Fourier series coefficients of g as

$$
\begin{aligned}
c_k &= \int_{\mathbb{T}^d} \mathrm{e}^{-2\pi \mathrm{i} k x} g(x) \, \mathrm{d}x \\
&= \int_0^1 \mathrm{e}^{-2\pi \mathrm{i} k_d x_d} \cdots \int_0^1 \mathrm{e}^{-2\pi \mathrm{i} k_1 x_1} \, g(x) \, \mathrm{d}x_1 \cdots \mathrm{d}x_d
\end{aligned}
$$

(9.8)

for $k \in \mathbb{Z}^d$, with $x = (x_1, \dots, x_d)$. Let us further assume that $g \in C^\infty(\mathbb{R}^d)$. By Theorem 8.1, this is a sufficient condition for the Fourier series

$$
(9.9) \qquad g(x) = \sum_{k \in \mathbb{Z}^d} c_k \, \mathrm{e}^{2\pi \mathrm{i} k x}
$$

to converge uniformly towards g.

Consider now a Schwartz function $\phi \in \mathcal{S}(\mathbb{R}^d)$, with its Fourier transform $\widehat{\phi}(k) = \int_{\mathbb{R}^d} \mathrm{e}^{-2\pi \mathrm{i} k x} \phi(x) \, \mathrm{d}x$ as defined in Section 8.3, so that also $\widehat{\phi} \in \mathcal{S}(\mathbb{R}^d)$.

Proposition 9.3 (PSF for functions). *For all $\phi \in \mathcal{S}(\mathbb{R}^d)$, one has*

$$
\sum_{m \in \mathbb{Z}^d} \phi(m) = \sum_{m \in \mathbb{Z}^d} \widehat{\phi}(m) .
$$

PROOF. Define $g(x) = \sum_{\ell \in \mathbb{Z}^d} \phi(x + \ell)$. Due to $\phi \in \mathcal{S}(\mathbb{R}^d)$, this is a uniformly convergent sum. Also, g is both infinitely differentiable and $\mathbb{Z}^d$-periodic. If c_m is the Fourier coefficient as defined in Eq. (9.8), the Fourier series of Eq. (9.9) implies

$$
\sum_{m \in \mathbb{Z}^d} c_m = g(0) = \sum_{m \in \mathbb{Z}^d} \phi(m).
$$

On the other hand, we have

$$
\begin{aligned}
c_m &= \int_{\mathbb{T}^d} \mathrm{e}^{-2\pi\mathrm{i}mx} g(x)\,\mathrm{d}x \;=\; \int_{\mathbb{T}^d} \sum_{\ell \in \mathbb{Z}^d} \mathrm{e}^{-2\pi\mathrm{i}mx} \phi(x+\ell)\,\mathrm{d}x \\
&= \sum_{\ell \in \mathbb{Z}^d} \int_{\mathbb{T}^d} \mathrm{e}^{-2\pi\mathrm{i}mx} \phi(x+\ell)\,\mathrm{d}x \;=\; \sum_{\ell \in \mathbb{Z}^d} \int_{\mathbb{T}^d} \mathrm{e}^{-2\pi\mathrm{i}m(x+\ell)} \phi(x+\ell)\,\mathrm{d}x \\
&= \sum_{\ell \in \mathbb{Z}^d} \int_{\ell+\mathbb{T}^d} \mathrm{e}^{-2\pi\mathrm{i}mx} \phi(x)\,\mathrm{d}x \;=\; \int_{\mathbb{R}^d} \mathrm{e}^{-2\pi\mathrm{i}mx} \phi(x)\,\mathrm{d}x \;=\; \widehat{\phi}(m),
\end{aligned}
$$

where $\mathrm{e}^{-2\pi\mathrm{i}m\ell} = 1$ for $m,\ell \in \mathbb{Z}^d$. The second last step follows from the decomposition $\mathbb{R}^d = \dot{\bigcup}_{\ell \in \mathbb{Z}^d}(\ell+\mathbb{T}^d)$, with $\dot{\cup}$ denoting the disjoint union of sets. Since all steps in the calculation are justified by the uniform convergence of our series, the claim follows. $\qquad\square$

9.2.2. PSF FOR DIRAC COMBS

Next, we have to transfer this result to certain (tempered) measures. To do so, we first formulate it for $\mathbb{Z}^d$-periodic Dirac combs. Recall that δ_S denotes the uniform Dirac comb of a locally finite point set S, as introduced in Example 8.6. Let us consider the Dirac comb for $S = \mathbb{Z}^d$ as an element of the space $\mathcal{S}'(\mathbb{R}^d)$ of tempered distributions.

Proposition 9.4 (PSF for lattice Dirac combs). *Interpreted as an equation in $\mathcal{S}'(\mathbb{R}^d)$, one has the identity $\widehat{\delta_{\mathbb{Z}^d}} = \delta_{\mathbb{Z}^d}$.*

PROOF. We have to verify that $(\widehat{\delta_{\mathbb{Z}^d}},\phi) = (\delta_{\mathbb{Z}^d},\phi)$ holds for all $\phi \in \mathcal{S}(\mathbb{R}^d)$. This is a simple calculation on the basis of Proposition 9.3,

$$
(\widehat{\delta_{\mathbb{Z}^d}},\phi) \;=\; (\delta_{\mathbb{Z}^d},\widehat{\phi}) \;=\; \sum_{m \in \mathbb{Z}^d} \widehat{\phi}(m) \;=\; \sum_{m \in \mathbb{Z}^d} \phi(m) \;=\; (\delta_{\mathbb{Z}^d},\phi),
$$

which proves the claim. $\qquad\square$

With this convention, the following calculation is based on the PSF and obtains a rigorous meaning as an equation between tempered distributions,

$$
(9.10) \qquad \sum_{x \in \mathbb{Z}^d} \mathrm{e}^{-2\pi\mathrm{i}xy} \;=\; \sum_{x \in \mathbb{Z}^d} \widehat{\delta_x} \;=\; \widehat{\delta_{\mathbb{Z}^d}} \;=\; \delta_{\mathbb{Z}^d} \;=\; \sum_{k \in \mathbb{Z}^d} \delta_k.
$$

Its validity (and meaning) is a direct consequence of Proposition 9.4.

What remains is the extension of our setting to general lattices, both for functions and for Dirac combs. By Definition 2.4, d linearly independent vectors $\{b_1,\ldots,b_d\}$ in $\mathbb{R}^d$ span the lattice

$$
\Gamma \;=\; \mathbb{Z}b_1 \oplus \cdots \oplus \mathbb{Z}b_d \;=\; \{m_1 b_1 + \cdots + m_d b_d \mid m_i \in \mathbb{Z}\}.
$$

This lattice is related to $\mathbb{Z}^d$ via $\Gamma = B\mathbb{Z}^d$, where $B \in \mathrm{GL}(d,\mathbb{R})$ is the invertible matrix that columnwise contains the basis vectors b_i. Observe that, if g is a Γ-periodic function on $\mathbb{R}^d$, $h = g \circ B$ is $\mathbb{Z}^d$-periodic, where $(g \circ B)(x) := g(Bx)$. One now has

$$(9.11) \qquad \sum_{m \in \Gamma} \phi(m) = \sum_{\ell \in \mathbb{Z}^d} (\phi \circ B)(\ell) = \sum_{\ell \in \mathbb{Z}^d} \widehat{(\phi \circ B)}(\ell)$$

by means of Proposition 9.3. A simple calculation shows that

$$(9.12) \qquad \widehat{(\phi \circ B)}(x) = |\det(B)|^{-1}\, \widehat{\phi}(B^* x),$$

where $B^* := (B^{-1})^T$ is the basis matrix of the dual lattice Γ^*, so $\Gamma^* = B^* \mathbb{Z}^d$; compare Eq. (3.4). Moreover, $\mathrm{dens}(\Gamma) := 1/|\det(B)|$ is the density of the lattice Γ. Putting this together, we obtain the general Poisson summation formula as follows.

Theorem 9.1 (General PSF). *If Γ is a lattice in $\mathbb{R}^d$, with dual lattice Γ^*, and if $\phi \in \mathcal{S}(\mathbb{R}^d)$ is an arbitrary Schwartz function, one has*

$$\sum_{m \in \Gamma} \phi(m) = \mathrm{dens}(\Gamma) \sum_{\ell \in \Gamma^*} \widehat{\phi}(\ell).$$

Moreover, in $\mathcal{S}'(\mathbb{R}^d)$, one has the following identity of lattice Dirac combs,

$$\widehat{\delta_\Gamma} = \mathrm{dens}(\Gamma)\, \delta_{\Gamma^*},$$

which simultaneously is an identity between translation bounded measures.

PROOF. The first claim follows immediately from Eqs. (9.11) and (9.12), together with the PSF for $\mathbb{Z}^d$ from Proposition 9.4. To show the identity of the Dirac combs, let ϕ be an arbitrary Schwartz function, and consider

$$(\widehat{\delta_\Gamma}, \phi) = (\delta_\Gamma, \widehat{\phi}) = \sum_{m \in \Gamma} \widehat{\phi}(m) = \mathrm{dens}(\Gamma) \sum_{\ell \in \Gamma^*} (\mathcal{F}^2 \phi)(\ell)$$

$$= \mathrm{dens}(\Gamma) \sum_{\ell \in \Gamma^*} \phi(-\ell) = \mathrm{dens}(\Gamma) \sum_{\ell \in \Gamma^*} \phi(\ell) = \mathrm{dens}(\Gamma)\, (\delta_{\Gamma^*}, \phi),$$

where we have used the relation $(\mathcal{F}^2 \phi)(x) = \phi(-x)$ and the inversion symmetry of lattices. Since lattice Dirac combs are translation bounded measures, the last claim is clear. $\qquad\square$

Remark 9.4 (*Radial PSF*). There is an analogue of the PSF for a situation with perfect radial symmetry that has applications to powder diffraction as well as pinwheel and related tilings [BFG07], and also occurs in analytic number theory [IK04].

  Let us formulate one version that is based on a general lattice $\Gamma \subset \mathbb{R}^d$ and its dual lattice Γ^*. Let the sets of radii for non-empty 'shells' $\partial B_r(0)$ in Γ and Γ^* be $\mathcal{D}_\Gamma$ and $\mathcal{D}_{\Gamma^*}$, with shelling numbers $\eta_\Gamma(r)$ and $\eta_{\Gamma^*}(r)$; compare [BG03]. The latter are the number of lattice points on spheres of radius r

around the origin. Now, the general PSF from Theorem 9.1 leads to the radial analogue

$$(9.13) \qquad \sum_{r \in \mathcal{D}_{\Gamma}} \eta_{\Gamma}(r)\, \widehat{\mu_r} \;=\; \mathrm{dens}(\Gamma) \sum_{r \in \mathcal{D}_{\Gamma^*}} \eta_{\Gamma^*}(r)\, \mu_r,$$

where μ_r denotes the uniform probability measure on $\partial B_r(0)$, with $\mu_0 = \delta_0$; see [BFG07] for a proof and various examples. $\qquad\qquad\diamond$

Various extensions are possible from here. For us, the most important consequence of the PSF is the diffraction of perfect crystals and, with one important additional ingredient, also that of perfect quasicrystals.

9.2.3. LATTICE PERIODIC MEASURES

If Γ is a lattice, a measure μ is called Γ-periodic if $\delta_x * \mu = \mu$ holds for all $x \in \Gamma$. For a function g, one has $(\delta_x * g)(y) = g(y - x)$, which shows the connection to our earlier considerations.

Proposition 9.5. *If $\Gamma \subset \mathbb{R}^d$ is a lattice, with fundamental domain* $\mathrm{FD}(\Gamma)$, *and if μ is a Γ-periodic measure on $\mathbb{R}^d$, there is a finite measure ϱ that is concentrated on* $\mathrm{FD}(\Gamma)$, *or on a subset of* $\mathrm{FD}(\Gamma)$, *so that $\mu = \varrho * \delta_{\Gamma}$.*

PROOF. Let C be a relatively compact, measurable fundamental domain of Γ, as in the proof of Proposition 3.1, which implies $\mathbb{R}^d = \bigcup_{t \in \Gamma}(t + C)$. If we now set $\varrho := \mu|_C$, this restriction is a well-defined finite measure. One can check that $\mu|_{t+C} = \delta_t * \varrho$ for $t \in \Gamma$, and thus

$$\mu = \sum_{t \in \Gamma} \mu|_{t+C} = \sum_{t \in \Gamma} \delta_t * \varrho = \varrho * \delta_{\Gamma},$$

which establishes the claim. $\qquad\qquad\square$

Example 9.2 (*Fourier transform of lattice periodic measures*). If μ is a Γ-periodic measure, with Γ a lattice in $\mathbb{R}^d$, it can be represented as $\mu = \varrho * \delta_{\Gamma}$, according to Proposition 9.5. Then, one can apply the convolution theorem for measures, see Theorem 8.5,

$$\widehat{\mu} = (\varrho * \delta_{\Gamma})\widehat{} = \widehat{\varrho} \cdot \widehat{\delta_{\Gamma}} = \mathrm{dens}(\Gamma)\, \widehat{\varrho}\, \delta_{\Gamma^*} = \mathrm{dens}(\Gamma) \sum_{y \in \Gamma^*} \widehat{\varrho}(y)\, \delta_y.$$

The second last step follows from the general PSF of Theorem 9.1. Since $\widehat{\varrho}$ is a bounded continuous function, $\widehat{\mu}$ is the pure point measure spelt out in the last step. Since ϱ has compact support within $\overline{\mathrm{FD}(\Gamma)}$, the function $\widehat{\varrho}$ is even analytic, by the Paley–Wiener theorem [RS80, Thm. IX.12]; see also Remark 8.4. $\qquad\qquad\diamond$

9.2.4. DIFFRACTION OF CRYSTALLOGRAPHIC MEASURES

The autocorrelation of the uniform lattice Dirac comb δ_Γ can be calculated from a density argument. This gives the coefficients $\eta(t) = \operatorname{dens}(\Gamma)$ for all $t \in \Gamma$, and $\eta(t) = 0$ otherwise. The result is $\gamma_\Gamma = \operatorname{dens}(\Gamma)\,\delta_\Gamma$, which also follows from Eq. (8.21). Another application of the general PSF of Theorem 9.1 thus leads to

$$(9.14) \qquad \widehat{\gamma_\Gamma} = \big(\operatorname{dens}(\Gamma)\big)^2 \delta_{\Gamma^*}.$$

In general, a Γ-periodic (complex) measure ω has the form $\omega = \varrho * \delta_\Gamma$ with a finite measure ϱ of compact support, according to Proposition 9.5. It is not difficult to show that the natural autocorrelation of ω exists and is given by

$$(9.15) \qquad \gamma_\omega = (\varrho * \widetilde{\varrho}) * \gamma_\Gamma = \operatorname{dens}(\Gamma)\,(\varrho * \widetilde{\varrho}) * \delta_\Gamma.$$

The diffraction measure of ω, which is the Fourier transform of γ_ω, can be calculated by means of Theorem 8.5 together with the PSF. It reads

$$(9.16) \qquad \widehat{\gamma_\omega} = \big(\operatorname{dens}(\Gamma)\big)^2 |\widehat{\varrho}\,|^2 \delta_{\Gamma^*},$$

which is the well-known result for perfect crystals [Cow95].

Note that the choice of ϱ in $\omega = \varrho * \delta_\Gamma$ is not unique — there are, in fact, infinitely many distinct possibilities. What finally enters the diffraction formula are the values of the continuous function $|\widehat{\varrho}\,|^2$ at the points of the dual lattice Γ^* only. All distinct choices share these values. Moreover, Eq. (9.16) also applies when ω is given as the convolution $\omega = \varrho * \delta_\Gamma$ with an arbitrary *finite* measure ϱ (so that $|\varrho|(\mathbb{R}^d) < \infty$). Let us summarise this as follows.

Theorem 9.2. *Let Γ be a lattice in $\mathbb{R}^d$, and ω a Γ-invariant measure, represented as $\omega = \varrho * \delta_\Gamma$ with ϱ a finite measure. Then, the autocorrelation γ_ω of ω is given by Eq. (9.15), with the diffraction measure $\widehat{\gamma_\omega}$ of Eq. (9.16). In particular, $\widehat{\gamma_\omega}$ is a positive pure point measure, with $\operatorname{supp}(\widehat{\gamma}) \subset \Gamma^*$ and $\widehat{\gamma_\omega}(\{0\}) = \big(\operatorname{dens}(\Gamma)\big)^2 |\widehat{\varrho}(0)|^2$.* $\qquad\square$

Note that the last claim follows constructively from Eq. (9.16), but also from the general result on the central intensity in Corollary 9.1. An important application of Theorem 9.2 is the diffraction of an idealised mono-atomic crystal, where the atomic positions are the points of the lattice Γ and ϱ describes the atomic scattering profile. More complicated situations with several types of atoms can also be modelled this way, as long as one has a stable density on the unit cell of Γ and thus no deviation from periodicity.

Example 9.3 ($\mathbb{Z}^2$-*periodic point measures*). Let $\Gamma = \mathbb{Z}^2$ be the square lattice. Place a unit Dirac measure at each lattice point, and a Dirac measure of complex weight κ at (a, b) and all its $\mathbb{Z}^2$-translates. We may thus, without loss of generality, assume that $0 \leqslant a, b < 1$. This constitutes a $\mathbb{Z}^2$-periodic weighted Dirac comb ω, which permits the representation

$$\omega = \varrho * \delta_{\mathbb{Z}^2} \quad \text{with} \quad \varrho = \delta_{(0,0)} + \kappa\, \delta_{(a,b)}\,.$$

The autocorrelation is $\gamma_\omega = (\varrho * \widetilde{\varrho}) * \delta_{\mathbb{Z}^2}$, with

$$\begin{aligned}
\varrho * \widetilde{\varrho} &= \big(\delta_{(0,0)} + \kappa\, \delta_{(a,b)}\big) * \big(\delta_{(0,0)} + \bar{\kappa}\, \delta_{-(a,b)}\big) \\
&= \big(1 + |\kappa|^2\big)\, \delta_{(0,0)} + \kappa\, \delta_{(a,b)} + \bar{\kappa}\, \delta_{-(a,b)}\,.
\end{aligned}$$

By Theorem 9.2, the corresponding diffraction measure is $\widehat{\gamma_\omega} = |\widehat{\varrho}\,|^2 \delta_{\mathbb{Z}^2}$ with

$$|\widehat{\varrho}\,|^2(k,\ell) = \big(1 + |\kappa|^2\big) + 2\operatorname{Re}\big(\kappa e^{-2\pi\mathrm{i}(ka+\ell b)}\big) = \big|1 + \kappa e^{-2\pi\mathrm{i}(ka+\ell b)}\big|^2$$

for $k, \ell \in \mathbb{Z}$. The diffraction measure is supported on $\mathbb{Z}^2$ (which is self-dual), but need *not* have any non-trivial period. In fact, when $\kappa \neq 0$, the diffraction measure $\widehat{\gamma_\omega}$ is crystallographic if both coordinates a and b are rational, and periodic (of rank 1) if precisely one of them is. $\Diamond$

Remark 9.5 (*Periodicity adjustment*). Let us consider the Lebesgue measure $\omega = \lambda$ on $\mathbb{R}$, which has autocorrelation $\gamma_\omega = \lambda \circledast \lambda = \lambda$ by Example 8.10, and hence diffraction $\widehat{\gamma_\omega} = \delta_0$; compare Example 8.5. Since λ is the normalised Haar measure on $\mathbb{R}$, it is also $\mathbb{Z}$-periodic, and we may represent it as $\lambda = \varrho * \delta_{\mathbb{Z}}$ with $\varrho = 1_{[0,1)}$. Starting from this representation, the autocorrelation is then $(\varrho * \widetilde{\varrho}) * \delta_{\mathbb{Z}}$ by an application of Eq. (9.15). This expression really is the Lebesgue measure in disguise. Now, Eq. (9.16) leads to

$$\widehat{\gamma_\omega} = |\widehat{\varrho}\,|^2\, \delta_{\mathbb{Z}} = \big(\operatorname{sinc}(\pi k)\big)^2 \delta_{\mathbb{Z}} = \delta_0 = \widehat{\lambda},$$

as it must. The penultimate step follows from $\operatorname{sinc}(\pi k) = 0$ for $k \in \mathbb{Z} \setminus \{0\}$; see Example 8.3. Consequently, using a lattice that perhaps does not capture all translation symmetries does not really matter — the amplitudes adjust the final diffraction formula by suitable extinctions, even in this somewhat extreme case. $\Diamond$

Let us add one example where we start from a measure that is not a Dirac comb itself, but still yields a pure point diffraction measure.

Example 9.4 (*Lattice comb with background*). Let $\alpha \in \mathbb{C}$ be fixed, and consider the measure $\omega_\alpha = \delta_{\mathbb{Z}} + \alpha \lambda$, which may be considered as a lattice Dirac comb with additional constant background. If we employ the Eberlein convolution and the relations from Example 8.10, we find

$$\gamma = \omega_\alpha \circledast \widetilde{\omega_\alpha} = \delta_{\mathbb{Z}} + \big(|\alpha|^2 + 2\operatorname{Re}(\alpha)\big)\lambda,$$

where we have used $\widetilde{\delta_{\mathbb{Z}}} = \delta_{\mathbb{Z}}$ and $\widetilde{\lambda} = \lambda$ together with Eq. (8.21). The corresponding diffraction measure is

$$\widehat{\gamma} = \delta_{\mathbb{Z}} + \big(|1 + \alpha|^2 - 1\big)\delta_0,$$

which is a pure point measure. If $a \in \mathbb{C}$ is an arbitrary complex number, the four choices with $\alpha \in \{a, \bar{a}, -2 - a, -2 - \bar{a}\}$ lead to the same $\hat{\gamma}$, so that the corresponding ω_α are homometric. One checks that $\hat{\gamma}(\{0\}) = |1 + \alpha|^2$, in line with Proposition 9.2. Note that the total contribution to δ_0 can take any non-negative value, with extinction for $\alpha = -1$.

Alternatively, since $\mathrm{per}(\omega) = \mathbb{Z}$, we can write $\omega = \varrho * \delta_{\mathbb{Z}}$ with the finite measure $\varrho = \delta_0 + \alpha\lambda|_{[0,1)}$, by an application of Proposition 9.5. Now, we can use Theorem 9.2 to calculate the diffraction as $\hat{\gamma} = |\hat{\varrho}(k)|^2\,\delta_{\mathbb{Z}}$, where $\hat{\varrho}(k) = 1 + \alpha\mathrm{e}^{-\pi \mathrm{i}k}\,\mathrm{sinc}(\pi k)$ from Example 8.3. Although this looks rather different from the previous formula, the result is the same since one has $\hat{\varrho}(0) = 1 + \alpha$ and $\hat{\varrho}(k) = 1$ for all $k \in \mathbb{Z} \setminus \{0\}$; compare Remark 9.5. $\Diamond$

An interesting class of examples is discussed in [Wit05], where certain structural elements are considered robust (or stiff), while the entire arrangement still permits some positional degree of freedom, for instance by relative rotations of the structural elements against one another. In general, such a scenario will produce a diffraction measure of mixed spectral type, but the following simple example stays within the realm of pure point systems.

Example 9.5 (*Planar σ-phases*). Consider the chequerboard and view it as a decoration of the square lattice (with lattice constant 1). Assume that the grey squares are stiff (or solid), while the white squares are empty. One can now rotate the grey squares alternately in opposite directions,

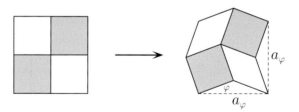

so that a new periodic structure emerges where the white squares are deformed into congruent rhombuses. This new structure can be seen as the lattice periodic repetition of the motif above, where the underlying lattice is a scaled copy of $\mathbb{Z}^2$. The result is an example of the class of uniform tilings, compare [GS87, Figs. 2.9.2, 4.1.3 and 4.2.2]. It has applications in materials science in the context of planar σ-phases and quasicrystal approximants [INF85]; see Figure 9.1 below for two examples. Let us now decorate the tilings by a normalised point (or Dirac) measure at each vertex point.

Introducing an angle $\varphi \in \left(-\frac{\pi}{4}, \frac{\pi}{4}\right)$ to parametrise the rotation, and the length scale $a_\varphi = 2\cos(\varphi)$, the new Dirac comb can be expressed as

$$\omega_\varphi = \delta_{R_\varphi S} * \delta_{a_\varphi \mathbb{Z}^2}.$$

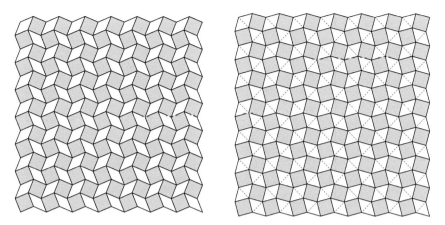

FIGURE 9.1. Planar σ-phases with angles $\varphi = \pi/8$ (left) and $\varphi = \pi/12$ (right), shown with the correct relative length scale. The rhombus on the right is composed of two equilateral triangles (indicated by dashed lines).

Here, $R_\varphi = \begin{pmatrix} \cos(\varphi) & -\sin(\varphi) \\ \sin(\varphi) & \cos(\varphi) \end{pmatrix}$, while $S = \{0, e_1, e_2, e_1 + e_2\}$ denotes the vertex set of the unit square $[0, 1]^2$. Note that $\omega_0 = \delta_{\mathbb{Z}^2}$, while

$$\omega_{\pm\frac{\pi}{4}} := \lim_{\varphi \to \pm\frac{\pi}{4}} \omega_\varphi = 2\delta_{R_{\frac{\pi}{4}}\mathbb{Z}^2}$$

by continuity of ω_φ in the parameter φ (even though the vertex set for these limits is $R_{\pi/4}\mathbb{Z}^2$). The corresponding autocorrelation γ_φ clearly exists and follows from Eq. (9.15), where $\mathrm{dens}(a_\varphi \mathbb{Z}^2) = a_\varphi^{-2}$. Invoking Eq. (9.16) and Theorem 9.2, a direct calculation shows that the diffraction measure reads

$$(9.17) \qquad \widehat{\gamma_\varphi} = \frac{1 + \cos\big(2\pi\langle R_\varphi e_1 | k\rangle\big)}{2\cos(\varphi)^2} \frac{1 + \cos\big(2\pi\langle R_\varphi e_2 | k\rangle\big)}{2\cos(\varphi)^2} \delta_{\mathbb{Z}^2/a_\varphi},$$

where the prefactor is written in terms of the wave vector k. In particular, only values $k \in \mathbb{Z}^2/a_\varphi$ contribute.

When $\varphi = 0$, this expression reduces to $\widehat{\gamma_0} = \delta_{\mathbb{Z}^2}$, as it must, while $\varphi = \pm\pi/4$ leads to $\widehat{\gamma_{\pm\pi/4}} = 4\delta_{R_{\pi/4}\mathbb{Z}^2}$, which reflects the double weight of the point measures at each vertex in this limit. For angles φ with $\tan(\varphi) \notin \mathbb{Q}$, one has extinctions precisely for all $k = \frac{1}{a_\varphi}(m_1, m_2)$ with $m_1 m_2 = 0$ and $m_1 + m_2 \in 2\mathbb{Z} + 1$. Further extinctions occur when $\tan(\varphi)$ is rational. These claims can be derived from the zeros of the intensity function in Eq. (9.17) on $\mathbb{Z}^2/a_\varphi$; compare Remark 9.10 for more details on the concept of extinctions.

The diffraction patterns for $\varphi = \pi/8$ and $\varphi = \pi/12$ are illustrated in Figure 9.2. Note that, due to incommensurate positions of the points in the fundamental cell, both patterns are non-periodic. This is in agreement with the comment made at the end of Example 9.3. All reflexes (shown as disks)

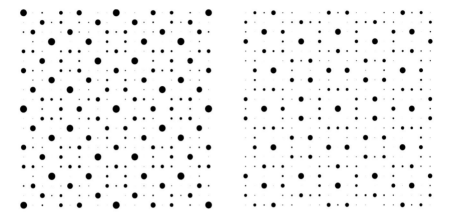

FIGURE 9.2. Diffraction patterns of the two σ-phases of Figure 9.1. Distances and intensities (areas of dots) are shown in the correct relative scale, as given by Eq. (9.17). Both patterns are non-periodic; see text for details.

are situated on the corresponding dual lattices, but display, to some degree, an approximate 8- or 12-fold symmetry. $\Diamond$

This example can be viewed as a planar 'frame' that possesses a macroscopic flexing degree of freedom. More precisely, it flexes while keeping lattice periodicity, with the lattice changing on the way. Such structures are used to model similar degrees of freedom in zeolites and related solids; compare [Pow14] and references therein.

9.2.5. EXPONENTIAL LATTICE SUMS

Although we derived the diffraction of crystallographic structures by means of the PSF, it is also possible and useful to derive the Fourier–Bohr coefficients (or amplitudes) directly. In view of the convolution structure of Proposition 9.5, it is sufficient to consider lattice Dirac combs.

Lemma 9.1. *For $n \in \mathbb{N}$ and $m \in \mathbb{Z}$, one has*

$$\frac{1}{n} \sum_{\ell=m}^{m+n-1} e^{-2\pi i k \ell} = \begin{cases} 1, & k \in \mathbb{Z}, \\ \mathcal{O}(n^{-1}), & \text{otherwise,} \end{cases}$$

where the implied constant of the estimate depends on k, but not on m.

PROOF. The claim is clear for $k \in \mathbb{Z}$. Otherwise, an explicit summation gives

$$\sum_{\ell=m}^{m+n-1} e^{-2\pi i k \ell} = e^{-2\pi i k m} \frac{1 - e^{-2\pi i k n}}{1 - e^{-2\pi i k}},$$

where the right-hand side is uniformly bounded by $2/|1 - e^{-2\pi i k}|$. $\square$

An immediate consequence is the convergence

$$(9.18) \qquad \lim_{n \to \infty} \frac{1}{2n+1} \sum_{\ell=m-n}^{m+n} \mathrm{e}^{-2\pi \mathrm{i} k \ell} = \begin{cases} 1, & k \in \mathbb{Z}, \\ 0, & \text{otherwise}, \end{cases}$$

uniformly in m, for each fixed value of k.

Remark 9.6 (*Rational versus irrational exponents*). The summation used in the proof of Lemma 9.1 'hides' two different mechanisms for $k \notin \mathbb{Z}$. Indeed, if $k \in \mathbb{R} \setminus \mathbb{Q}$, we know that the sequence $\big((kn) \bmod 1\big)_{n \in \mathbb{N}}$ is uniformly distributed in the unit interval in the sense of Weyl [Wey16]; compare Eq. (7.6) and the ensuing Example 7.1. Consequently, we have

$$\frac{1}{2n+1} \sum_{\ell=-n}^{n} \mathrm{e}^{-2\pi \mathrm{i} k \ell} \xrightarrow{n \to \infty} \int_0^1 \mathrm{e}^{-2\pi \mathrm{i} x} \, \mathrm{d}x = 0.$$

In comparison, when $k = \frac{p}{q} \notin \mathbb{Z}$ with coprime $p, q \in \mathbb{N}$ say, one finds

$$\sum_{\ell=0}^{nq-1} \mathrm{e}^{-2\pi \mathrm{i} k \ell} = \sum_{r=0}^{q-1} \sum_{m=0}^{n-1} \mathrm{e}^{-2\pi \mathrm{i} \frac{p}{q}(mq+r)} = n \sum_{r=0}^{q-1} \mathrm{e}^{-2\pi \mathrm{i} \frac{pr}{q}} = 0,$$

where the last sum vanishes because it is a sum over all qth roots of unity (with $q > 1$ by assumption). Also this case results in a uniform distribution of the exponents $\frac{pr}{q}$, however not on the unit interval, but only on the (finitely many) multiples of $\frac{1}{q}$ in it. $\diamond$

Lemma 9.1 can be lifted to $\mathbb{Z}^d$ for sums over hypercubic regions of side length n as follows.

Proposition 9.6. *Fix an arbitrary lattice point $x \in \mathbb{Z}^d$ and consider the cubic patch $C_n(x) := \mathbb{Z}^d \cap \big(x + [0, n-1]^d\big)$, with $n \in \mathbb{N}$. Then, with $k \in \mathbb{R}^d$, one has*

$$\frac{1}{n^d} \sum_{\ell \in C_n(x)} \mathrm{e}^{-2\pi \mathrm{i} k \ell} = \begin{cases} 1, & k \in \mathbb{Z}^d, \\ \mathcal{O}(n^{-1}), & \text{otherwise}, \end{cases}$$

where the implied constant of the asymptotic estimate is uniform in x.

PROOF. As in Lemma 9.1, the claim is clear for $k \in \mathbb{Z}^d$, as $\mathrm{card}\big(C_n(x)\big) = n^d$. Otherwise, at least one coordinate of k fails to be in $\mathbb{Z}$, so that

$$\sum_{\ell \in C_n(x)} \mathrm{e}^{-2\pi \mathrm{i} k \ell} = \mathcal{O}(n^{d-1})$$

by an application of Lemma 9.1 to each coordinate separately. $\square$

It is clear that one obtains stronger decay estimates if several coordinates of k are non-integral. Proposition 9.6 also shows that

$$(9.19) \qquad \lim_{n \to \infty} \frac{1}{n^d} \sum_{\ell \in C_n(x)} \mathrm{e}^{-2\pi \mathrm{i} k\ell} = \begin{cases} 1, & k \in \mathbb{Z}^d, \\ 0, & \text{otherwise,} \end{cases}$$

in line with the PSF for the lattice Dirac comb $\delta_{\mathbb{Z}^d}$.

Note that the result can again be understood from the point of view of uniform distribution, where the mechanisms explained in Remark 9.6 may differ for distinct directions. However, what finally counts here is the distribution of (kx) mod 1 with x running through the integer lattice. Indeed, the averaged exponential sum vanishes (as $n \to \infty$) as a result of (kx) mod 1 being uniformly distributed in various ways.

Corollary 9.2. *Let $(x_i)_{i \in \mathbb{N}}$ be a sequence obtained from ordering all elements of $\mathbb{Z}^d$ according to $\|x_i\| \leqslant \|x_{i+1}\|$ for all $i \in \mathbb{N}$, where $\|.\|$ is any norm of $\mathbb{R}^d$. If $k \notin \mathbb{Q}^d$, the induced sequence $\big((kx_i) \bmod 1\big)_{i \in \mathbb{N}}$ is uniformly distributed in the unit interval.*

PROOF. Since all norms in $\mathbb{R}^d$ are equivalent, we may choose the maximum norm $\|.\|_\infty$, which is best adapted to the cubic patches $C_n(0)$ used above. When $k \notin \mathbb{Q}^d$, $mk \in \mathbb{Z}^d$ with $m \in \mathbb{Z}$ is only possible for $m = 0$.

Let $y_i = (kx_i) \bmod 1$. Now, Proposition 9.6 implies that

$$\lim_{N \to \infty} \frac{1}{N} \sum_{j=1}^{N} \mathrm{e}^{2\pi \mathrm{i} m y_j} = 0$$

for every $m \in \mathbb{Z} \backslash \{0\}$, because $my_j = mkx_j$ and $mk \notin \mathbb{Z}^d$ by assumption. This gives the claimed uniform distribution via an application of Weyl's criterion [Wey16]. □

Remark 9.7 (*Compatible summation regions*). Since the estimates are uniform in the position of the patches, one can replace the cubes by suitable other compact regions, such as more general parallelotopes as well as spheres or ellipsoids. As long as the surface to volume ratio is $\mathcal{O}(s^{-1})$, where s is the diameter of the region, one obtains the same estimate. This asymptotic requirement defines a special class of van Hove sequences; see [Schl00] for a general discussion. ◇

Recall that a general lattice $\Gamma \subset \mathbb{R}^d$ is obtained from $\mathbb{Z}^d$ by a non-singular linear transformation. This transformation turns a cube into a parallelotope. Observing that the integrality condition above has now to be formulated in terms of the dual lattice Γ^*, one obtains the following consequence for the Fourier–Bohr coefficients (or amplitudes) of Γ.

Corollary 9.3. *Let $\Gamma \subset \mathbb{R}^d$ be a lattice with basis $\{b_1, \ldots, b_d\}$, and let Γ^* be the dual lattice, with dual basis $\{b_1^*, \ldots, b_d^*\}$. If $x \in \Gamma$ is arbitrary and $F_n(x) := \left\{ x + \sum_{i=1}^{d} m_i b_i \mid 0 \leqslant m_i \leqslant n-1 \right\}$, one has*

$$\frac{1}{n^d} \sum_{y \in F_n(x)} \mathrm{e}^{-2\pi i k y} = \begin{cases} 1, & k \in \Gamma^*, \\ \mathcal{O}(n^{-1}), & \text{otherwise,} \end{cases}$$

uniformly in $x \in \Gamma$, as $n \to \infty$, for any fixed k.

PROOF. By construction, $\mathrm{card}\big(F_n(x)\big) = n^d$, and $F_n(0) = B\,C_n(0)$ where B is the basis matrix of $\Gamma = B\,\mathbb{Z}^d$. In view of Proposition 9.6, the first situation emerges when $ky \in \mathbb{Z}$ holds for all $y \in \Gamma$, which means $k \in \Gamma^*$. The points of the dual lattice are the integer linear combinations of the b_i^*, while all other k still have a unique expansion in this basis, but with at least one non-integral coordinate. Since $b_i^*\, b_j = \delta_{i,j}$, an application of Proposition 9.6 and its proof produces the claimed estimate. $\qquad\square$

It is clear that Corollary 9.2 can also be adapted to the case of a general lattice. Here, the condition $k \notin \mathbb{Q}\Gamma^*$ implies the uniform distribution of $(kx) \bmod 1$ in the unit interval, where x now runs through the lattice Γ in a sequence that is $\|.\|$-ordered. Beyond lattices, it is a difficult problem to analyse the convergence behaviour of exponential sums; see [BT86] for a highly influential discussion in the diffraction context, and [BMRS03] and references therein for some answers.

9.2.6. INCOMMENSURATE STRUCTURES

An interesting extension emerges from the inclusion of incommensurability, which can be seen as a first step towards the structure of mathematical quasicrystals. Incommensurate systems were pioneered by de Wolff, Janner and Janssen [dWo74, JJ77], and continue to be an important part of crystallography. Here, we can only indicate some elementary examples, and refer to [vSm07] and references therein for a more detailed account, written from a crystallographer's perspective.

Example 9.6 (*Superposition of incommensurate Dirac combs*). Let $\alpha > 0$ be an irrational number and consider the Dirac comb

$$\omega_\alpha := \delta_{\mathbb{Z}} + \delta_{\alpha\mathbb{Z}}.$$

With Eqs. (8.21) and (8.22) from Example 8.10, one can check that the corresponding autocorrelation $\gamma_\alpha = \omega_\alpha \circledast \widetilde{\omega_\alpha}$ exists and reads

$$\gamma_\alpha = \delta_{\mathbb{Z}} + \frac{1}{\alpha} \delta_{\alpha\mathbb{Z}} + \frac{2}{\alpha} \lambda,$$

while it would be a periodic Dirac comb for rational α. Now, with Example 8.5 and the PSF of Theorem 9.1, the diffraction measure is

$$\widehat{\gamma_\alpha} = \delta_{\mathbb{Z}} + \frac{1}{\alpha^2}\,\delta_{\mathbb{Z}/\alpha} + \frac{2}{\alpha}\,\delta_0,$$

which is a pure point measure and satisfies $\widehat{\gamma_\alpha}(\{0\}) = \left(1 + \frac{1}{\alpha}\right)^2$, in line with the density result from Corollary 9.1.

One can also introduce a relative shift $u \in \mathbb{R}$ between the two combs, thus considering $\omega_{\alpha,u} = \delta_{\mathbb{Z}} + \delta_{u+\alpha\mathbb{Z}}$. It turns out that, independently of u, the autocorrelation is the measure γ_α from above. As before, this follows from calculating the Eberlein convolution $\omega_{\alpha,u} \circledast \widetilde{\omega_{\alpha,u}}$, where one notices $\delta_{u+\alpha\mathbb{Z}} = \delta_u * \delta_{\alpha\mathbb{Z}}$ together with $\delta_u * \delta_{-u} = \delta_0$ and $(\delta_u + \delta_{-u}) * \lambda = 2\lambda$. The diffraction measure is $\widehat{\gamma_\alpha}$. It is also independent of the shift u, as expected for irrational α.   ◇

This example is somewhat unrealistic in the sense that the superposition of the two incommensurate Dirac combs does not correspond to a Delone set. In view of possible applications, one is more interested in models that ensure a minimal distance between neighbouring points. An important variant (in higher dimensions) is the class of incommensurate composite systems, which also comprise Delone sets. We restrict ourselves to an explanation by means of a simple planar example.

Example 9.7 (*Incommensurate composite point set*). Let $\alpha > 0$ and consider the Dirac comb

$$\omega = \delta_{\mathbb{Z}^2} + \delta_{u+\Gamma} = \delta_{\mathbb{Z}^2} + \delta_u * \delta_\Gamma,$$

where $\Gamma = \alpha\mathbb{Z}\times\mathbb{Z} \subset \mathbb{R}^2$ is a lattice and $u \in \mathbb{R}^2$. If $\alpha \in \mathbb{Q}$, the underlying point set is still crystallographic, with $\mathbb{Z}^2 \cap \Gamma$ as its lattice of periods. Since this case is fully covered by our previous results, let us assume for the remainder of this example that α is irrational.

Observe that $\mathrm{dens}(\Gamma) = 1/\alpha$ and $\delta_u * \widetilde{\delta_u} = \delta_0$. By arguments similar to those used in Example 9.6, one finds $\delta_{\mathbb{Z}^2} \circledast \delta_\Gamma = \frac{1}{\alpha}(\lambda \otimes \delta_{\mathbb{Z}})$, where we write $\mu \otimes \nu$ for the product of the two measures μ and ν. With the formulas of Example 8.10, one then finds the autocorrelation

$$\gamma = \delta_{\mathbb{Z}^2} + \frac{1}{\alpha}\delta_\Gamma + \frac{1}{\alpha}(\delta_u + \delta_{-u}) * (\lambda \otimes \delta_{\mathbb{Z}}).$$

The Fourier transform can be calculated by the PSF (Theorem 9.1) and the convolution theorem (Theorem 8.5). The result reads

$$\widehat{\gamma} = \delta_{\mathbb{Z}^2} + \frac{1}{\alpha^2}\delta_{\Gamma^*} + \frac{2}{\alpha}\cos(2\pi ku)(\delta_0 \otimes \delta_{\mathbb{Z}})$$

$$= \delta_{\mathbb{Z}^2} + \frac{1}{\alpha^2}(\delta_{\mathbb{Z}/\alpha} \otimes \delta_{\mathbb{Z}}) + \frac{2}{\alpha}\cos(2\pi k_2 u_2)(\delta_0 \otimes \delta_{\mathbb{Z}}).$$

The product structure of Γ^* is obvious, while the simplification in the argument of the cosine term is due to the factor δ_0 in the last measure of the formula. Note that the first component of the shift u does not enter the diffraction. The composite structure is thus visible through the extra intensities added to the peaks on the vertical axis; compare the structure of Example 9.6. Notice that the total intensity at a point $(0, n)$ with $n \in \mathbb{Z}$ is

$$\hat{\gamma}(\{(0, n)\}) = 1 + \frac{1}{\alpha^2} + \frac{2}{\alpha} \cos(2\pi n u_2) \geqslant \left(1 - \frac{1}{\alpha}\right)^2 \geqslant 0.$$

The example can be made more realistic by modifying the positions according to the local neighbourhood; compare [vSm07, Ch. 4].   ◇

Another important class emerges via local displacements of a lattice (or a crystalline point set). Since this can be viewed as an intermediate step towards the general projection method for mathematical quasicrystals, we discuss one characteristic example in more detail. Consider $\mathbb{Z}$, and deform it by moving the points with a real-valued function h, so that

(9.20) $$\Lambda_h = \{x_n \mid n \in \mathbb{Z}\} \quad \text{with} \quad x_n = n + h(n)$$

is the new point set, and δ_{Λ_h} its Dirac comb. If we now assume that the limits

(9.21) $$\mu_m := \lim_{N \to \infty} \frac{1}{2N+1} \sum_{n=-N}^{N} \delta_{x_{n+m} - x_n}$$

exist for all $m \in \mathbb{Z}$ (with $\mu_0 = \delta_0$, which always holds in this case), the autocorrelation measure of δ_{Λ_h} is given by

$$\gamma = \sum_{m \in \mathbb{Z}} \mu_m,$$

where $\mu_{-m} = \mu_m$ by construction.

Example 9.8 (*Deterministic displacement model*). Consider the displacement function defined by $h(n) = \varepsilon \{\alpha n\}$ with $\alpha, \varepsilon \in \mathbb{R}$, where $\{x\} = x - [x]$ denotes the fractional part of x (and not a singleton set). To understand the set Λ_h of Eq. (9.20), it is advantageous to use an embedding in the plane that is known as the 'superspace approach' in crystallography;[1] compare [vSm07, Ch. 2]. Define the lattice $\Gamma = \left\langle \left(\begin{smallmatrix} 1 \\ -\alpha \end{smallmatrix}\right), \left(\begin{smallmatrix} 0 \\ 1 \end{smallmatrix}\right) \right\rangle_{\mathbb{Z}}$ and consider the line pattern obtained as the Γ-orbit of the line from the origin (included) to the point $\left(\begin{smallmatrix} \varepsilon \\ 1 \end{smallmatrix}\right)$ (not included). Then, Λ_h is the set of intersections of the horizontal axis with these line segments; see the left panel of Figure 9.3 for an illustration. This approach displays Λ_h as a deformed model set, in the formulation with tilted target lines; compare Figure 7.15.

---

[1]We thank Peter Zeiner for sharing his expertise on incommensurate systems.

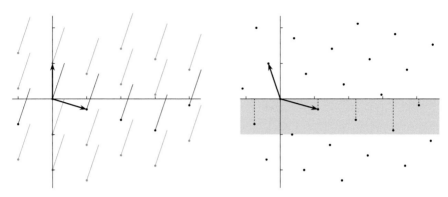

FIGURE 9.3. Illustration of the deterministic displacement model of Example 9.8, for $\varepsilon = 0.35$ and $\alpha \approx 0.2941$. On the left, the description is via tilted target lines that start at the points of the lattice Γ. Their intersection points with the horizontal axis are the points of Λ_h (black targets, giving $x_{-1}, \ldots, x_4$). The corresponding cut and project setting with the lattice $\mathcal{L}$ and the strip with cross-section $W = [0, -1)$ (shaded) is shown on the right; see text for details.

Let us reformulate this system as a cut and project set, to match the formalism introduced for the silver mean chain, which is based on orthogonal projections. Defining the matrix $A = \left(\begin{smallmatrix} 1 & -\varepsilon \\ 0 & 1 \end{smallmatrix} \right)$ and the lattice $\mathcal{L} = A\Gamma$, one finds $\mathcal{L} = \left\langle \left(\begin{smallmatrix} 1+\varepsilon\alpha \\ -\alpha \end{smallmatrix} \right), \left(\begin{smallmatrix} -\varepsilon \\ 1 \end{smallmatrix} \right) \right\rangle_{\mathbb{Z}}$. The corresponding dual lattice is

$$\mathcal{L}^* = \left\langle \begin{pmatrix} 1 \\ \varepsilon \end{pmatrix}, \begin{pmatrix} \alpha \\ 1 + \varepsilon\alpha \end{pmatrix} \right\rangle_{\mathbb{Z}},$$

as can easily be calculated with the methods from Chapter 3. The set Λ_h is now a model set for the CPS $(\mathbb{R}, \mathbb{R}, \mathcal{L})$, with the simple window $W = [0, -1)$, as illustrated in Figure 9.3 (right panel). If $\varepsilon = 0$, we get a (non-minimal) embedding of $\mathbb{Z}$, and if $\alpha = \frac{p}{q} \in \mathbb{Q}$ with p, q coprime, one obtains a periodic point set with lattice of periods $q\mathbb{Z}$.

Let us consider the case $\alpha \notin \mathbb{Q}$ in more detail. Here, the autocorrelation measure can be calculated via the measures μ_m from Eq. (9.21) as follows. When $m \in \mathbb{Z} \setminus \{0\}$, one has

$$x_{n+m} - x_n = m + \varepsilon\big(\{\alpha(n+m)\} - \{\alpha n\}\big),$$

which, for fixed m, takes precisely two different values, namely $m + \varepsilon\{\alpha m\}$ (when $0 \leqslant \{\alpha n\} < 1 - \{\alpha m\}$) or $m - \varepsilon(1 - \{\alpha m\})$ (which happens when $1 - \{\alpha m\} \leqslant \{\alpha n\} < 1$). Since $(\{\alpha n\})_{n \in \mathbb{Z}}$ is uniformly distributed in the unit interval (recall Example 7.1), the two cases occur in the sum for μ_m with frequencies $1 - \{\alpha m\}$ and $\{\alpha m\}$. Noting that $\mu_0 = \delta_0$, one finds the autocorrelation

$$\gamma_h = \sum_{m \in \mathbb{Z}} \mu_m = \sum_{m \in \mathbb{Z}} \Big((1 - \{\alpha m\}) \, \delta_{m + \varepsilon \{\alpha m\}} + \{\alpha m\} \, \delta_{m - \varepsilon (1 - \{\alpha m\})} \Big).$$

Since $L = \pi(\mathcal{L}) = \langle 1 + \varepsilon\alpha, -\varepsilon \rangle_{\mathbb{Z}}$, we can alternatively write the autocorrelation as $\gamma_h = \sum_{z \in L} \eta(z) \, \delta_z$, where $\eta(z) = f(z^\star)$ with the (continuous) tent-shaped function f given by $f(y) = 1 - |y|$ for $|y| \leqslant 1$ and by $f(y) = 0$ otherwise. The $\star$-map is defined by $\big(r(1 + \varepsilon\alpha) - s\varepsilon \big) \mapsto (-r\alpha + s)$, with $r, s \in \mathbb{Z}$. The structure behind this choice will be explained later.

The corresponding diffraction measure $\widehat{\gamma_h}$ can be calculated from the projection formalism, and reads

$$\widehat{\gamma_h} = \sum_{k \in \mathcal{M}} |A(k)|^2 \, \delta_k,$$

with the Fourier module $\mathcal{M} = \pi(\mathcal{L}^*) = \langle 1, \alpha \rangle_{\mathbb{Z}} = \mathbb{Z}[\alpha]$ and the amplitudes $A(k) = \widehat{1_W}(-k^\star) = \mathrm{e}^{-\pi \mathrm{i} k^\star} \operatorname{sinc}(\pi k^\star)$. Here, the $\star$-map is the induced $\star$-map for the dual CPS $(\mathbb{R}, \mathbb{R}, \mathcal{L}^*)$ and acts as $(r + s\alpha) \mapsto \big(r\varepsilon + s(1 + \varepsilon\alpha) \big)$, again with $r, s \in \mathbb{Z}$. Notice that $\widehat{\gamma_h}$ is a pure point measure. It is supported on a dense set, with the intensities being locally summable. The proof for this result and the diffraction formula are non-trivial. Both will follow from our later (and more general) treatment of regular model sets. ◇

This is an example of an incommensurately modulated structure, the latter being described in more details in [vSm07], including their superspace formulation. The formal calculation of the diffraction used in [vSm07] and other sources reproduces the pure point part correctly, while a proof of the absence of any continuous parts requires additional effort. One way to complete the argument employs the concept of deformed model sets; see below for a complete derivation.

Let us now move on to the diffraction theory of more general non-periodic systems. Since this requires a rather intricate series of arguments, we prefer to first develop the entire approach with a suitable example in one dimension, the silver mean chain of Section 7.1. We hope that the later general exposition will become more transparent this way.

9.3. Autocorrelation and diffraction of the silver mean chain

To continue our discussion of the silver mean chain, we use the point set $\Lambda = \lambda(W)$ with window $W = \left[-\frac{\sqrt{2}}{2}, \frac{\sqrt{2}}{2} \right]$, as constructed in Section 7.1 from the CPS (7.3). In other words, we identify the tiling with the point set of the left endpoints, which is justified by the MLD relation. We know that the (continuous) hull $\mathbb{X}(\Lambda)$ is strictly ergodic for the action of $\mathbb{R}$ by translation, wherefore the autocorrelation coefficients $\eta(z)$ exist as a limit, for all $z \in \mathbb{R}$.

Since $\Lambda - \Lambda \subset L = \mathbb{Z}[\sqrt{2}]$, one clearly has $\eta(z) = 0$ for all $z \notin L$. For $z \in L$, we set $z^{\star} = z'$, with $'$ denoting algebraic conjugation in $\mathbb{Q}(\sqrt{2})$.

Lemma 9.2. *For the silver mean chain Λ and an arbitrary $z \in L$, the autocorrelation coefficient $\eta(z)$ is given by*

$$\eta(z) = \mathrm{dens}(\Lambda)\, \frac{\mathrm{vol}\big(W \cap (z^{\star} + W)\big)}{\mathrm{vol}(W)} = \frac{1}{2\sqrt{2}} \int_{\mathbb{R}} 1_W(y)\, 1_{z^{\star}+W}(y)\, \mathrm{d}y.$$

In particular, one has $0 \leqslant \eta(z) \leqslant \mathrm{dens}(\Lambda)$, where $\mathrm{dens}(\Lambda) = \frac{1}{2}$.

PROOF. Due to the properties of Λ, we have to calculate the limit of Eq. (9.4) for $z \in L$ with weight function $w(x) = 1_W(x^{\star})$. This amounts to calculating the relative frequency of points $x, y \in \Lambda$ with $x - y = z$, similarly to Corollary 7.3. For $z \in L$, the points x and $y = x - z$ (both from L) are in Λ if and only if both $x^{\star}$ and $(x - z)^{\star}$ lie inside the window W. This is equivalent to $x^{\star} \in W \cap (z^{\star} + W)$. The relative frequency is again obtained from Weyl's lemma, with uniform distribution according to Proposition 7.3. This gives convergence of the sum in Eq. (9.4) to the integral as stated, noting that

$$\frac{1}{2r}\, \mathrm{card}\big(\Lambda \cap (-r, r)\big) \xrightarrow{\ r \to \infty\ } \mathrm{dens}(\Lambda) = \frac{1}{2}.$$

Since $\mathrm{vol}(W) = \sqrt{2}$, the alternative expression for $\eta(z)$ as well as the inequality for η are obvious. $\qquad\square$

Remark 9.8 (*Window covariogram*). Observe that the integral in the formula of Lemma 9.2 equals the *covariogram* of the window W, defined as

$$\mathrm{cvg}_W(v) := \big(1_W * \widetilde{1_W}\big)(v) = \int_{\mathbb{R}} 1_W(y)\, 1_{z^{\star}+W}(y)\, \mathrm{d}y.$$

This follows from a simple calculation using that $1_{v+W}(y) = 1_W(y - v)$ and that $\tilde{g}(v) = g(-v)$ holds for real-valued functions g. The covariogram of a compact interval (such as W) is a positive definite and symmetric continuous function with compact support $W - W$. Recall from Example 8.3 that, due to $\widehat{\tilde{g}} = \overline{\hat{g}}$, the covariogram has the positive Fourier transform

$$\widehat{\mathrm{cvg}_W}(k) = \big|\widehat{1_W}(k)\big|^2,$$

which will reappear frequently in what follows, as on page 392. $\qquad\Diamond$

Proposition 9.7. *The Dirac comb δ_Λ of the silver mean chain possesses the autocorrelation measure*

$$\gamma_\Lambda = \delta_\Lambda \circledast \widetilde{\delta_\Lambda} = \sum_{z \in \Lambda - \Lambda} \eta(z)\, \delta_z$$

with the coefficients $\eta(z)$ of Lemma 9.2. This is a translation bounded, pure point measure that is both positive and positive definite. Moreover, γ_Λ is also the autocorrelation measure for all other elements of the hull $\mathbb{X}(\Lambda)$.

PROOF. Since $\Lambda - \Lambda$ is uniformly discrete, the autocorrelation is a pure point measure of the form claimed, where $0 \leqslant \eta(z) \leqslant \frac{1}{2}$ was derived in Lemma 9.2. In particular, the coefficient function $\eta(z)$ vanishes for all $z \notin \Lambda - \Lambda$. The autocorrelation measure is always positive definite by construction.

Recall that the hull $\mathbb{X}(\Lambda)$ equals the LI class of Λ, so that the difference set equals $\Lambda - \Lambda$ for all elements of the hull. Since the latter is uniquely ergodic, the relative cluster frequencies are the same for all its members. This means that they share all autocorrelation coefficients with Λ. □

To continue, we need the dual lattice $\mathcal{L}^*$ of $\mathcal{L} = \{(x, x') \mid x \in \mathbb{Z}[\sqrt{2}]\}$, the latter being the Minkowski embedding of $\mathbb{Z}[\sqrt{2}]$ as given in Eq. (3.11) on page 60. The dual lattice is

$$\mathcal{L}^* = \{y \in \mathbb{R}^2 \mid xy \in \mathbb{Z} \text{ for all } x \in \mathcal{L}\} = \left\langle \frac{\sqrt{2}}{4}\begin{pmatrix} 1 \\ -1 \end{pmatrix}, \frac{1}{2}\begin{pmatrix} 1 \\ 1 \end{pmatrix} \right\rangle_{\mathbb{Z}}$$

(note the different star symbol), which has the projection

$$(9.22) \qquad L^{\circledast} := \pi(\mathcal{L}^*) = \left\{\frac{1}{2}\left(m + \frac{n}{\sqrt{2}}\right) \mid m, n \in \mathbb{Z}\right\} = \frac{\sqrt{2}}{4}\mathbb{Z}[\sqrt{2}].$$

Note that the $\star$-map is well-defined on the rational span of L, which includes the dense point set $L^{\circledast}$, and that $\mathcal{L}^* = \{(k, k^{\star}) \mid k \in L^{\circledast}\}$. This situation can be generalised by means of a dual CPS (though we will not need it here); see [Moo97a] for details.

To proceed to the diffraction measure of δ_Λ in a constructive manner, we need a technical result, which is somewhat subtle. We formulate this for the example at hand, but point out that it has an immediate generalisation to Euclidean model sets (with $\mathbb{R}^m$ as internal space of the CPS).

Lemma 9.3. *Let $\mathcal{L} = \{(x, x^{\star}) \mid x \in L\}$ be the lattice of the CPS of Eq. (7.3) from page 253. Consider $\mu = \sum_{x \in L} v(x^{\star})\,\delta_x$, where v is a bounded and integrable continuous function on $\mathbb{R}$ subject to the condition that μ is a tempered measure. Furthermore, let v be such that the tempered distribution $\widehat{\mu}$ is also a measure. Then, one has*

$$\widehat{\mu} = \mathrm{dens}(\mathcal{L})\sum_{k \in L^{\circledast}} \widehat{v}(-k^{\star})\,\delta_k.$$

PROOF. Define a measure ν on $\mathbb{R}^2$ by

$$\nu := \sum_{x \in L} v(x^{\star})\,\delta_x \otimes \delta_{x^{\star}} = \sum_{x \in L}(1 \otimes v)\,\delta_{(x,x^{\star})} = (1 \otimes v)\,\delta_{\mathcal{L}},$$

which is a translation bounded pure point measure, hence also tempered. We write $g \otimes h$ for the product function defined by $(x, y) \mapsto g(x)\,h(y)$, and $\delta_x \otimes \delta_y$ for the product measure. The relation of ν to μ is via marginalisation in the second coordinate, so that $\mu(g) = \nu(g \otimes 1)$ for all $g \in C_{\mathrm{c}}(\mathbb{R})$, which is well-defined due to the assumptions on the function v and the properties of μ and ν. Moreover, we also have (via dominated convergence)

$$\mu(f) = \lim_{\varepsilon \searrow 0} \nu(f \otimes \widehat{\psi_\varepsilon}),$$

with the Gaussian functions ψ_ε from Example 8.2 on page 310, where f can be any continuous function on $\mathbb{R}$ that is μ-measurable. Now, fix an arbitrary $g \in \mathcal{S}(\mathbb{R})$. Since $\widehat{\mu}$ is a tempered distribution, we know that

$$(9.23) \qquad \widehat{\mu}(g) = \mu(\widehat{g}) = \lim_{\varepsilon \searrow 0} \nu(\widehat{g} \otimes \widehat{\psi_\varepsilon})$$

as an equation of distributions. Recall that any $g \in C_{\mathsf{c}}(\mathbb{R})$ can be approximated arbitrarily well (in the $\|.\|_\infty$-topology) by Schwartz functions of compact support. Since μ is a well-defined linear functional on $\mathcal{S}(\mathbb{R})$ and $\widehat{\mu}$ is a measure by assumption, $\widehat{\mu}(g) = \mu(\widehat{g})$ also holds as an identity between measures, with $g \in C_{\mathsf{c}}(\mathbb{R})$, whenever $\widehat{g}$ is μ-integrable. This is automatically satisfied for $g \in C_{\mathsf{c}}(\mathbb{R}) \cap \mathcal{S}(\mathbb{R})$. Since ν is a tempered measure, the approximation identity of Eq. (9.23) holds on the level of measures in the same sense, where $C_{\mathsf{c}}(\mathbb{R}) \cap \mathcal{S}(\mathbb{R})$ is dense in $C_{\mathsf{c}}(\mathbb{R})$.

Observe next that

$$\nu(\widehat{g} \otimes \widehat{\psi_\varepsilon}) = \nu(\widehat{g \otimes \psi_\varepsilon}) = \widehat{\nu}(g \otimes \psi_\varepsilon).$$

The Fourier transform $\widehat{\nu}$ (which, due to our assumptions on μ, is also a measure) can now be calculated via the PSF (Theorem 9.1) and the convolution theorem (Theorem 8.5) for (tempered) measures, which gives

$$\widehat{\nu} = \big((1 \otimes v)\, \delta_\mathcal{L}\big)^{\widehat{}} = \big(\widehat{1 \otimes v}\big) * \widehat{\delta_\mathcal{L}} = \mathrm{dens}(\mathcal{L})\,(\delta_0 \otimes \widehat{v}) * \delta_{\mathcal{L}^*}.$$

Here, the regular measure $\widehat{v}$ is defined by $f \mapsto \int_\mathbb{R} f(y)\, \widehat{v}(y)\, \mathrm{d}y$, so $\delta_0 \otimes \widehat{v}$ is a finite measure on $\mathbb{R}^2$. With the dual lattice $\mathcal{L}^* = \big\{(k, k^\star) \mid k \in L^{\circledast}\big\}$, one has

$$(\delta_0 \otimes \widehat{v}) * \delta_{(k,k^\star)} = (\delta_0 * \delta_k) \otimes (\widehat{v} * \delta_{k^\star}) = \delta_k \otimes (\widehat{v} * \delta_{k^\star}),$$

where $\widehat{v} * \delta_{k^\star}$ is the regular measure defined by the function $y \mapsto \widehat{v}(y - k^\star)$. Consequently, we have

$$\widehat{\nu}(g \otimes \psi_\varepsilon) = \mathrm{dens}(\mathcal{L}) \sum_{k \in L^{\circledast}} g(k) \int_\mathbb{R} \widehat{v}(y - k^\star)\, \psi_\varepsilon(y)\, \mathrm{d}y.$$

By assumption, v is integrable, wherefore $\widehat{v}$ is a uniformly continuous and bounded function on $\mathbb{R}$. This implies $\int_\mathbb{R} \widehat{v}(y - k^\star)\, \psi_\varepsilon(y)\, \mathrm{d}y \xrightarrow{\varepsilon \searrow 0} \widehat{v}(-k^\star)$ via Example 8.2, uniformly in $k^\star$. Since $g(k) = \delta_k(g)$, the claim follows. $\qquad \square$

Remark 9.9 (*Niceness criterion*). One sufficient condition for continuity, boundedness and integrability of the function v in Lemma 9.3 is $v \in C_{\mathsf{c}}(\mathbb{R})$; another is that v is continuous with $v(y) = \mathcal{O}\big(|y|^{-\alpha}\big)$, as $|y| \to \infty$, for some $\alpha > 2$. In these cases, both measures μ and $\widehat{\mu}$ are translation bounded.

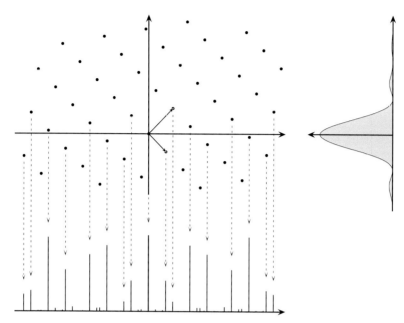

FIGURE 9.4. Sketch of the diffraction measure of the silver mean chain (bottom). As indicated by the arrows (for some points), each point $(k, k^\star)$ of the dual lattice $\mathcal{L}^*$ (dots) produces a point measure at k of intensity $0 \leqslant \left(\frac{1}{2} \operatorname{sinc}(\pi\sqrt{2}k^\star)\right)^2 \leqslant \frac{1}{4}$ (represented by a line). The intensity function is shown on the right. The complete diffraction pattern, comprising the contribution of the entire dual lattice $\mathcal{L}^*$, is symmetric in k.

Conversely, a function such as $v(y) = (1 + |y|)^{-1}$ is insufficient and will not lead to a measure; compare [Ric03] for a more detailed discussion of densely defined Dirac combs. ◇

Theorem 9.3. *The diffraction measure of the Dirac comb δ_Λ of the silver mean chain Λ is the Fourier transform $\widehat{\gamma_\Lambda}$ of the autocorrelation γ_Λ of Proposition 9.7. It is a translation bounded, pure point measure that is both positive and positive definite. It is explicitly given by*

$$\widehat{\gamma_\Lambda} = \sum_{k \in L^\circledast} |A(k)|^2 \, \delta_k,$$

with the Fourier module $L^\circledast$ of Eq. (9.22) and the amplitudes

$$A(k) = \frac{\operatorname{dens}(\Lambda)}{\operatorname{vol}(W)} \, \widehat{1_W}(-k^\star) = \frac{1}{2\sqrt{2}} \int_W e^{2\pi i k^\star y} \, dy = \frac{1}{2} \operatorname{sinc}(\pi\sqrt{2}\,k^\star),$$

where $\operatorname{sinc}(z) = \frac{\sin(z)}{z}$ is the function introduced in Example 8.3.

PROOF. The autocorrelation coefficient of Proposition 9.7 is

$$\eta(z) = \frac{\text{dens}(\Lambda)}{\text{vol}(W)} \, \text{cvg}_W(z^\star),$$

as follows from Remark 9.8. The right-hand side (viewed as a function on internal space) defines an element of $C_{\mathsf{c}}(\mathbb{R})$. By Remark 9.9, we are in the situation of Lemma 9.3 as soon as we know that $\widehat{\gamma_\Lambda}$ is a tempered measure. This is true by Proposition 8.6, because the autocorrelation is a positive definite measure by construction.

Lemma 9.3 then gives the explicit representation by a simple calculation, based on Remark 9.8 together with the relation $\text{dens}(\Lambda) = \text{dens}(\mathcal{L})\,\text{vol}(W)$ from Eq. (7.5). The coefficients $\eta(z)$ clearly make γ_Λ a pure point measure, with uniformly discrete support $\Lambda - \Lambda$. Positive definiteness and positivity of $\widehat{\gamma_\Lambda}$ follow from the corresponding properties of γ_Λ by the Bochner–Schwartz theorem, compare Remark 8.7, while translation boundedness is another consequence of Proposition 8.6. □

The diffraction of the silver mean chain is illustrated in Figure 9.4. Let us note that the existence of the diffraction measure $\widehat{\gamma_\Lambda}$ means that the intensities $I(k) = |A(k)|^2$, which are defined for $k \in L^\circledast$, are locally summable everywhere in $\mathbb{R}$. This follows from evaluating the measure with a suitable continuous function of compact support.

Let us briefly note that the silver mean chain is an example of a Pisot substitution with two letters. As a consequence of [HS03], the corresponding dynamical system is pure point, so that the pure point nature of the diffraction spectrum alternatively follows from the equivalence discussed in Appendix B as well as in [LMS02, BL04]; see also [SS02]. However, this approach does not provide the intensities of the point measures.

The explicit form of the diffraction measure indicates that one can give the notion of a diffraction *amplitude* a well-defined meaning, as in the lattice periodic case, even though the distribution $\widehat{\delta_\Lambda}$ is not a measure. This was first discussed in [Hof95a]. Let us expand on the meaning of the amplitudes (or Fourier–Bohr coefficients), viewed as limits of exponential sums.

Lemma 9.4. *Let Λ be the silver mean point set from above, and fix $k \in L^\circledast$. Then, with $\Lambda_r = \Lambda \cap B_r(0)$, one has*

$$\frac{1}{\text{vol}(B_r)} \sum_{x \in \Lambda_r} \text{e}^{-2\pi \mathrm{i} k x} \xrightarrow{\; r \to \infty \;} \frac{\text{dens}(\Lambda)}{\text{vol}(W)} \int_W \text{e}^{2\pi \mathrm{i} k^\star y} \, \mathrm{d}y = A(k).$$

PROOF. Observe that $(k, k^\star) \in \mathcal{L}^*$, so that, with $x \in \Lambda$ and hence $(x, x^\star) \in \mathcal{L}$, one has $kx + k^\star x^\star \in \mathbb{Z}$ and thus

$$\text{e}^{-2\pi \mathrm{i} k x} = \text{e}^{-2\pi \mathrm{i}(kx + k^\star x^\star)} \text{e}^{2\pi \mathrm{i} k^\star x^\star} = \text{e}^{2\pi \mathrm{i} k^\star x^\star}.$$

Inserting this into the left hand side of our claim, we obtain

$$\frac{1}{\mathrm{vol}(B_r)} \sum_{x \in \Lambda_r} \mathrm{e}^{-2\pi \mathrm{i} k x} = \frac{\mathrm{card}(\Lambda_r)}{\mathrm{vol}(B_r)} \frac{1}{\mathrm{card}(\Lambda_r)} \sum_{x \in \Lambda_r} \mathrm{e}^{2\pi \mathrm{i} k^\star x^\star}.$$

Since $\mathrm{card}(\Lambda_r)/\mathrm{vol}(B_r) \xrightarrow{r \to \infty} \mathrm{dens}(\Lambda)$, the convergence is a consequence of Weyl's lemma (note that $\mathrm{e}^{2\pi \mathrm{i} k^\star y}$ is continuous in y) and the uniform distribution of the points of $\Lambda^\star$ in W according to Proposition 7.3. $\square$

Remark 9.10 (*Extinction rules*). The diffraction formula of Theorem 9.3 involves an intensity function $I(k) = |A(k)|^2$, which is defined on $L^\circledast$ from Eq. (9.22). This $\mathbb{Z}$-module also satisfies

$$L^\circledast = \langle k \in \mathbb{R} \mid I(k) > 0 \rangle_{\mathbb{Z}}.$$

The intensity can still vanish on some subset of $L^\circledast$. If this happens, one speaks of an *extinction*. For the silver mean chain, one easily checks that the extinction subset is

$$\left\{ k \in L^\circledast \mid k^\star = \tfrac{m}{2}\sqrt{2} \text{ with } 0 \neq m \in \mathbb{Z} \right\},$$

which follows from the fact that $\mathrm{sinc}(y) = 0$ if and only if $y = m\pi$ with $m \in \mathbb{Z} \setminus \{0\}$. These extinctions reflect the special choice of the window, which also leads to the LIDS of the silver mean chain; compare [FR03] for a general approach to this kind of connection. More generally, any interval as a window is compatible with affine expansions that map the corresponding model set into itself. However, only intervals of certain lengths lead to hulls with *local* inflation rules. $\Diamond$

Remark 9.11 (*PV series of peaks*). Whenever $k \in L^\circledast$, also $sk \in L^\circledast$, with $s = \lambda_{\mathrm{sm}} = 1 + \sqrt{2}$. Since the amplitudes of the silver mean chain in Theorem 9.3 satisfy $A(sk) = \frac{\mathrm{dens}(\Lambda)}{\mathrm{vol}(W)} \widehat{1_W}(-s^\star k^\star)$ with $s^\star = 1 - \sqrt{2} \approx -0.4142$, the sequence $(s^m k)_{m \in \mathbb{N}}$ supports point measures of $\widehat{\gamma_\Lambda}$ with amplitudes

$$A(s^m k) = \frac{\mathrm{dens}(\Lambda)}{\mathrm{vol}(W)} \widehat{1_W}\bigl(-(s^\star)^m k^\star\bigr) \xrightarrow{m \to \infty} \frac{\mathrm{dens}(\Lambda)}{\mathrm{vol}(W)} \widehat{1_W}(0) = A(0).$$

Observing that $A(0) = \mathrm{dens}(\Lambda) = \sup\{|A(k)| \mid k \in L^\circledast\}$, this convergence property means that every peak of the diffraction measure $\widehat{\gamma_\Lambda}$ leads to a sequence of peaks at exponentially growing distance from 0 whose intensity converges (from below) to the intensity of the central peak at $k = 0$. $\Diamond$

Remark 9.12 (*Meyer property of intensity patterns*). As a consequence of Theorem 9.3, the supremum of $|A(k)|$ is $\mathrm{dens}(\Lambda)$, which is achieved at $k = 0$. Viewed as a function on internal space, $|A(k)|$ is continuous, in particular at

$k^\star = 0$. If we choose (a sufficiently small) $\varepsilon > 0$, there is a $\delta > 0$ such that

$$1 \geqslant \frac{|\widehat{1_W}(y)|}{\mathrm{vol}(W)} > 1 - \varepsilon \qquad \text{for all } y \in (-\delta, \delta).$$

This implies $(1 - \varepsilon)\,\mathrm{dens}(\Lambda) < |A(k)| \leqslant \mathrm{dens}(\Lambda)$ for any $k \in L^\circledast$ such that $k^\star \in (-\delta, \delta)$. The points selected this way form a regular model set and hence a Meyer set. In particular, peaks with almost the intensity of the central one populate a relatively dense set.

Moreover, given *any* $0 < \varepsilon < 1$, the set of all silver mean diffraction peaks with $(1 - \varepsilon)\,\mathrm{dens}(\Lambda) < |A(k)| \leqslant \mathrm{dens}(\Lambda)$ is still uniformly discrete. This follows from $|\widehat{1_W}(y)| = \mathcal{O}(1/|y|)$ as $|y| \to \infty$ (which is a consequence of the asymptotic behaviour of the sinc function). Consequently, the set

$$V = \left\{ y \in \mathbb{R} \mid |\widehat{1_W}(y)| > (1 - \varepsilon)\,\mathrm{vol}(W) \right\}$$

is bounded, so that $\{ k \in L^\circledast \mid k^\star \in V \}$ is uniformly discrete. This is the reason behind the *threshold discreteness* of the diffraction intensities; compare [Str13b] for an analogous property in a more general setting. ◇

One can extend Proposition 9.7 and Theorem 9.3 as follows. Fix a function $h \in C_{\mathsf{c}}(\mathbb{R})$ and define the weighted Dirac comb

$$\omega = \sum_{x \in L} h(x^\star)\,\delta_x.$$

One can then show that $\gamma_\omega = \sum_{z \in L} \eta(z)\,\delta_z$ is the unique autocorrelation measure of ω, where

$$\eta(z) = \mathrm{dens}(\Lambda)\,\frac{\big(h * \widetilde{h}\big)(z^\star)}{\big(h * \widetilde{h}\big)(0)} = \frac{\mathrm{dens}(\Lambda)}{\|h\|_2^2}\,\big(h * \widetilde{h}\big)(z^\star),$$

the proof of which is a variant of our previous argument. This brings us once again to the situation of Lemma 9.3, with the following immediate consequence for $L = \mathbb{Z}[\sqrt{2}\,]$.

Corollary 9.4. *The weighted Dirac comb* $\omega = \sum_{x \in L} h(x^\star)\,\delta_x$ *with a function* $h \in C_{\mathsf{c}}(\mathbb{R})$ *is pure point diffractive. The corresponding diffraction measure is the positive pure point measure given by* $\widehat{\gamma_\omega} = \sum_{k \in L^\circledast} |A(k)|^2\,\delta_k$, *where* $A(k) = \mathrm{dens}(\Lambda)\,\widehat{h}(-k^\star)/\|h\|_2$. □

Before we embark on a generalisation to higher dimensions, let us briefly comment on the structure of certain *deformations* of the silver mean chain. To be concrete, start with Λ as above and define

$$\Lambda_\vartheta := \{ x + \vartheta(x^\star) \mid x \in \Lambda \},$$

where $\vartheta \colon W \longrightarrow \mathbb{R}$ is assumed to be continuous. This defines a special class of deformations of Λ which preserve pure pointedness of the diffraction. Note that more general deformations may destroy this property. The set Λ_ϑ is called a *deformed model set*. The general theory of deformed model sets is developed in [Hof95a, BD00, BL05]. Here, we only give an example.

Example 9.9 (*Deformed silver mean chain*). A simple but interesting deformation of the silver mean chain Λ is provided by the deformation function

$$\vartheta(y) \,=\, \alpha y + \beta,$$

for $y \in W$, with some constants $\alpha, \beta \in \mathbb{R}$. Let us first restrict α in size so that the deformation does not change the order of the points along the line, and does not produce coinciding points. Then, the affine nature of ϑ on W has the effect of changing the relative length ratio of the a and b intervals, with β being a global translation. In particular, two points of Λ at distance t, say x and $x + t$, are mapped onto two points of Λ_ϑ at distance $t + \alpha t^\star$, independently of x. It is easy to check that the admissible values of α include $\alpha \in (-1, 3 + \sqrt{2}\,)$. The corresponding length ratio is

$$\kappa \,=\, \frac{\text{length}(a_\vartheta)}{\text{length}(b_\vartheta)} \,=\, 1 + \frac{1 - \alpha}{1 + \alpha}\sqrt{2},$$

where we use a_ϑ and b_ϑ for the intervals that result from the deformation. For a given ratio κ, the deformation parameter is $\alpha = (\sqrt{2} + 1 - \kappa)/(\sqrt{2} - 1 + \kappa)$. In particular, $\kappa = 1$ corresponds to $\alpha = 1$. The deformed model set Λ_ϑ is pure point diffractive; see [BD00, BL05] for a proof, and Remark 9.13 below. Note that, despite their close relation, the original and the deformed silver mean chains need *not* be MLD for more general deformation; see also [CS06].

Of particular interest is the fact that one also obtains an explicit formula for the diffraction measure. A detailed account for its calculation can be found in [BD00], which can also be derived explicitly via Weyl's lemma on uniform distribution; compare [Schl98, Moo02] for a formulation of the latter in the context of model sets. The result is

$$(9.24) \qquad \widehat{\gamma_{\Lambda_\vartheta}} \,=\, \sum_{k \in L^\circledast} |A_\vartheta(k)|^2\, \delta_k,$$

where the new amplitudes are obtained as

$$(9.25) \qquad A_\vartheta(k) \,=\, \frac{1}{2\sqrt{2}} \int_W e^{2\pi i (k^\star y - k\vartheta(y))}\, \mathrm{d}y$$

for all $k \in L^\circledast$, and as $A_\vartheta(k) = 0$ otherwise. The formula for the amplitudes $A_\vartheta(k)$ with $k \in L^\circledast$ can be derived by the same argument that was used in the proof of Lemma 9.4. Weyl's lemma applies since ϑ is continuous on W.

Note that $A_\vartheta(0) = \frac{1}{2} = \text{dens}(\Lambda)$, in agreement with the observation that the deformation defined by ϑ does not change the density of the set. For our special choice of ϑ, the amplitude formula simplifies to

$$A_{\alpha,\beta}(k) = \frac{1}{2}\, \mathrm{e}^{-2\pi\mathrm{i}\beta k}\, \mathrm{sinc}\big(\pi(\alpha k - k^\star)\sqrt{2}\,\big)$$

for all $k \in L^\circledast$. Depending on the value of α, there can be characteristic points of extinction, which can be calculated via the zeros of the sinc function.

Let us add that, if we use a formulation via measures, the previous formulas for diffraction and amplitudes remain valid for *any* (continuous) function ϑ, not just for those which preserve the Delone property. This is so because the Dirac comb $\delta_{\Lambda_\vartheta}$, unlike the point set Λ_ϑ itself, keeps track of coinciding points via multiple weights.   ◇

Remark 9.13 (*Deformations preserving the model set property*). Among the family of deformed silver mean model sets discussed in Example 9.9, the cases that led to Delone sets with two possible nearest neighbour distances (and thus to a tiling with two prototiles) seemed special. Indeed, one can show that they can be described as model sets again, via a suitable deformation of the embedding lattice. Consequently, these deformed model sets are Meyer sets, and represent a special case of a rather general robustness result; compare [LM08] for more.

For special values of α, the resulting tiling again has a local inflation symmetry. This is sometimes accompanied by extinctions, as mentioned in Example 9.9.   ◇

9.4. Autocorrelation and diffraction of regular model sets

Let us generalise the treatment of the silver mean chain to model sets in $\mathbb{R}^d$ with the more general CPS $(\mathbb{R}^d, H, \mathcal{L})$ of Eq. (7.11) on page 264. In what follows, we assume H to be equipped with the suitably normalised Haar measure μ_H.

Proposition 9.8. *Consider the general CPS $(\mathbb{R}^d, H, \mathcal{L})$ of Eq. (7.11), and let $\Lambda = \lambda(W)$ be a regular model set for it, with a compact window $W = \overline{W^\circ}$. The autocorrelation measure γ_Λ of Λ exists and is a positive and positive definite, translation bounded, pure point measure. It is explicitly given by*

$$\gamma_\Lambda = \sum_{z\in\Lambda-\Lambda} \eta(z)\,\delta_z,$$

with the autocorrelation coefficients

$$\eta(z) = \mathrm{dens}(\Lambda)\,\frac{\mu_H\big(W\cap(z^\star + W)\big)}{\mu_H(W)} = \frac{\mathrm{dens}(\Lambda)}{\mu_H(W)}\int_H 1_W(y)\,1_{z^\star + W}(y)\,\mathrm{d}\mu_H(y).$$

In particular, one has $\eta(0) = \mathrm{dens}(\Lambda)$.

SKETCH OF PROOF. We can step by step mimic the proof of Lemma 9.2. The crucial input is the uniform distribution of $\Lambda^\star$ in W as stated in Theorem 7.2, which implies the claim. $\square$

Note that the integral in the formula for $\eta(z)$ equals the convolution $(1_W * \widetilde{1_W})(z^\star)$. This observation will become useful in Sections 9.5 and 9.6.

Theorem 9.4. *Let $\Lambda = \curlywedge(W)$ be a regular model set for the CPS $(\mathbb{R}^d, H, \mathcal{L})$ of Eq. (7.11), with compact window $W = \overline{W^\circ}$ and autocorrelation γ_Λ according to Proposition 9.8. The diffraction measure $\widehat{\gamma_\Lambda}$ is a positive and positive definite, translation bounded, pure point measure. It is explicitly given by*

$$\widehat{\gamma_\Lambda} = \sum_{k \in L^\circledast} I(k)\, \delta_k,$$

where the diffraction intensities are $I(k) = |A(k)|^2$ with the amplitudes

$$A(k) = \frac{\mathrm{dens}(\Lambda)}{\mu_H(W)}\, \widehat{1_W}(-k^\star) = \frac{\mathrm{dens}(\Lambda)}{\mu_H(W)} \int_W \langle k^\star, y \rangle \, \mathrm{d}\mu_H(y).$$

SKETCH OF PROOF. For direct space $\mathbb{R}^d$ and internal space $H = \mathbb{R}^m$, one can follow the proof of Lemma 9.3 and extend it to this case. Consequently, Theorem 9.3 extends accordingly, which gives the claim for $H = \mathbb{R}^m$.

More generally, one needs the Fourier transform on H, with $\langle \ell, y \rangle$ for $\ell \in \widehat{H}$ denoting the character [Loo11]. One can now show that γ_Λ is a strongly almost periodic measure [BM04], wherefore its Fourier transform is a pure point measure by [GLA90, Thm. 11.2 and Cor. 11.1]. For the Euclidean case, a similar argument is employed in [Gou05]. $\square$

Remark 9.14 (*Diffraction of singular model sets*). In our discussion of the cut and project method, singular model sets (where $\partial W \cap L^\star \neq \varnothing$) played a special role. This is not so for diffraction, as long as the model set is regular. Indeed, the difference between $\curlywedge(\overline{W})$ and $\curlywedge(W^\circ)$ is then a set of density 0, which does not affect the autocorrelation. Consequently, the general diffraction formula holds irrespective of the position of the window relative to $L^\star$, provided that $\mu_H(\partial W) = 0$. $\Diamond$

An alternative proof (for the result in full generality) employs the theory of dynamical systems [Schl00], based on an argument due to Dworkin [Dwo93]. This approach goes back to Hof [Hof95a, Hof97a], who proved the result for Euclidean model sets with a restricted class of windows. In practice, there is often yet another approach via approximating pure point measures that norm converge to $\widehat{\gamma_\Lambda}$, thus giving both the pure point nature and its explicit form. We shall meet examples of this constructive procedure later.

Remark 9.15 (*Spherical approximation of highly symmetric windows*). If
the internal group for a regular model set Λ is Euclidean, with given window
$W \subset H = \mathbb{R}^m$, the amplitude $A(k)$ with $k \in L^{\circledast}$ is given by

$$A(k) \;=\; \frac{\mathrm{dens}(\Lambda)}{\mathrm{vol}(W)} \int_W \mathrm{e}^{2\pi \mathrm{i} k^{\star} y} \,\mathrm{d}y.$$

Since the diffraction intensity is $I(k) = |A(k)|^2$, any (global) phase factor of
$A(k)$ cancels out for $I(k)$, so that we may assume that the centre of mass
of W is the origin. If W is a compact set of high point symmetry (such as
a regular or semi-regular polytope), it is close to a ball of the same volume,
in the sense that the replacement of W by this ball in the definition of $A(k)$
results in a good approximation, paired with a significant simplification of
the computation. The latter claim is due to the resulting spherical symmetry
and the relation with the *Bessel functions* J_{ν} of the first kind [AAR99], called
'ordinary' for integer and 'spherical' for half-integer values of ν.

Indeed, using spherical coordinates and the ball $B_R = B_R(0)$, one finds

$$\frac{1}{\mathrm{vol}(B_R)} \int_{B_R} \mathrm{e}^{2\pi \mathrm{i} k^{\star} y} \,\mathrm{d}y \;=\; \frac{\Gamma(\frac{m}{2}+1)}{(\pi |k^{\star}| R)^{\frac{m}{2}}} \, J_{\frac{m}{2}}(2\pi |k^{\star}| R) \xrightarrow{\;|k^{\star}| \to 0\;} 1,$$

where Γ is the gamma function; compare Example 8.3. The limit property
follows from the asymptotic behaviour of $J_{\nu}(x)$ for small x. Some interesting
special cases, with the abbreviation $z = 2\pi |k^{\star}| R$, are

$$m = 1:\ \mathrm{sinc}(z), \quad m = 2:\ \frac{2 J_1(z)}{z}, \quad m = 3:\ \frac{3\big(\sin(z) - z \cos(z)\big)}{z^3}.$$

The cases with m odd can be expressed in terms of $\sin$ and $\cos$, the others in
terms of J_0 and J_1. The asymptotic behaviour for $|z| \to \infty$ is $\mathcal{O}\big(|z|^{-(m+1)/2}\big)$.

Given a window W, the appropriate radius R for an approximating spherical window derives from the formula for the volume of a ball and reads

$$(9.26) \qquad\qquad R \;=\; \Big(\frac{\mathrm{vol}(W)}{\pi^{\frac{m}{2}}}\, \Gamma(\tfrac{m}{2}+1)\Big)^{\frac{1}{m}},$$

where, for $n \in \mathbb{N}_0$, one has $\Gamma(n+1) = n!$ and $\Gamma\big(n+\tfrac{1}{2}\big) = \frac{(2n)!}{4^n \cdot n!}\sqrt{\pi}$. When used
for a regular dodecagon in the plane, or a triacontahedron in 3-space, the error
for $I(k)$ is of the order of 10^{-3} of the central intensity. The approximation
does not affect the Fourier module and shows the correct position of the
peaks. However, it is generally unsuitable to detect extinctions. $\diamond$

Let us come back to the diffraction amplitudes, where we now extend
Lemma 9.4 to the general case of Euclidean model sets.

Proposition 9.9. *Consider a regular model set $\Lambda = \curlywedge(W)$ for the Euclidean
CPS $(\mathbb{R}^d, \mathbb{R}^m, \mathcal{L})$, with compact window $W = \overline{W^{\circ}} \subset \mathbb{R}^m$ and Fourier module*

$L^\circledast = \pi(\mathcal{L}^*) \subset \mathbb{R}^d$. *Then, one has*

$$\frac{1}{\mathrm{vol}(B_r)} \sum_{x \in \Lambda_r} \mathrm{e}^{-2\pi \mathrm{i} kx} \xrightarrow{r \to \infty} \begin{cases} A(k), & \text{if } k \in L^\circledast, \\ 0, & \text{otherwise,} \end{cases}$$

where $\Lambda_r = \Lambda \cap \overline{B_r(0)}$ and $A(k)$ is the amplitude of Theorem 9.4 for the internal space $H = \mathbb{R}^m$. When $k \notin \mathbb{Q}L^\circledast$, the scalar products kx with $x \in \Lambda$ ordered according to increasing length are uniformly distributed mod 1.

PROOF. For $k \in L^\circledast$, we can copy the proof of Lemma 9.4, which was formulated in sufficient generality to apply here, too. It remains to consider the case $k \notin L^\circledast$, where the claim is a consequence of [Mey70, Prop. 4.5.1]; see [Mey12, Prop. 4.4] for the general model set formulation, or [Hof97a, Thm. 3.1] and [Len09, Thm. 5 and Sec. 7] for an alternative approach via dynamical systems. Two cases can be distinguished here, namely $k \in \mathbb{Q}L^\circledast \setminus L^\circledast$ and $k \notin \mathbb{Q}L^\circledast$. For the former, we give an independent direct argument.

If $k \in \mathbb{Q}L^\circledast \setminus L^\circledast$, define $n = \gcd\{p \in \mathbb{N} \mid pk \in L^\circledast\}$ together with $\kappa = nk$, which is a $\mathbb{Z}$-primitive element of the (free) $\mathbb{Z}$-module $L^\circledast$. Furthermore, consider $L = \pi(\mathcal{L})$ and $M = \frac{1}{n}L$, which are free $\mathbb{Z}$-modules with $[M : L] = n^q$, where q is the rank of L over $\mathbb{Z}$. By the elementary divisors theorem, see [Lan02, Thm. III.7.8], there is an integer $s \in \mathbb{N}$ together with elements $v_i \in M$ and numbers $n_i \in \mathbb{N}$ for $1 \leqslant i \leqslant s$ such that the set $\{n_i v_i \mid 1 \leqslant i \leqslant s\}$ is a $\mathbb{Z}$-basis of L and that $n_i | n_{i+1}$ for all $1 \leqslant i < s$. This means that we can express M as a disjoint union of cosets of L,

$$M = \dot{\bigcup_{j \in J}} (L + v_j),$$

where $\boldsymbol{j} = (j_1, \ldots, j_s)$ is a multi-index, $J = \times_{i=1}^{s} \{0, 1, \ldots, n_i - 1\}$, and $v_{\boldsymbol{j}} = \sum_{i=1}^{s} j_i v_i$. By construction, each n_i is the minimal integer such that $n_i v_i \in L$, and we have $n_1 \cdots n_s = n^q$.

For $x \in \Lambda_r$, we write $kx = \kappa \frac{x}{n}$, where $\frac{x}{n} \in M$ has a unique representation as $\frac{x}{n} = y + v$ with $y \in L$ and $v \in \{v_{\boldsymbol{j}} \mid \boldsymbol{j} \in J\}$. Observing the decomposition

$$\tfrac{1}{n}\Lambda_r = \left\{ \tfrac{x}{n} \mid x \in \Lambda_r \right\} = \dot{\bigcup_{j \in J}} \left\{ y + v_{\boldsymbol{j}} \mid y \in L \cap (\tfrac{1}{n}\Lambda_r - v_{\boldsymbol{j}}) \right\},$$

we can now calculate

$$\sum_{x \in \Lambda_r} \mathrm{e}^{-2\pi \mathrm{i} kx} = \sum_{x \in \Lambda_r} \mathrm{e}^{-2\pi \mathrm{i} \kappa x/n} = \sum_{j \in J} \sum_{y \in L \cap (\frac{1}{n}\Lambda_r - v_{\boldsymbol{j}})} \mathrm{e}^{-2\pi \mathrm{i} \kappa (y + v_{\boldsymbol{j}})}$$

$$= \sum_{j \in J} \mathrm{e}^{-2\pi \mathrm{i} \kappa v_{\boldsymbol{j}}} \sum_{y \in L \cap (\frac{1}{n}\Lambda_r - v_{\boldsymbol{j}})} \mathrm{e}^{-2\pi \mathrm{i} \kappa y}.$$

The inner sum is of the type we have dealt with before in the limit of large r. Here, one has (as $r \to \infty$)

$$\sum_{y \in L \cap (\frac{1}{n}\Lambda_r - v_j)} e^{-2\pi i \kappa y} = \sum_{y^\star \in W \cap (\frac{1}{n}\Lambda_r - v_j)^\star} e^{2\pi i \kappa^\star y^\star}$$

$$\sim \alpha \, \mathrm{vol}(B_r) \, e^{-2\pi i \kappa^\star v_j^\star} \int_{\frac{1}{n}W} e^{2\pi i \kappa^\star z} \, \mathrm{d}z \,,$$

where $\alpha = \mathrm{dens}(\Lambda) / \big(\mathrm{vol}(\frac{1}{n}W)\, n^q\big)$ is the same prefactor for all $j \in J$. Putting everything together, we find

$$\frac{1}{\mathrm{vol}(B_r)} \sum_{x \in \Lambda_r} e^{-2\pi i k x} \sim \alpha \int_{\frac{1}{n}W} e^{2\pi i \kappa^\star z} \, \mathrm{d}z \sum_{j \in J} e^{-2\pi i (\kappa v_j + \kappa^\star v_j^\star)},$$

where the inner sum equals

$$\prod_{i=1}^{s} \sum_{j_i=0}^{n_i-1} \big(e^{-2\pi i (\kappa v_i + \kappa^\star v_i^\star)} \big)^{j_i}.$$

By construction, $\kappa v_i + \kappa^\star v_i^\star = \frac{\ell_i}{n_i}$ with coprime ℓ_i and n_i, so each factor is a sum over a complete set of roots of unity, and hence vanishes.

Finally, if $k \notin \mathbb{Q}L^\circledast$, we can invoke the result from [Mey70, Mey12] for nk with any integer $n \neq 0$, which shows the uniform distribution of the numbers $(kx \bmod 1)_{x \in \Lambda}$ in the unit interval by an application of Weyl's criterion [KN74, Ch. 1, Thm. 2.1]; compare Example 7.1.   □

Note that, by Weyl's lemma, the uniform distribution of $(kx \bmod 1)_{x \in \Lambda}$ gives back

$$\frac{1}{\mathrm{vol}(B_r)} \sum_{x \in \Lambda_r} e^{-2\pi i k x} \xrightarrow{\;r \to \infty\;} \mathrm{dens}(\Lambda) \int_0^1 e^{-2\pi i y} \, \mathrm{d}y = 0,$$

as expected. This means that an independent proof of this uniform distribution property would provide a more direct approach to the exponential sums and their asymptotic behaviour. This is possible via suitable periodic approximants with the summation techniques from lattice exponential sums. For irrational projections in the square lattice, a concrete approach employs the continued fraction expansion of irrational numbers, which permits the control of the error of the estimate. In higher dimensions, one needs an analogous deformation of the projection matrix.

At this point, one can extend the treatment of model sets by suitable deformations, in analogy to the above silver mean example. One class of deformations is discussed in [Hof95a], while we refer to [SW06] for an interesting application to deformed Penrose tilings. For a more general treatment, we refer to [BD00, BL05].

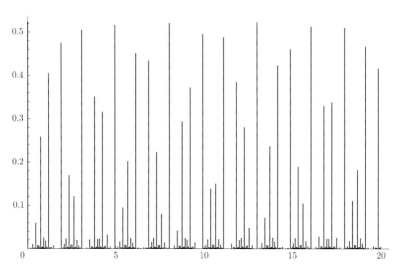

FIGURE 9.5. Diffraction image of the Fibonacci chain for $0 \leqslant k \leqslant 20$. A peak of this pure point measure is represented as a line with a height that equals its intensity. All peaks with at least $1/1000$ of the central intensity $I(0) = (\tau + 1)/5 \approx 0.5236$ are included.

9.4.1. DIFFRACTION OF THE FIBONACCI CHAIN

According to Example 7.3, the Fibonacci chain emerges from the CPS (7.11) with $d = 1$, $H = \mathbb{R}$, $L = \mathbb{Z}[\tau]$ and its Minkowski embedding $\mathcal{L}$ of Figure 3.3. This planar lattice (of density $\operatorname{dens}(\mathcal{L}) = 1/\sqrt{5}$) has basis matrix

$$B = \begin{pmatrix} \tau & 1 \\ 1 - \tau & 1 \end{pmatrix} \quad \text{with dual basis matrix} \quad B^* = \frac{1}{\sqrt{5}} \begin{pmatrix} 1 & \tau - 1 \\ -1 & \tau \end{pmatrix}.$$

One thus has $[\mathcal{L}^* : \mathcal{L}] = [L^\circledast : L] = 5$, and it is easy to verify that $L^\circledast = L/\sqrt{5}$. Since $L^\circledast \subset \mathbb{Q}(\sqrt{5})$, the $\star$-map is well-defined on the Fourier module.

The Fibonacci point set $\varLambda$ was constructed as the disjoint union $\varLambda_a \,\dot\cup\, \varLambda_b$ of the model sets $\varLambda_a = \curlywedge\big((\tau - 2, \tau - 1]\big)$ and $\varLambda_b = \curlywedge\big((-1, \tau - 2]\big)$, with $\operatorname{dens}(\varLambda_a) = \tau \operatorname{dens}(\varLambda_b) = 1/\sqrt{5}$. The corresponding Dirac combs $\delta_{\varLambda_a}$ and $\delta_{\varLambda_b}$ lead to diffraction amplitudes (compare Example 8.3)

$$A_a(k) = \frac{1}{\sqrt{5}} \int_{\tau-2}^{\tau-1} e^{2\pi i k^\star y} \, \mathrm{d}y \quad \text{and} \quad A_b(k) = \frac{1}{\sqrt{5}} \int_{-1}^{\tau-2} e^{2\pi i k^\star y} \, \mathrm{d}y$$

for $k \in L^\circledast$, by an application of Theorem 9.4, and $A_a(k) = A_b(k) = 0$ otherwise. If we start from the Dirac comb $\omega = \alpha \delta_{\varLambda_a} + \beta \delta_{\varLambda_b}$ with (possibly complex) weights α and β, the corresponding diffraction measure reads

$$\widehat{\gamma_\omega} = \sum_{k \in L/\sqrt{5}} \big| \alpha A_a(k) + \beta A_b(k) \big|^2 \, \delta_k.$$

The Dirac comb δ_Λ corresponds to the choice $\alpha = \beta = 1$, which is the Dirac comb of the Fibonacci point set Λ of density $\tau/\sqrt{5} = (\tau + 2)/5$. Here, the intensity for $k \in L/\sqrt{5}$ is given by

$$(9.27) \qquad I(k) \;=\; \big| A_a(k) + A_b(k) \big|^2 \;=\; \left(\frac{\tau}{\sqrt{5}} \operatorname{sinc}\big(\pi\tau k^\star\big) \right)^2,$$

in line with Example 8.3. An illustration is given in Figure 9.5. Note that the analogue of Remark 9.11 applies here as well. In particular, one can clearly see the phenomenon for the peaks at $\tau^i/\sqrt{5}$ with $0 \leqslant i \leqslant 7$, where $\tau^5/\sqrt{5} \approx 4.96$.

Inspecting the right hand side of Eq. (9.27), one notices that the intensity function $I(k)$ vanishes if and only if $\tau k^\star \in \mathbb{Z}\backslash\{0\}$. This corresponds to $k = \ell\tau$ with $\ell \in \mathbb{Z}\backslash\{0\}$. Since all these points are elements of the Fourier module $L/\sqrt{5}$, we have identified all extinctions. The absence of the potential peaks at these points accompanies the exact inflation symmetry of the Fibonacci point set, as in the previous example of the silver mean chain in Remark 9.10.

Let us now, after our detailed exposition of one-dimensional model sets, illustrate how the approach works in higher dimensions.

9.4.2. DIFFRACTION OF THE OCTAGONAL TILING

Using the description as a cyclotomic model set of Example 7.8 from page 273, the diffraction measure of the Dirac comb of the Ammann–Beenker point set Λ_{AB} reads

$$\widehat{\gamma_{\mathrm{AB}}} \;=\; \sum_{k\in\frac{1}{2}\mathbb{Z}[\xi_8]} \big| A(k) \big|^2 \, \delta_k.$$

It comprises point measures supported on the dual module $L^\circledast = \frac{1}{2}\pi(\mathcal{L}_8)$; compare Table 7.1. The amplitudes (or Fourier–Bohr coefficients) are

$$A(k) \;=\; \frac{1}{4}\,\widehat{1_{W_{\mathrm{AB}}}}(-k^\star)$$

because $\mathrm{dens}(\mathcal{L}_8) = \frac{1}{4}$.

For an approximate calculation of the amplitudes, we could have employed Remark 9.15; compare [BG11, Fig. 5] for an illustration. However, as it is not difficult to compute the Fourier transform of the window (the regular octagon W_{AB} of Example 7.8), it is instructive to derive the exact expression for the diffraction intensities for this example. Here we make use of the following observation. When calculating the Fourier transform of a polygon P in the plane at $k = (k_1, k_2)$ with $k_1 k_2 \neq 0$, one can profit from Stokes' theorem via

$$(9.28) \qquad \int_P e^{2\pi\mathrm{i}(k_1 x + k_2 y)} \, \mathrm{d}x\,\mathrm{d}y \;=\; \int_{\partial P} e^{2\pi\mathrm{i}(k_1 x + k_2 y)} \left(\frac{\mathrm{d}y}{4\pi\mathrm{i}k_1} - \frac{\mathrm{d}x}{4\pi\mathrm{i}k_2} \right),$$

where the right-hand side is a line integral over the positively oriented boundary ∂P of the polygon. The integral can be calculated in a suitable parametrisation, and also gives the correct result for $k_1 k_2 = 0$ via suitable limits. This formulation also shows the asymptotic behaviour of the amplitudes for large k. For the Ammann–Beenker tiling, the resulting intensities $I(k) = |A(k)|^2$ for $k \in \frac{1}{2}\mathbb{Z}[\xi_8] \subset \mathbb{Q}(\xi_8)$ are given by

$$
\begin{aligned}
I\big((k_1, k_2)\big) \;=\; & \frac{1}{\big(4\pi^2 (k_2' + k_1')(k_2' - k_1')\big)^2} \Big(\cos(k_2'\pi)\cos(\lambda_{\mathrm{sm}} k_1'\pi) \\
& - \cos(k_1'\pi)\cos(\lambda_{\mathrm{sm}} k_2'\pi) - \frac{k_1'}{k_2'}\sin(k_2'\pi)\sin(\lambda_{\mathrm{sm}} k_1'\pi) \\
& + \frac{k_2'}{k_1'}\sin(k_1'\pi)\sin(\lambda_{\mathrm{sm}} k_2'\pi)\Big)^2
\end{aligned}
$$

(9.29)

with $\lambda_{\mathrm{sm}} = 1 + \sqrt{2}$. Expressing $k \in L^\circledast$ in Cartesian coordinates (k_1, k_2) via $k = k_1 + k_2 \xi_8^2$, where k_1 and k_2 are then elements of $\mathbb{Q}(\sqrt{2})$, the $\star$-map acts as $k^\star = k_1' - k_2' \xi_8^2$, where $'$ denotes the algebraic conjugate in $\mathbb{Q}(\sqrt{2})$. The central intensity satisfies $I(0) = \lambda_{\mathrm{sm}}^2/4 = \mathrm{dens}(\Lambda_{\mathrm{AB}})^2$, in line with Corollary 9.1, and the intensities along the eight high symmetry directions read

$$
\begin{aligned}
I\big((\pm s, 0)\big) \;=\; & I\big((0, \pm s)\big) = I\big(\tfrac{\sqrt{2}}{2}(\pm s, \pm s)\big) \\
\;=\; & \left(\frac{\cos(s'\pi) - \cos(\lambda_{\mathrm{sm}} s'\pi) + s'\pi \sin(\lambda_{\mathrm{sm}} s'\pi)}{4\pi^2 (s')^2} \right)^2
\end{aligned}
$$

as follows by taking the appropriate limits. A central patch of the corresponding diffraction image is shown in Figure 9.6. In contrast to the previous one-dimensional examples, there are no extinctions on $L^\circledast$.

The diffraction measure has exact D_8 symmetry, which reflects the symmetry of the window W_{AB} (and that of $\mathrm{LI}(\Lambda_{\mathrm{AB}})$ as well), and possesses Dirac peaks on a dense point set. Nevertheless, since the peak intensities are locally summable, the positions of peaks with an intensity $I \geqslant \varepsilon > 0$ are supported on a locally finite point set. In fact, the latter is even a Meyer set, for any fixed value of ε; compare Remark 9.12 and [Str13b].

Note that every member of the Ammann–Beenker LI class has the same diffraction pattern, irrespective of its individual symmetry. This corresponds to the fact that the intensities are the same for all members of the LI class, while the amplitudes differ by phase factors. In particular, the diffraction result is not restricted to the representative with individual D_8 symmetry which is shown in Figures 7.4 and 6.41.

Other examples can easily be worked out in a similar manner, such as the diffraction of the tenfold symmetric Tübingen triangle tiling (TTT) or the twelvefold symmetric shield tiling. Both have a single window that is a regular polygon. More generally, one can also treat the rhombic Penrose

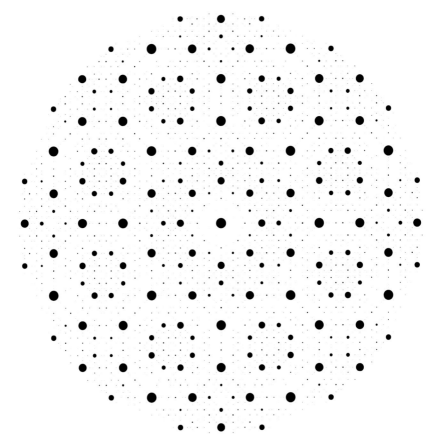

FIGURE 9.6. Diffraction image of the Ammann–Beenker model set Λ_{AB}. A peak of this pure point measure is represented as a disk with an area proportional to its intensity, centred at the peak position. The image has exact D_8 symmetry. The patch shows all peaks with at least $1/10000$ of the central intensity (within the circular area selected). This cut-off is the reason for the discrete appearance; see text for details.

tiling, where one starts from a description of its vertex point set as a four-component model set (meaning that the vertices are the disjoint union of four model sets); compare Example 7.11. In this formulation, the LTM of the tiling is $(1 - \xi)\mathbb{Z}[\xi]$, with $\xi = \xi_5$, so that the Fourier module (which is the corresponding dual module) turns out to be $\frac{2\sqrt{5}}{5}\mathbb{Z}[\xi]$. This follows from Table 7.1 together with the observation that $(1 - \xi)^2$ and $\sqrt{5}$ generate the same principal ideal of $\mathbb{Z}[\xi]$. Various diffraction patterns (in a slightly different, but equivalent formulation) are illustrated in [BKSZ90], wherefore we omit further details.

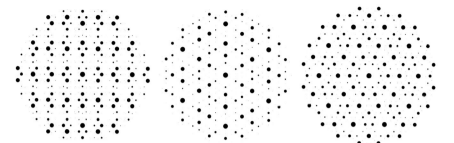

FIGURE 9.7. Planar sections of the three-dimensional, icosahedrally symmetric diffraction pattern of the primitive icosahedral model set of Section 7.4, in the spherical window approximation of Remark 9.15. The three sections are perpendicular to a twofold axis (left), a threefold axis (centre) and a fivefold axis (right), and coincide in the central peak.

9.4.3. DIFFRACTION OF ICOSAHEDRAL MODEL SETS

Here, from the diffraction point of view, the simplest example is the primitive icosahedral model set of Section 7.4. If $L = \mathcal{M}_\mathsf{P}$ is the primitive icosahedral module of Example 2.20, the corresponding Fourier module is

$$(9.30) \qquad L^\circledast = \mathcal{M}_\mathsf{P}^\circledast = \frac{1}{2(\tau + 2)}\,\mathcal{M}_\mathsf{P},$$

which follows from a simple calculation on the basis of Example 2.20 and Eq. (A.2). The general diffraction formula of Theorem 9.4 applies, with $H = \mathbb{R}^3$ and W a centred triacontahedron of edge length $\sqrt{\tau + 2}$, in a suitable orientation (as obtained from the projection of a 6D hypercube; compare [KN84]). The corresponding diffraction measure is a pure point measure concentrated on the Fourier module $L^\circledast$ of Eq. (9.30), which is dense in $\mathbb{R}^3$, and has full icosahedral symmetry. Choosing a small cut-off $\varepsilon > 0$ and restricting to intensities that exceed ε reduces the positions to a discrete set of spots, which show pronounced planar and linear subpatterns.

For our illustration, we employ the approximation of Remark 9.15. Since the volume of our triacontahedron is $20\tau^3$, Eq. (9.26) leads to the radius

$$R = \left(\frac{3\,\mathrm{vol}(W)}{4\pi}\right)^{1/3} = \left(\frac{15}{\pi}\right)^{1/3}\tau \approx 2.7246$$

for the approximating ball. Because the triacontahedral window is well approximated by this sphere, the difference between the approximate and the exact diffraction intensities is tiny, and irrelevant for our purpose. For a full calculation of the Fourier transform of the triacontahedron, we refer to [Els86, App. 2]. Figure 9.7 shows sections through the three-dimensional diffraction patterns orthogonal to the twofold, threefold and fivefold symmetry axes.

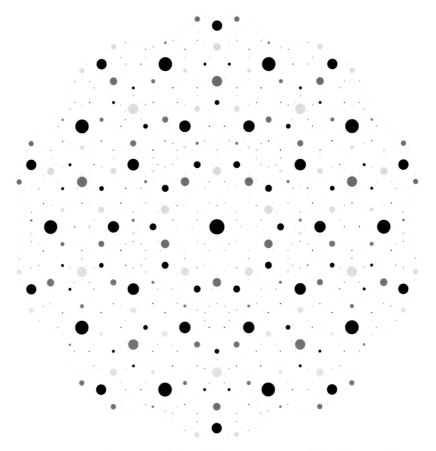

FIGURE 9.8. Illustration of the diffraction pattern of the Danzer tiling
along a fivefold direction. The black spots lie in the planar section through
the origin, and are located at points $k \in \frac{1}{2(\tau+2)} \mathcal{M}_\mathsf{P}$. The grey spots are
located in the coset $\frac{1}{2(\tau+2)} (\mathcal{M}_\mathsf{P} + u)$ and lie in parallel planes that contain
u (dark grey) or $-u$ (light grey); see text for details.

Let us also briefly describe the diffraction of Danzer's ABCK tiling. More
precisely, we focus on the vertices of the $\langle \text{ABCK} \rangle$ tiling, which were described
as a three-component model set in Example 7.13 on page 287. The formula-
tion was based on the module $\mathcal{M}_\mathsf{F}$, wherefore a calculation analogous to that
for Eq. (9.30) leads to the Fourier module

$$(9.31) \qquad \mathcal{M}_\mathsf{F}^\circledast = \frac{1}{2(\tau+2)} \mathcal{M}_\mathsf{B} = \frac{1}{2(\tau+2)} \left(\mathcal{M}_\mathsf{P} \cup (\mathcal{M}_\mathsf{P} + u) \right)$$

with $u = (1, 1, 1)$ as in Example 2.20. If we place a point measure of weight
α_ι on every vertex point of type $\iota \in \{\mathrm{I}, \mathrm{II}, \mathrm{III}\}$, the corresponding diffraction

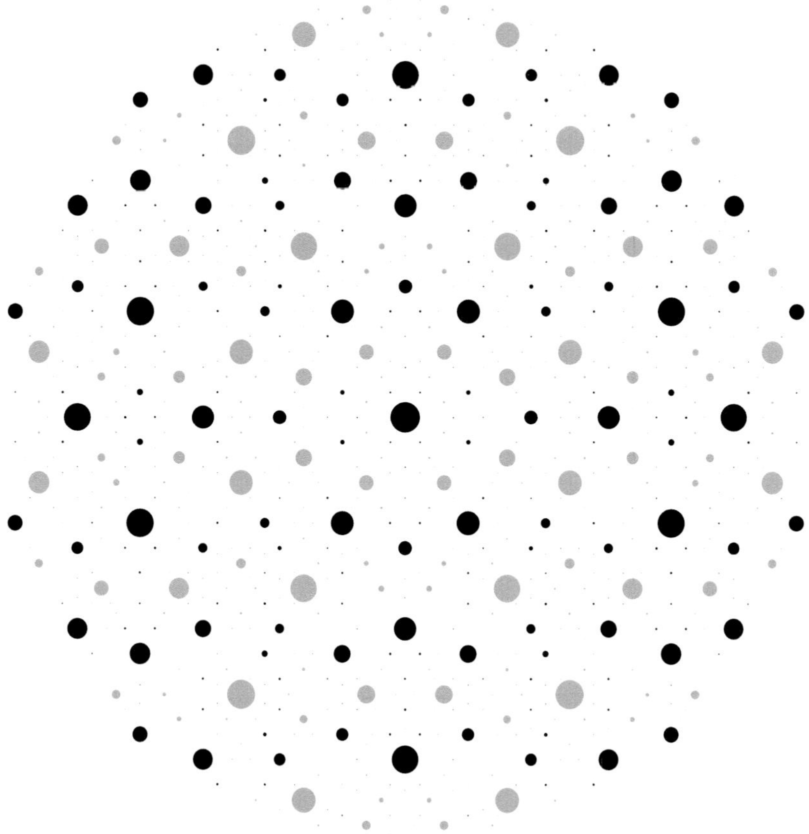

FIGURE 9.9. Illustration of the diffraction pattern of the Danzer tiling along a twofold direction. All spots lie in the planar section through the origin. The black spots belong to $\frac{1}{2(\tau+2)}\mathcal{M}_\mathsf{P}$, while the grey spots belong to the coset $\frac{1}{2(\tau+2)}(\mathcal{M}_\mathsf{P}+u)$. The pattern is reflection symmetric in the horizontal axis.

measure reads

$$(9.32) \qquad \widehat{\gamma} = \sum_{k \in \mathcal{M}_\mathsf{F}^\circledast} \left| \alpha_\mathrm{I} A_\mathrm{I}(k) + \alpha_\mathrm{II} A_\mathrm{II}(k) + \alpha_\mathrm{III} A_\mathrm{III}(k) \right|^2 \delta_k,$$

with the amplitudes $A_\iota(k) = \frac{7-4\tau}{400}\widehat{1_{W_\iota}}(-k^\star)$. For the choice $\alpha_\mathrm{I} = \alpha_\mathrm{III} = 0$ and $\alpha_\mathrm{II} = 1$, the diffraction along a fivefold direction is illustrated in Figure 9.8. As in our previous example, the spherical approximation of Remark 9.15 was used for the dodecahedral window W_II, here with radius $R = \left(\frac{3(\tau+2)}{\pi}\right)^{1/3} \approx 1.5118$.

A fivefold planar section through the origin only contains points from $\frac{1}{2(\tau+2)}\mathcal{M}_{\mathsf{P}}$, but not from the coset. The latter populate parallel planes. Figure 9.8 combines the spots from three different planes. It demonstrates that the overall rotational symmetry is fivefold (not tenfold), while the pattern is also inversion symmetric. The latter property accounts for the tenfold rotation symmetry of the section through the origin (black spots). Sections with threefold symmetry display the analogous phenomena.

In contrast, twofold sections through the origin comprise points from *both* cosets, as illustrated in Figure 9.9. This means that the distinction between the diffraction of a primitive and that of a face-centred icosahedral model set is immediately recognisable from the spot locations in such a twofold section. Further (practical) details and examples can be found in [SD09, Ch. 9]. The careful distinction between primitive (or P-type) and face-centred (or F-type) icosahedral models is important for practical structure analysis, where the latter appear to be prevalent.

All examples so far were Euclidean model sets with internal space $\mathbb{R}^m$, or a simple (finite) extension of such a space. Let us now discuss generalisations to other internal spaces, in particular to 2-adic numbers.

9.4.4. DIFFRACTION OF PERIOD DOUBLING AND PAPER FOLDING

Regular Toeplitz sequences are known to be pure point diffractive, which follows from [JK69, Thm. 6] together with the equivalence of pure point dynamical and pure point diffraction spectrum; see [LMS02, BM04, BL04] for more, and [vEHM92, vEdG11] for their appearance in a related context. However, this does not give the explicit form of the diffraction measure (or how to calculate it). Let us thus complete the picture for two limit-periodic examples by explicitly deriving $\widehat{\gamma}$.

A model set description of the period doubling chain of Section 4.5.1 as a two-colouring of $\mathbb{Z} = \Lambda_a \dot\cup \Lambda_b$ was given in Example 7.4, with the compact Abelian group $H = \mathbb{Z}_2$ of the 2-adic integers as internal space. Then, the general formulas for the autocorrelation coefficients (Proposition 9.8) and the diffraction intensities (Theorem 9.4) apply. This requires some techniques from 2-adic analysis, which would also establish the connection to the solenoid Σ_2 from Example 2.11; see [Sie03, BS07] for related results. In particular, one can show [BLM07] that the continuous hull of the period doubling chain is an almost everywhere $1 : 1$ cover of Σ_2.

Alternatively, since the period doubling chain is a limit periodic structure, one can replace the 2-adic approach by working along a converging sequence of periodic approximants [BM04]. The Fourier transforms of the latter can be calculated via the PSF. The norm convergence of the sequence

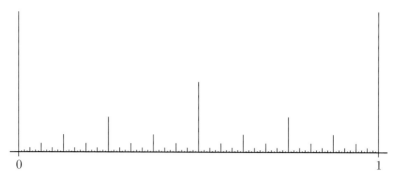

FIGURE 9.10. Sketch of the diffraction amplitudes $|A(k)|$ of Eq. (9.34) for the period doubling chain, with $k \in [0,1]$. The pattern, as well as that of the corresponding intensities, repeats $\mathbb{Z}$-periodically along the real line.

of approximating diffraction measures provides an independent proof of the pure point nature of the limit, by an application of Theorem 8.4.

Starting from Eq. (4.9) on page 96 and terminating the infinite unions at a finite integer N leads to two periodic point sets (neither of which contains the point -1) with Dirac combs $\omega_a^{(N)}$ and $\omega_b^{(N)}$. Their Fourier transforms follow directly from the PSF of Proposition 9.4,

$$\widehat{\omega_a^{(N)}} = \sum_{n=0}^{N} \frac{e^{-2\pi i k(4^n-1)}}{2 \cdot 4^n} \delta_{\mathbb{Z}/2 \cdot 4^n} \quad \text{and} \quad \widehat{\omega_b^{(N)}} = \sum_{n=1}^{N} \frac{e^{-2\pi i k(2 \cdot 4^{n-1}-1)}}{4^n} \delta_{\mathbb{Z}/4^n}.$$

All these distributions are of the form $f \delta_{L^\circledast}$, with the dense point set

$$\begin{aligned}
(9.33) \qquad L^\circledast &= \bigcup_{\ell \geqslant 1} \frac{\mathbb{Z}}{2^\ell} = \mathbb{Z} \,\dot\cup\, \bigcup_{\ell \geqslant 1} \frac{2\mathbb{Z}+1}{2^\ell} = \mathbb{Z}[\tfrac{1}{2}] \\
&= \left\{ \frac{m}{2^r} \mid (r=0,\, m \in \mathbb{Z}) \text{ or } (r \geqslant 1,\, m \text{ odd}) \right\},
\end{aligned}$$

which (in the second line) is written such that pairs (r,m) label the elements uniquely. The module $\mathbb{Z}[\tfrac{1}{2}]$ was mentioned in the context of almost periodic functions in Remark 8.2. With this explicit parametrisation, we can write

$$\widehat{\omega_a^{(N)}} = \sum_{k \in L^\circledast} A^{(N)}(k)\, \delta_k \quad \text{and} \quad \widehat{\omega_b^{(N)}} = \sum_{k \in L^\circledast} B^{(N)}(k)\, \delta_k$$

with amplitudes $A^{(N)}(k)$ and $B^{(N)}(k)$. As $N \to \infty$, they converge to

$$(9.34) \qquad A(k) = \frac{2}{3} \frac{(-1)^r}{2^r} e^{2\pi i k} \quad \text{and} \quad B(k) = \delta_{r,0} - A(k)$$

for any $k = \frac{m}{2^r} \in L^\circledast$; see [BM04] for details. This can be verified by standard geometric series identities. The amplitudes $A(k)$ and $B(k)$ are $\mathbb{Z}$-periodic. The values of $|A(k)|$ in the unit interval are shown in Figure 9.10.

The linear combination $w^{(N)} = \alpha w_a^{(N)} + \beta w_b^{(N)}$, with (possibly complex) weights α and β, can now be understood as a periodic approximant of the period doubling chain. Note that $\big(\widehat{w^{(N)}}\big)_{N \in \mathbb{N}}$ is not convergent as a sequence of measures (because $|\widehat{w^{(N)}}|(K)$ diverges for any compact set with non-empty interior). However, $w^{(N)}$ is still a periodic measure, with diffraction measure

$$\widehat{\gamma_{w^{(N)}}} = \sum_{k \in L^{\circledast}} \big| \alpha A^{(N)}(k) + \beta B^{(N)}(k) \big|^2 \delta_k$$

$$\xrightarrow{N \to \infty} \sum_{k \in L^{\circledast}} \big| \alpha A(k) + \beta B(k) \big|^2 \delta_k = \widehat{\gamma_w},$$

with convergence in the norm topology (which easily follows from the observation that each of the functions $|A(k)|^2$, $|A(k)B(k)|$ and $|B(k)|^2$ is summable over all $k \in [0,1] \cap L^{\circledast}$ by a geometric series calculation). By Theorem 8.4, the limit is thus a pure point measure. It is the diffraction measure $\widehat{\gamma_w}$ of the period doubling chain with weights α and β. Consequently, $L^{\circledast}$ is indeed the dual module of the CPS for the period doubling chain. When $\alpha = \beta$, the diffraction measure reduces to $\widehat{\gamma_w} = |\alpha|^2 \delta_{\mathbb{Z}}$, in line with $w = \alpha \delta_{\mathbb{Z}}$ in this case.

The autocorrelation can be derived by inverse Fourier transform of $\widehat{\gamma_w}$. Choosing $\alpha = 1$ and $\beta = 0$ leads to the diffraction measure $\widehat{\gamma_a}$ (only a-type points occupied). Then, one obtains [BM04] the corresponding autocorrelation $\gamma_a = \sum_{m \in \mathbb{Z}} \eta_a(m) \delta_m$ with the coefficients

$$(9.35) \qquad \eta_a(m) = \frac{2}{3} \cdot \begin{cases} 1, & \text{if } m = 0, \\ 1 - \frac{1}{2^{r+1}}, & \text{if } m = (2\ell + 1)2^r, \ r \geqslant 0, \ \ell \in \mathbb{Z}. \end{cases}$$

The expression for $\eta_a(0)$ coincides with the relative frequency of the letter a in the period doubling fixed point (it alternatively follows from the explicit calculation in Remark 9.16). To derive the formula for $m \neq 0$, one first observes that

$$\big(\delta_{(2\mathbb{Z}+1)/2^s}\big)^{\vee} = \mathrm{e}^{2\pi \mathrm{i} x/2^s} 2^{s-1} \delta_{2^{s-1}\mathbb{Z}}$$

for $s \geqslant 1$, with the first term on the right understood as a density function for the Dirac comb, written as a function of the variable x. Using the parametrisation $m = (2\ell + 1)2^r$ as above and collecting the contributions to $\eta_a(m)$ gives Eq. (9.35) by an application of standard geometric series identities.

Remark 9.16 (*Direct approach via substitution*)**.** It is also possible to compute the autocorrelation coefficients directly via the substitution rule. Consider the one-sided fixed point with $w_0 = 1$, $w_{2n} = 1$ and $w_{2n+1} = -w_n$ for $n \geqslant 0$, which corresponds to the choice $\alpha = 1$ and $\beta = -1$. Observe that 1 is twice as frequent as -1 in w, so that $\frac{1}{N} \sum_{i=0}^{N-1} w_i \xrightarrow{N \to \infty} \frac{1}{3}$. The recursion for w now implies a recursion for the corresponding coefficients η_{o}, namely

$$\eta_{\mathrm{o}}(2m) = \frac{1}{2}\bigl(1 + \eta_{\mathrm{o}}(m)\bigr) \quad \text{and} \quad \eta_{\mathrm{o}}(2m+1) = -\frac{1}{3},$$

which is valid for all $m \geqslant 0$. In particular, $\eta_{\mathrm{o}}(0) = 1$, and we also have $\eta_{\mathrm{o}}(-m) = \eta_{\mathrm{o}}(m)$. When $m = 2^r(2\ell+1)$ with $r, \ell \geqslant 0$, the recursion leads to $\eta_{\mathrm{o}}(m) = 1 - 1/(3 \cdot 2^{r-2})$, which is independent of ℓ.

Replacing 1 and -1 by general (real) weights α and β means to consider the sequence $\frac{1}{2}\bigl((\alpha+\beta)\mathbf{1} + (\alpha-\beta)w\bigr)$, which results in

$$\eta = \tfrac{1}{4}(\alpha-\beta)^2 \eta_{\mathrm{o}} + \tfrac{1}{12}\bigl(3(\alpha+\beta)^2 + 2(\alpha^2 - \beta^2)\bigr)\mathbf{1}.$$

The case $\alpha = 1$, $\beta = 0$ brings us back to Eq. (9.35). This can alternatively be calculated via the Eberlein convolution of Section 8.8. Indeed, if $\omega_{\mathrm{o}} = w\delta_{\mathbb{Z}}$, one finds $\delta_{\mathbb{Z}} \circledast \omega_{\mathrm{o}} = \delta_{\mathbb{Z}} \circledast \widetilde{\omega_{\mathrm{o}}} = \frac{1}{3}\delta_{\mathbb{Z}}$, wherefore $\omega = \frac{1}{2}(\alpha+\beta)\delta_{\mathbb{Z}} + \frac{1}{2}(\alpha-\beta)\omega_{\mathrm{o}}$ leads to the autocorrelation $\gamma = \omega \circledast \widetilde{\omega} = \eta\,\delta_{\mathbb{Z}}$. $\quad\diamond$

Let us also derive the remaining autocorrelation coefficients in our original approach. If $\widehat{\gamma_b}$ denotes the diffraction measure for the choice $\alpha = 0$, $\beta = 1$ (only b-type points occupied), one has the relation

$$\widehat{\gamma_b} = \widehat{\gamma_a} - \frac{1}{3}\delta_{\mathbb{Z}}$$

which immediately implies $\eta_b(m) = \eta_a(m) - \frac{1}{3}$ for all $m \in \mathbb{Z}$. It remains to calculate the cross-correlation coefficients $\eta_{ab}(m)$ and $\eta_{ba}(m)$. Since the choice $\alpha = \beta = 1$ corresponds to $\widehat{\gamma_\omega} = \delta_{\mathbb{Z}}$, the coefficients satisfy

$$\eta_a(m) + \eta_b(m) + \eta_{ab}(m) + \eta_{ba}(m) = 1$$

for all $m \in \mathbb{Z}$. By Proposition 4.8 from page 97, the period doubling chain is reflection symmetric, which implies that the frequency of any finite subword equals that of its reversed word. Consequently, the mixed autocorrelation coefficients must be equal for all m, which implies [BM04]

$$\eta_{ab}(m) = \eta_{ba}(m) = \frac{2}{3} - \eta_a(m),$$

again for all $m \in \mathbb{Z}$.

Finally, the autocorrelation coefficients $\eta(m)$ for the general period doubling Dirac comb $\omega = \alpha\omega_a + \beta\omega_b$ are given by

(9.36)
$$\begin{aligned}
\eta(m) &= |\alpha|^2 \eta_a(m) + |\beta|^2 \eta_b(m) + 2\,\mathrm{Re}(\alpha\bar{\beta})\eta_{ab}(m) \\
&= |\alpha-\beta|^2 \eta_a(m) - \frac{1}{3}|\beta|^2 + \frac{4}{3}\,\mathrm{Re}(\alpha\bar{\beta}),
\end{aligned}$$

where the above relations between the various coefficients were used. This expression properly extends the previous cases, which emerge via the appropriate choices of α and β.

Remark 9.17 (*Construction of internal space*). The autocorrelation coefficients $\eta_a(m)$ of Eq. (9.35) lead to a pseudo-metric d_a on $\mathbb{Z}$ via

$$\mathrm{d}_a(m,0) \,=\, \left|1 - \frac{\eta_a(m)}{\eta_a(0)}\right|^{\frac{1}{2}} = \begin{cases} 0, & \text{if } m = 0, \\ 2^{-(r+1)/2}, & \text{if } m = (2\ell+1)2^r, \end{cases}$$

again with $r \geqslant 0$ and $\ell \in \mathbb{Z}$. This pseudo-metric defines the 2-adic topology on $\mathbb{Z}$. The completion of $\mathbb{Z}$, which is the support of the autocorrelation γ, in this topology is $\mathbb{Z}_2$, the compact Abelian group we have used as internal space in the model set description of the period doubling chain in Example 7.4. For the general mechanism to construct the internal space from the autocorrelation measure, we refer to [BM04]. $\diamond$

Let us now briefly describe the diffraction of the paper folding chain, as introduced in Section 4.5.2; see also [BMRS03]. More precisely, we determine the amplitudes for the Dirac combs δ_{Λ_j} with $j \in \{0,1\}$ and the point sets Λ_j from Eq. (4.12) on page 98. The correct Fourier module is once again $L^\circledast$ as defined in Eq. (9.33) for the period doubling case. The amplitudes are

$$A_j(k) \,=\, \begin{cases} \frac{1}{2}, & \text{if } k \in \mathbb{Z}, \\ 0, & \text{if } k \in \frac{1}{2}(2\mathbb{Z}+1), \\ \frac{\mathrm{i}^{m(2j+1)}}{2^r}\,\mathrm{e}^{\pi\mathrm{i}m/2^{r-1}}, & \text{if } k = \frac{m}{2^r} \text{ with } m \text{ odd and } r \geqslant 2. \end{cases}$$

In particular, one has $A_0(k)+A_1(k) = 1$ for $k \in \mathbb{Z}$, while the sum vanishes for any other k. This is in agreement with $\Lambda_0 \,\dot{\cup}\, \Lambda_1 = \mathbb{Z}$, because $\widehat{\delta_\mathbb{Z}} = \delta_\mathbb{Z}$. The weighted Dirac comb $\omega = \alpha\delta_{\Lambda_0} + \beta\delta_{\Lambda_1}$ now has the diffraction measure

$$(9.37) \qquad\qquad \widehat{\gamma_\omega} \,=\, \sum_{k \in L^\circledast} \bigl|\alpha A_0(k) + \beta A_1(k)\bigr|^2 \delta_k,$$

again obtained as the limit of a norm-converging sequence of periodic pure point measures, wherefore it is pure point as well. All further quantities can be calculated in complete analogy to the period doubling case; compare [AMF95, Part IV.6] for an alternative derivation.

Remark 9.18 (*Autocorrelation coefficients for $\alpha = 1$, $\beta = -1$*). The strategy of Remark 9.16 can also be applied here. The one-sided fixed point (with weights ± 1) satisfies $w_{2n} = (-1)^n$ and $w_{2n+1} = w_n$, for all $n \geqslant 0$. To proceed, we observe $\frac{1}{N}\sum_{n=0}^{N-1}(-1)^n w_n \xrightarrow{N \to \infty} 0$, which follows from an application of the ergodic theorem to the underlying four-letter substitution rule, via the map $a, d \mapsto 1$ and $b, c \mapsto -1$. The limit is 0 because the four letters are equally frequent, as follows immediately from the substitution of Eq. (4.10).

The resulting recursion for the autocorrelation coefficients reads $\eta(0) = 1$, $\eta(2m+1) = 0$ and $\eta(2m) = \frac{1}{2}\bigl((-1)^m + \eta(m)\bigr)$ for $m \geqslant 0$ (together with

$\eta(-m) = \eta(m)$ as usual). When $m = 2^r(2\ell + 1)$, with $r \geqslant 1$ and $\ell \geqslant 0$, this gives $\eta(m) = 1 - 3/2^r$. The corresponding diffraction measure is

$$\hat{\gamma} = \sum_{r=1}^{\infty} \frac{1}{4^r} \delta_{(2\mathbb{Z}+1)/2^{r+1}},$$

which is what Eq. (9.37) reduces to for $\alpha = 1$ and $\beta = -1$. $\Diamond$

Analogous methods work for other limit periodic systems as well. Let us demonstrate this by an example in two dimensions from Section 6.4.

9.4.5. DIFFRACTION OF THE CHAIR TILING

For an explicit calculation of the chair tiling diffraction, it is advantageous to employ the coordinatisation that emerged from the block substitution discussed in Section 4.9. In particular, Eq. (4.30) shows the limit-periodic structure of the four constituting point sets $\Lambda_i \subset \mathbb{Z}^2$. We recall $\Gamma_\pm = \big\{(m,n) \in \mathbb{Z}^2 \mid m + n \overset{\text{even}}{\underset{\text{odd}}{}}\big\}$ and the relations

$$\Lambda_0 \,\dot\cup\, \Lambda_2 = \Gamma_+, \quad \Lambda_1 \,\dot\cup\, \Lambda_3 = \Gamma_- \quad \text{and} \quad \Gamma_+ \,\dot\cup\, \Gamma_- = \mathbb{Z}^2.$$

Here, we aim at calculating the diffraction measure of the Dirac comb

$$(9.38) \qquad \omega = \sum_{i=0}^{3} \alpha_i \, \delta_{\Lambda_i}$$

with the four point sets Λ_i from Eq. (4.30) on page 124, and arbitrary complex numbers α_i. Again, we make use of periodic approximants and the formulas for lattice periodic measures, employing the PSF of Theorem 9.1.

To do so, define $Q_r(x) := 2^{r+1}\Gamma_- + S_r x$ for $r \in \mathbb{N}_0$ and $x \in \mathbb{Z}^2$, with the sets $S_r = \{0, 1, \dots, 2^r-1\}$ as in Eq. (4.30). Recalling $\Gamma_- = \Gamma_+ + u$ with $u = (1,0)$ from Section 4.9 and observing $\Gamma_+^* = \tfrac{1}{2}\Gamma_+$, one can calculate

$$(9.39) \qquad \begin{aligned}
\widehat{\delta_{Q_r(x)}} &= \sum_{m=0}^{2^r-1} e^{-2\pi i m k x} \left(\delta_{2^{r+1}(\Gamma_+ + u)}\right)^{\widehat{}} \\
&= \frac{1 - e^{-2^{r+1}\pi i k x}}{1 - e^{-2\pi i k x}} \, e^{-2^{r+2}\pi i k u} \, \widehat{\delta_{2^{r+1}\Gamma_+}} \\
&= \frac{1 - e^{-2^{r+1}\pi i k x}}{1 - e^{-2\pi i k x}} \, \frac{e^{-2^{r+2}\pi i k u}}{2^{2r+3}} \, \delta_{\Gamma_+/2^{r+2}},
\end{aligned}$$

where we again write density factors as functions of k. Note that the first fraction evaluates as 2^r whenever $kx \in \mathbb{Z}$. The resulting Fourier module is

$$L^{\circledast} = \bigcup_{\ell \geqslant 2} \frac{\Gamma_+}{2^\ell} = \mathbb{Z}^2 \,\dot\cup\, \bigcup_{\ell \geqslant 1} \left\{ \frac{1}{2^\ell}(m,n) \mid m, n \in \mathbb{Z} \text{ with } \gcd(m,n,2) = 1 \right\}.$$

The diffraction measure of our weighted Dirac comb ω is of the form

$$(9.40) \qquad \widehat{\gamma_\omega} = \sum_{k \in L^\circledast} \left| \sum_{i=0}^3 \alpha_i A_i(k) \right|^2 \delta_k,$$

where $A_i(k)$ is the amplitude that belongs to Λ_i, based on the same kind of convergence result as in our previous examples.

The amplitudes can be calculated in the same way as for the period doubling and paper folding sequences above. For any $k \in L^\circledast$, there is a minimal $s \geqslant 0$ such that $k \in \Gamma_+/2^{s+2}$, and then $k \in \Gamma_+/2^{r+2}$ for all $r \geqslant s$. The amplitudes result from summing the corresponding contributions from Eq. (9.39), with $x = u + v$ for $i = 0$, $x = u - v$ for $i = 1$, $x = -u - v$ for $i = 2$, and $x = -u + v$ for $i = 3$, where $u = (1, 0)$ and $v = (0, 1)$. In addition, the results have to be multiplied by an overall phase factor that reflects the translation vectors on the left-hand sides of the four relations in Eq. (4.30). We omit the detailed calculations and simply state the result.

To this end, the integer and half-integer points of $L^\circledast$ are better treated separately. Whenever $k \in \mathbb{Z}^2$,

$$A_0(k) = A_1(k) = A_2(k) = A_3(k) = \frac{1}{4}.$$

If $k \in \frac{1}{2}\Gamma_+ \setminus \mathbb{Z}^2$, one has $k = \frac{1}{2}(m, n)$ with m and n odd, so that $kx \in \mathbb{Z}$. This gives

$$A_0(k) = A_2(k) = \frac{1}{4} \quad \text{and} \quad A_1(k) = A_3(k) = -\frac{1}{4},$$

while $k = \frac{1}{2}(m, n)$ with $m + n$ odd results in

$$A_0(k) = \frac{1}{8}, \quad A_2(k) = -\frac{1}{8}, \quad A_1(k) = \frac{(-1)^n}{8}, \quad \text{and} \quad A_3(k) = \frac{(-1)^m}{8}.$$

In particular, $A_0(k) + A_2(k) = 0$, and also $A_1(k) + A_3(k) = 0$ as $m + n$ is an odd integer here.

The remaining elements of $L^\circledast$ are of the form $k = (m, n)/2^\ell$ with $\ell \geqslant 2$ and $m, n \in \mathbb{Z}$ subject to the condition $\gcd(m, n, 2) = 1$. The amplitudes $A_0(k)$ and $A_2(k)$ depend on $m + n$. They satisfy $A_2(k) = -A_0(k)$ and

$$A_0(k) = \begin{cases} 0, & \text{if } 2^\ell \mid (m + n), \\[2ex] -\dfrac{2}{4^\ell} \dfrac{1 - (-1)^{(m+n)/2}}{1 - \varepsilon_\ell^{m+n}}, & \text{if } m + n \text{ even with } 2^\ell \nmid (m + n), \\[2ex] \dfrac{1}{4^\ell} \dfrac{1}{1 - \varepsilon_\ell^{m+n}}, & \text{if } m + n \text{ odd}, \end{cases}$$

FIGURE 9.11. Graphical representation of the D_4-symmetric diffraction measure for a 'balanced' version of the chair tiling, realised as a weighted decoration of $\mathbb{Z}^2$; see text for details. The patch shows the peaks with $k \in [-1,1]^2$, where $k = (0,0)$ is the centre of figure. Here, the Bragg peaks with $k \in \frac{1}{2}\Gamma_+$ are extinct, and the complete pattern is $\frac{1}{2}\Gamma_+$-periodic.

where $\varepsilon_\ell = \exp(-2\pi \mathrm{i}/2^\ell)$ is a root of unity. Similarly, one obtains the relation $A_3(k) = -A_1(k)$ together with

$$
A_1(k) = \varepsilon_\ell^{-n}
\begin{cases}
0, & \text{if } 2^\ell \mid (m-n), \\
-\dfrac{2}{4^\ell} \dfrac{1 - (-1)^{(m-n)/2}}{1 - \varepsilon_\ell^{m-n}}, & \text{if } m - n \text{ even with } 2^\ell \nmid (m-n), \\
\dfrac{1}{4^\ell} \dfrac{1}{1 - \varepsilon_\ell^{m-n}}, & \text{if } m - n \text{ odd.}
\end{cases}
$$

In particular, it is now easy to check that $\alpha_0 = \alpha_1 = \alpha_2 = \alpha_3 = 1$ leads to $\widehat{\gamma_\omega} = \delta_{\mathbb{Z}^2}$, as required. Moreover, setting $\alpha_0 = \alpha_2 = 1$ and $\alpha_1 = \alpha_3 = 0$ gives the diffraction $\widehat{\gamma_\omega} = \widehat{\delta_{\Gamma_+}} = \frac{1}{2}\delta_{\Gamma_+/2}$, while the alternative choice $\alpha_0 = \alpha_2 = 0$ together with $\alpha_1 = \alpha_3 = 1$ yields $\widehat{\gamma_\omega} = \widehat{\delta_{\Gamma_-}} = e^{-2\pi i k u} \widehat{\delta_{\Gamma_+}} = \frac{(-1)^m}{2}\delta_{\Gamma_+/2}$, where the last step uses the parametrisation of k as above.

An interesting case results from the choice $\alpha_j = i^j$, which produces extinctions for all $k \in \frac{1}{2}\Gamma_+$, as follows from our above calculations. The corresponding diffraction measure is illustrated in Figure 9.11. As in our previous figures, the intensity of a peak is proportional to the area of the disk centred at the peak position. This choice of weights results in a D_4-symmetric diffraction image, which is not the case in general; compare Eq. (9.40).

For a generic choice of the weights, our Dirac comb in Eq. (9.38) is MLD with the chair tiling. As it is based on four subsets of $\mathbb{Z}^2$, the diffraction measure $\widehat{\gamma_\omega}$ is always $\mathbb{Z}^2$-periodic. This is the generic maximal translation symmetry. However, since $\Lambda_0 \cup \Lambda_2 = \Gamma_+$, the diffraction of any weighted Dirac comb with these two components only is periodic under the dual lattice $\Gamma_+^* = \frac{1}{2}\Gamma_+$. Likewise, the corresponding statement holds for the diffraction of weighted Dirac combs with components Λ_1 and Λ_3. This is also the translation symmetry in Figure 9.11, which is a consequence of the particular choice of the weights in that example.

Note that any choice of the weights refers to a discrete structure that can be seen as a decoration of the chair tiling. In the converse direction, however, a typical decoration of the chair tiling will be MLD to our coloured point set, but need not be realisable as a simple decoration of the square-shaped building blocks. In this sense, our above results constitute only a first step of the complete diffraction analysis of the chair tiling. A similar analysis (which we omit here) is also possible for the pure point factor of the table tiling from Remark 4.17.

9.5. Pure point diffraction of weighted Dirac combs

So far, we have mainly concentrated on model sets and the diffraction of the associated Dirac combs with uniform weights. The methods employed in the proofs of Proposition 9.8 and Theorem 9.4 are powerful enough to accommodate a more general setting. In particular, one can treat weighted Dirac combs on model sets (provided that the weight function is sufficiently 'nice' [BM04]), in complete analogy to our discussion of the silver mean chain. Moreover, one can even extend the results to an interesting class of dense Dirac combs.

9.5.1. DIFFRACTION OF WEIGHTED MODEL SETS

Let $\Lambda = \curlywedge(W)$ be a regular model set for the general CPS of Eq. (7.11) on page 264, with compact window $W = \overline{W^\circ}$. Let $g \colon W \longrightarrow \mathbb{C}$ be a continuous function and define the *weighted Dirac comb*

$$(9.41) \qquad \omega_g = \sum_{x \in \Lambda} g(x^\star)\, \delta_x.$$

Since Λ is uniformly discrete and g is continuous on W and hence bounded, ω_g is a translation bounded measure. The uniform distribution of $\Lambda^\star$ in W is the key to the determination of the corresponding autocorrelation measure. Repeating the proof of Proposition 9.8 with (obvious) minor modifications establishes the following result.

Proposition 9.10. *Let ω_g of Eq. (9.41) be a weighted Dirac comb for a regular model set $\Lambda = \lambda(W)$ with the CPS $(\mathbb{R}^d, H, \mathcal{L})$ of Eq. (7.11) and compact window $W = \overline{W^\circ}$, with a function $g \colon H \longrightarrow \mathbb{C}$ which is continuous on W and vanishes on its complement. Then, the natural autocorrelation γ of ω_g exists and is given by*

$$\gamma = \sum_{z \in \Lambda - \Lambda} \eta(z)\, \delta_z \quad with \quad \eta(z) = \frac{\mathrm{dens}(\Lambda)}{\mu_H(W)}\,(g * \widetilde{g})(z^\star),$$

where μ_H is the Haar measure on H. □

As mentioned earlier in the silver mean example, the uniform distribution implies that $\mathrm{dens}(\Lambda) = \mu_H(W)\,\mathrm{dens}(\mathcal{L})$, so that we also have

$$\eta(z) = \mathrm{dens}(\mathcal{L})\,(g * \widetilde{g})(z^\star).$$

It is now clear that Theorem 9.4 generalises accordingly.

Theorem 9.5. *Under the assumptions of Proposition 9.10, the weighted Dirac comb of Eq. (9.41) has the positive, translation bounded diffraction measure*

$$\widehat{\gamma} = \sum_{k \in L^\circledast} |A(k)|^2\, \delta_k \quad with \quad A(k) = \mathrm{dens}(\mathcal{L})\,\widehat{g}(-k^\star),$$

where $L^\circledast = \pi(\mathcal{L}^*)$ is the Fourier module of the CPS of Eq. (7.11). □

Let us briefly mention that the continuity condition on g can be relaxed, for instance to a suitable notion of piecewise continuity. Note that some of our previous examples (such as period doubling, paper folding and chair) can be re-interpreted as weighted Dirac combs on a single model set (with finitely many distinct weights), which would give an alternative description to our direct approach. We leave it to the reader to reconsider other examples (such as the Fibonacci or the silver mean chain, or the vertex set of the rhombic Penrose tiling) with different weights on the different classes of points.

9.5.2. Dense Dirac combs in Euclidean space

Inspecting the diffraction formula of Theorem 9.5, it is tempting to assume that the same formula (with the corresponding autocorrelation) should also hold if g is no longer required to have compact support, but has a sufficiently rapid asymptotic decay instead. This would have important consequences for the theory of random tilings in three or more dimensions, and would also lead to a more symmetric model set analogue of the classic PSF. One has to start with a Dirac comb of the form

$$(9.42) \qquad \omega_g = \sum_{x \in L} g(x^\star)\, \delta_x$$

in analogy to Eq. (9.41), where we have used that $L - L = L$. An immediate obstacle is to give this object a proper meaning as a (translation bounded) measure and to extend the uniform distribution argument to this case. This was achieved in an important paper by Richard [Ric03] for the symmetric Euclidean CPS

$$\mathbb{R}^d \quad \xleftarrow{\ \pi\ } \quad \mathbb{R}^d \times \mathbb{R}^m \quad \xrightarrow{\ \pi_{\mathrm{int}}\ } \quad \mathbb{R}^m$$

$$\text{dense } \cup \qquad\qquad \cup \qquad\qquad \cup \text{ dense}$$

$$(9.43) \qquad \pi(\mathcal{L}) \quad \xleftarrow{\ 1-1\ } \quad \mathcal{L} \quad \xrightarrow{\ 1-1\ } \quad \pi_{\mathrm{int}}(\mathcal{L})$$

$$\| \qquad\qquad\qquad\qquad\qquad\qquad \|$$

$$L \quad \xleftrightarrow{\qquad\qquad \star \qquad\qquad} \quad L^\star$$

Note that the denseness and bijectivity conditions appear symmetrically, so that one can interchange the roles of direct and internal space. In particular, the $\star$-map is a bijection in this setting. It is precisely this symmetry that can be used to formulate a sufficient criterion for the measure property of ω_g.

Lemma 9.5. [Ric03, Thm. 6]. *Let $g \colon \mathbb{R}^m \longrightarrow \mathbb{C}$ be a continuous function that is $\mathcal{O}\bigl(|x|^{-m-1-\alpha}\bigr)$ for some $\alpha > 0$ as $|x| \to \infty$. Then, the weighted Dirac comb ω_g of Eq. (9.42) for the CPS of Eq. (9.43) defines a translation bounded complex measure on $\mathbb{R}^d$.* $\qquad\square$

To continue, we need the autocorrelation of the Dirac comb ω_g.

Proposition 9.11. *Under the assumptions of Lemma 9.5, the Dirac comb ω_g has the unique autocorrelation $\gamma = \sum_{z \in L} \eta(z)\, \delta_z$ with the autocorrelation coefficients $\eta(z) = \mathrm{dens}(\mathcal{L})\, \bigl(g * \tilde{g}\bigr)(z^\star)$. The measure γ is a translation bounded and positive definite pure point measure.*

SKETCH OF PROOF. The asymptotic condition on g is strong enough to work along the previous line of proof, provided we can still rely on Weyl's uniform distribution theorem, which gives the formula for $\eta(z)$. This is proved in [Ric03, Thm. 7]; compare the proof of [Ric03, Thm. 9] for the details of the somewhat technical steps involved. The underlying idea employs a sequence of functions g_n that are restrictions of g to compact regions, so that one can use Proposition 9.10 together with a convergence estimate.

The final claim on γ follows by the standard arguments used before. $\quad\square$

It is clear that the Fourier–Bohr coefficients (or amplitudes) of ω_g are well-defined. They read $A(k) = \mathrm{dens}(\mathcal{L})\,\widehat{g}(-k^\star)$ for $k \in L^\circledast = \pi(\mathcal{L}^*)$ as before, and enter the diffraction formula for ω_g of Eq. (9.42) as expected.

Theorem 9.6. *Under the assumptions of Lemma 9.5, the Dirac comb ω_g has the diffraction measure $\widehat{\gamma} = \sum_{k \in L^\circledast} |A(k)|^2\, \delta_k$ with $A(k) = \mathrm{dens}(\mathcal{L})\,\widehat{g}(-k^\star)$. The measure $\widehat{\gamma}$ is a translation bounded and positive pure point measure.*

SKETCH OF PROOF. As in the previous proposition, one can derive the result from Theorem 9.5 by a suitable norm converging sequence of approximations that are based on restricting g to compact supports of increasing volume; for the details, see the proof of [Ric03, Thm. 10]. $\quad\square$

Corollary 9.5. *If ω_g is a general Dirac comb that satisfies the conditions of Lemma 9.5 for the CPS of Eq. (9.43), one obtains the generalised PSF*

$$\left(\sum_{z \in L} (g * \widetilde{g})(z^\star)\, \delta_z \right)^{\!\widehat{}} = \mathrm{dens}(\mathcal{L}) \sum_{k \in L^\circledast} \left| \widehat{g}(-k^\star) \right|^2 \delta_k$$

with $L^\circledast = \pi(\mathcal{L}^*)$. $\quad\square$

Particularly appealing examples can be derived by suitable choices of g. For instance, if $g(x) = \mathrm{e}^{-\pi x^2/2}$, one has $(g * \widetilde{g})(x) = \mathrm{e}^{-\pi x^2} = |\widehat{g}(-x)|^2$, by Example 8.2. Choosing a lattice $\mathcal{L}$ with $\mathrm{dens}(\mathcal{L}) = 1$ results in self-dual Dirac combs. The key difference to the classic PSF of Theorem 9.1 is the fact that L is the projection of a lattice, and a dense set. The results of this section have been generalised in a systematic measure theoretic setting by Lenz and Richard in [LR07].

9.6. Homometric point sets

Due to the intermediate step with the autocorrelation, it is rather clear that distinct point sets may have the same diffraction measure. It is important to understand this phenomenon before one approaches the inverse problem of deriving a structure from its diffraction pattern. For a model set Λ which originates from a given CPS, the inverse problem is reduced to determining the window. Let us thus discuss how and to what extent the window is specified by the diffraction data.

A perfect diffraction image of a Euclidean model set Λ, as described by the positive measure $\widehat{\gamma_\Lambda}$, uniquely determines its inverse Fourier transform, which is the autocorrelation (or Patterson) measure γ_Λ. Our starting point is thus the (hypothetically complete) knowledge of γ_Λ. The remaining task is then to determine a window W from this information. This is an example of a class of inverse problems where one aims at the reconstruction of a finite or compact set in Euclidean space from its (possibly weighted) difference set. To describe the corresponding classes of point sets, we borrow the concept of homometry from crystallography, which is due to Patterson [Patt44].

9.6.1. HOMOMETRY OF FINITE POINT SETS

Given a finite point set $F \subset \mathbb{R}^d$, the *difference set* $F - F$ is a relevant quantity for the inverse problem. More precisely, one actually needs $F - F$ as a *multiset*, meaning that we also need to know the multiplicity of each element of $F - F$. As an intermediate step, one can also consider the *distance set* $\{|z| \mid z \in F - F\}$, again augmented by the multiplicities.

Definition 9.3. Two finite point sets in $\mathbb{R}^d$ are called *weakly homometric* when they share the same weighted distance set, meaning that all distances occur with the same cardinalities in each set.

In particular, weakly homometric point sets have the same cardinality. It is clear that congruent point sets (in the wider sense of including reflections) are weakly homometric. Weak homometry is an equivalence relation, both on finite points sets and on their congruence classes.

Example 9.10 (*Weakly homometric sets on the circle*). In his 1944 paper [Patt44], Patterson constructed various weakly homometric point sets, which he called cyclotomic sets due to their relations to roots of unity. The simplest non-trivial example of a weakly homometric pair, adapted from [Patt44, Fig. 2], consists of the two sets (with four points each) displayed in Figure 9.12. Both contain the points on the vertical bisector of the disk, while the remaining two points lie on an inscribed square (shaded). Moving the leftmost point of the left diagram to its antipode in the square produces the right diagram. It is straightforward to check that this comes with a bijection between the displayed edges, which represent the distances.

This example can be generalised to a one-parameter family of weakly homometric pairs by rotating the square relative to the vertical bisector; see [Patt44, Fig. 7] for details. $\Diamond$

The concept of weak homometry only captures distances and their statistics. A suitable refinement takes directions into account.

Definition 9.4. Two finite point sets in $\mathbb{R}^d$ are called *homometric* when they share the same weighted difference set, meaning that all difference vectors occur with the same cardinalities in each set.

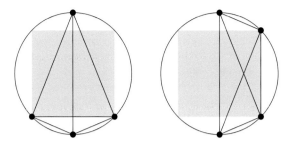

FIGURE 9.12. Patterson's construction of weakly homometric point sets.

Clearly, homometric point sets are weakly homometric, hence must have the same cardinality. Note that homometry is again an equivalence relation, which, in general, is not preserved by rotations. Also, the weakly homometric pair of Example 9.10 is not homometric. However, a point set $F \subset \mathbb{R}^d$ is homometric with $-F$, and with $t + F$ for arbitrary translations $t \in \mathbb{R}^d$. In Example 9.12 below, we shall see how weakly homometric point sets can be used to construct infinite homometric point sets that are periodic.

Example 9.11 (*A homometric pair in the plane*). A relatively simple homometric pair, realised by finite subsets $F_i \subset \mathbb{Z}^2$, was constructed in [GGZ05]. One choice of the coordinates results in

$$F_1 = \left\{ \binom{0}{0}, \binom{1}{0}, \binom{1}{1}, \binom{1}{2}, \binom{1}{3}, \binom{2}{1}, \binom{2}{2}, \binom{2}{3}, \binom{2}{4}, \binom{2}{5}, \binom{3}{3}, \binom{3}{4}, \binom{3}{5}, \binom{4}{4}, \binom{4}{5} \right\},$$

$$F_2 = \left\{ \binom{0}{0}, \binom{0}{1}, \binom{1}{0}, \binom{1}{1}, \binom{1}{2}, \binom{1}{3}, \binom{1}{4}, \binom{2}{2}, \binom{2}{3}, \binom{2}{4}, \binom{2}{5}, \binom{3}{3}, \binom{3}{4}, \binom{3}{5}, \binom{4}{5} \right\}.$$

Although $F_1 \neq F_2$, one can check that the two Minkowski difference sets $F_i - F_i$ are equal, not only as sets but also when considered as multisets (which are the sets including the multiplicities of their elements). In particular, $\binom{0}{0}$ occurs in $F_i - F_i$ with multiplicity $\mathrm{card}(F_i) = 15$.

Let us expand on the underlying structure in the spirit of [GGZ05, AL12]. Defining the sets

$$S = \left\{ \binom{0}{0}, \binom{1}{0}, \binom{1}{3}, \binom{2}{3}, \binom{3}{3} \right\} \quad \text{and} \quad T = \left\{ \binom{0}{0}, \binom{1}{1}, \binom{1}{2} \right\},$$

it is easy to verify that $F_1 = S \oplus T$ and $F_2 = \binom{1}{2} + S \oplus (-T)$, where $A \oplus B$ once again denotes the *direct* Minkowski sum of the point sets A and B. In particular, $\mathrm{card}(A \oplus B) = \mathrm{card}(A)\,\mathrm{card}(B)$. This representation immediately implies the homometry of the two sets. The direct sum structure is illustrated in Figure 9.13. $\diamond$

The structure as a direct Minkowski sum can be standardised by suitable affine transformations, and then generalised to an infinite family of examples of the same kind [AL12, Thm. 2.4 and Cor. 2.5]. At present, all known

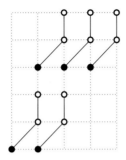 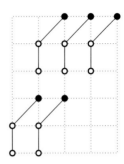

FIGURE 9.13. The two homometric point sets F_1 and F_2 of Example 9.11 and their decomposition as a direct Minkowski sum of S and $\pm T$. The set S is represented by the solid dots, with relative shift $\binom{1}{2}$.

examples of finite homometric subsets of the lattice $\mathbb{Z}^2$ seem to belong to this family; compare the discussion in [AL12].

9.6.2. HOMOMETRY OF INFINITE POINT SETS

So far, we have considered finite point sets only. In view of our general topic, we are mainly interested in *infinite* point sets, where the literal formulation of Definition 9.4 makes no sense. An appropriate generalisation needs the concept of density, for instance via the autocorrelation measure.

Definition 9.5. Two infinite, locally finite point sets in $\mathbb{R}^d$ are called *homometric* when their natural autocorrelation measures (in the sense of Definition 9.1) exist and are equal.

Two homometric point sets have the same density. Due to the volume averaging involved, adding or removing a point set of density 0 does not change the homometry class. As before, translations and inversion have no effect on the homometry class either.

Example 9.12 (*Homometric $\mathbb{Z}$-periodic point sets*). Let Λ be a locally finite point set that is $\mathbb{Z}$-periodic. By Proposition 3.1, we know that $\Lambda = F + \mathbb{Z}$, where F can be chosen as a finite subset of $[0,1)$, so that we actually have $\Lambda = F \oplus \mathbb{Z}$. An application of the results of Section 9.2.4 shows that Λ then has the natural autocorrelation $\gamma = \left(\delta_F * \widetilde{\delta_F}\right) * \delta_{\mathbb{Z}}$. Comparing this with a set $\Lambda' = F' \oplus \mathbb{Z}$, where we assume that $\mathrm{card}(F) = \mathrm{card}(F')$, homometry of Λ and Λ' is equivalent to $\gamma|_{[0,1)} = \gamma'|_{[0,1)}$. Note that $\delta_F * \widetilde{\delta_F} = \delta_{F'} * \widetilde{\delta_{F'}}$ is then sufficient for the homometry of Λ and Λ'.

Consider F and F' as subsets of $\mathbb{R}/\mathbb{Z}$ (hence with periodic boundary condition), so that Λ and Λ' are wrapped on a circle. The resulting sets are again finite. This representation reveals that Λ and Λ' are homometric as

$\mathbb{Z}$-periodic point sets if and only if their 'wrapped-up' versions are weakly homometric as finite subsets of the circle. This was Patterson's original approach [Patt44] for the homometry of crystals in one dimension. It led to the pair described in Example 9.10, which gives

$$\Lambda = \left\{0, \tfrac{3}{8}, \tfrac{1}{2}, \tfrac{5}{8}\right\} + \mathbb{Z} \quad \text{and} \quad \Lambda' = \left\{0, \tfrac{1}{8}, \tfrac{3}{8}, \tfrac{1}{2}\right\} + \mathbb{Z}.$$

With this approach, Patterson [Patt44] constructed many examples in $\mathbb{R}$; see also [Zob93a, Zob93b] and references therein for further cases. ◇

Remark 9.19 (*Homometry of translation bounded measures*). More generally, as already indicated in Remark 9.1, two translation bounded measures ω and ω' are called *homometric* when they share the same (natural) autocorrelation measure γ. The homometry of locally finite point sets Λ and Λ' as defined above is compatible with this extension by considering their Dirac combs δ_Λ and $\delta_{\Lambda'}$. ◇

Example 9.13 (*Weighted Dirac combs with homometry of higher order*). Let us consider a positively weighted $\mathbb{Z}$-periodic Dirac comb of the form $\omega = \varrho * \delta_{\mathbb{Z}}$ with $\varrho = \sum_{\ell=0}^{5} w_\ell \, \delta_{\ell/6}$. Choosing the weights $(w_\ell)_{0 \leqslant \ell \leqslant 5}$ as

$$(11, 25, 42, 45, 31, 14) \quad \text{or} \quad (10, 21, 39, 46, 35, 17)$$

results in two homometric Dirac combs [GM95]. Moreover, these two combs have the same correlation functions up to fifth order, and only differ beyond. This is the worst case scenario for $\mathbb{Z}$-periodic combs with rational weights at rational positions [GM95, Thm. 4].

The corresponding diffraction measure reads

$$\widehat{\gamma_\omega} = \left(28224\,\delta_0 + 2964\,(\delta_1 + \delta_{-1})\right) * \delta_{6\mathbb{Z}},$$

which is 6-periodic and has extinctions for all wave numbers $k \in \mathbb{Z}$ with $k \equiv 2, 3$ or $4 \bmod 6$. Note that the functions $\left|\hat{\varrho}(k)\right|^2$ for the two choices are distinct as functions on $\mathbb{R}$, but agree on $\mathbb{Z}$. ◇

9.6.3. COVARIOGRAM AND HOMOMETRY OF COMPACT SETS

There is an interesting connection between the homometry of Euclidean model sets (with internal space $H = \mathbb{R}^m$ of the CPS) and the covariogram problem for non-empty compact subsets $K \subset \mathbb{R}^m$. Recall from Remark 9.8 that the covariogram encapsulates the difference information. Given the covariogram of K, the corresponding inverse problem is the task to determine K; compare [Bia05, GGZ05]. This is also known as Matheron's problem, which was originally formulated as the question of whether the covariogram determines a convex body, among all convex bodies, up to translation and inversion; compare [Mat75, Mat86], and see [Bia05, BBD'E10] for nice and readable accounts of the history of the problem and its present state. Let us continue with a formal definition.

Definition 9.6. Let $K \subset \mathbb{R}^d$ be a non-empty, relatively compact set, which is Riemann measurable (meaning that $\lambda(\partial K) = 0$). The function

$$\mathrm{cvg}_K(x) := \mathrm{vol}\big(K \cap (x + K)\big),$$

defined for all $x \in \mathbb{R}^d$, is called the *covariogram* of the set K.

The covariogram appeared earlier in Remark 9.8. The real-valued function cvg_K on $\mathbb{R}^d$ is inversion symmetric, $\mathrm{cvg}_K(-x) = \mathrm{cvg}_K(x)$, and satisfies

$$\sup_{x \in \mathbb{R}^d} \mathrm{cvg}_K(x) = \mathrm{cvg}_K(0) = \mathrm{vol}(K).$$

The covariogram is continuous on $\mathbb{R}^d$ by [CJ94, Thm. 3.1], which remains true in the more general case where one defines cvg_B for a Borel set B of finite volume; see also [ReSt00, Lemma 3.6.3]. Here, cvg_K is a positive definite function with compact support. The latter is related to the inversion symmetric *difference body* $K - K$ via

(9.44) $\mathrm{supp}(\mathrm{cvg}_K) \subset \overline{K - K}$.

When K is compact with $\overline{K^\circ} = K$, one has $\mathrm{supp}(\mathrm{cvg}_K) = K - K$. The covariogram of K equals that of $-K$, as well as that of any translate $t+K$ with $t \in \mathbb{R}^d$. This means that cvg_K can determine K at best up to translations and inversion. As we shall see below, even this is generally not the case.

Definition 9.7. Two non-empty relatively compact and Riemann measurable sets $K, K' \subset \mathbb{R}^d$ are called *homometric* when $\mathrm{cvg}_K = \mathrm{cvg}_{K'}$.

With 1_K denoting the characteristic function of K, the function cvg_K is given by the convolution

(9.45) $\mathrm{cvg}_K(x) = \big(1_K * 1_{-K}\big)(x)$.

Its Fourier transform thus satisfies

(9.46) $\widehat{\mathrm{cvg}_K}(k) = \big|\widehat{1_K}(k)\big|^2$,

which is a positive and real analytic function on $\mathbb{R}$ that vanishes at $\pm\infty$. This relation is the reason why, if K is itself inversion symmetric in the sense that $-K = t + K$ for a suitable translation $t \in \mathbb{R}^d$, the function 1_K (and hence K) can be reconstructed from the knowledge of cvg_K, up to translation and inversion [CJ94].

If K is a convex polytope in dimension $d \leqslant 3$, it is determined by cvg_K, in the sense mentioned above; see [Bia02, Bia05, Bia09] and references therein. Other examples include convex bodies in the plane [AB09], though the general (unrestricted) claim is still open. For recent practical approaches, we refer

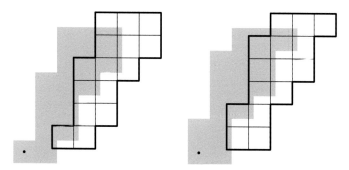

FIGURE 9.14. Example of two homometric polyominoes, each with an equally shifted copy overlayed. This case is constructed from the finite point sets of Example 9.11. The dot denotes the origin. The overlays illustrate how the two intersection areas coincide.

to [BGK11]. However, starting with dimension $d = 4$, uniqueness fails even within the class of convex polytopes; compare [Bia05, GGZ05]. Here, the direct products $S \times T$ and $S \times (-T)$ are homometric, where S and T are planar polytopes. This is a similar mechanism to the structure we saw above in the context of direct Minkowski sums in Example 9.11.

In general, the reconstruction of K from the knowledge of cvg_K is a difficult problem. In this context, an interesting example of two polyominoes with the same covariogram [BG07] follows from the point set pair F, F' of Example 9.11 by adding $A = \left[-\frac{1}{2}, \frac{1}{2}\right]^2$, so that $P = F + A$ and $P' = F' + A$. Their covariograms are equal as a consequence of the homometry of the finite point sets F and F', whence P and P' are homometric. The polyominoes P and P' are shown in Figure 9.14, and their joint covariogram is displayed in Figure 9.15.

Remark 9.20 (*Direct Minkowski sums and homometric polyominoes*). Any pair of point sets constructed via a direct Minkowski sum as described in Example 9.11 gives rise to a pair of polyominoes which are homometric as compact sets. In fact, this yields an infinite family of such pairs; compare [AL12] for a detailed exposition. ◇

Let us now apply the various concepts introduced so far to the homometry problem of Euclidean model sets.

9.6.4. HOMOMETRY OF MODEL SETS

Here, the general CPS of Eq. (7.11) is used with internal space $H = \mathbb{R}^m$. Recall that a regular model set is of the form $\Lambda = t + \curlywedge(W)$ with $t \in \mathbb{R}^d$ and a non-empty relatively compact window W with boundary of measure 0. For simplicity, we also assume that $W = \overline{W^\circ}$.

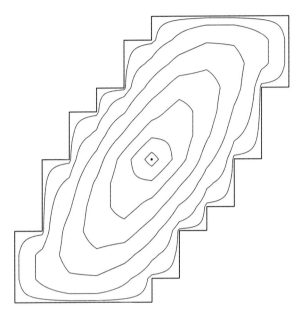

FIGURE 9.15. Contour plot of the covariogram of both P and P'. It is inversion symmetric with respect to the origin, the latter being marked by a dot. The outer polygonal line is the boundary of the difference body, which is the support of the covariogram.

By Proposition 9.8, the autocorrelation of the Dirac comb δ_Λ has the explicit form $\gamma_\Lambda = \sum_{z \in \Lambda - \Lambda} \eta(z)\,\delta_z$, with the autocorrelation coefficients

$$(9.47) \qquad \eta(z) \,=\, \mathrm{dens}(\Lambda)\,\frac{\mathrm{vol}\big(W \cap (W - z^\star)\big)}{\mathrm{vol}(W)} \,=\, \mathrm{dens}(\mathcal{L})\,\mathrm{cvg}_W(z^\star),$$

where $\mathrm{dens}(\Lambda)$ denotes the ordinary density of the model set $\Lambda \subset \mathbb{R}^d$ and $\mathrm{dens}(\mathcal{L})$ the density of the embedding lattice $\mathcal{L} \subset \mathbb{R}^d \times \mathbb{R}^m$. This formula expresses the autocorrelation of Λ in terms of the covariogram of the window.

Theorem 9.7. *Two Euclidean model sets obtained from the same CPS are homometric if and only if the defining windows share the same covariogram.*

PROOF. If two model sets, Λ and Λ' say, possess the same autocorrelation, the coefficient $\eta(z)$ is valid for both of them. Since the CPS is fixed, the factor $\mathrm{dens}(\mathcal{L})$ in Eq. (9.47) is the same for Λ and Λ'. Consequently, the covariograms of the windows coincide on all points $z \in \Lambda - \Lambda$. By the general setting of a CPS, $\Lambda^\star - \Lambda^\star$ is a dense subset of $W - W$, so that the covariogram, which is a continuous function [CJ94], is completely determined on $W - W$, while it vanishes on the complement. As the same argument applies to $W' - W'$, the covariograms of W and W' coincide. The converse claim is obvious from Eq. (9.47). □

Taking Fourier transforms, the following consequence is immediate.

Corollary 9.6. *Homometric model sets from the same CPS have the same diffraction measure. Conversely, kinematic diffraction cannot discriminate between homometric model sets.* □

Remark 9.21 (*Determination of the CPS*). So far, the formulation was based upon a fixed CPS. In general, however, there are different possibilities to generate a given model set. In particular, two regular model sets obtained from different cut and project schemes can still be homometric. If so, they have the same autocorrelation γ by definition, and thus the same diffraction measure $\hat{\gamma}$. Since regular model sets are pure point diffractive according to Theorem 9.4, $\hat{\gamma}$ comprises a countable sum of Dirac measures with positive intensity. The $\mathbb{Z}$-span of their positions is the Fourier module. Via minimal embedding, it defines an essentially unique CPS, whose dual in the sense of [Moo00] is a natural setting to describe both model sets in the same CPS. In this sense, the restriction in Theorem 9.7 is not essential. ◇

9.6.5. A SIMPLE PLANAR EXAMPLE

To turn the homometric pair of polyominoes into a pair of homometric model sets, we use them as windows for cyclotomic model sets in a CPS of Eq. (7.13), with lattice $\mathcal{L}_8 \subset \mathbb{R}^2 \times \mathbb{R}^2$, $L = \mathbb{Z}[\xi_8] \subset \mathbb{R}^2$ and the $\star$-map as defined in Table 7.1. We define the two model sets

$$\Lambda = \curlywedge(P) \quad \text{and} \quad \Lambda' = \curlywedge(P'),$$

where both windows are placed in internal space such that their lower left corner has coordinates $(-\frac{1}{2}, -\frac{1}{2})$. This way, the corresponding model sets are generic. They have density

$$\mathrm{dens}(\Lambda) = \mathrm{dens}(\Lambda') = \frac{15}{4}$$

and are homometric by construction, due to Theorem 9.7. Note that Λ and Λ' are *not* in the same LI class, and that they differ in points of positive density. In particular, the difference sets are model sets themselves, whose windows can be extracted from Figure 9.16. Their densities are

$$\mathrm{dens}(\Lambda' \setminus \Lambda) = \mathrm{dens}(\Lambda \setminus \Lambda') = \frac{2}{15}\,\mathrm{dens}(\Lambda) = \frac{1}{2}.$$

Note, however, that the sets $\Lambda' \setminus \Lambda$ and $\Lambda \setminus \Lambda'$ are not homometric. The same procedure is possible for any pair of homometric polyominoes according to Remark 9.20.

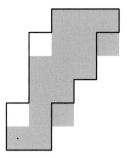

FIGURE 9.16. Comparison of the two windows P (grey area) and P' (bounded by the black line). The two white squares on the left (the two grey ones on the right) form the window for the points in $\Lambda' \setminus \Lambda$ (in $\Lambda \setminus \Lambda'$).

The diffraction measure $\widehat{\gamma}$ is the same for both Λ and Λ', and reads

$$\widehat{\gamma} \;=\; \sum_{k \in \frac{1}{2}L} I(k)\,\delta_k,$$

with intensity function $I(k) = |A_P(k)|^2$ derived from

$$A_P(k) \;=\; \mathrm{dens}(\mathcal{L}_8)\,\widehat{1_P}(-k^\star) \;=\; \frac{1}{4}\,\widehat{1_P}(-k^\star),$$

and similarly for P'. Note that the amplitudes depend on the window, but their absolute squares do not. One can now work out the explicit diffraction intensities by means of the formulas from Example 8.3; see [BG07] for details.

Remark 9.22 (*Homometry, local derivability and higher correlations*). The criteria for mutual local derivability (MLD) from above imply that the two model sets Λ and Λ' are MLD. This is due to the building blocks of the windows being unit squares in an orientation that is compatible with $\mathcal{L}_8$, and hence an untypical case. Rotating both polyominoes by the same angle or scaling them by the same factor (or both) clearly preserves homometry, but generically destroys the MLD property, by the criterion from Remark 7.6; see also [BSJ91].

Clearly, the two polyominoes are translationally inequivalent. By a result of Deng and Moody [DM09], compare also the discussion in [LM09], the corresponding model sets (which have the same diffraction and hence the same two-point correlation) must differ on the level of their three-point correlations. Indeed, it is not difficult to construct clusters of three points that occur in one set but not in the other.   ◇

At this point, we close our discussion of pure point diffractive point sets (and measures), and turn our attention to the other spectral types, with a similarly constructive approach.

CHAPTER 10

Beyond Model Sets

The result on the pure point diffraction of crystallographic systems and, more generally, of regular model sets is constructive in the sense that it also provides a closed formula for the diffraction measure and the Fourier-Bohr coefficients (or amplitudes). Such a situation cannot be expected in general, and concrete results are rather sparse outside the realm of model sets. However, there are some notable exceptions, and it is the purpose of the following paragraphs to discuss several paradigms as concretely and explicitly as possible.

Another important line of generalisation concerns the diffraction of subsets of model sets. We begin this part with a detailed treatment of lattice subsets, including the diffraction of visible lattice points and related structures, and afterwards summarise some rather remarkable properties of general subsets of Meyer sets. Further examples will follow in Chapter 11.

10.1. Diffraction of the Thue-Morse chain

Let us consider the symmetric bi-infinite fixed point of the Thue-Morse substitution, as constructed in Section 4.6, in the realisation as a symmetric sequence $w \in \{\pm 1\}^{\mathbb{Z}}$. Recall that the (discrete) hull $\mathbb{X}(w)$ is uniquely ergodic under the $\mathbb{Z}$-action of the shift. To continue, we follow the approach of [Kak72] and [AMF95, Part IV.4], and use the recursive definition of Eq. (4.14) together with Eq. (4.16) from Remark 4.8 on page 100; see also [BG08]. For the connection with diffraction, we consider the signed (or weighted) Dirac comb $\omega_{\mathrm{TM}} = w\delta_{\mathbb{Z}}$ with w as above. In this case, the extension to the corresponding continuous hull does not add any obstacles. In particular, one inherits unique ergodicity for the $\mathbb{R}$-action of translation. Consequently, an application of Birkhoff's ergodic theorem (compare Theorem 4.4) shows that the autocorrelation measure γ of ω_{TM} exists. Here, it is of the form

$$\gamma = \gamma_{\mathrm{TM}} = \sum_{m \in \mathbb{Z}} \eta(m)\, \delta_m.$$

Recall from Lemma 8.4 that γ is positive definite as a measure on $\mathbb{R}$ if and only if η is positive definite as a function on $\mathbb{Z}$. The coefficients clearly satisfy

TABLE 10.1. Some autocorrelation coefficients of the Thue–Morse chain. Note that $\eta(0) = 1$ and $\eta(-m) = \eta(m)$ for $m \in \mathbb{Z}$.

m	1	2	3	4	5	6	7	8	9	10
$\eta(m)$	$-\frac{1}{3}$	$-\frac{1}{3}$	$\frac{1}{3}$	$-\frac{1}{3}$	0	$\frac{1}{3}$	0	$-\frac{1}{3}$	$\frac{1}{6}$	0
$\eta(10+m)$	$-\frac{1}{6}$	$\frac{1}{3}$	$-\frac{1}{6}$	0	$\frac{1}{6}$	$-\frac{1}{3}$	$\frac{1}{12}$	$\frac{1}{6}$	$-\frac{1}{12}$	0
$\eta(20+m)$	$\frac{1}{12}$	$-\frac{1}{6}$	$-\frac{1}{12}$	$\frac{1}{3}$	$-\frac{1}{12}$	$-\frac{1}{6}$	$\frac{1}{12}$	0	$-\frac{1}{12}$	$\frac{1}{6}$
$\eta(30+m)$	$\frac{1}{12}$	$-\frac{1}{3}$	$\frac{1}{8}$	$\frac{1}{12}$	$-\frac{1}{8}$	$\frac{1}{6}$	$-\frac{1}{24}$	$-\frac{1}{12}$	$\frac{1}{24}$	0
$\eta(40+m)$	$-\frac{1}{24}$	$\frac{1}{12}$	$\frac{1}{24}$	$-\frac{1}{6}$	$\frac{1}{8}$	$-\frac{1}{12}$	$-\frac{1}{8}$	$\frac{1}{3}$	$-\frac{1}{8}$	$-\frac{1}{12}$
$\eta(50+m)$	$\frac{1}{8}$	$-\frac{1}{6}$	$\frac{1}{24}$	$\frac{1}{12}$	$-\frac{1}{24}$	0	$\frac{1}{24}$	$-\frac{1}{12}$	$-\frac{1}{24}$	$\frac{1}{6}$
$\eta(60+m)$	$-\frac{1}{8}$	$\frac{1}{12}$	$\frac{1}{8}$	$-\frac{1}{3}$	$\frac{5}{48}$	$\frac{1}{8}$	$-\frac{5}{48}$	$\frac{1}{12}$	$\frac{1}{48}$	$-\frac{1}{8}$

$\eta(0) = 1$ and, due to positive definiteness, $\eta(-m) = \eta(m)$ for all $m \in \mathbb{Z}$. Since our fixed point itself is inversion symmetric, we can use the one-sided sequence v from Remark 4.8 to obtain

$$(10.1) \qquad \eta(m) = \lim_{N \to \infty} \frac{1}{N} \sum_{i=0}^{N-1} v_i v_{i+m}$$

for all $m \in \mathbb{N}$ (alternatively, this follows from the existence of the autocorrelation coefficient as a limit, uniformly with respect to arbitrary translations of the averaging region). Using the relations from Eq. (4.14), one easily verifies from Eq. (10.1), by splitting the sum into one over even and one over odd indices, that the recursive identities [Kak72]

$$(10.2) \qquad \eta(2m) = \eta(m) \quad \text{and} \quad \eta(2m+1) = -\frac{1}{2}\big(\eta(m) + \eta(m+1)\big)$$

are valid for all $m \geqslant 0$. Given $\eta(0) = 1$, these relations uniquely specify $\eta(m)$ for all $m > 0$. In particular, $\eta(1) = -\frac{1}{3}$. The first values are shown in Table 10.1. It is routine to check that Eq. (10.2) also holds for all $m \in \mathbb{Z}$.

Since η is a positive definite function on $\mathbb{Z}$ by construction, the Herglotz–Bochner theorem (compare Theorem 8.6 or [Kat04, Thm. I.7.6]) guarantees the existence of a positive measure ϱ on $[0,1)$ with

$$(10.3) \qquad \eta(m) = \int_0^1 e^{2\pi i m y} \, d\varrho(y).$$

A slight modification of this result provides a connection with the diffraction measure of ω_{TM}. Since γ exists and is positive definite, the diffraction measure $\widehat{\gamma}$ is well-defined by Proposition 8.6.

Lemma 10.1. *The diffraction measure $\widehat{\gamma}$ of the weighted Thue–Morse comb ω_{TM} is a $\mathbb{Z}$-periodic positive measure. It can be written as $\widehat{\gamma} = \varrho * \delta_{\mathbb{Z}}$ with $\varrho = \widehat{\gamma}|_{[0,1)}$, the latter being a probability measure on $[0,1)$ (and also on $\mathbb{R}$). Moreover, the autocorrelation coefficients satisfy Eq. (10.3) with this ϱ.*

PROOF. Interpreting η as a real-valued function on $\mathbb{Z}$, the autocorrelation can be written as $\gamma = \eta\,\delta_{\mathbb{Z}}$. Moreover, whenever h is a complex-valued function on $\mathbb{R}$ with $h(m) = 1$ for all $m \in \mathbb{Z}$, one has $\gamma = h\,\gamma$ as an equality of measures on $\mathbb{R}$. Choosing $h(x) = \mathrm{e}^{2\pi\mathrm{i}x}$, one obtains $\widehat{\gamma} = \delta_1 * \widehat{\gamma}$ by a (backwards) application of the convolution theorem (Theorem 8.5) and hence the $\mathbb{Z}$-periodicity of $\widehat{\gamma}$. The latter is positive by the Bochner–Schwartz theorem; see [RS80, Thm. IX.10] or Proposition 8.6. Consequently, by Proposition 9.5 and the argument used in its proof, $\widehat{\gamma} = \varrho*\delta_{\mathbb{Z}}$ with the finite positive measure ϱ as claimed, which vanishes on the complement of $[0,1)$.

Since γ is a pure point measure with uniformly discrete support, we can calculate the coefficients $\eta(m)$ by inverse Fourier transform, the (inverse) convolution theorem, and the PSF (Proposition 9.4) as follows,

$$\eta(m) = \gamma(\{m\}) = \big(\varrho * \delta_{\mathbb{Z}}\big)^{\vee}(\{m\}) = \big(\check{\varrho}\,\delta_{\mathbb{Z}}\big)(\{m\}) = \check{\varrho}(m)$$

$$= \int_{\mathbb{R}} \mathrm{e}^{2\pi\mathrm{i}my}\,\mathrm{d}\varrho(y) = \int_0^1 \mathrm{e}^{2\pi\mathrm{i}my}\,\mathrm{d}\varrho(y).$$

This establishes the claim about the coefficients. In particular, one has the normalisation $1 = \eta(0) = \varrho\big([0,1)\big) = \varrho(\mathbb{R})$. $\qquad\qquad \square$

Our next aim is to show that ϱ is a purely singular continuous measure (and so is then $\widehat{\gamma}$ by construction). This is done by first excluding pure points, and then by showing the absence of any absolutely continuous part. For the first step, recall Wiener's result (Theorem 8.6) stating that the discrete part of a measure can be recovered from its Fourier–Stieltjes coefficients.

Lemma 10.2. *The probability measure ϱ from Lemma 10.1 is a continuous measure on $[0,1]$, meaning that $\varrho = \varrho_{\mathsf{cont}}$.*

PROOF. Due to the construction, we may view ϱ as a measure on $\mathbb{T} \simeq [0,1)$. By Wiener's criterion from Proposition 8.9, the continuity of ϱ is equivalent to the condition $\lim_{N\to\infty} \frac{1}{2N+1}\Sigma(N) = 0$ with

$$\Sigma(N) := \sum_{m=-N}^{N} \big(\eta(m)\big)^2,$$

because $\widehat{\varrho}(m) = \eta(-m) = \eta(m) \in \mathbb{R}$.

When $N \geqslant 1$, we can use the recursion relations of Eq. (10.2) to derive

$$
\begin{aligned}
\Sigma(4N) &= \sum_{m=-2N}^{2N} \big(\eta(2m)\big)^2 + \sum_{m=-2N}^{2N-1} \big(\eta(2m{+}1)\big)^2 \\
&= \sum_{m=-2N}^{2N} \big(\eta(m)\big)^2 + \frac{1}{4} \sum_{m=-2N}^{2N-1} \big(\eta(m) + \eta(m{+}1)\big)^2 \\
&= \frac{3}{2}\,\Sigma(2N) - \frac{1}{4}\Big(\big(\eta(2N)\big)^2 + \big(\eta(-2N)\big)^2\Big) + \frac{1}{2} \sum_{m=-2N}^{2N-1} \eta(m)\,\eta(m{+}1) \\
&= \frac{3}{2}\,\Sigma(2N) - \frac{1}{2}\big(\eta(2N)\big)^2 - \frac{1}{4} \sum_{m=-N}^{N-1} \big(\eta(m) + \eta(m{+}1)\big)^2 \leqslant \frac{3}{2}\,\Sigma(2N),
\end{aligned}
$$

since $\eta(2m)\,\eta(2m{+}1) + \eta(2m{+}1)\,\eta(2m{+}2) = -\frac{1}{2}\big(\eta(m) + \eta(m{+}1)\big)^2$ holds for any $m \in \mathbb{Z}$. This implies $\Sigma(2^{k+1}) \leqslant \big(\frac{3}{2}\big)^k \Sigma(2)$ and thus $\Sigma(N) \leqslant C\big(\frac{3}{2}\big)^{\log_2(N)}$ for some positive constant C. With $\alpha = \log_2(3/2) < 1$, one then obtains the estimate $\frac{1}{N}\Sigma(N) = \mathcal{O}(1/N^{1-\alpha})$, which proves the claim. $\square$

At this point, we thus know that

$$
\varrho = \varrho_{\mathsf{cont}} = \varrho_{\mathsf{sc}} + \varrho_{\mathsf{ac}}.
$$

The Radon–Nikodym theorem [RS80, Thm. I.19] implies that $\varrho_{\mathsf{ac}} = g\,\lambda$, where g is a real-valued L^1-function on $[0,1]$. To continue, it is advantageous to define the distribution function $F = F_{\mathrm{TM}}$ of ϱ as

$$
(10.4) \qquad\qquad F(x) := \varrho\big([0,x)\big) = \varrho\big([0,x]\big),
$$

which is now a continuous non-decreasing function on $[0,1]$ (which will later be extended to a skew-symmetric function on $\mathbb{R}$ via $F(x) = \hat{\gamma}\big([0,x]\big)$ for $x \geqslant 0$). At this point, F is a continuous function of bounded variation on $[0,1]$. In fact, the proof of Lemma 10.2 shows a bit more: By a result of Last, see [Kni98, Thm. 3.7] for a formulation that fits the present context, the asymptotic estimate $\frac{1}{N}\Sigma(n) = \mathcal{O}(1/N^{1-\alpha})$ also implies that F is uniformly Hölder continuous, with exponent $\sqrt{1-\alpha} > 0$.

Lemma 10.3. *The Radon–Nikodym density g of ϱ vanishes, hence $\varrho_{\mathsf{ac}} = 0$.*

PROOF. Observe that each coefficient $\eta(m)$ in Eq. (10.3) may contain contributions from both continuous components of ϱ. We now show that the absolutely continuous contribution is absent.

Viewing ϱ as a Lebesgue–Stieltjes measure with distribution function F (see [Car00, Ch. 14], [Lan93, Ch. X] or [Bil95, Sec. 12] for background), the two recursion relations of Eq. (10.2) imply the identities

(10.5)
$$\mathrm{d}F(\tfrac{y}{2}) + \mathrm{d}F(\tfrac{y+1}{2}) = \mathrm{d}F(y),$$
$$\mathrm{d}F(\tfrac{y}{2}) - \mathrm{d}F(\tfrac{y+1}{2}) = -\cos(\pi y)\,\mathrm{d}F(y),$$

for all $y \in [0,1]$. The left-hand sides are obtained from the corresponding sides of the recursions in Eq. (10.2) by a change of variables, followed by a split of the new integration region into two parts, which results in one pair of relations for each $m \in \mathbb{Z}$. The actual equality of the measures follows because we may now take linear combinations of these relations to obtain equality of the integrals over arbitrary trigonometric polynomials, whence the Fourier uniqueness theorem (Proposition 8.7) applies. The relations in Eq. (10.5) must also hold for the two continuous components of the measure separately, because $\varrho_{\mathsf{sc}} \perp \varrho_{\mathsf{ac}}$ by Proposition 8.4.

Rewriting the absolutely continuous part of ϱ in terms of its Radon–Nikodym density g, the identities of Eq. (10.5) result in

$$\tfrac{1}{2}\big(g(\tfrac{y}{2}) + g(\tfrac{y+1}{2})\big) = g(y) \quad \text{and} \quad \tfrac{1}{2}\big(g(\tfrac{y}{2}) - g(\tfrac{y+1}{2})\big) = -\cos(\pi y)\,g(y),$$

this time for almost all $y \in [0,1]$. If we now set

$$\eta_{\mathsf{ac}}(m) = \int_0^1 \mathrm{e}^{2\pi i m y}\,\mathrm{d}\varrho_{\mathsf{ac}}(y) = \int_0^1 \mathrm{e}^{2\pi i m y}\,g(y)\,\mathrm{d}y,$$

the Riemann–Lebesgue lemma (Proposition 8.8) implies that

(10.6)
$$\lim_{m \to \pm\infty} \eta_{\mathsf{ac}}(m) = 0.$$

On the other hand, the previous splitting argument (together with its implication for the recursions of η_{ac}) shows that we also have

$$\eta_{\mathsf{ac}}(2m) = \eta_{\mathsf{ac}}(m) \quad \text{and} \quad \eta_{\mathsf{ac}}(2m+1) = -\frac{1}{2}\big(\eta_{\mathsf{ac}}(m) + \eta_{\mathsf{ac}}(m+1)\big)$$

for all $m \geqslant 0$, together with $\eta_{\mathsf{ac}}(-m) = \eta_{\mathsf{ac}}(m)$. As before, all $\eta_{\mathsf{ac}}(m)$ are uniquely determined from $\eta_{\mathsf{ac}}(0)$, with $\eta_{\mathsf{ac}}(1) = -\frac{1}{3}\eta_{\mathsf{ac}}(0)$. Consequently, Eq. (10.6) forces $\eta_{\mathsf{ac}}(0) = 0$, which implies $\eta_{\mathsf{ac}}(m) = 0$ for all $m \in \mathbb{Z}$. By the Fourier uniqueness theorem (Proposition 8.7), this implies $g(y) = 0$ almost everywhere, and hence $\varrho_{\mathsf{ac}} = 0$ as claimed. $\square$

The three Lemmas 10.1, 10.2 and 10.3 imply that $\varrho = \varrho_{\mathsf{sc}} \neq 0$, and hence give the following result, which was (somewhat implicitly) proved by Mahler [Mah27], based on work by Wiener [Wie27], and later (in the form presented here) by Kakutani in the context of dynamical systems [Kak72].

Theorem 10.1. *Let ω_{TM} be the weighted Dirac comb of the Thue–Morse chain in its realisation as a sequence in $\{\pm 1\}^{\mathbb{Z}}$. This signed measure possesses* *a unique autocorrelation γ_{TM} whose Fourier transform $\widehat{\gamma_{\mathrm{TM}}}$ is a purely singular continuous, positive, $\mathbb{Z}$-periodic measure. The latter is also the diffraction measure for the signed measure derived from any other element of the Thue–Morse LI class.*   □

Note that the last claim of Theorem 10.1 follows, as before, from the fact that the autocorrelation measure is the same for all elements of the hull. Since $\hat{\gamma} = \widehat{\gamma_{\mathrm{TM}}}$ is another paradigm of a singular continuous measure (beyond the Cantor measure from Figure 8.5), it is worthwhile to also calculate it explicitly. We employ a method that can be used more generally.

Adding the two equations for $\mathrm{d}F$ in the proof of Lemma 10.3 gives the relation $\mathrm{d}F(\frac{y}{2}) = \frac{1}{2}\bigl(1 - \cos(\pi y)\bigr)\,\mathrm{d}F(y)$, and thus the integral equation

$$(10.7) \qquad F(x) = \frac{1}{2}\int_0^{2x}\bigl(1 - \cos(\pi y)\bigr)\,\mathrm{d}F(y)$$

for the continuous function F, valid for $x \in [0, \frac{1}{2}]$. Noting that the diffraction measure $\widehat{\gamma_{\mathrm{TM}}}$ is inversion symmetric and $\mathbb{Z}$-periodic, the function F satisfies $F(1 - x) = 1 - F(x)$ for all $x \in [0, 1]$, so it suffices to know F on $[0, \frac{1}{2}]$. In particular, $F(\frac{1}{2}) = \frac{1}{2}$, and $\int_0^1 \sin(2\pi m x)\,\mathrm{d}F(x) = 0$ for $m \in \mathbb{Z}$.

Since F is a continuous and non-decreasing function on $[0, 1]$, the difference $\psi(x) = F(x) - x$ defines a continuous function ψ of bounded variation, with $\psi(0) = 0$ and $\psi(x) + \psi(1 - x) = 0$ for all $x \in [0, 1]$. By Proposition 8.1, ψ possesses a uniformly converging Fourier series of the form

$$(10.8) \qquad \psi(x) = F(x) - x = \sum_{m \geqslant 1} b_m \sin(2\pi m x),$$

with $b_m = 2\int_0^1 \sin(2\pi m x)\,\psi(x)\,\mathrm{d}x$. From Eqs. (10.3) and (10.4), we find

$$\eta(m) = \int_0^1 \mathrm{e}^{2\pi \mathrm{i} m x}\,\mathrm{d}F(x) = \int_0^1 \cos(2\pi m x)\,\mathrm{d}F(x) = \int_0^1 \cos(2\pi m x)\,\mathrm{d}\psi(x)$$

$$= \Bigl[\cos(2\pi m x)\,\psi(x)\Bigr]_0^1 + 2\pi m \int_0^1 \sin(2\pi m x)\,\psi(x)\,\mathrm{d}x = \pi m\, b_m,$$

using $\mathrm{d}F(x) = \mathrm{d}\psi(x) + \mathrm{d}x$ in the first line (with $\int_0^1 \cos(2\pi m x)\,\mathrm{d}x = 0$ for all $m \in \mathbb{N}$), and integration by parts in the second. Consequently,

$$(10.9) \qquad F(x) = x + \sum_{m \geqslant 1} \frac{\eta(m)}{m\pi} \sin(2\pi m x)$$

is a uniformly converging Fourier series representation of the distribution function F. Note that Eq. (10.9) can also be obtained from the tempered distribution $\gamma = \eta\, \delta_{\mathbb{Z}}$ by formal Fourier transform, followed by a term by term integration (which is well-defined in the context of tempered distributions).

As explained above, the distribution function F satisfies the symmetry relation $F(x) + F(1 - x) = 1$ for all $x \in [0,1]$. Moreover, F is indeed a solution of the integral equation (10.7), which can now be checked via a term by term integration using standard trigonometric formulas. In fact, within the space

$$(10.10) \qquad D = \left\{ G \in C\big([0,1], \mathbb{R}\big) \, \middle| \, \begin{array}{l} G(0) = 0, \, G \text{ non-decreasing and} \\ G(x) + G(1 - x) = 1 \text{ on } [0,1] \end{array} \right\}$$

of continuous, non-decreasing distribution functions with the required symmetry, F of Eq. (10.9) is the *only* solution of Eq. (10.7). For reasons that will become clear below, we equip the space D with the topology induced by uniform convergence of continuous functions. Our functional equation (10.7) suggests to define a mapping $G \mapsto \varPhi(G)$ with

$$(10.11) \qquad \big(\varPhi(G)\big)(x) = \begin{cases} \frac{1}{2} \int_0^{2x} \big(1 - \cos(\pi y)\big) \, \mathrm{d}G(y), & \text{if } 0 \leqslant x \leqslant \frac{1}{2}, \\ 1 - \big(\varPhi(G)\big)(1-x), & \text{if } \frac{1}{2} < x \leqslant 1. \end{cases}$$

It follows from elementary observations that $\varPhi$ maps D into itself. If a function $G \in D$ is extended to a function on $\mathbb{R}$ via $G(x + 1) = G(x) + 1$ (and hence $G(x+n) = G(x)+n$ for all $n \in \mathbb{Z}$ by induction), an explicit calculation shows that the representation

$$\big(\varPhi(G)\big)(x) = \frac{1}{2} \int_0^{2x} \big(1 - \cos(\pi y)\big) \, \mathrm{d}G(y)$$

now holds for all $x \in \mathbb{R}$. Note that the extended function G satisfies $G(z) = z$ for all $z \in \frac{1}{2}\mathbb{Z}$.

Corollary 10.1. *The distribution function F of the Thue–Morse diffraction measure on $[0,1]$ is the limit of the uniformly converging series in Eq. (10.9). It is also the unique solution of the functional equation $G = \varPhi(G)$ in the space D of Eq. (10.10).*

PROOF. The first claim follows as outlined around Eq. (10.8). By construction, F satisfies Eq. (10.7) for $x \in [0, \frac{1}{2}]$, while the validity on $[\frac{1}{2}, 1]$ follows via the symmetry relation $F(x) + F(1 - x) = 1$. For the uniqueness, observe that each function $G \in D$ has a uniformly converging series of the form

$$G(x) = x + \sum_{m=1}^{\infty} \frac{\beta_m}{m\pi} \sin(2\pi mx).$$

A direct calculation shows that $G = \varPhi(G)$ implies that $\beta_1 = -\frac{1}{2}(1 + \beta_1)$, hence $\beta_1 = -\frac{1}{3}$, together with

$$\beta_{2m} = \beta_m \quad \text{and} \quad \beta_{2m+1} = -\frac{1}{2}\big(\beta_m + \beta_{m+1}\big)$$

for all $m \geqslant 1$, hence $\beta_m = \eta(m)$ for all $m \in \mathbb{N}$ by comparison with Eq. (10.2). This means that $G = F$. $\qquad\qquad\qquad\qquad\qquad\qquad\qquad\qquad\qquad\qquad$ □

To continue, it is advantageous to employ a converging Volterra-type iteration within D. Starting from $F_0(x) = x$, where $F_0 \in D$, we set

$$(10.12) \qquad\qquad F_{n+1} = \Phi(F_n),$$

for $n \geqslant 0$. This defines a sequence of functions in D, explicitly given by

$$(10.13) \qquad\qquad F_n(x) = x + \sum_{m=1}^{\infty} \frac{c_m^{(n)}}{m\pi} \sin(2\pi m x),$$

where $c_m^{(0)} = 0$ for all $m \geqslant 1$ from our initial condition. The coefficients $c_m^{(n)}$ satisfy the recursions $c_1^{(n+1)} = -\frac{1}{2}(1 + c_1^{(n)})$ together with

$$(10.14) \qquad c_{2m}^{(n+1)} = c_m^{(n)} \quad \text{and} \quad c_{2m+1}^{(n+1)} = -\frac{1}{2}\left(c_m^{(n)} + c_{m+1}^{(n)}\right)$$

for $n \geqslant 0$ and $m \geqslant 1$. They approach the autocorrelation coefficients η as

$$(10.15) \qquad\qquad \lim_{n\to\infty} c_m^{(n)} = \eta(m)$$

for arbitrary $m \in \mathbb{N}$. To prove this, observe first that the affine mapping $x \mapsto -\frac{1}{2}(1 + x)$ is a contraction on $\mathbb{R}$ (with Lipschitz constant $\frac{1}{2}$). The sequence $\left(c_1^{(n)}\right)_{n\in\mathbb{N}}$ thus converges exponentially fast by Banach's contraction mapping principle, with limit $\eta(1) = -\frac{1}{3}$. The general convergence now is a consequence of Eq. (10.14), which fixes the other limits inductively. By construction, each F_n is in fact a finite sum, and can be written as

$$(10.16) \qquad\qquad F_n(x) = x + \sum_{m=1}^{2^n-1} \frac{c_m^{(n)}}{m\pi} \sin(2\pi m x).$$

The advantage of this approach for numerical calculations (over truncating the Fourier series of F, say) is that each F_n is in D, and hence a distribution function with the correct symmetry.

Corollary 10.2. *The iteration sequence from Eqs. (10.12) and (10.13) converges uniformly on $\mathbb{R}$ to the Thue–Morse distribution function F.*

PROOF. We first show pointwise convergence of the sequence $(F_n)_{n\in\mathbb{N}_0}$ to F on $\mathbb{R}$, which is equivalent to $F_n(x) - x \xrightarrow{n\to\infty} F(x) - x$ for all $x \in [0,1]$. We know the representation of $F_n(x) - x$ as a uniformly converging Fourier series given in Eq. (10.16), with $c_m^{(n)} \xrightarrow{n\to\infty} \eta(m)$ from Eq. (10.15). Hence

$$\int_0^1 (F_n(x) - x)\, h(x)\, \mathrm{d}x \xrightarrow{n\to\infty} \int_0^1 (F(x) - x)\, h(x)\, \mathrm{d}x$$

for all trigonometric polynomials h and hence for all continuous functions on $\mathbb{T}$ (by the Stone–Weierstraß theorem), which means weak convergence of the sequence. Since F and all F_n are continuous, we obtain the claimed pointwise convergence.

Due to the 1-periodicity of the functions $F_n - F$, uniform convergence on $\mathbb{R}$ follows from that on $[0,1]$. The latter can be shown via the 'stepping-stone' argument from the proof of [Bau01, Thm. 30.13]: Given $\varepsilon > 0$, there are numbers $0 = x_0 < x_1 < \cdots < x_m = 1$ such that

$$|F(x_i) - F(x_{i-1})| = F(x_i) - F(x_{i-1}) < \varepsilon$$

for $1 \leqslant i \leqslant m$. Also, for all sufficiently large $n \in \mathbb{N}$, one has

$$|F_n(x_i) - F(x_i)| < \varepsilon$$

for all $0 \leqslant i \leqslant m$. Since $F_n(1) = F(1) = 1$, consider now an arbitrary $x \in [0,1)$, so that $x_{i-1} \leqslant x < x_i$ for precisely one $i \in \{1, 2, \ldots, m\}$. Using monotonicity, this implies the inequalities

$$F(x_{i-1}) \leqslant F(x) \leqslant F(x_i) < F(x_{i-1}) + \varepsilon \quad \text{and}$$

$$F(x_{i-1}) - \varepsilon < F_n(x_{i-1}) \leqslant F_n(x) \leqslant F_n(x_i) < F(x_i) + \varepsilon < F(x_{i-1}) + 2\varepsilon.$$

Together, they give $|F_n(x) - F(x)| < 2\varepsilon$, which holds for all $x \in [0,1]$, and then uniformly for all $x \in \mathbb{R}$. $\qquad \square$

By construction, the distribution functions F_n represent absolutely continuous measures (although their limit does not). Writing $\mathrm{d}F_n(x) = f_n(x)\,\mathrm{d}x$, one finds $f_0(x) = 1$, and the functional equation (10.7) induces the recursion

$$f_{n+1}(x) = \big(1 - \cos(2\pi x)\big) f_n(2x) = 2\big(\sin(\pi x)\big)^2 f_n(2x)$$

for $n \geqslant 0$. This gives the well-known explicit representation

$$(10.17) \qquad f_n(x) = \prod_{\ell=0}^{n-1} \big(1 - \cos(2^{\ell+1}\pi x)\big) = 2^n \prod_{\ell=0}^{n-1} \big(\sin(2^\ell \pi x)\big)^2$$

of the Thue–Morse measure as a Riesz product; compare [Que95, Sec. 8.2.1].

Remark 10.1 (*A shortcut to the Thue–Morse Riesz product*)**.** Let us point out that, once the uniform convergence properties are established, there is a rather quick way to derive Eq. (10.17) from the substitution structure. If $v^{(n+1)} = v^{(n)}\overline{v^{(n)}}$ with $v^{(0)} = 1$ is the concatenation recursion for the one-sided fixed point $v \in \{\pm 1\}^{\mathbb{N}_0}$ of Remark 4.8, where $\overline{\pm 1} = \mp 1$, we define the corresponding exponential sums as

$$g_n(x) = \sum_{\ell=0}^{2^n-1} v_\ell^{(n)}\,\mathrm{e}^{-2\pi \mathrm{i}\ell x} = \sum_{\ell=0}^{2^n-1} v_\ell\,\mathrm{e}^{-2\pi \mathrm{i}\ell x}.$$

This leads to $g_0 = 1$ and $g_{n+1}(x) = (1 - \mathrm{e}^{-2\pi \mathrm{i}2^n x}) g_n(x)$ for $n \geqslant 0$. For the absolute squares, this gives

$$|g_{n+1}(x)|^2 = 2\big(1 - \cos(2^{n+1}\pi x)\big)\,|g_n(x)|^2,$$

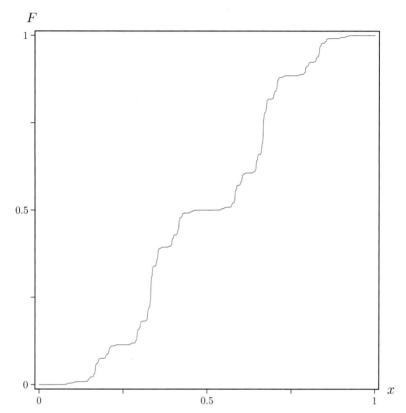

FIGURE 10.1. Distribution function F of the purely singular continuous diffraction measure of the Thue–Morse chain in the interval $[0, 1]$. Note that F is strictly increasing, despite the appearance of plateau-like regions.

so that $|g_n(x)|^2 = 2^n f_n(x)$ for $n \in \mathbb{N}_0$. In particular, $2^{-n} |g_n(x)|^2$ vaguely converges to the diffraction measure. This often used derivation hides the fact that convergence needs to be considered in the sense of measures, while pointwise convergence of the sequence of functions is a subtle issue. $\lozenge$

One natural way to deal with the difficulties of singular continuous measures in a numerical way employs the distribution function.

Remark 10.2 (*Distribution functions versus densities*). The Volterra-type iteration leads to a sequence $(F_n)_{n \in \mathbb{N}_0}$ of continuous distribution functions that converge uniformly (on $[0, 1]$) to F, which is continuous as well. The latter is shown in Figure 10.1 and resembles the classic middle-thirds Cantor measure of Figure 8.5 in various aspects, though it has full support (see below). Note that the corresponding sequence $(\mathrm{d}F_n)_{n \in \mathbb{N}_0}$ of absolutely continuous measures is only vaguely convergent, with the limit being purely singular continuous. Therefore, it is somewhat misleading to show a density for the Thue–Morse measure, as is often encountered in the literature. Still, it may be instructive to inspect the sequence of densities f_n in order to get some intuition on the singular nature of the Thue–Morse measure, or to study some of its scaling properties; see [CSM88, Zak02] and references therein. Note, however, that one has to deal with uncountably many potential peak locations, which is impossible in practice. In particular, scaling exponents can either be calculated for countably many positions explicitly, or for almost all positions via results from uniform distributions. $\diamondsuit$

On initial inspection, the distribution function F seems to have a plateau around $x = \frac{1}{2}$, with exact value $\frac{1}{2}$. More generally, as suggested by the Riesz products in Eq. (10.17), one might expect a plateau around any

$$x \in \left(\{0,1\} \cup \{m/2^k \mid k, m \in \mathbb{N} \text{ with } m \text{ odd}\}\right) \cap [0,1],$$

because the densities f_n vanish at x for all sufficiently large n, with the order of this zero increasing linearly with n. However, these potential gaps are all closed (see below). The corresponding values of F can be calculated with the series expansion of Eq. (10.9). Unlike the situation of the gap labelling theorem for one-dimensional Schrödinger operators [BHZ00], where one knows the values of the integrated density of states (IDOS) on the (always non-overlapping) plateaux but not their positions, we know the possible locations of the plateaux, but cannot see a topological constraint for the corresponding values of the distribution function (which can be determined from Eq. (10.9)).

It is interesting to note that the set of potential plateau locations coincides with the set of potential (but in our case extinct) Bragg peak positions, so the (extinct) Bragg peaks appear to 'repel' the continuous diffraction spectrum. Nevertheless, since the support is always closed by definition, one has

$$\mathrm{supp}\big(\mathrm{d}F|_{[0,1]}\big) = \mathrm{supp}(\varrho) = [0,1]$$

by [BBK05, Prop. 28], which also implies that F is a strictly increasing function. In particular, as one can quickly see from Eq. (10.7), $F(x) = 0$ forces $F(2x) = 0$, which (when repeated) contradicts $F(1/2) = 1/2$ unless $x = 0$. This shows that there is no gap around 0 (and none around 1 by symmetry). The general argument uses the fact that ϱ is a regular Borel measure; see [BBK05] and references therein for further details.

Remark 10.3 (*Thue–Morse comb with general weights*). When we replace the weights ± 1 in ω_{TM} by general (possibly complex) weights $h_\pm$, we obtain the new comb

$$\omega = \frac{h_+ + h_-}{2}\, \delta_{\mathbb{Z}} + \frac{h_+ - h_-}{2}\, \omega_{\mathrm{TM}}.$$

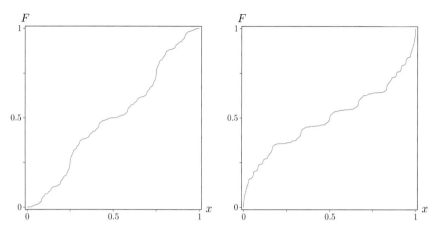

FIGURE 10.2. Distribution functions of the purely singular continuous diffraction measure of the generalised Thue–Morse chains for $\ell = 1$, with $k = 2$ (left) and $k = 5$ (right). Both are strictly increasing functions.

Its autocorrelation γ_ω can directly be calculated by volume-weighted (or Eberlein) convolution, which (using Example 8.10) gives

$$\gamma_\omega = \omega \circledast \widetilde{\omega} = \left| \frac{h_+ + h_-}{2} \right|^2 \delta_{\mathbb{Z}} + \left| \frac{h_+ - h_-}{2} \right|^2 \gamma_{\mathrm{TM}}.$$

This simple expression follows from $\omega_{\mathrm{TM}} \circledast \delta_{\mathbb{Z}} = 0 = \delta_{\mathbb{Z}} \circledast \widetilde{\omega_{\mathrm{TM}}}$. The Fourier transform $\widehat{\gamma_\omega}$ is then a singular measure, with a pure point and a singular continuous component. The pure point part (the singular continuous part) vanishes if and only if $h_+ + h_- = 0$ (respectively $h_+ = h_-$).   ◇

Example 10.1 (*Diffraction of generalised TM sequences*). Our above discussion can be extended (in the spirit of [Kea68]) to the family of substitutions $\varrho_{k,\ell}$ defined in Remark 4.9 on page 102. Here, we again use the binary representation with the alphabet $\{\pm 1\}$. In particular, given $k, \ell \in \mathbb{N}$, the corresponding distribution function F has a uniformly converging expansion precisely as in Eq. (10.9), however with the appropriate autocorrelation coefficients. The latter satisfy $\eta(0) = 1$ and $\eta(-m) = \eta(m)$ together with the recursion [BGG12a]

$$\eta\big((k+\ell)m + r\big) = \frac{1}{k+\ell}\big(\alpha(k,\ell,r)\eta(m) + \alpha(k,\ell,k+\ell-r)\eta(m+1)\big)$$

for arbitrary $m \in \mathbb{Z}$ and $0 \leqslant r \leqslant k + \ell - 1$, with the integral coefficients $\alpha(k,\ell,r) = k + \ell - r - 2\min(k,\ell,r,k+\ell-r)$. In particular, taking $m = 0$ and $r = 1$, one gets $\eta(1) = (k+\ell-3)/(k+\ell+1)$.

The Riesz product of Eq. (10.17) possesses the generalisation

$$f(x) = \prod_{n=0}^{\infty} \vartheta\big((k+\ell)^n x\big) \quad \text{with} \quad \vartheta(x) = 1 + \frac{2}{k+\ell} \sum_{r=1}^{k+\ell-1} \alpha(k,\ell,r)\cos(2\pi r x),$$

which is to be understood as a limit in the vague topology. The product represents a purely singular continuous measure with continuous (and also Hölder continuous) distribution function. Two further cases of corresponding distribution functions are shown in Figure 10.2; we refer to [BGG12a] for a detailed exposition. $\quad\Diamond$

The contrast to the Cantor measure of Figure 8.5 is interesting, as the Cantor and the (generalised) TM measures represent the opposite extremes of singular continuous measures on $[0,1]$ as far as their supporting set is concerned.

It is clear that the methods explained in this section also work in higher dimensions, though it might become technically more demanding to apply them explicitly. Let us briefly indicate how it works.

Example 10.2 (*Diffraction of the squiral tiling*). If we describe the hull of the squiral tiling via the symbolic version of Example 6.7 from page 215, it is a compact subset $\mathbb{X} \subset \{1,\bar{1}\}^{\mathbb{Z}^2}$. Due to unique ergodicity, each element of $\mathbb{X}$ defines a weighted Dirac comb with the same autocorrelation measure $\gamma = \eta\delta_{\mathbb{Z}^2}$, where the coefficients satisfy $\eta(0,0) = 1$ together with

$$\eta(3m+r, 3n+s) = \sum_{k=0}^{1}\sum_{\ell=0}^{1} \alpha_{k,\ell}^{(r,s)} \eta(m+k, n+\ell)$$

for $m,n \in \mathbb{Z}$. Here, the coefficients $\alpha_{k,\ell}^{(r,s)}$ are given by the following table.

$\genfrac{}{}{0pt}{}{(r,s)}{(k,\ell)}$	$(0,0)$	$(0,1)$	$(0,2)$	$(1,0)$	$(1,1)$	$(1,2)$	$(2,0)$	$(2,1)$	$(2,2)$
$(0,0)$	1	$-\frac{2}{9}$	$\frac{1}{3}$	$-\frac{2}{9}$	0	$-\frac{2}{9}$	$\frac{1}{3}$	$-\frac{2}{9}$	$\frac{1}{9}$
$(0,1)$	0	$\frac{1}{3}$	$-\frac{2}{9}$	0	$-\frac{2}{9}$	0	0	$\frac{1}{9}$	$-\frac{2}{9}$
$(1,0)$	0	0	0	$\frac{1}{3}$	$-\frac{2}{9}$	$\frac{1}{9}$	$-\frac{2}{9}$	0	$-\frac{2}{9}$
$(1,1)$	0	0	0	0	$\frac{1}{9}$	$-\frac{2}{9}$	0	$-\frac{2}{9}$	0

The coefficients for $-2 \leqslant m,n \leqslant 2$ follow from $\eta(0,0)$ by solving linear equations, while all others are determined by simple recursions. They satisfy $(-1)^{m+n}\eta(m,n) > 0$ for all $m,n \in \mathbb{Z}$.

Defining $\Sigma(N) = \sum_{m,n=0}^{N-1} \eta(m,n)^2$, one can show that $\Sigma(3N) < 4\Sigma(N)$. This allows the application of Wiener's criterion (in the formulation of Theorem 8.6) to establish that the diffraction measure is continuous. The relation $\eta(3m,3n) = \eta(m,n)$, together with the Riemann–Lebesgue lemma, implies that the diffraction measure is singular. Consequently, the diffraction is purely singular continuous; see [BG14] for further details and an illustration of the diffraction measure with its Riesz product structure. Once more, the corresponding distribution function is continuous, which need not hold for general singular continuous measures in dimensions $d > 1$.   ◊

It is clear that this method works for other (binary) bijective substitutions of constant length as well, and it can also be applied in higher dimensions. The key observation is that there is no need to calculate the recursion coefficients for the determination of the spectral type. Indeed, the mere existence of the recursion for the η-coefficients (with some mild and easy to verify conditions) implies that the spectral type of the diffraction measure is pure. From the analysis of the underlying dynamical system [Fra05], one can then extract the existence of a singular continuous component, whence the diffraction is purely singular continuous. For the actual calculation of the diffraction measure, via a multi-dimensional Riesz product, the explicit knowledge of the recursion coefficients is advantageous. Fortunately, they can conveniently be determined from the substitution by means of a computer algebra program.

The TM diffraction measure, even with general weights, does not explore the entire pure point part of the dynamical spectrum of the Thue–Morse chain, which is $\mathbb{Z}[\frac{1}{2}]$. We have seen earlier that the period doubling chain (which is pure point diffractive and of Toeplitz type) is a factor of the Thue–Morse chain. Its diffraction spectrum recovers the 'missing' part, and is actually an almost everywhere one-to-one cover of the Kronecker factor of the Thue–Morse chain. For the finiteness of the number of non-periodic factors, we refer to [Dur00, CDP10].

Let us briefly mention another (but rather different) example with singular diffraction, which is based on Section 5.7.7.

Example 10.3 (*Diffraction of SCD tilings*)**.** Consider an SCD tiling $\mathcal{T}$, as introduced in Eq. (5.7) on page 173, and place a unit point measure at an arbitrary (but fixed) control point inside each tile. By an application of [BF05, Thm. 2.1], the diffraction measure $\hat{\gamma}$ of the corresponding Dirac comb ω is a singular measure, so that $\hat{\gamma} = (\hat{\gamma})_{\mathsf{pp}} + (\hat{\gamma})_{\mathsf{sc}}$. If $\Gamma = \mathbb{Z}a + \mathbb{Z}b$ is the planar lattice of Section 5.7.7 and Γ^* its dual, the measure $\hat{\gamma}$ has support in the set

$$\overline{\left(\bigcup_{m \in \mathbb{Z}} R^m \Gamma^*\right)} \times \mathbb{R},$$

where $\overline{(.)}$ denotes the closure of a set and $\mathbb{R}$ represents the z-axis. Moreover, the restriction of $\hat{\gamma}$ to the z-axis is a pure point measure that is proportional to $\delta_{\mathbb{Z}/|c|}$, where $|c|$ is the distance between neighbouring layers. This result is indicated in [Dan95] and spelt out in [BF05, Thm. 2.2].

If $\mathcal{T}$ is incommensurate, but has a screw axis as in Eq. (5.7), the measure $\widehat{\gamma}$ is singular continuous on the complement of the z-axis; see [BF05, Thm. 2.5] for further details. In the commensurate case, an SCD tiling may be crystallographic or not. The former case is equivalent to the existence of a coincidence site sublattice of Γ for the rotation R, as mentioned in Remark 5.12. For further details, we refer to the discussion in [BF05].   ◇

10.2. Diffraction of the Rudin–Shapiro chain

Let us now consider the binary Rudin–Shapiro chain introduced in Section 4.7.1. More precisely, we start with the Dirac comb $\omega_{\mathrm{RS}} = w\delta_{\mathbb{Z}}$ where the function $w : \mathbb{Z} \longrightarrow \{\pm 1\}$ is indexed in steps of 4 and defined by the recursion

$$(10.18) \qquad w(4m + \ell) = \begin{cases} w(m), & \text{for } \ell \in \{0,1\}, \\ (-1)^{m+\ell}\, w(m), & \text{for } \ell \in \{2,3\}, \end{cases}$$

together with the initial conditions $w(0) = 1$ and $w(-1) = -1$. This corresponds (via $a \leftrightarrow 1$ and $b \leftrightarrow -1$) to the bi-infinite fixed point under the staggered substitution rule ϱ_{even} and ϱ_{odd} of Lemma 4.9 from page 105 with seed $b|a$. Since the staggered substitution rule has constant length 4, the letters at position $4m + \ell$, with $0 \leqslant \ell \leqslant 3$, are obtained by either ϱ_{even} or ϱ_{odd} applied to the letter at position m, which yields the recursion relations from Eq. (10.18).

Proposition 10.1. *Let* $\omega_{\mathrm{RS}} = w\delta_{\mathbb{Z}}$ *be the signed Dirac comb of the binary Rudin–Shapiro chain in its realisation as a sequence* $w \in \{\pm 1\}^{\mathbb{Z}}$, *as given by Eq. (10.18). Then, the autocorrelation coefficients*

$$\eta(m) = \lim_{N \to \infty} \frac{1}{2N+1} \sum_{i=-N}^{N} w(i)\, w(i+m)$$

exist for all $m \in \mathbb{Z}$ *and are given by* $\eta(m) = \delta_{m,0}$.

PROOF. Consider the autocorrelation coefficients $\eta(m)$ as given above, and additionally introduce the signed coefficients $\vartheta(m)$ by

$$\vartheta(m) = \lim_{N \to \infty} \frac{1}{2N+1} \sum_{i=-N}^{N} (-1)^i\, w(i)\, w(i+m).$$

The limits exist for all $m \in \mathbb{Z}$ as a consequence of Birkhoff's ergodic theorem (see Theorem 4.4), applied to the fixed point of the underlying 4-letter

substitution rule (which is uniquely ergodic). Indeed, consider the functions
$f_\pm\colon \{0,1,2,3\}^{\mathbb{Z}} \longrightarrow \{\pm 1\}$ defined by

$$s \longmapsto f_\pm(s) := (\mp 1)^{s_0}(-1)^{s_m}$$

for arbitrary, but fixed $m \in \mathbb{Z}$. These functions are continuous, as are their
restrictions to the original, quaternary Rudin–Shapiro hull of Section 4.7.1.
Recalling the alternating structure of the fixed point (with 0 or 2 on even and
1 or 3 on odd positions), the orbit averages of f_+ and f_- give the expressions
for η and ϑ. Continuity of $f_\pm$ together with the unique ergodicity now imply
that the limits indeed exist for all $m \in \mathbb{Z}$, via Theorem 4.4.

Clearly, one has $\eta(0) = 1$ and $\vartheta(0) = 0$, because $w(i)^2 = 1$ for all $i \in \mathbb{Z}$.
Then, considering m modulo 4, and splitting the sums in the definition of
$\eta(m)$ and $\vartheta(m)$ accordingly, the recursion relations of Eq. (10.18) imply

$$\eta(4m) = \frac{1 + (-1)^m}{2}\,\eta(m),$$

$$\eta(4m{+}1) = \frac{1 - (-1)^m}{4}\,\eta(m) + \frac{(-1)^m}{4}\,\vartheta(m) - \frac{1}{4}\vartheta(m{+}1),$$

$$\eta(4m{+}2) = 0,$$

$$\eta(4m{+}3) = \frac{1 + (-1)^m}{4}\,\eta(m{+}1) - \frac{(-1)^m}{4}\,\vartheta(m) + \frac{1}{4}\vartheta(m{+}1),$$

together with

$$\vartheta(4m) = 0,$$

$$\vartheta(4m{+}1) = \frac{1 - (-1)^m}{4}\,\eta(m) - \frac{(-1)^m}{4}\,\vartheta(m) + \frac{1}{4}\,\vartheta(m{+}1),$$

$$\vartheta(4m{+}2) = \frac{(-1)^m}{2}\,\vartheta(m) + \frac{1}{2}\,\vartheta(m{+}1),$$

$$\vartheta(4m{+}3) = -\frac{1 + (-1)^m}{4}\,\eta(m{+}1) - \frac{(-1)^m}{4}\,\vartheta(m) + \frac{1}{4}\,\vartheta(m{+}1).$$

It is a remarkable property that the introduction of ϑ suffices to obtain a
closed set of recursion relations.

Starting with the initial data $\eta(0) = 1$ and $\vartheta(0) = 0$, the sixth equation
(with $m = 0$) forces $\vartheta(1) = 0$, which then implies $\eta(1) = \vartheta(2) = 0$, as
well as $\eta(3) = \vartheta(3) = 0$. From here, the recursion relations imply that
$\eta(m) = \vartheta(m) = 0$ for all $m > 0$. Similarly, the last equation (with $m = -1$)
implies $\vartheta(-1) = 0$, hence $\eta(-1) = 0$ from the fourth equation. The recursions
now give $\eta(m) = \vartheta(m) = 0$ for all $m < 0$, thus establishing the claim.   □

This result also follows immediately from [PF02, Prop. 2.2.6], where a
concrete growth estimate is proved for certain finite sums that approximate

$\eta(m)$. Both approaches are rather short in comparison to the original accounts; compare [Que10] and references therein, and [BH00, BG09] for the diffraction context.

Theorem 10.2. *Let $\omega_{\mathrm{RS}} = w\delta_{\mathbb{Z}}$ be the weighted Dirac comb of the binary Rudin–Shapiro chain in its realisation as a sequence $w \in \{\pm 1\}^{\mathbb{Z}}$. This signed measure possesses the unique autocorrelation and diffraction measures*

$$\gamma_{\mathrm{RS}} = \delta_0 \quad and \quad \widehat{\gamma_{\mathrm{RS}}} = \lambda.$$

In particular, the diffraction measure is purely absolutely continuous.

PROOF. By Proposition 10.1, one has $\gamma = \sum_{m \in \mathbb{Z}} \eta(m)\,\delta_m = \delta_0$, so that the relation $\widehat{\delta_0} = \lambda$ from Example 8.4 completes the argument. $\square$

Let us mention that, as in the case of the Thue–Morse system, the Rudin–Shapiro chain is the best known, but by no means the only example of this kind. In fact, one can construct families of substitution systems in arbitrary dimension with analogous properties, for instance by means of Hadamard matrices; see [Fra03] for the construction method.

Remark 10.4 (*Autocorrelation and subword frequencies*). Let us come back to the frequency calculation of Example 4.13 on page 120. Let $m \geqslant 1$ be fixed. Then, with the subword cluster frequencies of Example 4.13, one finds

$$\eta(m) = \lim_{N \to \infty} \frac{1}{2N+1} \sum_{i=-N}^{N} w(i)w(i-m)$$

$$= \left(\nu_{a.a}^{(m+1)} + \nu_{b.b}^{(m+1)}\right) - \left(\nu_{a.b}^{(m+1)} + \nu_{b.a}^{(m+1)}\right) = 0,$$

where the last step follows from Proposition 10.1. Together with the symmetries discussed in Example 4.13, one arrives at Eq. (4.26). $\diamond$

Remark 10.5 (*Rudin–Shapiro comb with general weights*). As above in Remark 10.3 for the Thue–Morse chain, we can consider a comb with general (complex) weights $h_{\pm}$ as $\omega = \frac{1}{2}(h_+ + h_-)\delta_{\mathbb{Z}} + \frac{1}{2}(h_+ - h_-)\omega_{\mathrm{RS}}$. The corresponding autocorrelation is $\gamma_{\omega} = \omega \circledast \widetilde{\omega} = \frac{1}{4}|h_+ + h_-|^2\delta_{\mathbb{Z}} + \frac{1}{4}|h_+ - h_-|^2\delta_0$, due to $\gamma_{\mathrm{RS}} = \delta_0$ and the observation (along the lines of Example 8.10) that the volume-weighted convolutions of the cross terms vanish. This gives the diffraction measure

$$\widehat{\gamma_{\omega}} = \left|\frac{h_+ + h_-}{2}\right|^2 \delta_{\mathbb{Z}} + \left|\frac{h_+ - h_-}{2}\right|^2 \lambda,$$

which is of mixed type, with a pure point and an absolutely continuous component. $\diamond$

Let us point out that, once again, even the diffraction formula with general weights does *not* reveal the full pure point part of the dynamical spectrum of the RS system, which is given by $\mathbb{Z}[\frac{1}{2}]$; see Appendix B for background. However, the factor $\mathbb{Y}^{(2)} = \psi\big(\mathbb{X}_{\mathrm{RS}}^{(2)}\big)$ from Section 4.7.1 leads to Dirac combs with pure point diffraction spectra that are supported on $\mathbb{Z}[\frac{1}{2}]$. The explicit amplitudes can be calculated from Eq. (4.20) along the lines explained for Toeplitz sequences such as period doubling or paper folding in Section 9.4.4.

Let us now turn our attention to the diffraction theory of lattice subsets and their generalisations.

10.3. Diffraction of lattice subsets

So far, we have concentrated on point sets that are rather regular, in the sense that they are either crystallographic point sets, model sets, or arise from inflation rules. There are clearly many other structures of interest, displaying various types of deviations from the perfect order considered above. A natural first step consists of the analysis of certain subsets of our previous classes of examples. Here, it is an interesting question to consider to what extent the order of the underlying full point set is preserved.

Let us first investigate subsets $S \subset \Gamma$ of a lattice $\Gamma \subset \mathbb{R}^d$ or, more generally, weighted lattice Dirac combs

$$\omega = \sum_{x \in \Gamma} w(x)\, \delta_x$$

with a bounded function $w \colon \Gamma \longrightarrow \mathbb{C}$, as in Example 8.6. Clearly, one has $w(x) = \omega(\{x\})$ and ω is then a translation bounded measure. The lattice subset case $S \subset \Gamma$ is contained via the weight function $w = 1_S$, where 1_S is the characteristic function of the set S.

It is clear that such measures ω will not be crystallographic in general. In fact, generically, they do not have any non-trivial period at all. The diffraction measure $\widehat{\gamma_\omega}$ need not be pure point, but is typically of mixed type. Nevertheless, under rather weak assumptions, $\widehat{\gamma_\omega}$ is Γ^*-periodic, irrespective of its spectral type. This setting includes our previous examples as well as the large class of *lattice gases*, both with and without (stochastic) interactions.

Let us begin by sketching a heuristic idea of why this periodicity shows up; compare Lemma 10.1. Assume that we can find a 'nice' continuous function h such that we can rewrite the weighted Dirac comb ω as $\omega = h\,\delta_\Gamma$, where h satisfies $h(x) = w(x)$ for all $x \in \Gamma$. Then, a formal application of the convolution theorem, together with the PSF (Theorem 9.1), gives

$$\widehat{\omega} = \widehat{h} * \widehat{\delta_\Gamma} = \big(\mathrm{dens}(\Gamma)\,\widehat{h}\big) * \delta_{\Gamma^*}\,,$$

which would represent a Γ^*-periodic measure — provided all quantities are well-defined and all operations justifiable, which is generally not the case. Consequently, we need a more careful analysis via the autocorrelation.

10.3.1. INTERPOLATION OF THE AUTOCORRELATION

Let $\omega = w\,\delta_\Gamma$ be a weighted lattice Dirac comb with bounded weight function w. Consider its natural autocorrelation $\gamma_\omega = \sum_{z\in\Gamma} \eta(z)\,\delta_z$, which we assume to exist for the moment, with the autocorrelation coefficients

$$
\begin{aligned}
\eta(z) &= \lim_{r\to\infty} \frac{1}{\mathrm{vol}(B_r)} \sum_{\substack{x,x'\in\Gamma_r \\ x-x'=z}} w(x)\,\overline{w(x')} \\
&= \lim_{r\to\infty} \frac{1}{\mathrm{vol}(B_r)} \sum_{x\in\Gamma_r} w(x)\,\overline{w(x-z)}
\end{aligned}
$$

(10.19)

where $\Gamma_r = \Gamma \cap \overline{B_r(0)}$ as before. Note that the last step is correct because the sums, prior to taking the limit, differ only by a surface term that vanishes as $r \to \infty$; compare Example 9.1 and see [Schl00, Baa02b] for details.

If we now define $\omega_r = \sum_{x\in\Gamma_r} w(x)\,\delta_x$, we obtain a family $\{\gamma_r \mid r > 0\}$ of finite (or approximating) autocorrelations

$$
\gamma_r := \frac{\omega_r * \widetilde{\omega_r}}{\mathrm{vol}(B_r)}\,,
$$

which is, by construction, precompact in the vague topology. Consequently, by Proposition 9.1, this family has at least one limit point. There is thus a sequence of radii along which the finite autocorrelations converge, towards some γ say. Let us look at this specific limit point in detail.

Following [Lan93, Sec. VIII.4], let ϕ be the C^∞ 'hat function'

(10.20)
$$
\phi(x) := \begin{cases} \exp\!\left(\frac{|x|^2}{|x|^2-1}\right), & \text{if } |x| < 1, \\ 0, & \text{otherwise,} \end{cases}
$$

with compact support $\overline{B_1(0)}$. Define h by $h(x) = h_0\,\phi(x/\varepsilon)$ with some $h_0 > 0$ and some $\varepsilon > 0$ that is smaller than half the packing radius of the lattice Γ. Clearly, one has $\|h\|_\infty = h(0) = h_0 \geqslant h(x) \geqslant 0$, and h_0 is now chosen so that

$$
\bigl(h * \tilde{h}\bigr)(0) = \int_{\mathbb{R}^d} h(x)\,\tilde{h}(-x)\,\mathrm{d}x = \int_{\mathbb{R}^d} |h(x)|^2\,\mathrm{d}x = \|h\|_2^2 = 1.
$$

This way, h is an infinitely smooth bump function that is concentrated on $B_\varepsilon(0)$, with maximum value h_0 at the origin. Such a function is globally *Lipschitz continuous*, meaning that there is a constant L_h such that

$$
|h(x) - h(y)| \leqslant L_h\,|x - y|
$$

for all $x, y \in \mathbb{R}^d$.

Lemma 10.4. *If h is an integrable Lipschitz function with global Lipschitz constant L_h, the functions $\tilde{h}$ and $h * \tilde{h}$ are again globally Lipschitz, with $L_{\tilde{h}} = L_h$ and the estimate $L_{h*\tilde{h}} \leqslant \|h\|_1\, L_h$.*

PROOF. The Lipschitz continuity of $\widetilde{h}$ with $L_{\widetilde{h}} = L_h$ is clear. For the second claim, consider

$$\left| \left(h * \widetilde{h} \right)(x) - \left(h * \widetilde{h} \right)(y) \right| = \left| \int_{\mathbb{R}^d} h(z) \left(\widetilde{h}(x-z) - \widetilde{h}(y-z) \right) \mathrm{d}z \right|$$

$$\leqslant \int_{\mathbb{R}^d} |h(z)| \left| \widetilde{h}(x-z) - \widetilde{h}(y-z) \right| \mathrm{d}z$$

$$\leqslant L_h \int_{\mathbb{R}^n} |h(z)| \, |x-y| \, \mathrm{d}z = L_h \, \|h\|_1 \, |x-y|,$$

which proves both the Lipschitz property and the estimate.   □

Let us continue with the argument around the interpolating bump function. Defining $f_r = h * \omega_r$, one finds $f_r(z) = \sum_{x \in \Gamma_r} w(x) h(z-x)$ and

$$\left(f_r * \widetilde{f}_r \right)(z) = \sum_{u,v \in \Gamma_r} w(u) \overline{w(-v)} \left(h * \widetilde{h} \right)(z - u - v).$$

The special choice of ε implies that, for fixed $z \in \Gamma$ and arbitrary $u, v \in \Gamma$, $\left(h * \widetilde{h} \right)(z - u - v) = 1$ if and only if $u + v = z$, while it takes the value 0 otherwise. With

$$g_r := \frac{1}{\mathrm{vol}(B_r)} \left(f_r * \widetilde{f}_r \right),$$

one can now check that $\lim_{r \to \infty} g_r(z) = \eta(z)$ holds for all $z \in \Gamma$.

Lemma 10.5. *The family of functions $\{g_r \mid r > 0\}$ is uniformly Lipschitz, equicontinuous, and uniformly bounded.*

PROOF. Define $W = \sup_{x \in \Gamma} |w(x)|$ and observe that

$$\frac{\mathrm{card}(\Gamma_r)}{\mathrm{vol}(B_r)} = \mathrm{dens}(\Gamma) + \mathcal{O}(\tfrac{1}{r}) \quad \text{as } r \to \infty.$$

If $z \in \mathbb{R}^d$ is arbitrary (but fixed), one has $f_r(z) = w(x_z) h(z - x_z)$, where x_z is a point from Γ_r with minimal distance to z. If there is more than one point in Γ_r with minimal distance to z, a and b say, it does not matter which one we choose, as then $h(z-a) = h(z-b) = 0$ due to the condition on the support of h. Consider now $z, z' \in \mathbb{R}^d$. If $x_z = x_{z'}$, one clearly has $|f_r(z) - f_r(z')| \leqslant W L_h |z - z'|$. Otherwise, if $x_z \neq x_{z'}$, we know that the function h satisfies $h(z - x_{z'}) = h(z' - x_z) = 0$, and thus obtain

$$|f_r(z) - f_r(z')| \leqslant |w(x_z) h(z - x_z) - w(x_{z'}) h(z' - x_{z'})|$$

$$\leqslant |w(x_z)| \, |h(z - x_z) - h(z' - x_z)|$$

$$+ |w(x_{z'})| \, |h(z - x_{z'}) - h(z' - x_{z'})| \leqslant 2 W L_h |z - z'|.$$

This implies the inequalities

$$L_{f_r} \leqslant 2 \, W L_h \quad \text{and} \quad L_{f_r * \widetilde{f}_r} \leqslant 2 \, \mathrm{card}(\Gamma_r) \, W^2 \, \|h\|_1 \, L_h,$$

where the latter follows from Lemma 10.4 via the estimate

$$\|f_r\|_1 \leqslant \sum_{x \in \Gamma_r} |w(x)| \int_{\mathbb{R}^d} |h(z-x)| \, \mathrm{d}x = \mathrm{card}(\Gamma_r) W \|h\|_1.$$

Transferring this to g_r results in

$$L_{g_r} \leqslant 2 \frac{\mathrm{card}(\Gamma_r)}{\mathrm{vol}(B_r)} W^2 \|h\|_1 L_h = 2 \, \mathrm{dens}(\Gamma) W^2 \|h\|_1 L_h + \mathcal{O}(\tfrac{1}{r})$$

as $r \to \infty$, which shows uniform Lipschitz continuity and hence also equicontinuity of our family of functions.

In a similar fashion, since $\|h * \widetilde{h}\|_\infty = 1$ by construction, one derives

$$|g_r(z)| \leqslant \|g_r\|_\infty \leqslant \mathrm{dens}(\Gamma) W^2 + \mathcal{O}(\tfrac{1}{r}) \quad \text{as } r \to \infty,$$

from which uniform boundedness follows immediately. $\qquad\square$

By Ascoli's theorem, see [Lan93] for an exposition that matches our situation, we know that the set $\{g_r \mid r > 0\}$ is relatively compact with respect to $\|.\|_\infty$, so that, on any compact set $K \subset \mathbb{R}^d$, a uniformly converging subsequence is contained in $\{g_r \mid r > 0\}$, with a limit function g that is Lipschitz, positive definite, and satisfies $g(x) = \eta(x)$ for all $x \in \Gamma \cap K$. Since $\mathbb{R}^d$ has a countable base for its topology (a property also called separability), we can extend this to a general function g on all of $\mathbb{R}^d$, with compact convergence of the subsequence. So, we have proved the following result.

Proposition 10.2. *Let $\Gamma \subset \mathbb{R}^d$ be a lattice and consider the weighted Dirac comb $\omega = \sum_{x \in \Gamma} w(x) \, \delta_x$, with bounded function w. Let γ_ω be any limit point of the family $\{\gamma_{\omega_r} \mid r > 0\}$ of approximating autocorrelations. Then, there is a representation of the form $\gamma_\omega = g \, \delta_\Gamma$ with a bounded, positive definite Lipschitz function g on $\mathbb{R}^d$.* $\qquad\square$

10.3.2. Application to lattice subsets

The advantage of the above derivation is that we are now in the position to start with a well-defined autocorrelation (which can be any limit point of the set of finite autocorrelations, if more than one limit point exists) and to derive its Fourier transform.

Note that the bounded and continuous interpolation function g from Proposition 10.2 is positive definite, so that, by Bochner's theorem (see Theorem 8.6), its Fourier transform $\widehat{g}$ is a *finite* positive measure. So, the following calculation is perfectly justified,

$$\widehat{\gamma_\omega} = \big(g \, \delta_\Gamma\big)^{\widehat{}} = \widehat{g} * \widehat{\delta_\Gamma} = \mathrm{dens}(\Gamma) \, \widehat{g} * \delta_{\Gamma^*},$$

where the convolution theorem was used backwards, followed by another application of the PSF from Theorem 9.1.

Proposition 10.3. *Under the assumptions of Proposition* 10.2, *any diffraction measure* $\widehat{\gamma_\omega}$ *of the weighted Dirac comb* ω *is* (*at least*) Γ^*-*periodic.* □

Recall that a diffraction measure is always a positive measure. Employing Proposition 9.5, we may conclude that a finite positive measure ϱ exists that is supported in a fundamental domain and satisfies $\widehat{\gamma_\omega} = \varrho * \delta_{\Gamma^*}$. This is the precise version of the heuristic idea sketched above.

Theorem 10.3. *Let the assumptions be as in Proposition* 10.2, *and let* γ_ω *be any of the autocorrelations of the weighted Dirac comb* ω. *Then, the following properties hold.*

(1) $\gamma_\omega = \Phi \delta_\Gamma$, *with* Φ *an analytic function.*

(2) $\widehat{\gamma_\omega} = \varrho * \delta_{\Gamma^*}$, *with a finite positive measure* ϱ *of compact support.*

PROOF. Most steps have already been derived. That the interpolation function g from above can be replaced by an analytic function Φ is a simple consequence of the Paley–Wiener theorem; compare [RS80, Thm. IX.12]. □

Here, we have explicitly used the underlying lattice structure. One consequence is that the set of Bragg peak positions is either relatively dense or empty. This type of result is robust in the sense that the more general class of Meyer sets permits a similar consideration; see Section 10.5 below.

10.3.3. COMPLEMENTARY LATTICE SUBSETS

A particularly interesting situation emerges in the comparison of a lattice subset $S \subset \Gamma$ with its complement $S' = \Gamma \setminus S$. Here, the Dirac combs to be compared are δ_S and $\delta_{S'}$. Note that the Dirac comb δ_S results in an autocorrelation with coefficients

$$(10.21) \quad \eta_S(z) = \lim_{r \to \infty} \frac{\mathrm{card}\big(S \cap (z + S) \cap B_r(0)\big)}{\mathrm{vol}(B_r(0))} = \mathrm{dens}\big(S \cap (z + S)\big)$$

for all $z \in \mathbb{R}^d$, provided the limits exist. Otherwise, one restricts to a suitable unbounded sequence of increasing radii, in order to define a specific accumulation point. This can easily be derived from Eq. (10.19) and the comments following it.

Next, recall from Definition 9.5 that two infinite point sets are called *homometric* when they share the same (natural) autocorrelation. As we know from Section 9.6, such sets cannot be distinguished by (kinematic) diffraction.

Theorem 10.4. *Let* $\Gamma \subset \mathbb{R}^d$ *be a lattice, and let* $S \subset \Gamma$ *be a subset that possesses the* (*natural*) *autocorrelation coefficients* $\eta_S(z) = \mathrm{dens}\big(S \cap (z+S)\big)$ *for* $z \in \Gamma$ (*and* $\eta_S(z) = 0$ *otherwise*). *Then, the following properties hold.*

(1) The coefficients $\eta_{S'}(z)$ of the complement set $S' = \Gamma \setminus S$ also exist. They are $\eta_{S'}(z) = 0$ for all $z \notin \Gamma$ and otherwise, for $z \in \Gamma$, satisfy the relation

$$\eta_{S'}(z) - \mathrm{dens}(S') = \eta_S(z) - \mathrm{dens}(S).$$

(2) The diffraction spectra of the sets S and S' are related by

$$\widehat{\gamma_{S'}} = \widehat{\gamma_S} + \big(\mathrm{dens}(S') - \mathrm{dens}(S)\big)\,\mathrm{dens}(\Gamma)\,\delta_{\Gamma^*}.$$

In particular, $\widehat{\gamma_{S'}} = \widehat{\gamma_S}$ if $\mathrm{dens}(S') = \mathrm{dens}(S)$, in which case S and S' are homometric.

(3) The diffraction measure $\widehat{\gamma_{S'}}$ is pure point if and only if $\widehat{\gamma_S}$ is pure point.

PROOF. In what follows, each term involving a density may be viewed as the limit along a fixed increasing and unbounded sequence of radii, as in Eq. (10.21) and below it. Due to the decomposition $\Gamma = S \,\dot\cup\, S'$, we get $\mathrm{dens}(S') = \mathrm{dens}(\Gamma) - \mathrm{dens}(S)$, so that the natural density of S' exists because $\mathrm{dens}(S) = \eta_S(0)$. Since $S' \subset \Gamma$, we also have $\eta_{S'}(z) = 0$ whenever $z \notin \Gamma$.

So, let $z = x \in \Gamma$ from now on. Next, observe that $\Gamma \cap (x + \Gamma) = \Gamma$ and thus, using $\Gamma = S \,\dot\cup\, S'$, we obtain

$$\begin{aligned}
\mathrm{dens}(\Gamma) &= \mathrm{dens}\big(\Gamma \cap (x + \Gamma)\big) \\
&= \eta_{S'}(x) + \eta_S(x) + \mathrm{dens}\big(S \cap (x + S')\big) + \mathrm{dens}\big(S' \cap (x + S)\big).
\end{aligned}$$

Since $S' = \Gamma \setminus S$, it is easy to verify that

$$\begin{aligned}
\mathrm{dens}\big(S' \cap (x + S)\big) &= \mathrm{dens}\big(\Gamma \cap (x + S)\big) - \mathrm{dens}\big(S \cap (x + S)\big) \\
&= \mathrm{dens}(S) - \eta_S(x)
\end{aligned}$$

because $(x + S) \subset \Gamma$ and $\mathrm{dens}(x + S) = \mathrm{dens}(S)$. Similarly,

$$\mathrm{dens}\big(S \cap (x + S')\big) = \mathrm{dens}(S) - \eta_S(-x),$$

by first shifting (by $-x$) and then using the previous formula. Since η_S is a positive definite real function, we have $\eta_S(-x) = \eta_S(x)$, and obtain

$$\mathrm{dens}(\Gamma) = 2\,\mathrm{dens}(S) + \eta_{S'}(x) - \eta_S(x)$$

from which the first assertion follows with $\mathrm{dens}(\Gamma) = \mathrm{dens}(S) + \mathrm{dens}(S')$.

Since $S \subset \Gamma$, its autocorrelation is $\gamma_S = \sum_{x \in \Gamma} \eta_S(x)\delta_x$, and analogously for S', the complement set in Γ. From the first assertion, we then infer

$$\gamma_{S'} = \gamma_S + c\,\delta_\Gamma$$

with $c = \mathrm{dens}(S') - \mathrm{dens}(S)$. Assertion (2) now follows from taking the Fourier transform and applying the PSF of Theorem 9.1 to the lattice Dirac comb δ_Γ. When $c = 0$, one has $\eta_{S'}(z) = \eta_S(z)$ for all z by property (1), and hence the claimed homometry.

Finally, the difference between $\widehat{\gamma_{S'}}$ and $\widehat{\gamma_S}$ in the third assertion is a multiple of δ_{Γ^*}, which is a uniform lattice Dirac comb and hence a pure point measure, whence the last claim is obvious. □

Let us now apply these findings to some interesting lattice subsets.

10.4. Visible lattice points

Various point sets emerge from problems in elementary number theory. A versatile example is provided by the visible points of a lattice in $\mathbb{R}^d$ with $d \geqslant 2$. They are the points with coprime coordinates in a given lattice basis. Note that this characterisation does not depend on the choice of the basis. Here, we mainly focus on the lattice $\mathbb{Z}^2$; see [Apo76] for the case of $\mathbb{Z}^2$ and [BMP00, BM04, PH13] for the general theory. Let us define

$$ \mathcal{V} = \mathcal{V}_{\mathbb{Z}^2} = \left\{ x \in \mathbb{Z}^2 \mid \gcd(x) = 1 \right\}, $$

where $\gcd(x) = \gcd(x_1, x_2)$ for $x = (x_1, x_2)$. The term 'visible' refers to the geometric interpretation that the points of $\mathcal{V}$ are the ones visible from the origin, in the sense that no other lattice point lies between the origin and any of the points of $\mathcal{V}$. Due to their characterisation with the gcd, they are also known as the *primitive points* of $\mathbb{Z}^2$. A central (square-shaped) patch of $\mathcal{V}$ is shown in Figure 10.3.

Proposition 10.4. *The set $\mathcal{V}$ of visible lattice points of $\mathbb{Z}^2$ has the following properties.*

  (1) *The set $\mathcal{V}$ is uniformly discrete, but not relatively dense. In particular, $\mathcal{V}$ contains holes of arbitrary size that repeat lattice-periodically.*
  (2) *The group $\mathrm{GL}(2, \mathbb{Z})$ acts transitively on $\mathcal{V}$, and one has the partition $\mathbb{Z}^2 = \dot{\bigcup}_{m \in \mathbb{N}_0} m\mathcal{V}$ of $\mathbb{Z}^2$ into $\mathrm{GL}(2, \mathbb{Z})$-invariant sets.*
  (3) *The difference set is $\mathcal{V} - \mathcal{V} = \mathbb{Z}^2$.*
  (4) *The natural density of $\mathcal{V}$ exists and is given by $\mathrm{dens}(\mathcal{V}) = \frac{1}{\zeta(2)} = \frac{6}{\pi^2}$, where ζ denotes Riemann's zeta function.*

SKETCH OF PROOF. The uniform discreteness of the first claim follows from $\mathcal{V} \subset \mathbb{Z}^2$. The existence of holes comprising s lattice points follows from the Chinese remainder theorem, applied to simultaneous congruences modulo $m_1 \mathbb{Z}^2, \ldots, m_s \mathbb{Z}^2$ with pairwise coprime integers m_i. The holes then repeat with $m_1 \cdots m_s \mathbb{Z}^2$ as lattice of periods; see [BMP00, Prop. 4] for details.

Property (2) follows from the observation that $\mathrm{GL}(2, \mathbb{Z})$ transformations map primitive points to primitive points, hence preserve the gcd of the coordinates. Since $(1, 0)$ can be mapped to any other primitive point this way, transitivity follows. The partition property is obvious.

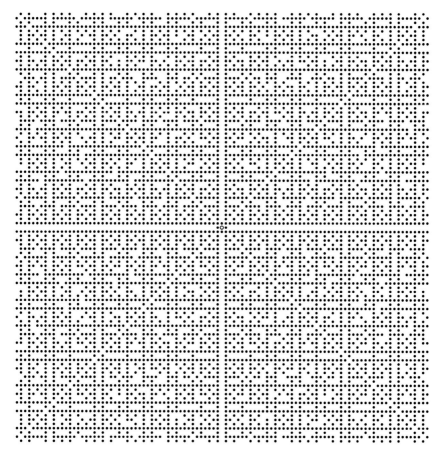

FIGURE 10.3. Visible points of the square lattice, as seen from the origin (circle). The point set $\mathcal{V}$ is D_4-symmetric, and invariant under $\mathrm{GL}(2, \mathbb{Z})$.

To establish property (3), observe that any element of $\mathbb{Z}^2$ is the difference $(x_1, x_2) = (x_1 + 1, 1) - (1, 1 - x_2)$ of two primitive points.

Claim (4) about the density is a standard example of the use of Möbius inversion, see [Apo76, Thm. 3.9], where the average is taken over squares. It is not difficult to see that the average over centred disks gives the same result; see [BMP00, Appendix] for details. □

Remark 10.6 (*Holes in $\mathcal{V}$*). Square-shaped holes of increasing size are exponentially rare, but repeat lattice-periodically. The 2×2 holes closest to the origin are those with lower left corner $(20, 14)$ and its symmetry related counterparts, eight altogether. Since the primes involved in the congruences are 2, 3, 5 and 7, each of these eight holes repeats lattice periodically, with lattice $210\,\mathbb{Z}^2$.

A 3×3 hole begins at $(1308, 1274)$, with lattice of periods $39\,270\,\mathbb{Z}^2$, or at $(3794, 1000)$, with periods $30\,030\,\mathbb{Z}^2$. A 4×4 example starts at the corner $(13\,458\,288, 13\,449\,225)$. ◊

To develop this example further, we need to analyse the natural autocorrelation γ of $\mathcal{V}$. Provided that γ exists, part (3) of Proposition 10.4 implies that it must be of the form $\gamma = \sum_{x\in\mathbb{Z}^2} \eta(x)\,\delta_x$. The following result emerges from a less trivial application of the Möbius inversion principle, as derived in [BMP00, Thm. 2]. The results are expressed as Euler products over rational primes p (either finite products or absolutely convergent ones).

Lemma 10.6. *For each $x \in \mathbb{Z}^2$, the natural autocorrelation coefficient $\eta(x)$ of the visible points $\mathcal{V}$ exists, and is given by*

$$\eta(x) \;=\; \xi \prod_{p\,|\,\gcd(x)} \left(1 + \frac{1}{p^2 - 2}\right)$$

with $\xi = \prod_p (1 - 2\,p^{-2}) \approx 0.3226$. In particular, with $\gcd(0) = 0$, this also gives the density $\eta(0) = \prod_p (1 - p^{-2}) = 1/\zeta(2)$. □

Next, we have to determine the diffraction measure $\widehat{\gamma}$. Observe that

$$
\begin{aligned}
\prod_p \left(1 + \frac{1}{p^2 - 2}\right) &= 1 + \sum_p \frac{1}{p^2 - 2} + \sum_{p<q} \frac{1}{p^2 - 2}\,\frac{1}{q^2 - 2} + \cdots \\
&= \sum_{\substack{m\geqslant 1 \\ m\ \text{squarefree}}} \prod_{p\,|\,m} \frac{1}{p^2 - 2}
\end{aligned}
$$

(10.22)

where the last sum is over all integers that are not divisible by a square $\neq 1$. The corresponding identity also holds under the additional constraints $p\,|\,k$ and $m\,|\,k$. In view of Lemma 10.6, and the partition of $\mathbb{Z}^2$ into subsets $m\mathcal{V}$ of fixed gcd m according to part (2) of Proposition 10.4, the autocorrelation measure γ of $\mathcal{V}$ can thus be expressed as

$$\gamma \;=\; \xi \sum_{\substack{m\geqslant 1 \\ m\ \text{squarefree}}} \left(\prod_{p\,|\,m} \frac{1}{p^2 - 2}\right) \delta_{m\mathbb{Z}^2}\,.$$

In particular, one can verify that the overall contribution to δ_0 is indeed $\eta(0) = \mathrm{dens}(\mathcal{V})$. A termwise application of the PSF of Theorem 9.1 leads to

$$\widehat{\gamma} \;=\; \xi \sum_{\substack{m\geqslant 1 \\ m\ \text{squarefree}}} \left(\prod_{p\,|\,m} \frac{1}{p^2\,(p^2 - 2)}\right) \delta_{\mathbb{Z}^2/m}\,,$$

(10.23)

which is justified by proper convergence. Let us recall that we write a pure point diffraction measure as $\sum_k I(k)\,\delta_k$ with intensities $I(k)$, where the sum runs over a finite or a countable set. We can now summarise the spectral properties of the visible points as follows.

Theorem 10.5. *The natural diffraction measure of the visible points $\mathcal{V}$ of the square lattice $\mathbb{Z}^2$ exists. It is the positive pure point measure of Eq. (10.23), with ξ as in Lemma 10.6. The measure $\widehat{\gamma}$ is concentrated on the points of $\mathbb{Q}^2$ with squarefree denominator. In particular, $I(0) = \left(1/\zeta(2)\right)^2 = 36/\pi^4$. When $0 \neq k \in \mathbb{Q}^2$ has squarefree denominator $\mathrm{den}(k) := \gcd\{n \in \mathbb{N} \mid nk \in \mathbb{Z}^2\}$, the corresponding intensity is given by*

$$I(k) \;=\; \left(\frac{6}{\pi^2} \prod_{p \mid \mathrm{den}(k)} \frac{1}{p^2 - 1} \right)^2,$$

while $I(k) = 0$ for all $k \in \mathbb{R}^2 \setminus \mathbb{Q}^2$ and for all $k \in \mathbb{Q}^2$ whose denominator $\mathrm{den}(k)$ is divisible by a square $\neq 1$.

PROOF. The expression in Eq. (10.23) represents a norm converging series of positive pure point measures and is thus a positive pure point measure, by an application of Theorem 8.4. The pure points obviously form a subset of $\mathbb{Q}^2$.

The formula for $k = 0$ follows from Eq. (10.23) by an explicit calculation. Let now $0 \neq k \in \mathbb{Q}^2$, with denominator $\mathrm{den}(k)$. If $\mathrm{den}(k)$ is divisible by a square $\neq 1$, Eq. (10.23) gives $\widehat{\gamma}(\{k\}) = 0$, so let $d = \mathrm{den}(k)$ be squarefree. The intensity $I(k)$ is the sum of various contributions, yielding

$$I(k) = \xi \sum_{\substack{\ell \geqslant 1 \\ \gcd(\ell,d)=1 \\ \ell \text{ squarefree}}} \prod_{p \mid \ell d} \frac{1}{p^2 \left(p^2 - 2\right)} = \xi \prod_{p \mid d} \frac{1}{p^2 \left(p^2 - 2\right)} \sum_{\substack{\ell \geqslant 1 \\ \gcd(\ell,d)=1 \\ \ell \text{ squarefree}}} \prod_{p \mid \ell} \frac{1}{p^2 \left(p^2 - 2\right)}$$

$$= \prod_{p} \frac{p^2 - 2}{p^2} \prod_{p \mid d} \frac{1}{p^2 \left(p^2 - 2\right)} \prod_{p \nmid d} \left(1 + \frac{1}{p^2 \left(p^2 - 2\right)} \right)$$

$$= \prod_{p \mid d} \frac{1}{p^4} \prod_{p \nmid d} \left(\frac{p^2 - 1}{p^2} \right)^2 = \frac{1}{\zeta(2)^2} \prod_{p \mid d} \frac{1}{\left(p^2 - 1\right)^2},$$

where an identity of the type of Eq. (10.22) with the additional constraints $\gcd(p,d) = 1$ and $\gcd(\ell,d) = 1$ was used in the third step, while the last equality follows from $1/\zeta(2) = \prod_p (1 - p^{-2}) = 6/\pi^2$. Since $\mathrm{den}(0) = 1$, the general formula also reproduces the correct intensity for $k = 0$. $\qquad \square$

Let us analyse some of the properties of $\widehat{\gamma}$. The result of Theorem 10.5 implies the relation $I(k) = I(0)\,f(\mathrm{den}(k))^2$ with

$$(10.24) \qquad f(d) := \begin{cases} \prod_{p \mid d} \frac{1}{p^2 - 1}, & d \text{ squarefree}, \\ 0, & \text{otherwise}, \end{cases}$$

for all $k \in \mathbb{Q}^2$, where $f(1) = 1$. This permits a straightforward calculation of the diffraction image, which is shown in Figure 10.4; see also [BGW94] which contains a comparison with an optical diffraction experiment. As before,

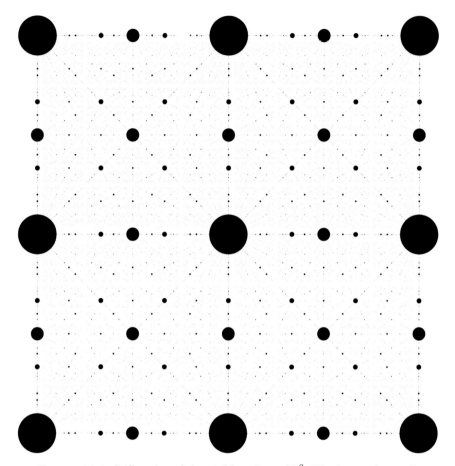

FIGURE 10.4. Diffraction of the visible points of $\mathbb{Z}^2$. The image shows all intensities with $I(k)/I(0) \geqslant 10^{-6}$ and $k \in \mathbb{Q}^2 \cap [0,2]^2$; see text for details, in particular on the $\mathrm{GL}(2,\mathbb{Z})$ invariance.

a point measure at k is represented by a disk, centred at k, with an area proportional to $I(k)$. Since $f(d) = 0$ if and only if d fails to be squarefree, we have a complete description of the extinction rules in this case. This is particularly visible for the denominator $d = 4$, which leads to 'white lines' in Figure 10.4.

Moreover, one explicitly sees that the diffraction image is fully translation invariant, with $\mathbb{Z}^2$ as its lattice of periods, in accordance with Proposition 10.3. Finally, as $\mathrm{GL}(2,\mathbb{Z})$-transformations do not change the denominator of a rational point, we inherit complete $\mathrm{GL}(2,\mathbb{Z})$-invariance of the diffraction measure from that of the set $\mathcal{V}$ itself. Our summary reads as follows.

Corollary 10.3. *The diffraction measure $\hat{\gamma}$ of the visible lattice points $\mathcal{V}$ of $\mathbb{Z}^2$ is a pure point measure with positive intensities precisely on the rational points with squarefree denominator. Moreover, the measure $\hat{\gamma}$ has the symmetry group $\mathbb{Z}^2 \rtimes \mathrm{GL}(2, \mathbb{Z})$.* □

The above results, with minor modifications, also hold for visible points in arbitrary planar lattices, and similarly for lattices in higher dimensions; see [BMP00, BM04, PH13] for further details. Since our previous examples with pure point diffraction were regular model sets, it is an obvious question to ask how visible points are connected to model sets.

Proposition 10.5. *Let Γ be a lattice in $\mathbb{R}^d$ with $d \geqslant 2$. The visible points $\mathcal{V}_\Gamma$ of Γ cannot be described as a regular model set. Moreover, they cannot be turned into one by adding or removing point sets of zero density.*

PROOF. All regular model sets are Meyer sets by Proposition 7.5, and hence certainly Delone. The first claim is clear because $\mathcal{V}_\Gamma$ fails to be Delone by part (1) of Proposition 10.4, which easily extends to all lattices under consideration. Because the holes of a given size in $\mathcal{V}_\Gamma$ repeat lattice periodically, turning $\mathcal{V}_\Gamma$ into a Delone set requires the addition of a point set of positive density, which establishes the second claim. □

In spite of this negative result, the visible lattice points can be realised as weak model sets, with an adelic internal space and a window W that coincides with its own boundary, $W = \partial W$. Moreover, ∂W has positive measure, so that the general theory does not apply. This leads to several interesting questions, including the patch counting entropy of these sets; see [BMP00, Sin06, BLR07, PH13] and references therein for additional material.

One immediate question in this context concerns the (discrete) hull of $\mathcal{V}$,

$$\mathbb{X}_0(\mathcal{V}) := \overline{\{t + \mathcal{V} \mid t \in \mathbb{Z}^2\}},$$

where we can view $\mathcal{V}$ and its translates simultaneously as subsets of $\mathbb{Z}^2$ and as configurations (meaning subsets of $\{0, 1\}^{\mathbb{Z}^2}$). It is clear from the existence of holes of unbounded size that $\mathbb{X}_0(\mathcal{V})$ contains the empty set (the configuration of 0 on every lattice point). Less obvious is the fact that it also contains *all* finite subsets of $\mathcal{V}$ (and their translates). In general, one has the following result [PH13].

Theorem 10.6. *Let $\mathcal{V}$ denote the visible points of $\mathbb{Z}^2$. Its discrete hull is then given by*

$$\mathbb{X}_0(\mathcal{V}) = \left\{ \Lambda \subset \mathbb{Z}^2 \mid \mathrm{card}\big(\Lambda \bmod p\mathbb{Z}^2\big) \leqslant p^2 - 1 \text{ for all } p \in \mathcal{P} \right\},$$

where $\mathcal{P}$ denotes the set of rational primes. □

When viewing a (finite) subset as a configuration, its cardinality is the number of positions that carry a 1. Deciding whether a finite subset of $\mathbb{Z}^2$ belongs to $\mathbb{X}_0(\mathcal{V})$ amounts to verifying only finitely many modulo conditions.

Let us expand on some possible extensions of the above results.

Example 10.4 (*Squarefree numbers*). Obviously, the set of visible lattice points in one dimension is finite and hence not of interest in this context. However, the set of *squarefree integers*

$$\mathcal{V} := \{n \in \mathbb{Z} \mid n \text{ not divisible by a square} \neq 1\}$$

is a point set of density $6/\pi^2$ and shows striking similarities with the set of visible lattice points. In particular, it has holes of arbitrary size, and it is pure point diffractive. In fact, by [BMP00, Thm. 5], the diffraction measure is the positive pure point measure supported on the rational numbers with cubefree denominator q. The diffraction intensity at $k = m/q$ with $\gcd(m, q) = 1$ and q cubefree reads $I(k) = I(0)f(q)^2$, with $I(0) = 1/\zeta(2)^2$ and the function f from Eq. (10.24), used only for cubefree q.

Similarly to Theorem 10.6, one can characterise the discrete hull of the squarefree numbers as

$$\mathbb{X}_0(\mathcal{V}) = \{\Lambda \subset \mathbb{Z} \mid \mathrm{card}(\Lambda \bmod p^2\mathbb{Z}) \leqslant p^2 - 1 \text{ for all } p \in \mathcal{P}\},$$

again with $\mathcal{P}$ the set of rational primes; see [Sar11, PH13] for two rather different proofs of this statement, and [CS13] for dynamical aspects. For further developments, see [CV13, HuB14] and references therein.   ◇

Remark 10.7 (*Invisible lattice points*). Above, it was shown that the set $\mathcal{V}$ of visible lattice points is pure point diffractive. The last assertion of Theorem 10.4 then tells us that their complement, the set of *invisible* points, is pure point diffractive as well. Similarly, the set of kth power-free integers, a subset of $\mathbb{Z}$, has pure point diffraction [BMP00], and then so does its complement, the set of integers divisible by the kth power of some integer $\geqslant 2$; see also Example 10.4 above. This indicates that many more pure point diffractive point sets of independent interest exist. A general criterion based on the almost periodicity of the autocorrelation measure is derived in [BM04]; see [GLA90] for general background material.   ◇

At this point, one natural question is whether there exists a unifying framework for the above examples, and the answer is affirmative [PH13]. Indeed, one can consider the set of points in a d-dimensional lattice Γ that are k-free. This means that their coordinates in a lattice basis are not divisible by any non-trivial kth power of an integer. This leads to a systematic exposition both of the entropy and the diffraction questions; see [PH13] for the details. Also, assuming the lattice $\Gamma \subset \mathbb{R}^d$ to be unimodular ($|\det(\Gamma)| = 1$), the hull

of the k-free points in Γ can be characterised in this generality as
$$\mathbb{X}_0(\mathcal{V}) = \big\{ \Lambda \subset \Gamma \mid \operatorname{card}(\Lambda \bmod p^k \Gamma) \leqslant p^{kd} - 1 \text{ for all } p \in \mathcal{P} \big\},$$
with $\mathcal{P}$ as above; compare [PH13, Thm. 6].

An obvious next question concerns an extension to visible points in a model set. This, however, is a more difficult problem, and only preliminary work has been done. There are two obstacles to overcome. On the one hand, the visibility condition is more involved than in the lattice case, where it simply reduces to primitivity of the coordinates in a lattice basis. For instance, for the individually eightfold symmetric Ammann–Beenker model set of Example 7.8, the subset of visible points (with respect to its symmetry centre) is given by[1]
$$\mathcal{V}_{\mathrm{AB}} = \big\{ x = \alpha + \beta \xi_8 \mid x^\star \in (W_{\mathrm{AB}} \setminus \lambda^\star W_{\mathrm{AB}}) \text{ and } \alpha, \beta \in \mathbb{Z}[\sqrt{2}] \text{ coprime} \big\},$$
where $\lambda = 1 + \sqrt{2}$ as before. On the other hand, visibility relative to different points of a model set might (and will) give rather different sets. It is not clear whether a suitable averaging is the best way to proceed, or whether a maximally symmetric model set is preferable.

The above examples fail to be Meyer sets. However, there are also interesting lattice subsets that are Meyer themselves, such as those in Example 2.4 on page 15. Let us consider the coin tossing case in more detail.

Example 10.5 (*Diffraction of Meyer sets with entropy*). We start from the Dirac comb $\omega = \sum_{n \in \mathbb{Z}} w(n) \delta_n$, where $w(n) = 1$ for all $n \in 2\mathbb{Z}$. For $n \in 2\mathbb{Z}+1$, the weight $w(n)$ takes values 0 or 1 with equal probability, and independently of each other. The autocorrelation coefficients almost surely exist in the probabilistic sense (which will be made precise in Chapter 11) and are given by $\eta(0) = \frac{3}{4}$, $\eta(2m+1) = \frac{1}{2}$ for $m \in \mathbb{Z}$ and $\eta(2m) = \frac{5}{8}$ for $0 \neq m \in \mathbb{Z}$. The autocorrelation and diffraction measures thus read
$$\gamma = \frac{1}{2} \delta_{\mathbb{Z}} + \frac{1}{8} \delta_{2\mathbb{Z}} + \frac{1}{8} \delta_0 \quad \text{and} \quad \widehat{\gamma} = \Big(\frac{1}{16} + \frac{1 + \cos(2\pi k)}{4} \Big) \delta_{\mathbb{Z}/2} + \frac{1}{8} \lambda,$$
both equations being almost surely true in the probabilistic sense. $\diamond$

A characteristic feature of the last example, despite its mixed spectrum, is the appearance of relatively dense patterns of Bragg peaks of equal intensity. A similar phenomenon turns out to be true for the diffraction of all Meyer sets, as we will now briefly summarise.

10.5. Extension to Meyer sets

Our exposition above shows that the diffraction of any lattice subset S of positive density must comprise a non-trivial pure point component. Since

---
[1]We thank Bernd Sing for interesting discussions on this problem.

Meyer sets are natural generalisations of lattices, with a high degree of intrinsic order, one might expect that a similar property holds for Meyer sets and their subsets. Although we know that Meyer sets are always relatively dense subsets of model sets [Mey72, Moo97a], the Dirac combs of such sets are generally *not* pure point diffractive. This is clear due to the existence of Meyer sets with entropy, such as those of Examples 2.4 and 10.5, or those of Section 11.2.3 below.

This question has been investigated by Strungaru [Str05a, Str05b, Str11, Str13a, Str13b] by methods from the theory of almost periodic measures [GLA90]. For Meyer sets, the result reads as follows.

Theorem 10.7. *Let $\Lambda \subset \mathbb{R}^d$ be a Meyer set and $\omega = \delta_\Lambda$ the corresponding Dirac comb. If γ is any autocorrelation of ω, its Fourier transform $\widehat{\gamma}$ comprises a non-trivial pure point part. In particular, for any $\varepsilon > 0$, the set $\left\{ k \in \mathbb{R}^d \mid \widehat{\gamma}(\{k\}) \geqslant (1 - \varepsilon)\, \widehat{\gamma}(\{0\}) \right\}$ is relatively dense.*

SKETCH OF PROOF. Any autocorrelation of ω is obtained as a limit along a suitably chosen averaging sequence $\mathcal{A}$ of van Hove type. The first claim is now an application of [Str05b, Prop. 3.12], which also yields that the positions of the Bragg peaks are relatively dense. The stronger second claim is obtained as a special case of [Str13b, Thm. 5.1]. □

In general, the diffraction spectrum is of mixed type. If so, also the continuous part is extended in the sense that its support is relatively dense. This also follows from [Str05b, Prop. 3.12]. Various possibilities exist to derive similar results for certain classes of Delone sets. For instance, any Delone set that contains a Meyer set has the property that each of its autocorrelations possesses an infinite set of Bragg peaks [Str05b, Prop. 3.14].

As a further step, one can consider subsets of Meyer sets of positive density. Any such set, S say, still has the property that both S and $S - S$ are uniformly discrete, though they need not be relatively dense. This case can conveniently be viewed as a weighted Dirac comb with Meyer set support, with weights in $\{0, 1\}$. In general, one obtains the following result from [Str05b, Prop. 4.1].

Theorem 10.8. *Let $\omega = \sum_{x \in M} w(x)\, \delta_x$ be a translation bounded measure with Meyer set support M, and let γ be any autocorrelation of ω. Then, the support of $\left(\widehat{\gamma}\right)_{\mathsf{pp}}$ is non-empty if and only if it is relatively dense.* □

This is the adequate generalisation of Proposition 10.3 to Meyer sets. It is thus impossible to have $\left(\widehat{\gamma}\right)_{\mathsf{pp}} = \widehat{\gamma}(\{0\})\, \delta_0$ with $\widehat{\gamma}(\{0\}) \neq 0$, which once more demonstrates the high intrinsic degree of coherence of Meyer sets. In particular, if the average scattering strength is non-zero, one immediately inherits a relatively dense set of Bragg peaks in this class.

Let us add a heuristic argument for the validity and structure of Theorems 10.7 and 10.8. Consider a Meyer set $M \subset \mathbb{R}^d$, and let $\Lambda \subset \mathbb{R}^d$ be a suitably chosen model set such that $M \subset \Lambda$ (which we know to exist from the general theory [Moo97a, Moo00]). Any weighted Dirac comb with support M can then be written as $\omega = w\delta_\Lambda$, with a suitable weight function w. Let us assume that its autocorrelation $\gamma_\omega = \sum_{z \in \Lambda - \Lambda} \eta(z)\delta_z$ exists, possibly defined along a special averaging sequence.

Let us now write $\gamma_\omega = h\gamma_\Lambda$ with $\gamma_\Lambda = \delta_\Lambda \circledast \widetilde{\delta_\Lambda}$, and assume that the function h is sufficiently 'nice' (as a function on $\mathbb{R}^d$), in the sense that it is bounded and that $\widehat{h\lambda}$ exists. The latter must then be a finite measure, and one obtains

$$\widehat{\gamma_\omega} = \widehat{h} * \widehat{\gamma_\Lambda}.$$

The case $h \equiv 1$ brings us back, via $\widehat{h} = \delta_0$, to the diffraction of the model set Λ itself, while other cases with $\widehat{h}$ of mixed spectral type lead to the distribution of these components over $\mathbb{R}^d$. This (heuristic) argument is not suitable for a formal proof of Theorem 10.8 (at least not in its generality), but it nevertheless illustrates the result in a way that is analogous to our treatment of the special subclass of (weighted) lattice subsets.

CHAPTER 11

Random Structures

In the absence of deterministic order, hence beyond the systems discussed so far, one can still expect interesting regimes of (partial) order. This point of view is supported by results from statistical physics, in particular from the theory of equilibrium systems, which are ensembles that are governed by Gibbs measures (to model thermal disorder). In the interest of readability, we restrict to simple cases and mainly focus on geometric properties or independent disorder. In particular, the notions of *stochastic order* and *symmetry* need some development. Although random structures are of growing importance, relatively little is known about such systems to date. This chapter attempts to give a preliminary account of some of the relevant structures. Our initial examples demonstrate the increased level of complexity.

11.1. Probabilistic preliminaries

The (sometimes implicit) starting point is a *probability space* (Ω, Σ, P), where Ω is some locally compact space, Σ is a σ-algebra of measurable subsets of Ω (often the σ-algebra of Borel sets of Ω), and P is a (regular) probability measure on Ω relative to Σ. The *probability* $\mathbb{P}$ of an event $A \in \Sigma$ is

$$\mathbb{P}(\text{event } A \text{ happens}) = \int_{\Omega} 1_A \, \mathrm{d}P,$$

though this integral is also simply written as $P(A)$. In standard situations, it is common to immediately use the probability measure P, while the notation with the symbol $\mathbb{P}$ is more practical for complicated events, as we shall see later. We refer to [Bil95, Geo12] for general background.

A real-valued *random variable* X is a measurable mapping from a probability space (Ω, Σ, P) into the measure space $(\mathbb{R}, \mathcal{B})$, where $\mathcal{B}$ usually is the σ-algebra of Borel sets of $\mathbb{R}$. The *expectation value* (or *mean*) of the random variable X (assuming that it exists) is denoted as

$$\mathbb{E}_P(X) = \int_{\Omega} X \, \mathrm{d}P = \int_{\mathbb{R}} x \, \mathrm{d}\mu(x),$$

where x is the value taken by X and μ is the induced probability measure on $\mathbb{R}$, as defined by the relation $\mu(B) = P(X^{-1}(B))$ for $B \in \mathcal{B}$. The measure

μ is called the *distribution* (or the *law*) of X. By using the triple $(\mathbb{R}, \mathcal{B}, \mu)$, one can often avoid any explicit reference to the underlying probability space. Also, when the context is clear, we simply write $\mathbb{E}$ instead of $\mathbb{E}_P$.

Similarly, provided the quantity exists, $\mathbb{E}(X^r)$ with $r \geqslant 0$ is called the rth *moment* of X, while

$$\operatorname{var}(X) = \mathbb{E}\big((X - \mathbb{E}(X))^2\big) = \mathbb{E}(X^2) - \big(\mathbb{E}(X)\big)^2$$

is the *variance* of X. For $r = 0$, one has $\mathbb{E}(X^0) = 1$, which is just the normalisation condition. In complete analogy, one can define random variables taking values in more general measure spaces, such as $\mathbb{R}^d$ or $\mathbb{C}$. In the case of a complex random variable Z, it is advantageous to properly include complex conjugation into the definition of moments. In particular, as $\mathbb{E}(\bar{Z}) = \overline{\mathbb{E}(Z)}$, the variance for this complex case is

$$(11.1) \qquad \operatorname{var}(Z) = \mathbb{E}\big(|Z - \mathbb{E}(Z)|^2\big) = \mathbb{E}\big(|Z|^2\big) - \big|\mathbb{E}(Z)\big|^2.$$

Example 11.1 (*Gamma distributed random variables*)**.** An interesting family of non-negative random variables emerges from the (absolutely continuous) gamma law, which is described by the two-parameter density function

$$g(x; \alpha, \beta) = \begin{cases} \frac{\beta^\alpha}{\Gamma(\alpha)} \, x^{\alpha-1} \, \mathrm{e}^{-\beta x}, & x > 0, \\ 0, & x \leqslant 0. \end{cases}$$

Here, Γ denotes the (standard) gamma function, and both parameters may be arbitrary positive numbers. This distribution has mean α/β and variance α/β^2. The case $\alpha = 1$ is the exponential distribution with mean $1/\beta$. $\diamond$

Let X and Y be two real-valued random variables (with σ-algebras $\mathcal{B}_X$ and $\mathcal{B}_Y$), with distributions μ and ν. They are called *independent* when

$$\mathbb{P}(x \in A, \, y \in B) = \mathbb{P}(x \in A) \, \mathbb{P}(y \in B) = \mu(A) \, \nu(B)$$

holds for all $A \in \mathcal{B}_X$ and $B \in \mathcal{B}_Y$. In this situation, the new random variable $Z = X + Y$ has distribution $\mu * \nu$ (and σ-algebra $\mathcal{B}_X \cap \mathcal{B}_Y$). The independence of more than two random variables is defined analogously, leading to a stronger condition than pairwise independence; see [Geo12] for details. Of particular interest are countable families $(X_i)_{i \in \mathbb{N}}$ of independent random variables with identical distribution μ, called i.i.d. for short.

For families of i.i.d. random variables, one often needs to know the behaviour of the random variable $\frac{1}{N} \sum_{i=1}^{N} X_i$ for large N. This is covered by the *strong law of large numbers* (SLLN) as follows, where we adopt the beautiful (and slightly more general) approach of Etemadi [Ete81].

Theorem 11.1. *Let $(X_i)_{i \in \mathbb{N}}$ be a family of pairwise independent, identically distributed, real random variables with common distribution μ, subject to the* *integrability condition* $\mathbb{E}\big(|X_1|\big) < \infty$. *Then,*

$$\frac{1}{N}\sum_{i=1}^{N} X_i \xrightarrow[a.s.]{N\to\infty} \mathbb{E}(X_1) = \int_{\mathbb{R}} x \, \mathrm{d}\mu(x).$$

The corresponding result also holds for families of random variables that take values in a fixed Banach space.

SKETCH OF PROOF. The SLLN is a kind of ergodic theorem for independent variables. In detail, the first assertion is [Ete81, Thm. 1], while the extension follows from [Ete81, Rem. 2]. □

The usual formulation for i.i.d. random variables is a special case, as pairwise independence is a weaker assumption than independence.

In a stochastic setting, our original (topological) concept of aperiodicity in Definition 4.13 is of limited value. This is already apparent for the classic coin tossing system. Its hull is the full shift $\{0,1\}^{\mathbb{Z}}$, though the canonical invariant measure on it has the remarkable property that all elements with a non-trivial period are in a set of measure 0. This motivates the following counterpart to our previous concept.

Definition 11.1. Consider a measure-theoretic dynamical system $(\mathbb{X}, G, \mu)$ with some compact space $\mathbb{X}$, a group $G \in \{\mathbb{Z}^d, \mathbb{R}^d\}$ and a G-invariant probability measure μ. The system is called *aperiodic* in the measure-theoretic sense when $\mu\big(\{x \in \mathbb{X} \mid \mathrm{per}(x) \neq \{0\}\}\big) = 0$.

Clearly, aperiodicity in the topological sense implies that in the measure-theoretic sense, but not vice versa. Examples of measure-theoretic aperiodicity will be provided by Bernoulli systems, certain random tiling ensembles, and by equilibrium systems such as the Ising lattice gas at positive temperature. Note that the non-periodic set from Example 4.3 still fails to be aperiodic in this slightly weaker sense, as there is no invariant probability measure on the hull that 'hides' its periodic element; compare Remark 4.5.

Other concepts such as repetitivity can be extended in a similar fashion. Definition 11.1 is also motivated by the observation that, in the typical stochastic setting, the notions of unique and strict ergodicity become irrelevant. This applies both to systems with independent random variables and to equilibrium systems (in the Gibbs measure sense); see [MvE90, vEZ97, vEMZ98, Mię93] for some examples that we cannot cover in this volume.

11.2. Bernoulli systems

Let us start with the most elementary stochastic system, based upon the classic coin tossing (or Bernoulli) experiment. This will lead to one-dimensional examples with quite different types of (dis)order, and later even to homometric structures with different entropies. The result will also indicate that the inverse problem of structure determination becomes more involved in the presence of mixed spectra.

11.2.1. BERNOULLI COMB ON THE LINE

Our first model is a Bernoulli system (or shift) on $\mathbb{Z}$. In our setting, the structure is represented by the stochastic Dirac comb

$$(11.2) \qquad \omega_{\mathrm{B}} = \sum_{m \in \mathbb{Z}} W_m \delta_m$$

where $(W_m)_{m \in \mathbb{Z}}$ is a family of i.i.d. random variables, the latter taking the values 1 and -1 with probabilities p and $1 - p$, where $0 \leqslant p \leqslant 1$. The autocorrelation, if it exists, will be a pure point measure that is supported on $\mathbb{Z}$. It thus can directly be analysed via the autocorrelation coefficients

$$(11.3) \qquad \eta_{\mathrm{B}}(m) = \lim_{N \to \infty} \frac{1}{2N+1} \sum_{j=-N}^{N} W_j W_{j-m}$$

for $m \in \mathbb{Z}$, since $\gamma_{\omega_{\mathrm{B}}} = \eta_{\mathrm{B}} \delta_{\mathbb{Z}}$ if all limits exist.

Lemma 11.1. *The autocorrelation coefficients* $\eta_{\mathrm{B}}(m)$ *of Eq. (11.3) almost surely exist for all* $m \in \mathbb{Z}$, *and they satisfy*

$$\eta_{\mathrm{B}}(m) = \begin{cases} 1, & m = 0, \\ (2p-1)^2, & m \neq 0. \end{cases}$$

PROOF. The statement is obviously always true for $m = 0$, so let some $m \neq 0$ be fixed. The products $Z(i) := W_i W_{i-m}$ form a family $\left(Z(i) \right)_{i \in \mathbb{Z}}$ of identically distributed random variables, which take the values 1 and -1 with probabilities $p^2 + (1-p)^2$ and $2p(1-p)$, respectively. However, these new random variables fail to be pairwise independent, because $Z(i)$ and $Z(i+m)$ share the factor W_i, wherefore the SLLN is not immediately applicable.

This small complication can be removed by splitting the sum in Eq. (11.3) into two sums as follows. Note first that, for fixed N, there are finitely many contributing terms of the form $Z(-N + km)$, with $0 \leqslant k \leqslant \left[\frac{2N}{m} \right]$. Those with k even (with k odd) are pairwise independent and go to the first (to the second) sum, so that each arithmetic progression is distributed alternately to the two sums. All remaining terms are grouped into similar chains and distributed analogously to the two sums. Since different chains have no overlap, both sums now satisfy the conditions of Theorem 11.1 separately. As $N \to \infty$, each sum almost surely converges to $\frac{1}{2}(2p-1)^2$ by the SLLN, because the individual terms are evenly distributed to the two sums. Their addition thus gives the formula for $\eta(m)$ for fixed m. The claim is now obvious, as $\mathbb{Z}$ is countable. □

Proposition 11.1. *The stochastic Dirac comb ω_{B} of Eq. (11.2) almost surely possesses the autocorrelation and diffraction measures*

$$\gamma_{\omega_{\mathrm{B}}} = (2p-1)^2 \delta_{\mathbb{Z}} + 4p(1-p)\,\delta_0,$$
$$\widehat{\gamma_{\omega_{\mathrm{B}}}} = (2p-1)^2 \delta_{\mathbb{Z}} + 4p(1-p)\,\lambda.$$

In particular, one has $\widehat{\gamma_{\omega_{\mathrm{B}}}} = \lambda$ for $p = \frac{1}{2}$ and $\widehat{\gamma_{\omega_{\mathrm{B}}}} = \delta_{\mathbb{Z}}$ for $p = 0$ or $p = 1$.

PROOF. The formula for the autocorrelation $\gamma_{\omega_{\mathrm{B}}}$ is an immediate consequence of Lemma 11.1. Its Fourier transform follows from the PSF of Proposition 9.4 together with the relation $\widehat{\delta_0} = \lambda$ from Example 8.5. □

The special choices $p = 0$ and $p = 1$ correspond to the deterministic limiting cases $\omega_{\mathrm{B}} = \pm\delta_{\mathbb{Z}}$, while $p = \frac{1}{2}$ describes the stochastic comb of a 'fair coin' with balanced weights (they have average 0, for almost all realisations). The latter case is an example of a structure with (almost surely) a purely absolutely continuous diffraction measure.

One can extend this result to arbitrary (possibly complex) weights $h_{\pm}$ by means of i.i.d. random variables H_i which take values $h_{\pm}$ rather than ± 1. The new stochastic comb can be re-expressed as

$$(11.4) \qquad \omega_{\mathrm{B}}^{(H)} = \frac{h_+ + h_-}{2}\,\delta_{\mathbb{Z}} + \frac{h_+ - h_-}{2}\,\omega_{\mathrm{B}}$$

with ω_{B} of Eq. (11.2). To calculate the autocorrelation of the new measure $\omega_{\mathrm{B}}^{(H)}$, one can repeat the previous argument (taking into account complex conjugates). The new autocorrelation coefficients turn out to be (this time, almost surely for all $m \in \mathbb{Z}$)

$$\eta_{\mathrm{B}}^{(H)}(m) = \begin{cases} \mathbb{E}(|H|^2), & m = 0, \\ |\mathbb{E}(H)|^2, & m \neq 0. \end{cases}$$

Alternatively, one can calculate the new autocorrelation from Eq. (11.4) by a volume-weighted convolution, in the spirit of Example 8.10. This needs the autocorrelation of $\delta_{\mathbb{Z}}$ (which is $\delta_{\mathbb{Z}}$), that of ω_{B} (which is given in Proposition 11.1) and the two cross-correlations involving $\delta_{\mathbb{Z}}$, $\widetilde{\delta_{\mathbb{Z}}} = \delta_{\mathbb{Z}}$, ω_{B} and $\widetilde{\omega_{\mathrm{B}}}$. These cross-correlations contribute the measure $(p - \frac{1}{2})(|h_+|^2 - |h_-|^2)\delta_{\mathbb{Z}}$ to the new autocorrelation (a proof of this claim follows from another application of the SLLN). The overall result is a scalar multiple of $\delta_{\mathbb{Z}}$ plus a contribution to δ_0 with coefficient $p(1-p)\,|h_+ - h_-|^2$. The outcome of this derivation can be summarised as follows.

Corollary 11.1. *The Bernoulli comb on $\mathbb{Z}$ with weights according to a binary random variable H almost surely has the diffraction measure*

$$\widehat{\gamma_{\omega_{\mathrm{B}}^{(H)}}} = |\mathbb{E}(H)|^2 \delta_{\mathbb{Z}} + \mathrm{var}(H)\,\lambda,$$

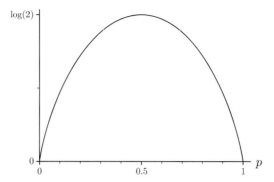

FIGURE 11.1. Entropy function $-\bigl(p\log(p)+(1-p)\log(1-p)\bigr)$ of the binary Bernoulli shift for probability parameter $0 \leqslant p \leqslant 1$. This strictly concave function has a unique maximum at $p = 1/2$, with value $\log(2) \approx 0.693147$.

where $\operatorname{var}(H) = \mathbb{E}\bigl(|H|^2\bigr) - \bigl|\mathbb{E}(H)\bigr|^2$ is the variance of the (complex) random variable H. In particular, the diffraction measure is purely absolutely continuous whenever $\mathbb{E}(H) = 0$.   □

Example 11.2 (*Lattice gas on* $\mathbb{Z}$). A special case of interest emerges from the choice $h_+ = 1$ (with probability p) and $h_- = 0$, which is a lattice gas on $\mathbb{Z}$ with occupation probability p. An application of Corollary 11.1 gives the diffraction measure

$$(11.5) \qquad\qquad \widehat{\gamma_{\omega_{\mathrm{B}}^{(H)}}} = p^2\, \delta_{\mathbb{Z}} + p(1-p)\, \lambda.$$

The coefficient of the absolutely continuous component is maximal for $p = \frac{1}{2}$, which is also the point of maximum entropy (in the parameter p), as follows from the expression $-\bigl(p\log(p) + (1-p)\log(1-p)\bigr)$ for the measure-theoretic (or metric) entropy of a binary Bernoulli system [Wal00, Thm. 4.26]; see also the first section in [Khi57] as well as [GS85] for general background on entropy, and Figure 11.1 for an illustration of the entropy function. We will return to this observation in the context of symmetry in Remark 11.13.   ◇

Remark 11.1 (*General Bernoulli combs*). The diffraction formula of Corollary 11.1 holds for more general Bernoulli combs on $\mathbb{Z}$. In fact, one can replace the binary random variable by any complex-valued random variable subject to the condition that its first and second moments exist; see [BH00, BM98] for details. As the argument only relies on the SLLN, analogous results hold for all lattices and model sets [BM98, BBM10] and, with minor modifications, for most locally finite point sets [Kül03a, Kül03b].

More precisely, Proposition 11.1 and Corollary 11.1 are robust in the sense that they easily generalise to arbitrary lattices $\Gamma \subset \mathbb{R}^d$. Indeed, the

SLLN permits us to analyse the Dirac comb $\omega_{\mathrm{B}}^{(H)} = \sum_{x \in \Gamma} H_x \delta_x$ with i.i.d. random variables $(H_x)_{x \in \Gamma}$, now (almost surely) giving the autocorrelation $\gamma_{\omega_{\mathrm{B}}} = \mathrm{dens}(\Gamma)\big(|\mathbb{E}(H)|^2 \delta_\Gamma + \mathrm{var}(H) \delta_0\big)$ and thus (via the PSF and $\widehat{\delta_0} = \lambda$) the diffraction measure

$$\widehat{\gamma_{\omega_{\mathrm{B}}^{(H)}}} = \big(\mathrm{dens}(\Gamma)\big)^2 \big|\mathbb{E}(H)\big|^2 \delta_{\Gamma^*} + \mathrm{dens}(\Gamma) \, \mathrm{var}(H) \, \lambda.$$

An interesting special case emerges for a random variable H with values in $\mathbb{S}^1$ and $\mathbb{E}(H) = 0$, hence $\mathrm{var}(H) = 1$. This includes the case of equally likely values ± 1 as well as that of uniformly distributed random phases. When also $\mathrm{dens}(\Gamma) = 1$, one (almost surely) finds λ as the diffraction measure. This provides a large class of homometric examples. $\diamond$

11.2.2. BERNOULLISATION OF BINARY SEQUENCES

The Bernoulli chain discussed above is an example of a completely random and interaction-free system. In view of realistic examples, it is interesting to explore what happens if one imposes the influence of coin tossing on the order of a deterministic system. This can be realised in many different ways. Here, we focus on binary sequences and modify them by an i.i.d. family of Bernoulli variables.

Consider a bi-infinite binary sequence $S \in \{\pm 1\}^{\mathbb{Z}}$, which we assume to be uniquely ergodic (in the sense that its hull under the action of the shift map is a uniquely ergodic dynamical system; compare Chapter 4). Then, the corresponding Dirac comb $\omega_S = \sum_{i \in \mathbb{Z}} S_i \, \delta_i$ possesses the (natural) autocorrelation measure $\gamma_S = \sum_{m \in \mathbb{Z}} \eta_S(m) \, \delta_m$ with the autocorrelation coefficients $\eta_S(m)$, where we have $\eta_S(0) = 1$ by construction.

Let $(W_i)_{i \in \mathbb{Z}}$ be an i.i.d. family of random variables, each taking values $+1$ and -1 with probabilities p and $1-p$. The 'Bernoullisation' of ω_S is the random Dirac comb

$$(11.6) \qquad \omega := \sum_{i \in \mathbb{Z}} S_i W_i \, \delta_i,$$

which emerges from ω_S by independently changing the sign of each S_i with probability $1-p$. This procedure can be interpreted as a 'model of second thought'. Setting $Z_i := S_i W_i$ defines a new family of independent (though in general not identically distributed) random variables, with values in $\{\pm 1\}$. Despite this modification, the autocorrelation γ of ω almost surely exists and can be determined via its autocorrelation coefficients $\eta(m)$. Since one always has $\eta(0) = \eta_S(0) = 1$, let $m \neq 0$ and consider, for large N, the sum

$$\frac{1}{2N+1} \sum_{i=-N}^{N} Z_i Z_{i-m} = \frac{1}{2N+1} \Big(\sum_{(+,+)} + \sum_{(-,-)} - \sum_{(+,-)} - \sum_{(-,+)} \Big) W_i W_{i-m},$$

which is split according to the value of (S_i, S_{i-m}). Each of the four sums can be handled by the SLLN in the same way as in the proof of Lemma 11.1, thus contributing $(2p - 1)^2$ times the frequency of the corresponding sign pair. Observing that the overall signs are the products $S_i S_{i-m}$, it is clear that, as $N \to \infty$, one obtains

$$\eta(m) = (2p - 1)^2 \eta_S(m)$$

for all $m \neq 0$.

Lemma 11.2. *Let $S \in \{\pm 1\}^{\mathbb{Z}}$ be a uniquely ergodic sequence, and let ω be the Bernoullisation of ω_S as introduced in Eq. (11.6). The autocorrelation γ of ω then exists almost surely, and is given by*

$$\gamma = (2p - 1)^2 \gamma_S + 4p(1 - p) \delta_0,$$

where γ_S is the unique autocorrelation of ω_S.

PROOF. Since S is uniquely ergodic, the frequencies of all clusters exist uniformly. Consequently, our above derivation is well-defined and γ follows from the knowledge of the coefficients $\eta(m)$ for $m \in \mathbb{Z}$. The almost sure convergence on the level of measures follows because $\mathbb{Z}$ is countable. $\square$

Example 11.3 (*Bernoullisation of the Thue–Morse chain*). Consider the Thue–Morse chain of Sections 4.6 and 10.1, realised in $\{\pm 1\}^{\mathbb{Z}}$ by turning a and b into 1 and -1. Its Bernoullisation leads to the new autocorrelation given by Lemma 11.2, and hence to the new diffraction measure

$$\widehat{\gamma} = (2p - 1)^2 \widehat{\gamma_{\mathrm{TM}}} + 4p(1 - p)\lambda.$$

Since $\widehat{\gamma_{\mathrm{TM}}}$ is purely singular continuous by Theorem 10.1, we have thus obtained (for $p \neq \frac{1}{2}$) an example of a continuous measure of mixed type.

Taking this example one step further, let us start from this Bernoullisation ω, and replace all weights -1 by 0, thus changing the underlying sequence S into an element of $\{0, 1\}^{\mathbb{Z}}$. The new comb ω' satisfies $\omega' = \frac{1}{2}(\delta_{\mathbb{Z}} + \omega)$, and almost surely has the autocorrelation

$$\gamma' = \frac{1}{4}(\delta_{\mathbb{Z}} + \gamma),$$

because 1 and -1 are equally frequent in the original sequence, so that the cross-terms (such as $\delta_{\mathbb{Z}} \circledast \omega$) in the Eberlein convolution for γ vanish; compare Example 8.10. Put together, this gives the diffraction measure

$$\widehat{\gamma'} = \frac{1}{4}\delta_{\mathbb{Z}} + (p - \frac{1}{2})^2 \widehat{\gamma_{\mathrm{TM}}} + p(1 - p)\lambda,$$

which is an interesting example of a mixed spectrum with all three types of contributions. Note that $p = \frac{1}{2}$ is the diffraction of the Bernoulli lattice gas (for weights 1 and 0), while $p = 0$ and $p = 1$ bring us back to the Thue–Morse diffraction for weights 1 and 0; compare Remark 10.3. $\diamond$

Remark 11.2 (*Bernoullisation and entropy*). The Thue–Morse chain is deterministic, with linear word complexity, and thus topological and measure-theoretic entropy 0. In contrast, the binary Bernoulli chain of Eq. (11.2) with parameter p has measure-theoretic entropy $-\bigl(p\log(p) + (1-p)\log(1-p)\bigr)$. The latter is also the measure-theoretic entropy of the Bernoullisation in Example 11.3, which vanishes if and only if the absolutely continuous diffraction component vanishes, in line with the results of [BLR07]. The entropy is maximal for $p = \frac{1}{2}$, where the singular continuous component vanishes and the absolutely continuous one has maximal intensity.   ◇

Let us apply the Bernoullisation procedure to the Rudin–Shapiro sequence. By Theorem 10.2, the autocorrelation and diffraction measures of the binary Rudin–Shapiro chain in its realisation as a sequence in $\{\pm 1\}^{\mathbb{Z}}$ are

$$\gamma_{\mathrm{RS}} = \delta_0 \quad \text{and} \quad \widehat{\gamma_{\mathrm{RS}}} = \lambda.$$

This is an example with a purely absolutely continuous diffraction despite the fact that the Rudin–Shapiro chain is deterministic and has entropy 0.

Proposition 11.2. *Let ω be the random Dirac comb obtained from the Bernoullisation (with parameter p) of the binary Rudin–Shapiro chain, realised as a subset of $\{\pm 1\}^{\mathbb{Z}}$. Then, the autocorrelation measure of ω almost surely exists and reads $\gamma = \delta_0$, independently of p. This means that the random Dirac combs ω, even for different values of p, are almost surely homometric, and share the purely absolutely continuous diffraction measure $\widehat{\gamma} = \lambda$.*

PROOF. Since the binary Rudin–Shapiro (RS) sequence is uniquely ergodic, the first claim follows from Lemma 11.2 by a simple calculation; compare [BG09]. The remainder of the statement is obvious.   □

Remark 11.3 (*Homometry and entropy*). The RS Bernoullisation explores the full entropy range: The Bernoulli case (with $p = \frac{1}{2}$) has entropy $\log(2)$, the maximal value for a binary system, while RS has entropy 0, and the parameter p interpolates continuously between the two limiting cases. The solution of the corresponding inverse problem is thus highly degenerate. Unless additional information is available, a possible strategy employs an optimisation approach, for instance via maximising the entropy, which singles out the Bernoulli comb here. For a general discussion of the maximum entropy method, we refer to [Jay82, GS85] and references therein.   ◇

The applicability of the SLLN is not limited to simple Bernoulli systems. Let us demonstrate this with another example of a slightly more complex nature, which was originally suggested by van Enter [BvE11].

Example 11.4 (*Random dimers on $\mathbb{Z}$*). Consider a partition of $\mathbb{Z}$ into 'dimers' (pairs of neighbouring numbers), for which there are two possibilities. Next, give each dimer a random orientation by decorating it with either $(+, -)$ or $(-, +)$, with equal probability, resulting in configurations such as

$$\ldots [+ \; -][- \; +][- \; +][+ \; -][- \; +][- \; +][- \; +][+ \; -][+ \; -] \ldots$$
$$\ldots [- \; +][+ \; -][+ \; -][- \; +][+ \; -][+ \; -][+ \; -][- \; +][- \; +][+ \; -] \ldots$$

Identifying $\pm$ with ± 1, this defines the (closed and compact) shift space

$$\mathbb{X} = \left\{ w \in \{\pm 1\}^{\mathbb{Z}} \mid M(w) \subset 2\mathbb{Z} \text{ or } M(w) \subset 2\mathbb{Z} + 1 \right\},$$

where $M(w) := \{i \in \mathbb{Z} \mid w_i = w_{i+1}\}$. Note that $M(w)$ can be empty, which happens precisely for the 2-periodic sequences $u_+ = \cdots + - | + - \cdots$ and $u_- = \cdots - + | - + \cdots$. One has a natural invariant measure on $\mathbb{X}$, which emerges from the stochastic process via the probability of the possible finite patches (which define the generating cylinder sets as usual).

Turning a configuration $w \in \mathbb{X}$ into a signed Dirac comb with weights $w_i \in \{\pm 1\}$, an application of the SLLN shows that the corresponding autocorrelation γ_w almost surely exists. It is not difficult to derive [BvE11] that $\gamma_w = \delta_0 - \frac{1}{2}(\delta_1 + \delta_{-1})$, so that the corresponding diffraction measure is

$$\widehat{\gamma_w} = \big(1 - \cos(2\pi k)\big)\lambda.$$

This is another example of a purely absolutely continuous diffraction measure. The Radon–Nikodym density relative to λ is written as a function of k.

The dynamical spectrum of this system contains eigenvalues, wherefore this is an analogue of the comparison between the Thue–Morse and period doubling systems in Theorem 4.7, this time in the presence of absolutely continuous spectra. A similar situation also occurred in the discussion of the Rudin–Shapiro system in Section 4.7.1. As before, a sliding block map ϕ can be defined by $\phi(w)_i = -w_i w_{i+1}$ for $i \in \mathbb{Z}$. This maps w to a new sequence u (with $u_i = -1$ on $\mathbb{Z} \setminus M(w)$), which almost surely has the diffraction measure

$$\widehat{\gamma_u} = \frac{1}{4}\delta_{\mathbb{Z}/2} + \frac{1}{2}\lambda$$

of mixed type. In particular, it displays the entire point part of the original dynamical spectrum, namely $\frac{1}{2}\mathbb{Z}$.

Going one step further, consider the finite subspace $\mathbb{X}_0$ of $\mathbb{X}$ defined by $\mathbb{X}_0 = \{w \in \mathbb{X} \mid M(w) = \varnothing\} = \{u_+, u_-\}$, which contains the two periodic sequences mentioned above. Define a mapping on $\mathbb{X}$ that is the identity when restricted to $\mathbb{X}_0$ and sends all sequences $w \in \mathbb{X}$ with $\varnothing \neq M(w) \subset 2\mathbb{Z}$ (with $\varnothing \neq M(w) \subset 2\mathbb{Z} + 1$) to u_+ (to u_-). This mapping is continuous and establishes $\mathbb{X}_0$ as a topological factor of $\mathbb{X}$, both seen as systems under the $\mathbb{Z}$-action of the shift. A simple calculation shows that $\mathbb{X}_0$ has a pure point spectrum that exhausts the point part of the spectrum of $\mathbb{X}$. In fact, the suspension of $\mathbb{X}_0$ for the action of the group $\mathbb{R}$ is a topological Kronecker factor of the corresponding suspension of $\mathbb{X}$; see [BvE11] for details. $\Diamond$

11.2.3. RANDOMISED INFLATION RULES AND POSITIONAL DISORDER

Yet another way to introduce randomness emerges from the mixture of different substitution rules at a local level. Since little is known about this class so far, we illustrate the approach with the two Fibonacci substitutions ϱ and ϱ' of Example 4.6 on page 90 and the subsequent Remark 4.6. Viewed as individual substitutions, they define the same discrete hull (by Proposition 4.6). Moreover, any mixture of these two rules, when applied *globally*, still produces the same hull. This claim follows from the observation that the set of finite legal words is closed under the application of either rule, as is (obviously) the entire hull.

Matters change dramatically when the two rules are randomly mixed *locally*, as originally suggested in [GL89]. Here, a letter b is always replaced by an a in a substitution step. The extra degree of freedom now emerges from replacing a by either ab or ba with equal probability (or, more generally, with probabilities p and $1 - p$). We are interested in this process on the level of the corresponding geometric inflation rules, where a and b represent intervals of lengths τ and 1, respectively, as in the deterministic case. Indeed, this is consistent with the substitution matrix, which remains unchanged. Due to the randomness, the atlas of legal words increases considerably. Since we can view infinite sequences as the image of a coin tossing sequence, one might expect an exponentially growing complexity function. On the level of exact substitution words, this was calculated in [GL89], leading to the entropy

$$h = \sum_{n=2}^{\infty} \frac{\log(n)}{\tau^{n+2}} \approx 0.444398725.$$

A proof on the basis of the full complexity function, via a somewhat intricate estimate, was presented more recently in [Nil12a], in the context of the corresponding problem for the randomised versions of the noble mean substitutions of Remark 4.7.

In the random Fibonacci case, as well as in the randomised noble means substitutions, a key observation is that *every* element of the hull (viewed as the point set obtained from the left endpoints of the intervals of the tiling) is a Meyer set. This can be proved by showing that each such set is a relatively dense subset of a model set. The latter emerges from the CPS of the underlying deterministic inflation, with a larger window. By an application of Strungaru's result, see Theorem 10.7, each element thus has a non-trivial pure point part to its diffraction spectrum, which is altogether of mixed type [BM13, Mol14]. In fact, its continuous part is absolutely continuous, which was (heuristically) derived already in [GL89]. As an important consequence, a dynamical system defined by this type of random inflation rule cannot be weakly mixing because it possesses non-trivial eigenvalues; compare [Wal00] and [BL04]. The dynamical system comes with a natural invariant measure, defined via Perron–Frobenius theory for the induced (probabilistic) substitution on legal words of arbitrary (finite) length, which is ergodic.

This class of examples, which has an analogue in higher dimensions as well [GL89], is interesting in many respects. In particular, it provides a rather natural family of Meyer sets with entropy, as well as a class of dynamical systems that are ergodic, but not weakly mixing. Moreover, such sets seem relevant for applications. Little is known so far about further extensions, for instance about mixtures of substitutions that define distinct hulls; compare [GM13] and references therein for recent progress.

Before we continue with our general discussion, let us briefly consider positional disorder of Bernoulli type. More precisely, we are interested in the random displacement of a given point set by uncorrelated movements of the individual points. This situation can be formulated via independent random variables, and thus does not require the Gibbs measure formalism for equilibrium systems. Let us have a detailed look at the lattice case.

Example 11.5 (*Random displacement model for lattices*). Let $\Gamma \subset \mathbb{R}^d$ be a lattice, and let $(X_t)_{t \in \Gamma}$ be a family of i.i.d. random variables that take values in $\mathbb{R}^d$. Their common distribution is given by the probability measure ν. By the SLLN, we know that

$$\frac{1}{\mathrm{vol}(B_r(0))} \sum_{y \in \Gamma_r} \delta_{X_y} \xrightarrow{\; r \to \infty \;} \mathrm{dens}(\Gamma)\, \nu,$$

with almost sure convergence in the vague topology. Here, $\Gamma_r = \Gamma \cap \overline{B_r(0)}$ as before, and δ_{X_y} is the random point measure of total mass 1 at X_y.

If we now consider the random point set $\{y + X_y \mid y \in \Gamma\}$, we can employ the approach of Section 9.2.6 to calculate the (random) autocorrelation as $\gamma = \sum_{z \in \Gamma} \mu_z$, where

(11.7) $$\mu_z = \lim_{r \to \infty} \frac{1}{\mathrm{vol}(B_r(0))} \sum_{y \in \Gamma_r} \delta_{z + X_{y+z} - X_y},$$

provided that these limits exist. Clearly, $\mu_0 = \mathrm{dens}(\Gamma)\delta_0$, so we can focus on $0 \neq z \in \Gamma$. Observing that

$$\delta_{z + X_{y+z} - X_y} = \delta_z * \left(\delta_{X_{y+z}} * \widetilde{\delta_{X_y}} \right),$$

an application of the SLLN (after splitting the sum into two sums over independent random variables, similar to the argument used in the proof of Lemma 11.1) shows almost sure convergence in Eq. (11.7), with limit

$$\mu_z = \mathrm{dens}(\Gamma)\, \delta_z * (\nu * \widetilde{\nu}).$$

Consequently, the autocorrelation almost surely reads

$$\gamma = (\nu * \tilde{\nu}) * \gamma_\Gamma + \mathrm{dens}(\Gamma)\,(\delta_0 - \nu * \tilde{\nu}),$$

where $\gamma_\Gamma = \mathrm{dens}(\Gamma)\delta_\Gamma$ is the (deterministic) autocorrelation of the lattice Γ. The diffraction (by our standard calculations) is then

(11.8) $$\hat{\gamma} = |\hat{\nu}|^2 \, \widehat{\gamma_\Gamma} + \mathrm{dens}(\Gamma)\big(1 - |\hat{\nu}|^2\big)\lambda,$$

which holds, like the previous formula, almost surely; see [Hof95b, BBM10] for the mathematical details. ◊

This random displacement model is a probabilistic approximation to the thermal movement of atoms [Hof95b]. One would assume that the result of Eq. (11.8) holds for more general point sets, and this is indeed the case. Under rather mild assumptions on the underlying point set Λ, the formula remains literally true, with γ_Γ simply replaced by the autocorrelation γ_Λ. We refer to [Hof95b, BBM10] for two different proofs and further details.

11.3. Renewal processes on the line

An interesting class of examples with some form of effective interaction is provided by the classic *renewal process* on the real line, defined by a probability measure ϱ of finite mean on $\mathbb{R}_+ := \{x > 0\}$ as follows. Starting from some initial point, at an arbitrary position, a machine moves to the right with constant speed and drops a point on the line with a random waiting time that is distributed according to ϱ. When this happens, the clock is reset and the process resumes. In what follows, we assume that both the velocity of the machine and the expectation value of ϱ are 1, so that we end up (in the limit that we let the initial point move to $-\infty$) with realisations that are almost surely point sets in $\mathbb{R}$ of density 1.

Clearly, the process just described defines a stationary process. It can thus be analysed by considering all realisations which contain the point 0. Moreover, there is a clear (distributional) symmetry around this point, so that we can determine the autocorrelation (in the sense of Definition 9.1) of almost all realisations by studying what happens on $\mathbb{R}_+$. Indeed, if we want to know what the frequency per unit length of the occurrence of two points with distance x is (or the corresponding density), we need to sum the contributions from x being the first point after 0, the second point, the third, and so on. In other words, we almost surely obtain the autocorrelation

(11.9) $$\gamma = \delta_0 + \nu + \tilde{\nu}$$

with $\nu = \varrho + \varrho * \varrho + \varrho * \varrho * \varrho + \cdots$, provided that this sum converges properly. The corresponding vague convergence of $\tilde{\nu}$ then follows from $\tilde{\nu}(g) = \overline{\nu(\tilde{g})}$ with $\tilde{g}(x) = \overline{g(-x)}$ for continuous functions of compact support. Note that the point measure at 0 simply reflects that the almost sure density of the resulting point set is 1, in line with Corollary 9.1. In the slightly more general case of a probability measure ϱ on $\mathbb{R}_+ \cup \{0\}$, one has the following convergence result. It is essentially a measure-theoretic reformulation of the main lemma in [Fel72, Sec. XI.1], but we prefer to give a complete proof that is adjusted to our setting; see [BBM10] for more. The n-fold convolution of a finite measure ϱ with itself it abbreviated as ϱ^{*n}, where $\varrho^{*1} = \varrho$.

Lemma 11.3. *If ϱ is a probability measure on $\mathbb{R}_+ \cup \{0\}$, with $\varrho(\mathbb{R}_+) > 0$, $\nu_n := \varrho + \varrho * \varrho + \cdots + \varrho^{*n}$ defines a sequence $(\nu_n)_{n \in \mathbb{N}}$ of positive measures that converges to a translation bounded measure ν in the vague topology.*

PROOF. Note that the condition $\varrho(\mathbb{R}_+) > 0$ implies that $0 \leqslant \varrho(\{0\}) < 1$, hence excludes the case $\varrho = \delta_0$. When $\varrho = \delta_a$ for some $a > 0$, one obtains $\nu_n = \sum_{m=1}^n \delta_{ma}$ by a simple convolution calculation, and the claim is obvious. In all remaining cases, it is possible to choose some $a \in \mathbb{R}_+$ with $\varrho(\{a\}) = 0$ and $0 < \varrho([0,a)) = p < 1$, so that we also have $\varrho([a,\infty)) = 1 - p < 1$. Since the sequence ν_n is monotonically increasing, the claimed vague convergence follows from showing that $\limsup_{n \to \infty} \nu_n([0,x))$ is bounded by $C_1 + C_2 x$ for some constants C_1 and C_2. In a second step, we then demonstrate that $\sum_{n=1}^\infty \varrho^{*n}([b, b+x))$ is bounded by $1 + C_1 + C_2 x$, independently of b, which establishes translation boundedness.

If $(X_i)_{i \in \mathbb{N}}$ denotes a family of i.i.d. random variables, with common distribution according to ϱ (and thus values in $\mathbb{R}_+ \cup \{0\}$), one has

$$\mathbb{P}\big(X_1 + \cdots + X_m < x\big) = \varrho^{*m}([0,x)).$$

On the other hand, for the a chosen above, one has the inequality

$$\mathbb{P}(X_1 + \cdots + X_m < x) \leqslant \mathbb{P}\big(\operatorname{card}\{1 \leqslant i \leqslant m \mid X_i \geqslant a\} \leqslant x/a\big)$$

$$= \sum_{\ell=0}^{[x/a]} \binom{m}{\ell} (1-p)^\ell \, p^{m-\ell},$$

where $\binom{m}{\ell} = 0$ whenever $\ell > m$, and where the Gauss bracket $[x/a]$ denotes the largest integer $\leqslant x/a$. Observing $\sum_{m=1}^\infty p^m = p/(1-p)$ and

$$\sum_{m=1}^\infty \binom{m}{\ell} (1-p)^\ell \, p^{m-\ell} = (1-p)^\ell \frac{1}{\ell!} \frac{\mathrm{d}^\ell}{\mathrm{d}p^\ell} \sum_{m=0}^\infty p^m = \frac{1}{1-p}$$

for all $\ell \geqslant 1$, the previous inequality implies that, for arbitrary $n \in \mathbb{N}$,

$$\nu_n\big([0,x)\big) \leqslant \sum_{m=1}^\infty \sum_{\ell=0}^{[x/a]} \binom{m}{\ell} (1-p)^\ell \, p^{m-\ell} = \frac{p + [x/a]}{1-p}$$

$$\leqslant \frac{p}{1-p} + \frac{1}{a(1-p)}\, x,$$

which establishes the first claim.

For the second estimate, we choose $b > 0$, $x > 0$ and define the summatory random variable $\Sigma_{k,\ell} := X_k + \cdots + X_\ell$, with $\Sigma_{k,\ell} := 0$ for $k > \ell$. We then observe

$$
\sum_{n=1}^{\infty} \varrho^{*n}\big([b, b+x)\big) = \sum_{n=1}^{\infty} \mathbb{P}\big(b \leqslant \Sigma_{1,n} < b+x\big)
$$

$$
= \sum_{n=1}^{\infty} \sum_{k=1}^{n} \mathbb{P}\big(\Sigma_{1,k-1} < b \leqslant \Sigma_{1,k} \text{ and } b \leqslant \Sigma_{1,n} < b+x\big)
$$

$$
\leqslant \sum_{n=1}^{\infty} \sum_{k=1}^{n} \mathbb{P}\big(\Sigma_{1,k-1} < b \leqslant \Sigma_{1,k}\big)\, \mathbb{P}\big(\Sigma_{k+1,n} < x\big)
$$

$$
= \sum_{k=1}^{\infty} \mathbb{P}\big(\Sigma_{1,k-1} < b \leqslant \Sigma_{1,k}\big) \sum_{n=k}^{\infty} \mathbb{P}\big(\Sigma_{k+1,n} < x\big)
$$

$$
= 1 + \sum_{m=1}^{\infty} \mathbb{P}\big(\Sigma_{1,m} < x\big) = 1 + \lim_{n \to \infty} \nu_n\big([0, x)\big),
$$

with $\mathbb{P}(0 < x) = 1$. The first inequality follows from an estimate via conditional probabilities. The last step used the i.i.d. property of the random variables (wherefore $\mathbb{P}(X_{k+1} + \cdots + X_{k+m} < x) = \mathbb{P}(X_1 + \cdots + X_m < x)$ for $m \geqslant 1$) together with $\sum_{k=1}^{\infty} \mathbb{P}\big(\Sigma_{1,k-1} < b \leqslant \Sigma_{1,k}\big) = 1$. In conjunction with our previous estimate, this completes the proof. $\square$

When $\varrho(\{0\}) > 0$, we are outside the realm of standard renewal point processes, and Eq. (11.9) for the autocorrelation no longer applies. This case can still be analysed with methods from point process theory; see [BBM10] for more. For the remainder of this section, we assume $\varrho(\{0\}) = 0$, so that ϱ is a measure on $\mathbb{R}_+$. For an alternative approach via random counting measures, we refer to [Fel72, Ch. XI.9] or [BBM10, Rem. 14].

Proposition 11.3. *Consider a renewal process on the real line, defined by a probability measure ϱ of mean 1 on $\mathbb{R}_+$. This defines a stationary stochastic process, whose realisations are point sets that almost surely possess the autocorrelation measure $\gamma = \delta_0 + \nu + \tilde{\nu}$ of Eq. (11.9).*

Here, $\nu = \sum_{n=1}^{\infty} \varrho^{*n}$ is a translation bounded positive measure. It satisfies the renewal equations

$$
\nu = \varrho + \varrho * \nu \quad \text{and} \quad (1 - \hat{\varrho})\hat{\nu} = \hat{\varrho},
$$

where $\hat{\varrho}$ is a uniformly continuous function on $\mathbb{R}$. In this setting, the measure γ is both positive and positive definite.

PROOF. The renewal process is a classic stochastic process on the real line which is known to be stationary and ergodic; compare [Fel72, Ch. VI.6] for details. Consequently, the measure of occurrence of a pair of points at distance $x + \mathrm{d}x$ (or the corresponding density) can be calculated by fixing one point at 0 (due to stationarity) and then determining the ensemble average for another point at $x + \mathrm{d}x$ (due to ergodicity). This is the justification for the heuristic reasoning given above, prior to Eq. (11.9).

By Lemma 11.3, ν is a translation bounded measure, so that the convolution $\varrho * \nu$ is well defined by Theorem 8.5. The first renewal identity is then clear from the structure of ν as a limit, while the second follows by Fourier transform and the convolution theorem. The claim about $\widehat{\varrho}$ is standard; compare [Pin02, Prop. 5.2.1]. The autocorrelation is a positive definite measure by construction (which also implies translation boundedness of γ), though this is not immediate here on the basis of its form as a sum; see [Ata75] for a related discussion. How to determine the autocorrelation in detail follows from our previous argument. □

Let us now consider the spectral type of the resulting diffraction measure for the class of point sets generated by a renewal process. This requires a distinction on the basis of the support of the distribution ϱ. To this end, the second identity of Proposition 11.3 is helpful, because one has

$$(11.10) \qquad\qquad \widehat{\nu}(k) \;=\; \frac{\widehat{\varrho}(k)}{1 - \widehat{\varrho}(k)}$$

at all positions k with $\widehat{\varrho}(k) \neq 1$. This is in line with summing $\widehat{\nu}$ as a geometric series, which gives the same formula for $\widehat{\nu}(k)$ for all k with $|\widehat{\varrho}(k)| < 1$ and has Eq. (11.10) as the unique continuous extension to all k with $|\widehat{\varrho}(k)| = 1 \neq \widehat{\varrho}(k)$. In fact, one sees that $\widehat{\nu}(k)$ is a continuous function on the complement of the set $S = \{k \in \mathbb{R} \mid \widehat{\varrho}(k) = 1\}$. For most ϱ, one has $S = \{0\}$.

In general, a probability measure μ on $\mathbb{R}$ is called *lattice-like* when its support is a subset of a translate of a lattice; see [Gne98] for details. Since the origin will play a special role, we need a slightly stronger property here, and call μ *strictly lattice-like* (called *arithmetic* in [Fel72]) when its support is a subset of a lattice. So, the difference is that we do not allow any translates here; see [Baa02b] for related results.

Lemma 11.4. *If μ is a probability measure on $\mathbb{R}$, its Fourier transform $\widehat{\mu}(k)$ is a uniformly continuous and positive definite function on $\mathbb{R}$, which satisfies $|\widehat{\mu}(k)| \leqslant \widehat{\mu}(0) = 1$. Moreover, the following three properties are equivalent.*

 (1) $\operatorname{card}\{k \in \mathbb{R} \mid \widehat{\mu}(k) = 1\} > 1$;

 (2) $\operatorname{card}\{k \in \mathbb{R} \mid \widehat{\mu}(k) = 1\} = \infty$;

 (3) $\operatorname{supp}(\mu)$ *is contained in a lattice.*

If one of these properties holds and $\mu \neq \delta_0$, the set $\{k \in \mathbb{R} \mid \widehat{\mu}(k) = 1\}$ is a lattice in $\mathbb{R}$.

PROOF. One has $\widehat{\mu}(k) = \int_{\mathbb{R}} e^{-2\pi i kx} \, d\mu(x)$, whence the first claims are standard consequences of Fourier analysis; compare [Pin02, Prop. 5.2.1] and [Rud62, Sec. 1.3.3].

If $\mu = \sum_{x \in \Gamma} p(x) \delta_x$ for a lattice $\Gamma \subset \mathbb{R}$, with $p(x) \geqslant 0$ and the normalisation $\sum_{x \in \Gamma} p(x) = 1$, one has

$$(11.11) \qquad \widehat{\mu}(k) - \sum_{x \in \Gamma} p(x) \, e^{-2\pi i k r},$$

so that $\widehat{\mu}(k) = 1$ for any $k \in \Gamma^*$. In particular, $\Gamma^* \subset \{k \in \mathbb{R} \mid \widehat{\mu}(k) = 1\}$, so that we have the implications $(3) \implies (2) \implies (1)$.

If $\widehat{\mu}(k) = 1$ for some $k \neq 0$, one has $\int_{\mathbb{R}} e^{-2\pi i kx} \, d\mu(x) = 1$ and hence

$$(11.12) \quad \int_{\mathbb{R}} \big(1 - \cos(2\pi kx)\big) \, d\mu(x) = \int_{\mathrm{supp}(\mu)} \big(1 - \cos(2\pi kx)\big) \, d\mu(x) = 0,$$

where $\mathrm{supp}(\mu)$, the support of the probability measure μ, is a closed subset of $\mathbb{R}$ and measurable. The integrand is a continuous non-negative function which, due to $k \neq 0$, vanishes precisely on the set $\frac{1}{k}\mathbb{Z}$, which is a lattice.

Write $\mathrm{supp}(\mu) = A \,\dot\cup\, B$ as a disjoint union of two measurable sets, with $A = \mathrm{supp}(\mu) \cap \frac{1}{k}\mathbb{Z}$ and $B = \mathrm{supp}(\mu) \cap (\mathbb{R} \setminus \frac{1}{k}\mathbb{Z})$. We can now split the second integral in Eq. (11.12) into an integral over A, which vanishes because the integrand does, and one over the set B, which would give a positive contribution by standard arguments, unless $B = \varnothing$. But this means that $\mathrm{supp}(\mu) = A \subset \frac{1}{k}\mathbb{Z}$, so that $(1) \implies (3)$, which establishes the result.

For the final claim, since $\mu \neq \delta_0$, we know that $\langle \mathrm{supp}(\mu) \rangle_{\mathbb{Z}} = \Gamma$ is a lattice in $\mathbb{R}$. By construction, Γ is the coarsest lattice that contains $\mathrm{supp}(\mu)$. Note that the right-hand side of Eq. (11.11) can only sum to 1 if $e^{-2\pi i kx} = 1$ for all points $x \in \Gamma$ with $p(x) > 0$, which include a set of points that generates Γ. This implies that $k \in \Gamma^*$, so that $\{k \in \mathbb{R} \mid \widehat{\mu}(k) = 1\} = \Gamma^*$, which is a lattice. $\qquad \square$

Theorem 11.2. *Let ϱ be a probability measure on $\mathbb{R}_+$ with mean 1, and assume that a moment of ϱ of order $1 + \varepsilon$ exists for some $\varepsilon > 0$. Then, the point sets obtained from the stationary renewal process based on ϱ almost surely have a diffraction measure of the form*

$$\widehat{\gamma} = \widehat{\gamma}_{\mathsf{pp}} + (1 - h) \, \lambda,$$

where h is a locally integrable function on $\mathbb{R}$ that is continuous except for at most countably many points (namely those of the set $S = \{k \mid \widehat{\varrho}(k) = 1\}$). On $\mathbb{R} \setminus S$, the function h is given by

$$h(k) = \frac{2 \left(|\widehat{\varrho}(k)|^2 - \mathrm{Re}(\widehat{\varrho}(k)) \right)}{|1 - \widehat{\varrho}(k)|^2}.$$

Moreover, the pure point part is

$$\left(\widehat{\gamma}\right)_{\mathsf{pp}} = \begin{cases} \delta_0, & \text{if } \mathrm{supp}(\varrho) \text{ is not a subset of a lattice}, \\ \delta_{\mathbb{Z}/b}, & \text{otherwise}, \end{cases}$$

where $b\mathbb{Z}$ *is the coarsest lattice that contains* $\mathrm{supp}(\varrho)$.

PROOF. The process has a well-defined autocorrelation γ by Proposition 11.3, in the sense that almost every realisation of the process is a point set Λ with this autocorrelation. Since γ is a positive definite measure, it is Fourier transformable by Proposition 8.6, with $\widehat{\gamma}$ being a positive measure on $\mathbb{R}$.

If $\mathrm{supp}(\varrho)$ is not contained in a lattice, we may invoke Lemma 11.4 to see that $\widehat{\varrho}(k) \neq 1$ whenever $k \neq 0$, so that Lemma 11.3 together with Proposition 11.3 implies pointwise convergence

$$\widehat{\nu}_n(k) \xrightarrow{\;n\to\infty\;} \widehat{\nu}(k) = \frac{\widehat{\varrho}(k)}{1 - \widehat{\varrho}(k)}$$

on $\mathbb{R} \setminus \{0\}$, and similarly for $\widehat{\widetilde{\nu}}$. Since $\widehat{\varrho}$ is uniformly continuous on $\mathbb{R}$ and $\widehat{\varrho}(k) \neq 1$ on $\mathbb{R} \setminus \{0\}$, both ν and $\widetilde{\nu}$ are represented, on $\mathbb{R} \setminus \{0\}$, by continuous Radon–Nikodym densities. Writing $(\nu + \widetilde{\nu})^{\widehat{\;}} = (1 - h)\lambda$, the formula for h follows from $\widehat{\widetilde{\nu}} = \overline{\widehat{\nu}}$. In the remaining cases, an analogous argument applies to all $k \in \mathbb{R} \setminus S$. From Lemma 11.4, we know that S is a lattice (and hence countable).

We next show that $1 - h$ is locally integrable. This is clear on any compact set in the complement of S due to the continuity of h, wherefore it needs to be shown around all points $k \in S$. If ϱ is strictly lattice-like, with $\langle \mathrm{supp}(\varrho) \rangle_{\mathbb{Z}} = \Gamma$ say, each realisation of the process is a subset of Γ, so that $\widehat{\gamma}$ is Γ^*-periodic by Proposition 10.3, with $\Gamma^* = S$ (as follows from the proof of Lemma 11.4). It thus suffices to show integrability around $k = 0$, which then covers all cases.

Let X be a random variable with distribution ϱ. Since the latter has mean 1 and $\langle X^{1+\varepsilon} \rangle = \int_0^\infty x^{1+\varepsilon}\, \mathrm{d}\varrho(x) < \infty$ due to our moment assumption, we have the Taylor series expansion

$$\widehat{\varrho}(k) = 1 - 2\pi\mathrm{i}k + \mathcal{O}\big(|k|^{1+\varepsilon}\big), \quad \text{as } |k| \to 0,$$

by an application of [Ush99, Thm. 1.5.4]. Inserting this into the expression for h results in

$$h(k) = 2 + \mathcal{O}\big(|k|^{-1+\varepsilon}\big), \quad \text{as } |k| \to 0,$$

which establishes integrability around 0, and thus absolute continuity of the measure $(1 - h)\lambda$.

The presence of a Dirac measure with intensity 1 at the origin is a consequence of Corollary 9.1. It reflects the fact that the resulting point set Λ almost surely has density 1. If ϱ is not strictly lattice-like, this is the only contribution to the pure point part, as follows from the above convergence argument for all $k \in \mathbb{R} \setminus S$. Otherwise, the lattice generated by the support of ϱ is $\Gamma = b\mathbb{Z}$ for a unique $b > 0$, with dual lattice $\Gamma^* = \mathbb{Z}/b = S$. The claim on the pure point part is then clear by the Γ^*-periodicity of $\widehat{\gamma}$. $\qquad\square$

Remark 11.4 (*Asymptotic behaviour of the background function h*)**.** When, under the general assumptions of Theorem 11.2, also the second moment of ϱ exists, one obtains from [Ush99, Thm. 1.5.3] the slightly stronger expansion

$$\widehat{\varrho}(k) = 1 - 2\pi \mathrm{i} k - 2\pi^2 \langle X^2 \rangle \, k^2 + o\big(|k|^2\big), \quad \text{as } |k| \to 0.$$

This leads to the asymptotic behaviour

$$h(k) = 2 - \langle X^2 \rangle + o(1), \quad \text{as } |k| \to 0,$$

which implies that h is bounded and can continuously be extended to 0 via $h(0) = 2 - \langle X^2 \rangle = 1 - \sigma^2$, where σ^2 is the variance of ϱ. Clearly, the existence of higher moments implies stronger smoothness properties. $\qquad\Diamond$

Remark 11.5 (*Renewal process and random tilings*)**.** When $\mathrm{supp}(\varrho)$ is a finite set, one is in the situation that the corresponding renewal process produces random tilings of the real line with finitely many prototiles. A more detailed discussion, together with an explicit calculation of h for this case, is given in Section 11.6.1 below and in [BH00, Thm. 2]; see also Example 11.10 as well as Remark 11.9. $\qquad\Diamond$

Let us turn to some examples, for which we need the Heaviside function,

$$(11.13) \qquad \Theta(x) := \begin{cases} 1, & \text{if } x > 0, \\ \frac{1}{2}, & \text{if } x = 0, \\ 0, & \text{if } x < 0. \end{cases}$$

This formulation of Θ has some advantage for formal calculations around generalised functions (or distributions) and their Fourier transforms.

Example 11.6 (*Poisson process on the real line*)**.** The probably best-known stochastic point process is the classic (homogeneous) Poisson process on the line, with intensity 1, where $\varrho = f\lambda$ is given by the density

$$f(x) = \mathrm{e}^{-x}\,\Theta(x),$$

which is a special case of Example 11.1 (with $\alpha = \beta = 1$). It is easy to check that the convolution of $n + 1$ copies of this function yields $\mathrm{e}^{-x}x^n\Theta(x)/n!$, which results in $\nu = \Theta\lambda$. As the intensity is 1, this results in the autocorrelation

$$\gamma = \delta_0 + \nu + \widetilde{\nu} = \delta_0 + \lambda$$

and thus in the diffraction $\widehat{\gamma} = \gamma$. This follows from $(\delta_0 + \lambda)^{\widehat{}} = \delta_0 + \lambda$; see Example 8.5 on page 317. $\qquad\Diamond$

Remark 11.6 (*Universal characterisation of Poisson processes*). Let N denote a homogeneous Poisson process on the real line, so that, for any measurable $A \subset \mathbb{R}$, $N(A)$ is the number of renewal points that fall into A. It is well-known that $N(A)$ is then Poisson-$(\lambda(A))$-distributed, meaning that

$$\mathbb{P}\left(N(A) = k\right) = \frac{e^{-\lambda(A)} \left(\lambda(A)\right)^k}{k!}$$

with $k \in \mathbb{N}_0$, and that, for any collection $A_1, A_2, \ldots, A_m$ of pairwise disjoint sets, the random numbers $N(A_1), \ldots, N(A_m)$ are independent. In fact, this property characterises the Poisson process (compare [DV-J88, Ch. 2.1]), and it can serve as a definition in higher dimensions or in more general measure spaces, to which the renewal process cannot be extended. This approach also explains why the Poisson process is often considered as a model for an ideal gas. A finite patch was used in Figure 2.3 on page 21 for the illustration of random Voronoi and Delone cells. ◊

Example 11.7 (*Renewal process with repulsion*). A perhaps more interesting example in this spirit is given by the density

$$f(x) = 4x \, e^{-2x} \, \Theta(x),$$

which corresponds to the choice $\alpha = \beta = 2$ in Example 11.1. It is normalised with mean 1, as in Example 11.6, but models a repulsion of points for small distances. This distribution can also be realised via Example 11.6 by taking only every second point (known as 'thinning'), followed by a rescaling of time.

By induction (or by using well-known properties of the gamma distributions; compare [Fel72, Sec. II.2]), one can verify that

$$f^{*n}(x) = \frac{4^n}{(2n-1)!} \, x^{2n-1} \, e^{-2x} \, \Theta(x),$$

which results in the autocorrelation

$$\gamma = \delta_0 + (1 - e^{-4|x|}) \, \lambda = \delta_0 + \lambda - e^{-4|x|} \, \lambda$$

and in the diffraction measure

$$\widehat{\gamma} = \delta_0 + \frac{2 + (\pi k)^2}{4 + (\pi k)^2} \, \lambda = \delta_0 + \lambda - \frac{2\lambda}{4 + (\pi k)^2}.$$

The absolutely continuous part is illustrated in Figure 11.2 (case $\alpha = 2$). The 'dip' around 0, and thus the deviation from the previous example ($\alpha = 1$), reflects the effectively repulsive nature of the stochastic process when viewed from the perspective of neighbouring points. ◊

Example 11.8 (*Renewal process with gamma law of mean 1*). The previous two examples are special cases of the gamma family of measures. For fixed

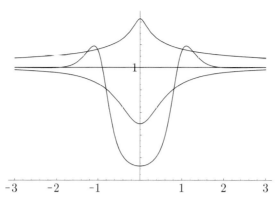

FIGURE 11.2. Absolutely continuous part of the diffraction measure from Example 11.8, for $\alpha = 0.7$ (upmost curve), $\alpha = 1$ (horizontal line, which also represents Example 11.6), $\alpha = 2$ (see also Example 11.7) and $\alpha = 8$ (overshooting curve). This is also known as the Bartlett spectrum in the theory of point processes; see [DV-J88, Ch. 8.2] and [BBM10, Rem. 15].

mean 1 (hence $\alpha = \beta$ in Example 11.1), they are parametrised with a real number $\alpha > 0$ via $\varrho_\alpha = f_\alpha \lambda$ and the density

$$(11.14) \qquad f_\alpha(x) \, = \, \frac{\alpha^\alpha}{\Gamma(\alpha)} \, x^{\alpha-1} \, \mathrm{e}^{-\alpha x} \, \Theta(x).$$

While $\alpha = 1$ is the 'interaction-free' Poisson process of Example 11.6, the density implies an effectively attractive (repulsive) nature of the process for $0 < \alpha < 1$ (for $\alpha > 1$).

Observing $f_\alpha^{*n}(x) = \frac{\alpha^{n\alpha}}{\Gamma(n\alpha)} \, x^{n\alpha-1} \, \mathrm{e}^{-\alpha x} \, \Theta(x)$, for $n \in \mathbb{N}$, this leads to the measure

$$(11.15) \qquad \nu_\alpha \, = \, g_\alpha \, \Theta \, \lambda \quad \text{with} \quad g_\alpha(x) \, = \, \alpha \, \mathrm{e}^{-\alpha x} \sum_{n=1}^{\infty} \frac{(\alpha x)^{n\alpha-1}}{\Gamma(n\alpha)}.$$

Note that, for fixed α, one has $\lim_{x \to \infty} g_\alpha(x) = 1$. The calculations result in the autocorrelation

$$\gamma_\alpha \, = \, \delta_0 + g_\alpha(|x|) \, \lambda$$

and in the diffraction $\widehat{\gamma_\alpha} = \delta_0 + (1 - h_\alpha) \, \lambda$, where h_α is the symmetric function defined by

$$h_\alpha(k) \, = \, \frac{2 \left(1 - \mathrm{Re}\big((1 + 2\pi \mathrm{i} k/\alpha)^\alpha \big) \right)}{\left| 1 - (1 + 2\pi \mathrm{i} k/\alpha)^\alpha \right|^2}.$$

The latter follows from the general form of h in Theorem 11.2, together with the observation that $\widehat{f_\alpha}(k) = (1 + 2\pi \mathrm{i} k/\alpha)^{-\alpha}$.

It is easy to see that $\lim_{k \to \pm\infty} h_\alpha(k) = 0$, for any fixed $\alpha > 0$, which makes the role of h_α as the deviation from the Poisson process diffraction more transparent, where $\alpha = 1$ and $h_1 \equiv 0$. Note also that $\lim_{\alpha \to \infty} \widehat{\gamma_\alpha} = \delta_{\mathbb{Z}}$

in the vague topology, in line with the limits mentioned before. The underlying transition from the Poisson case (completely random) to the lattice case (which is periodic) can nicely be studied in a series of plots of the diffraction with growing values of the parameter α. Figure 11.2 shows the onset of this phenomenon for small values of α. ◇

The interpolating property (as a function of the parameter α) is remarkable, as it shows that one can construct rather simple models for the transition from perfect order to complete disorder.

Remark 11.7 (*Delone sets from renewal processes*). Of particular interest in many applications are Delone sets, because points (representing atoms, say) should neither be too close nor too far apart. Such sets can also arise from a renewal process. In fact, if one considers a probability measure ϱ on $\mathbb{R}_+$, the resulting point sets are always Delone sets when $\mathrm{supp}(\varrho) \subset [a, b]$ with $0 < a \leqslant b < \infty$, and conversely. In this case, $a \leqslant r \leqslant R \leqslant b$ in the terminology of Section 2.1. This equivalence does not depend on the nature of ϱ on $[a, b]$, while the local complexity of the resulting point sets does. In particular, if ϱ is absolutely continuous, the point sets will (almost surely) *not* have finite local complexity (FLC). ◇

It is clear that, in the generic case, ϱ is not lattice-like (and hence certainly not strictly lattice-like). Nevertheless, the remaining cases are of interest, because they form a link to FLC tilings. Let us consider some examples.

Example 11.9 (*Deterministic lattice case*). The simplest case is $\varrho = \delta_1$. From $\delta_1 * \delta_1 = \delta_2$, one sees that $\nu = \delta_{\mathbb{N}}$ and hence

$$\gamma = \delta_0 + \delta_{\mathbb{N}} + \delta_{-\mathbb{N}} = \delta_{\mathbb{Z}},$$

which is a lattice Dirac comb, with Fourier transform

$$\widehat{\gamma} = \delta_{\mathbb{Z}}$$

by means of the Poisson summation formula from Eq. (9.14), or according to Theorem 11.2, where one finds $h \equiv 1$ and $b = 1$. This is the deterministic case of the integer lattice, covered in this setting. ◇

Remark 11.8 (*Concentration property for $\alpha \to \infty$*). Example 11.9 can also be seen as a deterministic limit of the measure ϱ_α defined by Eq. (11.14). In particular, one has $\lim_{\alpha \to \infty} \varrho_\alpha = \delta_1$ and $\lim_{\alpha \to \infty} \nu_\alpha = \delta_{\mathbb{N}}$, with ν_α as in Eq. (11.15) and both limits to be understood in the vague topology. This can also be seen by means of the SLLN. For each $n \in \mathbb{N}$, by well-known divisibility properties of the family of gamma distributions, ϱ_n is the distribution of

$$\frac{1}{n} \sum_{i=1}^{n} X_i,$$

where the X_i are independent and exponentially distributed random variables with mean 1. This sum then concentrates around 1, with a standard deviation of order $1/\sqrt{n}$. $\diamond$

Example 11.10 (*Random tilings with finitely many prototiles*). Consider the probability measure

$$\varrho = \alpha \delta_a + (1-\alpha)\delta_b,$$

with $\alpha \in (0,1)$ and $a, b > 0$, subject to the restriction $\alpha a + (1-\alpha)b = 1$ to ensure density 1, and $a \neq b$. Each realisation of the corresponding renewal process results in a point set that can also be viewed as a random tiling on the line with two prototiles, of lengths a and b. As before, place a normalised point measure at each point of the realisation. Then, the diffraction (almost surely) has a pure point and an absolutely continuous part, but no singular continuous one. The pure point part can be just δ_0 (when b/a is irrational) or a lattice comb. The details are given in [BH00], including an explicit formula for the absolutely continuous part (see also Theorem 11.6 below, where the result is spelt out without the restriction to density 1).

This example has a straightforward generalisation to any finite number of prototiles, with a similar result. Also in this case, there is an explicit formula for the diffraction measure, which was derived in [BH00] by a direct method, without using the renewal process. $\diamond$

Remark 11.9 (*Spikiness of absolutely continuous diffraction measures*). Looking back at Lemma 11.4, one realises that Example 11.10 revolves around the lattice condition in an interesting way. Namely, even if ϱ is *not* strictly lattice-like, supp(ϱ) for a random tiling example with finitely many prototiles is a finite set, and thus a subset of a Meyer set. We then know from the harmonic analysis of Meyer sets, compare [Moo00] and references therein, that $\widehat{\varrho}(k)$ will come ε-close to 1 with bounded gaps in k; compare Section 10.5.

This means that the diffraction measure of a typical realisation of the process, though it is absolutely continuous apart from the central peak at $k = 0$, will develop sharp 'needles' that are close to point measures in the vague topology — a phenomenon that was also observed in [BH00] on the basis of the explicit solution; compare Figure 11.7 on page 470. $\diamond$

11.4. Point processes from random matrix theory

It is possible to define interesting random point sets by stationary ergodic point processes that emerge as scaling limits of the eigenvalue distributions of certain random matrix ensembles; see [AGZ10] for general background. Below, we briefly discuss two classic examples that go back to Dyson [Dys62] and Ginibre [Gin65]. For the underlying derivations, we refer to the original

articles, and to further references given in [BK11]. This section assumes
some familiarity with elementary results from random matrix theory, but is
not required for our later results.

11.4.1. RANDOM POINT SETS ON THE LINE

The global eigenvalue distribution of random orthogonal, unitary or sym-
plectic matrix ensembles is known to asymptotically follow the classic semi-
circle law of Wigner. More precisely, this law (with a density function of the
form of a semi-circle, hence the name) describes the eigenvalue distribution
of the underlying ensembles of symmetric, Hermitian and (symplectically)
self-dual $N \times N$-matrices with Gaussian distributed entries, in the limit as
$N \to \infty$. The corresponding random matrix ensembles are called GOE, GUE
and GSE, with attached β-parameters 1, 2 and 4, respectively. They permit
an interpretation as a Coulomb gas, where β is the power in the central potential; see [AGZ10, Meh04, For10] for general background and [Dys62] for
the results that are relevant to our point of view.

For matrices of dimension N, the semi-circle has radius $\sqrt{2N/\pi}$ and area
N. Note that, in comparison to [Meh04], we have rescaled the density by a
factor $1/\sqrt{\pi}$ here, so that we really have a semi-circle and not a semi-ellipse.
To study the local eigenvalue distribution with our application in mind, we
rescale the central region (between ± 1, say) by $\sqrt{2N/\pi}$. This leads, in the
limit as $N \to \infty$, to a new ensemble of point sets on the line that can be
interpreted as a stationary, ergodic point process of intensity 1; for $\beta = 2$,
see [AGZ10, Ch. 4.2] or [Sos00] and references therein for details. Since the
process is simple (meaning that, almost surely, no point is occupied twice),
almost all realisations are point sets of density 1.

It is possible to calculate the autocorrelation of these processes, on the
basis of Dyson's correlation functions [Dys62]. Though the latter originally
apply to the circular ensembles, they have been adapted to the other ensembles by Mehta [Meh04]; see also [For10]. For all three ensembles mentioned
above, this leads to an autocorrelation of the form

(11.16) $$\gamma = \delta_0 + \big(1 - f(|x|)\big)\lambda,$$

where f is a locally integrable function that depends on β. Defining the
short-hand notation $s(r) = \mathrm{sinc}(\pi r) = \frac{\sin(\pi r)}{\pi r}$, one obtains (with $r \geqslant 0$)

$$f(r) = \begin{cases} s(r)^2 + s'(r) \int_r^\infty s(t)\,\mathrm{d}t, & \text{if } \beta = 1, \\ s(r)^2, & \text{if } \beta = 2, \\ s(2r)^2 - 2s'(2r) \int_0^r s(2t)\,\mathrm{d}t, & \text{if } \beta = 4. \end{cases}$$

The diffraction measure is the Fourier transform of γ, which has also
been calculated in [Dys62, Meh04, For10] in a slightly different formulation.

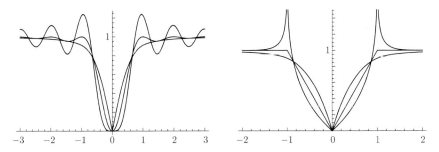

FIGURE 11.3. Absolutely continuous part of the autocorrelation (left) and the diffraction (right) for the three Dyson ensembles on the line, with $\beta \in \{1, 2, 4\}$. On the left, the oscillatory behaviour increases with β. On the right, $\beta = 2$ corresponds to the piecewise linear function with corners at 0 and ± 1, while $\beta = 4$ shows a locally integrable singularity at ± 1. The latter reflects the slowly decaying oscillations in the left diagram.

Recalling $\widehat{\delta_0} = \lambda$ and $\widehat{\lambda} = \delta_0$, the result is always of the form

$$(11.17) \qquad \widehat{\gamma} = \delta_0 + \big(1 - b(k)\big)\lambda = \delta_0 + h(k)\,\lambda,$$

where $b = \widehat{f}$. The Radon–Nikodym density h for $\beta = 1$ reads

$$h_1(k) = \begin{cases} |k|\big(2 - \log(2|k| + 1)\big), & \text{if } |k| \leqslant 1, \\ 2 - |k|\log\frac{2|k|+1}{2|k|-1}, & \text{if } |k| > 1, \end{cases}$$

where $k \in \mathbb{R}$. The result for $\beta = 2$ is simpler and reads

$$h_2(k) = \begin{cases} |k|, & \text{if } |k| \leqslant 1, \\ 1, & \text{if } |k| > 1, \end{cases}$$

while $\beta = 4$ leads to

$$h_4(k) = \begin{cases} \frac{1}{4}|k|\big(2 - \log\big|1 - |k|\big|\big), & \text{if } |k| \leqslant 2, \\ 1, & \text{if } |k| > 2. \end{cases}$$

Figure 11.3 illustrates the three cases. Let us summarise as follows.

Theorem 11.3. *The eigenvalues of the Dyson random matrix ensembles for $\beta \in \{1, 2, 4\}$, in the scaling of the local region around 0 as used above, almost surely give rise to point sets of density 1, with autocorrelation and diffraction measures as specified in Eqs. (11.16) and (11.17).* $\qquad\square$

Note that h_4 is smooth at $k = \pm 2$, but has integrable singularities at $k = \pm 1$. The latter are a consequence of the stronger oscillatory behaviour of the function f_4 at integer values, as was already noticed in [Dys62]. When extrapolating to other values of β (in particular to $\beta > 4$), this is the onset of another point measure. In fact, for $0 < \beta < \infty$, one finds another family of processes that interpolate between the Poisson process ($\beta \to 0$) and the integer lattice ($\beta \to \infty$); see [KS09, For10] for more.

Note that the circular random matrix ensembles (COE, CUE, CSE) asymptotically give the same local correlations [Dys62, Meh04], and hence the same autocorrelation and diffraction (after appropriate rescaling).

11.4.2. RANDOM POINT SETS IN THE PLANE

The above examples were derived from matrix ensembles with real eigenvalues, and thus lead to point processes in $\mathbb{R}$. There is also one ensemble, due to Ginibre [Gin65] (see also [Meh04, AGZ10]), of general complex matrices with Gaussian distributed entries that will give rise to a stationary point process in $\mathbb{R}^2$. Again, this emerges (by proper rescaling) from the eigenvalues (now seen as elements of the plane), which approach the uniform distribution in a circle of radius $\sqrt{N/\pi}$ (and hence area N) as $N \to \infty$.

As before, the system can be interpreted as a Coulomb gas, with a potential parameter $\beta = 2$. Other matrix ensembles permit this interpretation, too, but do not seem to correspond to interesting stationary processes from our point of view, wherefore we stick to Ginibre's example here.

Following the original approach of [Gin65], the limit $N \to \infty$ leads to a stationary and ergodic, simple point process of intensity 1, so that almost every realisation is a point set in $\mathbb{R}^2$ of density 1. Using complex variables $z_i \in \mathbb{C} \simeq \mathbb{R}^2$, the 2-point correlation function is of determinantal form,

$$\rho(z_1, z_2) \;=\; \mathrm{e}^{-\pi(|z_1|^2 + |z_2|^2)} \begin{vmatrix} \mathrm{e}^{\pi|z_1|^2} & \mathrm{e}^{\pi z_1 \overline{z_2}} \\ \mathrm{e}^{\pi \overline{z_1} z_2} & \mathrm{e}^{\pi|z_2|^2} \end{vmatrix} \;=\; 1 - \mathrm{e}^{-\pi|z_1 - z_2|^2};$$

see [Gin65] or [Meh04] for a derivation.[1] Note that, despite using complex coordinates here, the expression is calculated relative to the volume element of real coordinates (hence relative to Lebesgue measure, as in [Meh04]). The result is translation invariant and only depends on the distance $r = |z_1 - z_2|$ between the two points; see [BKM14] for related examples.

As a consequence, the autocorrelation of a realisation almost surely reads

$$(11.18) \qquad \gamma = \delta_0 + (1 - \mathrm{e}^{-\pi r^2})\lambda,$$

which is radially symmetric, with r as above. By a standard calculation, compare Remark 8.3, the Fourier transform of γ results in

$$(11.19) \qquad \widehat{\gamma} = \delta_0 + (1 - \mathrm{e}^{-\pi|k|^2})\lambda,$$

---

[1]The result is a reasonable approximation for matrices of dimension 10^3 or larger. We thank Thorsten Hüls and Holger Kösters for an interesting discussion on this point.

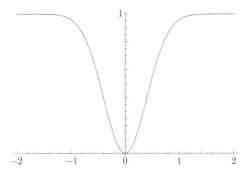

FIGURE 11.4. Radial dependence (section) of the absolutely continuous part both of the autocorrelation and the diffraction measure for the planar point set ensemble, as derived from Ginibre's matrix ensemble.

so that we obtain a self-dual pair of measures under Fourier transform (as in the Poisson process of density 1). The radial dependence is illustrated in Figure 11.4, while the general result reads as follows.

Theorem 11.4. *The Ginibre complex random matrix ensemble, in the scaling used above, almost surely results in point sets of density* 1*, with the autocorrelation of Eq.* (11.18) *and the diffraction of Eq.* (11.19). $\qquad\square$

So far, all examples relied on independence or on some effective stochastic interaction. In general, an understanding of equilibrium systems requires an explicit treatment of the interaction between the points. This can be done via the theory of Gibbs measures, which is beyond the scope of this book; see [Geo11] and references therein for a comprehensive exposition. Below, we restrict ourselves to an illustration by means of some classic examples.

11.5. Lattice systems with interaction

In Example 11.2, we have seen the interaction-free lattice gas on $\mathbb{Z}$, which readily generalises to $\mathbb{Z}^d$. We now turn our attention to the diffraction measure of certain lattice gas models on $\mathbb{Z}^d$ with translation invariant interaction. An obvious candidate in one dimension is the binary Ising lattice gas, compare [Wel04], which we formulate as an ergodic Markov chain.

Example 11.11 (*Reversible Markov lattice gas*). Let $p, q \in [0, 1]$ be two probabilities subject to the condition $0 < p + q < 2$, and consider the Markov chain on $\mathbb{Z}$ defined by

$$
\text{with matrix} \quad M = \begin{pmatrix} p & 1-p \\ 1-q & q \end{pmatrix}.
$$

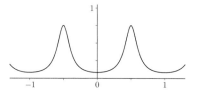 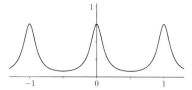

FIGURE 11.5. Absolutely continuous part of the Markov chain diffraction $\hat{\gamma}$ of Example 11.11. The left panel shows the case $p = q = \frac{1}{4}$, which is effectively repulsive, while the right panel is the corresponding attractive case for $p = q = \frac{3}{4}$. In both cases, the pure point part (not shown) is $\frac{1}{4}\delta_{\mathbb{Z}}$.

Due to our restrictions, M is primitive with eigenvalues 1 and $r = p + q - 1$, where $|r| < 1$. The unique stationary state is defined by the probability vector $v = \frac{1}{2-p-q}(1 - q, 1 - p)$, which is the left PF eigenvector of M in statistical normalisation. When using weights 0 (for 'empty') and 1 (for 'occupied'), the density of occupied sites of $\mathbb{Z}$ almost surely is $\rho = \frac{1-p}{2-p-q}$.

Since our conditions imply ergodicity [Pet83], we can calculate the auto-correlation of a typical realisation by the ensemble average, which is an application of the ergodic theorem; see Theorem 4.2. Almost surely, by application of standard methods from [Pet83], we thus get the autocorrelation coefficients $\eta(m) = (0,1)DM^{|m|}(0,1)^T$, where $D = \text{diag}(v_1, v_2)$. The validity for negative values of m follows from the reversibility of our Markov chain [Bil95]. A straightforward calculation gives

$$\eta(m) = \rho^2 + \frac{(1-p)(1-q)}{(2-p-q)^2} r^{|m|}$$

for all $m \in \mathbb{Z}$. Since $|r| < 1$, we can calculate the Fourier transform of $\gamma = \eta\,\delta_{\mathbb{Z}}$ by means of the PSF and the geometric series. Almost surely, the 1-periodic diffraction measure finally reads

$$\hat{\gamma} = \rho^2\delta_{\mathbb{Z}} + \rho h \lambda \quad \text{with} \quad h(k) = \frac{(1-q)(1+r)}{1 - 2r\cos(2\pi k) + r^2},$$

which is a special case of the more general treatment in [BH00]. We illustrate two examples in Figure 11.5. We will come back to the qualitative interpretation of $\hat{\gamma}_{\text{ac}}$ below in Corollary 11.2. Note that the choice $q = 1 - p$ (which means $r = 0$ and $\rho = 1 - p$) brings us back to the Bernoulli comb of Example 11.2.   ◊

Before we discuss the planar lattice gas based on the classic Ising model, we illustrate the emergence of new phenomena via a planar system of algebraic origin which has an internal group structure.

Example 11.12 (*Ledrappier's model on $\mathbb{Z}^2$*). A prominent example of algebraic origin in the plane is due to Ledrappier [Led78]. In multiplicative notation, it can be defined as the shift space

$$\mathbb{X}_{\mathrm{L}} = \left\{ w \in \{\pm 1\}^{\mathbb{Z}^2} \mid w_x w_{x+e_1} w_{x+e_2} = 1 \text{ for all } x \in \mathbb{Z}^2 \right\},$$

where e_1 and e_2 denote the standard Euclidean basis vectors in the plane. It is a subspace of $\{\pm 1\}^{\mathbb{Z}^2}$, but also an Abelian group (under pointwise multiplication in our formulation, which follows [BW10]). As a dynamical system, it is thus equipped with the corresponding Haar measure μ_{L}, which is positive and normalised so that $\mu_{\mathrm{L}}(\mathbb{X}_{\mathrm{L}}) = 1$. For an alternative formulation with a Gibbs (or equilibrium) measure, we refer to [Sla81]. Obviously, the system has no entropy, because the knowledge of a configuration along one horizontal line determines everything above it. However, it is clearly not deterministic. In fact, essentially along any given lattice direction, it looks like a one-dimensional Bernoulli system [Led78]. It is thus said to have rank-1 entropy, which means that the number of circular patches of a given size grows exponentially with its diameter, but not with its area.

Given an element $w \in \mathbb{X}_{\mathrm{L}}$, the corresponding Dirac comb

$$\omega = \sum_{x \in \mathbb{Z}^2} w_x \, \delta_x$$

possesses μ_{L}-almost surely the autocorrelation γ and the diffraction measure $\widehat{\gamma}$ given by [BW10]

$$\gamma = \delta_0 \quad \text{and} \quad \widehat{\gamma} = \lambda \, .$$

The system is thus homometric with the planar Bernoulli system with $p = \frac{1}{2}$ (coin tossing on $\mathbb{Z}^2$), and also with the direct product of two binary Rudin–Shapiro sequences. The similarity with the Bernoulli system goes a lot further, in the sense that many other correlation functions also agree, although the systems differ for certain 3-point correlations; see [BW10] for details.

This system is meant to indicate that higher-dimensional symbolic dynamics is always good for a surprise, as is well-known from [Schm95]. It is thus clear that the corresponding inverse problem becomes more complicated. Another famous example is the $(\times 2, \times 3)$ dynamical system [Rud90], which shares almost all correlation functions with a lattice Bernoulli system with continuous degrees of freedom [BW10]. $\diamond$

Let us continue with the discussion of lattice gases on $\mathbb{Z}^d$. Here, each lattice site $x \in \mathbb{Z}^d$ is either occupied ($n_x = 1$) or empty ($n_x = 0$), so that a *configuration* is an element of $\{0, 1\}^{\mathbb{Z}^d}$. We are then interested in the diffraction of a 'typical' set of occupied sites.

More specifically, we consider models that derive from the *Ising model* of magnetism, with single spin space $\{-1, 1\}$ and pair potentials which can be described by a symmetric real function $J(x) = J(-x)$ for $x \in \mathbb{Z}^d$. Then, the corresponding Hamiltonian for the total energy can formally be written as $H(\boldsymbol{\sigma}) = -\sum_{x,y\in\mathbb{Z}^d} J(x-y)\,\sigma_x\sigma_y$, where $\boldsymbol{\sigma} \in \{\pm 1\}^{\mathbb{Z}^d}$ and σ_x denotes the spin variable at $x \in \mathbb{Z}^d$. The pair potential models a translation invariant two-body interaction. If J is a positive function, the interaction is called *ferromagnetic* in reference to the meaning for the description of magnetism.

The lattice gas interpretation is achieved via $n_x = \frac{1}{2}(1+\sigma_x)$. The ferromagnetic interaction then translates into an effectively attractive interaction. The description of the infinite system is made rigorous via a limit over finite systems of increasing size; see below for some details. For convenience and compatibility with the literature, we mainly use the spin formulation.

11.5.1. GENERAL SETTING

For a finite subsystem $T \subset \mathbb{Z}^d$ (with periodic boundary conditions, which mean that the system is wrapped on a d-torus), the *partition function* in the spin formulation is

$$Z_\beta = \sum_{\boldsymbol{\sigma}\in\{\pm 1\}^T} \exp\big(-\beta H(\boldsymbol{\sigma})\big) = \sum_{\boldsymbol{\sigma}\in\{\pm 1\}^T} \exp\Big(\beta \sum_{x,y\in T} J(x-y)\,\sigma_x\sigma_y\Big),$$

where $\beta = 1/(k_\mathrm{B}T)$ is the inverse temperature, with Boltzmann's constant k_B, and the first sum runs over all configurations $\boldsymbol{\sigma} \in \{\pm 1\}^T$. Via

$$(11.20) \qquad \mu_\beta\big(\{\boldsymbol{\sigma}\}\big) = \frac{1}{Z_\beta}\exp\Big(\beta \sum_{x,y\in T} J(x-y)\,\sigma_x\sigma_y\Big),$$

one defines a probability measure on the (finite) configuration space. The *expectation* of a (measurable) function $f\colon \{\pm 1\}^T \longrightarrow \mathbb{C}$ then is

$$\langle f\rangle_\beta := \mu_\beta(f) = \sum_{\boldsymbol{\sigma}\in\{\pm 1\}^T} f(\boldsymbol{\sigma})\,\mu_\beta\big(\{\boldsymbol{\sigma}\}\big).$$

In the infinite volume (or thermodynamic) limit, Eq. (11.20) defines the corresponding *Gibbs measures* as the closure of the set of accumulation points in the vague topology, details of which are suppressed here. The Gibbs measures form a non-empty simplex (see [Geo11, Thm. 7.26]), which need not be a singleton set. If its nature changes as a function of β (for instance, from a singleton to a 1-simplex), the system undergoes a *phase transition*. The point at which this happens is $\beta_\mathrm{c} = 1/(k_\mathrm{B}T_\mathrm{c})$, the so-called *inverse critical temperature*. In the following, we mainly consider cases where the Gibbs measure is unique. Since extremal Gibbs measures are ergodic (see [Geo11, Thm. 14.15]), the unique Gibbs measure is ergodic, and quantities obtained as an average over the ensemble (such as the correlation functions which we consider next) almost surely apply to each realisation of the underlying stochastic process (with respect to the Gibbs measure). For a suitable exposition of the ergodic theorem for such $\mathbb{Z}^2$-actions, we refer to [Kel98].

Assuming uniqueness of the (translation invariant) Gibbs measure, the *pair correlation function* $\langle n_0 n_x \rangle_\beta$ in the lattice gas interpretation of the models considered can be deduced from the spin-spin correlations via the relation $n_x = \frac{1}{2}(1 + \sigma_x)$. The assumed uniqueness of the Gibbs measure, by a simple symmetry argument, implies that $\langle \sigma_x \rangle_\beta = 0$, and thus $\langle n_x \rangle_\beta = \frac{1}{2}$. This means that, on average, half the sites are occupied, and this leads to the following relationship between the correlation functions,

$$(11.21) \qquad \langle n_0 n_x \rangle_\beta = \frac{1}{4} + \frac{\langle \sigma_0 \sigma_x \rangle_\beta}{4}.$$

Note that the interaction between two sites only depends on their distance, which results in translation invariant expectations. In particular, one has $\langle n_y n_{x+y} \rangle_\beta = \langle n_0 n_x \rangle_\beta$. Since the autocorrelation coefficient $\eta(x)$ is the volume average of these pair correlation functions, one almost surely gets $\eta(x) = \langle n_0 n_x \rangle_\beta$. This yields the following positive definite autocorrelation measure

$$\gamma = \sum_{x \in \mathbb{Z}^d} \langle n_0 n_x \rangle_\beta \, \delta_x \qquad \text{(a.s.)}.$$

As before, we are interested in the (positive) diffraction measure $\widehat{\gamma}$ and its Lebesgue decomposition.

The constant part of Eq. (11.21) results in the pure point part

$$(11.22) \qquad \widehat{\gamma}_{\mathsf{pp}} = \frac{1}{4} \sum_{k \in \mathbb{Z}^d} \delta_k = \frac{1}{4} \delta_{\mathbb{Z}^d} \qquad \text{(a.s.)}$$

of the diffraction measure (recall from Example 3.1 that $\mathbb{Z}^d$ is self-dual), as a consequence of the PSF for distributions; see Proposition 9.4. Strictly speaking, the validity of Eq. (11.22) is not clear at this stage, as we have only shown that $\widehat{\gamma}_{\mathsf{pp}}$ 'contains' $\frac{1}{4} \delta_{\mathbb{Z}^d}$. However, the argument is complete if the measure $\frac{1}{4} \sum_{x \in \mathbb{Z}^d} \langle \sigma_0 \, \sigma_x \rangle_\beta \delta_x$ is null weakly almost periodic; see [GLA90, Ch. 11]. This is true of the present example, and also of our later ones. The reason is that $|\langle \sigma_0 \, \sigma_x \rangle_\beta|$ decays sufficiently rapidly as $|x| \to \infty$. In our above argument, we have used that $\langle \sigma_x \rangle_\beta = 0$, which led to Eq. (11.21). More generally, assuming only translation invariance, one finds

$$(11.23) \qquad \langle n_0 n_x \rangle_\beta = \rho^2 + \frac{1}{4} \big(\langle \sigma_0 \sigma_x \rangle_\beta - \langle \sigma_0 \rangle_\beta \langle \sigma_x \rangle_\beta \big),$$

where $\rho = \langle n_0 \rangle_\beta = \langle n_x \rangle_\beta$ is the density of the lattice gas. Note that the second term on the right hand side has the form of a covariance.

Let us now show that, in addition to this pure point part, almost surely only an absolutely continuous part (and no singular continuous one) is present in the diffraction measure of models on $\mathbb{Z}^d$ with finite-range ferromagnetic (or attractive) two-body interactions. This property holds for all temperatures above T_c. For the analogous situation in more general models, deep in the

Dobrushin uniqueness regime (which implies uniqueness of the Gibbs measure for sufficiently high temperatures), we refer to Remark 11.10 below.

Similar observations have also been made in [vEM92, BH00]. In general, when the interaction is sufficiently long-ranged (such as in the Thue–Morse ground state [MvE90, Mię93]), one should expect (on the basis of [Hal44, Kni98], say) the appearance of systems with singular spectral components (at least in the ground state [Rad87]). Systems with singular continuous diffraction may even be generic, in analogy to Simon's wonderland theorem [Sim95] from the theory of Schrödinger operators. Note, however, that such systems might still be negligible in a measure-theoretic sense. Indeed, in view of the widespread application of lattice gas models with short-ranged interactions to disordered phenomena in solids (at positive temperatures), the Ising model results (and their generalisations discussed below) provide an indication of why singular continuous spectra are usually not considered in classical crystallography; compare [BZ08] and references therein.

11.5.2. FOURIER SERIES OF DECAYING CORRELATIONS

Let us assume that the correlation coefficients of the model under consideration are exponentially decaying as

$$(11.24) \qquad |\langle \sigma_0 \sigma_x \rangle_\beta| \leqslant C \,\mathrm{e}^{-\varepsilon \|x\|},$$

where C and ε are positive constants depending only on β and the model considered, and $\|x\|^2 = x_1^2 + \cdots + x_d^2$ denotes the (squared) Euclidean norm. Let us now deduce from the absolute convergence of $\sum_{x \in \mathbb{Z}^d} \mathrm{e}^{-\varepsilon \|x\|}$ that the inequality in Eq. (11.24) for the correlation coefficients results in an absolutely continuous part of the diffraction measure.

Lemma 11.5. *If the correlation coefficients are bounded as in Eq. (11.24), their sum $\sum_{x \in \mathbb{Z}^d} \langle \sigma_0 \sigma_x \rangle_\beta$ is absolutely convergent.*

PROOF. Recall the classic inequality

$$|x_1| + \cdots + |x_d| \leqslant \sqrt{d}\,\|x\|,$$

which, with $\varepsilon > 0$, implies

$$\sum_{x \in \mathbb{Z}^d} \mathrm{e}^{-\varepsilon \|x\|} \leqslant \sum_{x \in \mathbb{Z}^d} \mathrm{e}^{-\frac{\varepsilon}{\sqrt{d}}(|x_1| + \cdots + |x_d|)} = \left(\sum_{n \in \mathbb{Z}} \mathrm{e}^{-\frac{\varepsilon}{\sqrt{d}}|n|} \right)^d$$

$$= \left(\frac{\mathrm{e}^{\varepsilon/\sqrt{d}} + 1}{\mathrm{e}^{\varepsilon/\sqrt{d}} - 1} \right)^d = \left(\coth\left(\frac{\varepsilon}{2\sqrt{d}} \right) \right)^d.$$

Due to the inequality (11.24), absolute convergence of the sum follows. □

A similar result can also be proved for correlations with suitable algebraic (power law) decay; see [BS04b] for details. Here, we formulate one consequence of Lemma 11.5 for *ergodic lattice gas* models on $\mathbb{Z}^d$, by which we mean a closed subset of $\{0,1\}^{\mathbb{Z}^d}$ that is equipped with an ergodic and translation invariant equilibrium measure.

Proposition 11.4. *The diffraction measure of an ergodic lattice gas model on $\mathbb{Z}^d$, with density ρ and correlation coefficients bounded as in Eq.* (11.24), *almost surely exists, is $\mathbb{Z}^d$-periodic and consists of the pure point part $\rho^2 \delta_{\mathbb{Z}^d}$ and an absolutely continuous part with smooth density. No singular continuous part is present.*

PROOF. Lattice gas models can also be treated as weighted lattice Dirac combs (with weight 1 if a site is occupied and weight 0 otherwise). So, due to the assumed ergodicity, the diffraction measure $\widehat{\gamma}$ almost surely can be represented as

$$\widehat{\gamma} = \mu * \delta_{\mathbb{Z}^d}$$

with a finite positive measure μ that is supported on a fundamental domain of $\mathbb{Z}^d$, by an application of Theorem 10.3. This yields the $\mathbb{Z}^d$-periodicity, which is also implied by the following more explicit arguments.

We have already treated the pure point part in Eq. (11.22) for a situation with $p = \frac{1}{2}$, which readily generalises via Eq. (11.23) and our assumptions on the invariant measure. Lemma 11.5 implies that the positive measure $\sum_{x \in \mathbb{Z}^d} |\langle \sigma_0 \sigma_x \rangle_\beta| \, \delta_x$ is a finite measure, so has vanishing volume mean.

Since the sum $\sum_{x \in \mathbb{Z}^d} \langle \sigma_0 \sigma_x \rangle_\beta$ is absolutely convergent, we can view the pair correlation coefficients $\langle \sigma_0 \sigma_x \rangle_\beta$ as functions in $L^1(\mathbb{Z}^d)$. Their Fourier transforms are uniformly converging Fourier series (by an application of the Weierstraß M-test) and are therefore continuous functions on $\mathbb{R}^d / \mathbb{Z}^d$, see [Rud62, Thm. 1.2.4(a)], which are then also in $L^1(\mathbb{R}^d / \mathbb{Z}^d)$. Applying the Radon–Nikodym theorem completes the proof. In fact, exponential decay of the pair correlation further implies that the Radon–Nikodym density of the measure $\widehat{\gamma}_{\mathsf{ac}}$ is a C^∞-function. $\qquad \square$

11.5.3. LATTICE GASES WITH SHORT-RANGE INTERACTIONS

The crucial step is now to find a large and relevant class of models with exponentially decaying correlations. One is provided by lattice gases with finite-range attractive two-body interaction, where the assumptions of Lemma 11.5 are satisfied, so that the diffraction can be analysed by means of Proposition 11.4. Another is given by the large class of systems in the Dobrushin uniqueness regime. We summarise the situation for the former class, where the average occupation rate of the lattice sites is $\rho = \frac{1}{2}$.

Theorem 11.5 ([BS04b]). *For $\beta < \beta_c$ $(T > T_c)$, the diffraction measure of an ergodic lattice gas model on $\mathbb{Z}^d$ with finite-range attractive two-body interaction almost surely exists, is $\mathbb{Z}^d$-periodic and consists of the pure point part $\widehat{\gamma}_{pp} = \frac{1}{4}\delta_{\mathbb{Z}^d}$ and an absolutely continuous part whose Radon–Nikodym density is C^∞. No singular continuous part is present.*

SKETCH OF PROOF. The claim follows from Proposition 11.4, provided we know that the pair correlations decay sufficiently rapidly for all $\beta < \beta_c$. While this is generally only known for small β, compare Remark 11.10 below, one can derive sufficient, explicit bounds for the class stated in the theorem; see [CIV03, Thm. A] and [BS04b, Thm. 2 and Lemma 5] for details. In particular, one always has $\langle \sigma_x \rangle_\beta = 0$ for $\beta < \beta_c$. ☐

Remark 11.10 (*Lattice gases in the Dobrushin uniqueness regime*). A decomposition of the form $\widehat{\gamma} = \widehat{\gamma}_{pp} + \widehat{\gamma}_{ac}$, and thus the absence of any singular continuous diffraction part, also holds for many systems in the Dobrushin uniqueness regime, because the correlation functions decrease sufficiently fast with increasing distance. This includes all systems with finite local state space and short-range interaction for sufficiently high temperatures [Geo11, BS04b, BZ08], and many more. This is an indication that singular continuous spectra are rather atypical for lattice systems with true short-range stochastic interaction. ◇

One further qualitative property of the absolutely continuous component can be extracted without making additional assumptions. In our setting, we have the inequality in Eq. (11.24) together with

$$\eta(x) = \langle \sigma_0 \sigma_x \rangle_\beta = \langle \sigma_0 \sigma_{-x} \rangle_\beta = \eta(-x),$$

which follows from the translation invariance of the invariant measure (or from the positive definiteness of the autocorrelation). Consequently, one has

(11.25) $$\Big(\sum_{x \in \mathbb{Z}^d} \eta(x)\, \delta_x \Big)^{\widehat{}}(k) = \sum_{x \in \mathbb{Z}^d} \eta(x) \cos(2\pi k x),$$

where the right-hand side is the uniformly converging Fourier series of a $\mathbb{Z}^d$-periodic continuous function, f say, as a consequence of Lemma 11.5. In fact, in our setting of exponential decay of $\eta(x)$, the function f is C^∞. Since $\eta(x) \geqslant 0$ for all $x \in \mathbb{Z}^d$, f has absolute maxima at $k \in \mathbb{Z}^d$ (viewed as the dual lattice of $\mathbb{Z}^d$). This can be summarised as follows.

Corollary 11.2. *Under the assumptions of Theorem 11.5, the absolutely continuous component of the diffraction measure is represented by a smooth function that assumes its maximal value at positions $k \in \mathbb{Z}^d$.* ☐

This result reflects the qualitative property that the continuous component (the so-called diffuse background) concentrates around the point measures (Bragg peaks) if the effective stochastic interaction is attractive. Otherwise, the two components 'repel' each other. We previously observed this behaviour in the context of Markov systems in Example 11.11 and Figure 11.5. We will meet it again in the dimer models discussed later; see [Cow95, Wel04, BH00, Höf00] for further examples.

Remark 11.11 (*Extension to general lattices*). All results also hold — *mutatis mutandis* — for an arbitrary lattice $\Gamma \subset \mathbb{R}^d$, since there exists a bijective linear map $\Gamma \longrightarrow \mathbb{Z}^d$, $x \mapsto Ax$ with $A \in \mathrm{GL}(d, \mathbb{R})$. This means that one can interpret a finite-range model on Γ with range R as a finite-range model on $\mathbb{Z}^d$ with range (bounded by) $\|A\|_2 R$, where $\|.\|_2$ denotes the spectral norm of a matrix. The attractive two-body interaction $\tilde{J}(x) = \tilde{J}(-x) \geqslant 0$ on Γ changes to $J(y) = \tilde{J}(A^{-1}y) = \tilde{J}(-A^{-1}y) = J(-y) \geqslant 0$ for $y \in \mathbb{Z}^d$. $\Diamond$

It is also possible to extend the results to model sets. Since one loses the underlying group structure of the point set, the analysis of the corresponding Gibbs measures becomes rather technical and involved. For first results in this direction, we refer to [Zin09, Mat10].

11.5.4. Classical Ising model as lattice gas

Let us illustrate the above findings with one of the best analysed models in statistical physics, the 2D Ising model without external field, in the lattice gas interpretation with scatterers of strength $n_x \in \{0, 1\}$ for $x \in \mathbb{Z}^2$. The partition function in the spin formulation ($\sigma_x \in \{\pm 1\}$) reads

$$(11.26) \qquad Z_\beta = \sum_{\boldsymbol{\sigma}} \exp\Big(\sum_x \sigma_x \big(K_1 \sigma_{x+e_1} + K_2 \sigma_{x+e_2} \big) \Big),$$

where we sum over all configurations $\boldsymbol{\sigma}$ and the e_i, $i \in \{1, 2\}$, are the standard Euclidean unit vectors. This is to be understood as explained above, by first restricting to a torus and then taking the thermodynamic limit to obtain a Gibbs measure. We consider the ferromagnetic case with coupling constants $K_\ell = \beta J_\ell > 0$, $\ell \in \{1, 2\}$, where β is the inverse temperature as before. The model undergoes a second order phase transition at $\kappa = 1$, where $\kappa := (\sinh(2K_1) \sinh(2K_2))^{-1}$. It is common knowledge (compare [Geo11, Sec. 6.2]) that, in the regime with coupling constants smaller than the critical ones (corresponding to $T > T_c$), the ergodic equilibrium state (with vanishing magnetisation) is unique, whereas otherwise (meaning $T < T_c$), there exist two extremal translation invariant equilibrium states $\mu_\beta^{(\pm)}$, which are thus ergodic. In this case, we assume to be in the extremal state with positive magnetisation $m = \langle \sigma_x \rangle_\beta^{(+)} = (1 - \kappa^2)^{1/8}$.

The diffraction properties of the Ising model can be extracted from the known asymptotic behaviour [WMTB76] of the autocorrelation coefficients. We first state the result for the isotropic case ($K_1 = K_2 = K$) and comment on the general case later.

Proposition 11.5. *Away from the critical point, the diffraction measure $\widehat{\gamma}$ of the Ising lattice gas almost surely exists, is $\mathbb{Z}^2$-periodic and consists of a pure point and an absolutely continuous part with continuous density. The pure point part reads, for $T \neq T_{\mathsf{c}}$,*

$$\widehat{\gamma}_{\mathsf{pp}} = \rho^2 \sum_{k \in \mathbb{Z}^2} \delta_k \qquad (a.s.),$$

where the density ρ is the ensemble average of the number of scatterers per unit area. It is related to the magnetisation m via $\rho = \frac{m+1}{2}$, so that $\rho = \frac{1}{2}$ for all $T > T_{\mathsf{c}}$.

PROOF. Recall that $n_x = \frac{1}{2}(1 + \sigma_x)$, and thus $\langle \sigma \rangle_\beta = m = 2\rho - 1$, so that $\rho = \frac{m+1}{2}$ varies between 1 and $\frac{1}{2}$. The asymptotic correlation function of two spins at distance $\|x\|$ (as $\|x\| \to \infty$) is [WMTB76]

$$(11.27) \qquad \langle \sigma_0 \sigma_x \rangle_\beta \sim \begin{cases} c_1 \dfrac{\mathrm{e}^{-\|x\|/c_2}}{\sqrt{\|x\|}}, & T > T_{\mathsf{c}}, \\ m^2 + c_3 \dfrac{\mathrm{e}^{-2\|x\|/c_2}}{\|x\|^2}, & T < T_{\mathsf{c}}, \end{cases}$$

where the constants c_1, c_2 and c_3 only depend on K and T. The pure point part $\widehat{\gamma}_{\mathsf{pp}}$ directly results from the Fourier transform of the constant part of the autocorrelation γ (by means of the PSF), as derived from the asymptotics of $\langle n_0 n_x \rangle_\beta = \frac{1}{4}\big(\langle \sigma_0 \sigma_x \rangle_\beta + 2m + 1\big)$; compare Eq. (11.23).

Observe further that the two sums

$$\sum_{x \in \mathbb{Z}^2} \frac{\mathrm{e}^{-\|x\|/c_2}}{\sqrt{\|x\|}} \quad \text{and} \quad \sum_{x \in \mathbb{Z}^2} \frac{\mathrm{e}^{-2\|x\|/c_2}}{\|x\|^2}$$

converge absolutely, so we can view the corresponding correlation coefficients as functions in $L^1(\mathbb{Z}^2)$. Their Fourier transforms (which are uniformly converging Fourier series) are continuous functions on $\mathbb{R}^2/\mathbb{Z}^2$, see [Rud62, Sec. 1.2.3], which are then also in $L^1(\mathbb{R}^2/\mathbb{Z}^2)$. Applying the Radon–Nikodym theorem completes the proof. $\qquad \square$

Remark 11.12 (*Behaviour at criticality*). At the critical point, the correlation function $\langle \sigma_0 \sigma_x \rangle_{\beta_{\mathsf{c}}}$ is asymptotically proportional to $\|x\|^{-1/4}$ as $\|x\| \to \infty$ [WMTB76]. Again, first taking out the constant part of γ, we get the same pure point part of $\widehat{\gamma}$ as in Proposition 11.5 for $T > T_{\mathsf{c}}$. However, for the remaining part of $\widehat{\gamma}$, our previous convergence arguments fail. Nevertheless, using a theorem of Hardy [Bro65, p. 97], one can show that the corresponding Fourier series of Eq. (11.25) still converges for $k \notin \mathbb{Z}^2$ (a natural order of

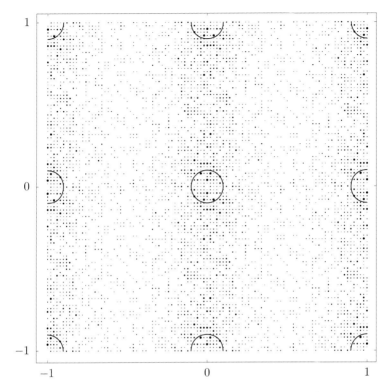

FIGURE 11.6. Illustration of the diffraction image of a (slightly) non-isotropic Ising lattice gas on the square lattice. Point measures are indicated by open circles (centred at their location), while the small black disks represent (via halftoning) the absolutely continuous background, calculated numerically by means of fast Fourier transform from a (large) periodic approximant.

summation is given by shells of increasing radii). In particular, this remaining part of the autocorrelation measure is still null weakly almost periodic, so that its Fourier transform is a continuous measure, by an application of [GLA90, Thm. 11.2].

For $k \in \mathbb{Z}^2$, which are the locations of the point measures, the series diverges, while the function it defines is still locally integrable. Consequently, these locations can neither result in further contributions to the pure point part (the constant part of γ had already been taken care of) nor in singular continuous contributions (because the points of divergence form a uniformly discrete set). So, even though the series diverges for $k \in \mathbb{Z}^2$, it still represents (we know that $\widehat{\gamma}$ exists) a function in $L^1(\mathbb{R}^2/\mathbb{Z}^2)$ and hence the Radon–Nikodym density of an absolutely continuous background.   ◇

On the diffraction image, we can thus see, for any temperature, point measures on the square lattice and a $\mathbb{Z}^2$-periodic, absolutely continuous background concentrated around the peaks (the interaction is attractive); see Figure 11.6 for an illustration, where the intensity variation in the horizontal direction is more pronounced than in the vertical one. At the critical point, the intensity of the diffuse scattering diverges when approaching the lattice positions of the peaks, but the qualitative picture remains the same. In particular, one still has local integrability, which is clear in view of the continuity of the pure point part in dependence of β.

Analogous arguments hold in the anisotropic case. The asymptotic behaviour still conforms to Eq. (11.27) and the above discussion, provided that $\|x\|$ is replaced by the corresponding term given in [WMTB76, Eq. 2.6]. The pure point part is again that of Proposition 11.5 with fourfold symmetry, while (as in the case of the domino tiling; see [BH00]) the continuous background breaks this symmetry if $K_1 \neq K_2$; see Figure 11.6. Let us finally remark that a different choice of the scattering strengths, such as ± 1 rather than 1 and 0, may result in the extinction of the point spectrum in the disordered phase $(T > T_c)$, but no choice does so in the ordered phase $(T < T_c)$.

11.6. Random tilings

Another large and important class of models can be formulated by means of *random tilings*. Here, starting from a finite number of prototiles (the 'building blocks'), one considers all gapless and overlap-free coverings of larger and larger domains by copies of these prototiles. In the infinite volume limit, if the number of translationally inequivalent coverings grows exponentially with the volume of the region, one speaks of a *random tiling ensemble*.

Such random tilings have many interesting features; see [Hen99] for a comprehensive review. In particular, one has to carefully set up appropriate symmetry concepts and to relate them to entropic aspects of the ensemble [RHHB98]. Here, we only consider some particularly simple cases from the diffraction point of view.

11.6.1. BINARY RANDOM TILINGS IN ONE DIMENSION

The simplest setting is that of a random tiling of the line with two intervals of length u and v, with prescribed frequencies p and q, respectively, where $p + q = 1$. We consider such a structure, and place a normalised Dirac measure on the left endpoints of all intervals. The key parameter then is the length ratio $\alpha = u/v$. The diffraction of 1D random tilings has been investigated in [BH00], but also follows as a simple extension from Theorem 11.2 via Remark 11.5 and Example 11.10. A 1D binary random tiling has a non-trivial (extended) pure point part if and only if α is rational.

Theorem 11.6. *Consider a random tiling of $\mathbb{R}$, built from two intervals of lengths u and v with corresponding probabilities p and q. Let Λ denote the point set obtained from the left endpoints of the intervals of the tiling.*

The natural density ζ of Λ exists with probability 1, and is given by $\zeta = (pu + qv)^{-1}$. If $\omega = \delta_\Lambda$ denotes the corresponding random Dirac comb, the autocorrelation γ_ω of ω exists almost surely, and is a positive definite pure point measure. The diffraction measure almost surely consists of a pure point part and an absolutely continuous part, so that $\widehat{\gamma_\omega} = (\widehat{\gamma_\omega})_{\mathsf{pp}} + (\widehat{\gamma_\omega})_{\mathsf{ac}}$.

If $\alpha = u/v$, the pure point part is

$$(\widehat{\gamma_\omega})_{\mathsf{pp}} = \zeta^2 \cdot \begin{cases} \delta_0, & \text{if } \alpha \notin \mathbb{Q}, \\ \delta_{\mathbb{Z}/\xi}, & \text{if } \alpha \in \mathbb{Q}, \end{cases}$$

where, if $\alpha \in \mathbb{Q}$, we set $\alpha = a/b$ with coprime $a, b \in \mathbb{Z}$ and define the number $\xi = u/a = v/b$.

The absolutely continuous part $(\widehat{\gamma_\omega})_{\mathsf{ac}}$ can be represented by the continuous Radon–Nikodym density

$$g(k) = \frac{\zeta \, pq \, \sin^2(\pi k \, (u - v))}{p \, \sin^2(\pi k \, u) + q \, \sin^2(\pi k \, v) - pq \, \sin^2(\pi k \, (u - v))},$$

which is well-defined for $k \, (u - v) \notin \mathbb{Z}$. It has a smooth continuation to the excluded points. If α is irrational, this continuation is $g(k) = 0$ for $k(u - v) \in \mathbb{Z}$ with $k \neq 0$ and

$$g(0) = \frac{\zeta \, pq \, (u - v)^2}{p \, u^2 + q \, v^2 - pq \, (u - v)^2} = \zeta \, \frac{pq \, (u - v)^2}{(p \, u + q \, v)^2}.$$

For $\alpha = a/b \in \mathbb{Q}$ as above, it is $g(k) = 0$ for $k \, (u - v) \in \mathbb{Z}$, but $k \, u \notin \mathbb{Z}$ (or, equivalently, $k \, v \notin \mathbb{Z}$), and

$$g(k) = \zeta \, \frac{pq \, (a - b)^2}{(p \, a + q \, b)^2}$$

for the case that also $k \, u \in \mathbb{Z}$.

PROOF. The random tiling process is stationary, so that we may assume (without loss of generality) that 0 is contained in the point set realisation Λ. As mentioned above, we may view the underlying process as a renewal process, with $\varrho = p\delta_u + q\delta_v$. The claim on the density of Λ is obvious.

Let us now assume that $pu + qv = 1$ so that the density is 1. Theorem 11.2 gives the claim on the structure of $\widehat{\gamma_\omega}$. If $\alpha \in \mathbb{Q}$, there is a unique $\xi \in \mathbb{R}_+$ such that $\xi\mathbb{Z}$ is the coarsest lattice to contain u and v (and thus $\mathrm{supp}(\varrho)$), which is the lattice stated above. This proves the formula for the pure point part (with density $\zeta = 1$).

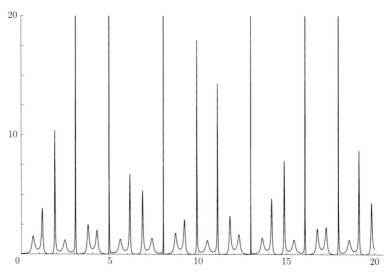

FIGURE 11.7. Absolutely continuous part of the diffraction of a typical Fibonacci random tiling. To illustrate the smoothness of the Radon–Nikodym density, the y-axis is truncated at a value of 20. This affects the six largest maxima of this figure.

For the absolutely continuous part of the diffraction, observe the relation $\widehat{\varrho}(k) = p\mathrm{e}^{-2\pi\mathrm{i}ku} + q\mathrm{e}^{-2\pi\mathrm{i}kv}$, wherefore

$$g = 1 - h = \frac{1 - |\widehat{\varrho}|^2}{|1 - \widehat{\varrho}|^2}$$

with the h from Theorem 11.2. The stated expression for g (with $\zeta = 1$) follows by a straightforward calculation. The points requiring special attention are the elements of the set $S = \{k \in \mathbb{R} \mid \widehat{\varrho}(k) = 1\}$, where $S = \{0\}$ if $\alpha \notin \mathbb{Q}$ and $S = \mathbb{Z}/\xi$ otherwise. The function g is clearly continuous on $\mathbb{R} \setminus S$, and the continuous extensions follow via l'Hospital's rule.

The case for an arbitrary value of the density ζ follows from the observation that the pure point part scales with ζ^2, while the absolutely continuous one scales with ζ; compare also our previous examples in this section. $\qquad\square$

The most prominent 1D random tiling is the Fibonacci random tiling, where $u = \tau = (1 + \sqrt{5})/2$ and $v = 1$, with the occupation probabilities $p = 1/\tau$ and $q = 1 - p = 1/\tau^2$ of the intervals (almost surely). Each interval endpoint of any realisation of a Fibonacci random tiling belongs to the module $\mathbb{Z}[\tau] = \{m + n\tau \mid m, n \in \mathbb{Z}\}$, which we have encountered many times by now. All perfect Fibonacci tilings (as obtained from the projection method) also appear in the random tiling ensemble, but they are atypical in the sense that their appearance has probability 0.

According to Theorem 11.6, the diffraction measure comprises, with probabilistic certainty, a trivial point measure at the origin (of intensity $\rho^2 = \frac{\tau+2}{5}$) and an absolutely continuous background; see Figure 11.7 for the latter. The absolutely continuous background, despite being smooth, shows localised, bell-shaped needles of increasing height at sequences of points scaling with the golden ratio τ. This is reminiscent of the perfect Fibonacci tiling, see Figure 9.5 on page 369, where the point measures of the diffraction line up in a similar fashion; compare Remark 11.9.

One can also understand the diffraction result for the Fibonacci random tiling from a fluctuation point of view, via comparing the random version with the perfect one. In fact, one can consider the $\star$-image of the set of vertex points of a typical random tiling realisation, which is no longer bounded in internal space (in contrast to the random inflation setting of Section 11.2.3). Indeed, the fluctuations about the average direction (as defined by the perfect tiling) diverge linearly with the system size. Moreover, one can define a distribution of the lifted points in internal space, which does not have compact support; see [BMRS03] for an illustration.

Clearly, the discussion can be extended to one-dimensional random tilings with more than two interval types as prototiles, and control point set Λ, say. One obtains expressions with multinomial coefficients in return, and the appearance of point measures beyond the trivial one at $k = 0$ depends on the mutual commensurability of all interval lengths present. As soon as only one length ratio is irrational, only the trivial peak $\big(\mathrm{dens}(\Lambda)\big)^2 \delta_0$ survives [BH00].

11.6.2. Random tilings in two dimensions

Beyond one dimension, the picture for random tilings is less complete. For various exactly solved planar models of statistical mechanics, one knows the autocorrelation sufficiently well, at least asymptotically, to determine the diffraction, again almost surely in the ensemble sense. This strategy works analogously to the case of the Ising model, and can be applied to various dimer models; see [Wu09, Ken97] and references therein.

One classic example is the planar random tiling built from a rhombic prototile with unit edge length and an opening angle of $\pi/3$, which is known as a *lozenge*. Here, it is available in three distinct orientations. An example is shown in Figure 11.8. The corresponding ensemble is in one-to-one correspondence with the ensemble of fully packed dimer configurations on the hexagonal or honeycomb packing (which is the periodic repetition of a regular hexagon). This ensemble is known to have positive entropy density, compare [Hen99, RHHB98], with the maximal contribution from the realisations with sixfold (and hence maximal) symmetry, the latter to be understood in the statistical (or ensemble) sense; compare Section 5.6.

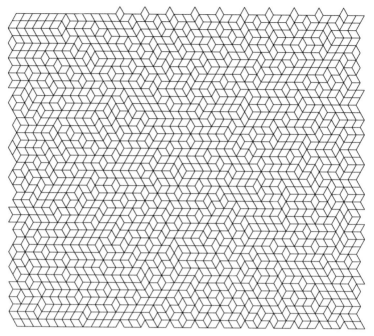

FIGURE 11.8. Typical patch of a rhombus (or lozenge) random tiling, with periodic boundary conditions. The frequencies of the three orientations differ (the vertical rhombus is less frequent than the other two types), so that this example breaks (statistical) three- or sixfold symmetry.

The dimer model on the honeycomb packing is a much studied, exactly solved model of statistical mechanics; see [Kas63, DG72, Ken97, Höf01] for background and applications in our context. The equivalent lozenge ensemble is equipped with a unique Gibbs measure, which is parametrised by the frequencies of the tiles in the three orientations; compare [Ken97] for an alternative formulation. This measure is thus an ergodic measure, and its correlations are rather well understood. With similar methods as used above for the diffraction of lattice gases, one can thus formulate and prove the following result [BH00]; see also [BS04b] for its more general embedding.

Theorem 11.7. *Choose a random lozenge tiling, with unit edge length and with prescribed prototile frequencies* ρ_i, $i \in \{1, 2, 3\}$, *and consider the Dirac comb built from the centres of the tiles. Then, its diffraction measure* $\widehat{\gamma}$ *exists with probability* 1, *and consists of a pure point and an absolutely continuous part,* $\widehat{\gamma} = \widehat{\gamma}_{\mathsf{pp}} + \widehat{\gamma}_{\mathsf{ac}}$, *with*

$$\widehat{\gamma}_{\mathsf{pp}} = \frac{4}{3} \sum_{(k_1, k_2) \in \Gamma^*} \left((-1)^{k_1} \rho_1 + (-1)^{k_2} \rho_2 + \rho_3 \right)^2 \delta_{(k_1, k_2)},$$

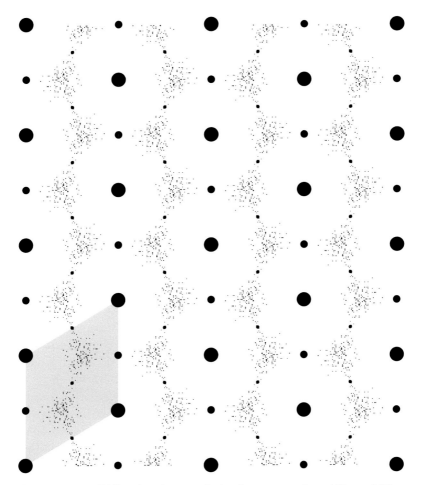

FIGURE 11.9. Diffraction image of the lozenge random tiling of Figure 11.8. The point measures are represented by the three types of large spots, while the smaller spots represent a sketch of the absolutely continuous background, as calculated numerically by fast Fourier transform of the periodic approximant. A fundamental domain of Γ^* is indicated (shaded). The (lack of) symmetry corresponds to that of the tiling.

where Γ^* is the dual lattice of the triangular lattice, so that Γ^* is spanned by $v_1 = \left(1, -\frac{1}{\sqrt{3}}\right)$ and $v_2 = \left(0, \frac{2}{\sqrt{3}}\right)$, as in Example 3.1. Here, we use the notation $(k_1, k_2) = k_1 v_1 + k_2 v_2$. There is no singular continuous part, and $\widehat{\gamma}$ is crystallographic, with lattice of periods $2\Gamma^*$.

SKETCH OF PROOF. The autocorrelation γ is a positive and positive definite pure point measure (hence translation bounded), supported on the union of the triangular lattice Γ and three of its cosets. For almost all realisations of the random tiling, by ergodicity of the underlying Gibbs measure, γ is of the form obtained by the usual ensemble average. Then, its coefficients can be split into a finite sum of constant parts (which are combinations of density factors) and a covariance type part, the latter decaying algebraically with the distance from 0; see [Höf01, BH00] and references given therein.

The pure point part of the diffraction measure, $\widehat{\gamma}_{\mathsf{pp}}$, is now simply the Fourier transform of the constant parts of γ emerging that way, again calculated by means of the PSF from Theorem 9.1 for the triangular lattice Γ and its translates (the latter simply giving additional phase factors); see also Proposition 9.5 and Theorem 9.2.

What remains once again gives a Fourier series which converges to a function that is square integrable on a fundamental domain of $2\Gamma^*$, the latter also being its lattice of periods. By Hölder's inequality, this is then also an integrable function, and thus the Radon–Nikodym density of an absolutely continuous measure; see [BH00] for further details.   □

Figure 11.9 illustrates the diffraction image, which displays the correct symmetry already on the level of the pure point part alone. In the closely related square lattice dimer model, however, one would extract the correct symmetry only from the diffuse background, as in our earlier Ising lattice gas example. Further examples can be analysed along similar lines; compare [Höf00] for a more complex case. As before, one can expand on the properties of the background, which turns out to have a continuous Radon–Nikodym density [BS04b]. Also, its 'repulsive' nature is clearly visible, and lines up with the repulsive nature of the effective interaction created by the stochastic process [Höf01, BS04b]; compare Corollary 11.2 and Example 11.11.

In a certain sense, this picture seems rather satisfactory. Unfortunately, things change and get more involved when one aims at the analogous results for genuinely non-crystallographic random tiling ensembles, such as the randomised version of the Ammann–Beenker tiling. A typical patch of the random octagonal tiling is shown in Figure 11.10. It is obtained from a patch of the Ammann–Beenker tiling by repeated random applications of the *simpleton flip*

(11.28)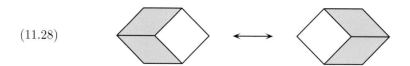

and its various rotated versions.

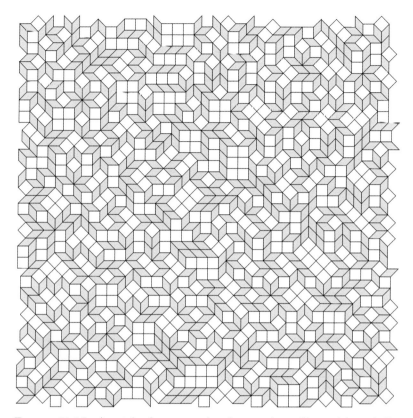

FIGURE 11.10. A patch of a square rhombus random tiling, with periodic boundary conditions and approximate statistical eightfold symmetry.

For this tiling, it is still not completely clear what happens — and the planar case apparently represents the demarcation line between purely absolutely continuous and the onset of pure point contributions, wherefore one expects a singular continuous component. As we have seen above, a generic one-dimensional random tiling produces an absolutely continuous diffraction image (apart from the trivial peak at $k = 0$). On the other hand, it is a long-standing conjecture that relevant random tilings in 3-space, such as the icosahedrally symmetric random tiling built from the two classic Kepler rhombohedra of Figure 7.11, will produce a mixture of pure point and absolutely continuous parts; see [Els85, Hen99] for the underlying scaling arguments. The pure point part of the diffraction coincides with the diffraction measure of a weighted dense Dirac comb of the kind treated in Section 9.5.2. In particular, the function in internal space is similar to a Gaussian profile, wherefore we meet the generalised PSF type formula of Corollary 9.5; see [Ric03, LR07] for further details.

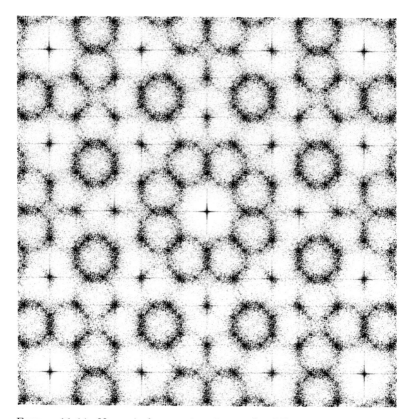

FIGURE 11.11. Numerical approximation to the diffraction pattern of the finite patch of the octagonal random tiling from Figure 11.10. A comparison with Figure 9.6 on page 372 shows remarkable similarities in the sense that the spiky structure of the random tiling diffraction resembles the location and the intensity of the strong peaks for the projection tiling.

Based upon heuristic scaling arguments [Hen99] and various numerical calculations, one expects the square rhombus random tiling with statistical eightfold symmetry to show only the trivial peak at the origin, but otherwise a mixture of singular and absolutely continuous parts. A numerical snapshot is shown in Figure 11.11. From the geometric insight, due to a fluctuation argument in embedding space [Hen99], it is also plausible that 2 really is the critical dimension for this phenomenon to happen, but it remains a challenge to prove (or disprove) this claim, in particular in view of the fact that scaling properties, in general, are not conclusive; compare [Hof97b] for a discussion.

Let us note that Figure 11.11 shows the diffraction of the *single* finite patch of Figure 11.10. Assuming ergodicity, the system is self-averaging in the sense that the diffraction of patches of increasing size almost surely converges to the diffraction of the infinite system (in the measure-theoretic sense). However, this convergence is really slow, which manifests itself in strong local fluctuations. In view of the (assumed) ergodicity, a better approximation can usually be achieved by averaging the diffraction measure over independent realisations; see [STJ89, Tan90, Hen99] for details.

Remark 11.13 (*Entropy of random tiling ensembles*). The random tiling ensembles mentioned above are systems of finite local complexity. As such, their topological entropy coincides with the combinatorial (or patch counting) entropy [BLR07]. The latter is easy to calculate for many one-dimensional examples, but only known numerically in dimensions $d \geqslant 3$. In two dimensions, some exact results are known via models of statistical mechanics.

The ensemble of the planar lozenge tiling is equivalent to the antiferromagnetic Ising model on the triangular lattice; see [Ken97, Ken00] and references therein. The latter ensemble was studied by Wannier [Wan50], who determined the entropy (per vertex) as

$$h = \frac{2}{\pi} \int_0^{\pi/3} \log\bigl(2\cos(x)\bigr) \, \mathrm{d}x \approx 0.323066.$$

Note that the numerical value given in [Wan50] in imprecise, while the integral is correct. This is an early example of the occurrence of a Mahler measure in entropy calculations; see [LSW90] for more recent developments and [SV09] for other models with similar entropy expressions.

One can refine the entropy picture by considering sub-ensembles with prescribed densities for the three distinct lozenge orientations (such as the example shown in Figure 11.8). The emerging entropy function then has a unique maximum (with the above value) that coincides with the point of maximal (statistical sixfold) symmetry [Ric99a, Ric99b]. The proof of this entropic mechanism for maximal symmetry is group-theoretic in nature, and applies to many other random tiling ensembles as well, including those with 8-, 10- or 12-fold symmetry, and those with icosahedral symmetry. This amounts to a proof of the first random tiling hypothesis of [Hen99] in a rather general setting. The result can be viewed as the generalisation of the maximum in Figure 11.1, which corresponds to a binary random tiling in one dimension with (statistical) reflection symmetry.

There are some other planar cases where the entropy has been calculated exactly by means of Bethe ansatz techniques. They include the random square-triangle tiling ensemble [Wid93, Kal94, dGN97b] as well as rectangle-triangle ensembles with 8- and 10-fold symmetry [dGN97a, dGN98]. For instance, the entropy per vertex for the random square-triangle ensemble is [Kal94]

$$h = \log(108) - 2\sqrt{3}\log(2 + \sqrt{3}\,) \approx 0.120055.$$

The results for the other ensembles have a similar form, but we are not aware of a connection to a Mahler measure. Although there is no doubt about the correctness of these results, proofs in the mathematical sense are still missing, and the ensembles themselves are not yet well understood from the point of view of equilibrium (or Gibbs) measures.   ◇

For the random tiling ensembles that emerge from the rhombic tilings with fivefold (Penrose) and eightfold (Ammann–Beenker) symmetry, the exact values of the entropy are not known. For the occurrence of the random analogue of the TTT of Section 6.2.2 in an atomic model with realistic pair potentials, we refer to [KET12].

Remark 11.14 (*Generation of random tilings*). The simpleton flip for the octagonal tiling from Eq. (11.28), and its various analogues in similar ensembles, provides a standard approach for the preparation of random tiling samples. It works well also for other tilings with rhombic prototiles, where one might have different types of simpletons to consider (for instance, there are, up to similarity, two such configurations in the rhombic Penrose tiling). One usually starts from a periodic approximant (to minimise boundary effects) to a perfect tiling, which is not difficult to construct, and runs the simpleton flip thermalisation until correlations have decayed. In such ensembles, the process can be shown to be topologically transitive, so that the entire ensemble compatible with these boundary conditions is accessible [Hen98, Höf01].

Note, however, that there are other important ensembles, such as the random square triangle tilings, where no such local flip mechanism exists. Here, one needs alternative methods, such as the well-studied 'zipper' move [OH93]. The latter temporarily creates some new (auxiliary) tiles that enable a (finite) randomisation path, until the created tiles annihilate themselves again and leave a modified square triangle tiling behind.   ◇

Beyond the ground covered so far, rigorous extensions become difficult. Determinantal, permanental and related point processes (such as those discussed in [HKPV09, BKM14]) can provide various generalisations of the Ginibre process of Section 11.4.2, while the highly developed theory of point processes [DV-J88, Kar91] can also help to make further progress. Here, the connection emerges via the observation that the intensity measure of the Palm measure of an ergodic, stationary point process coincides with the autocorrelation of the process [Gou03, BBM10]. Whether this will also lead to concrete insight into new classes of examples is presently not obvious to us, although the relevance of this connection for the inverse problem (through inference methods) is undeniable.

APPENDIX A

The Icosahedral Group

In this appendix, we informally summarise some basic properties of the (proper) icosahedral group Y and its representations. A comprehensive exposition, tailored to the setting of icosahedral quasicrystals, can be found in [KH89]. For material on the symmetric group, we refer to [JK81]; for general background on group and representation theory, see [JL01].

Recall from [CM80] (or from Section 2.3.1) that the icosahedral group Y can be presented as

$$Y = \langle r, u \mid r^3 = u^2 = (ru)^5 = e \rangle \, .$$

This is the rotation symmetry group of the icosahedron (as well as the dodecahedron, the icosidodecahedron and Kepler's triacontahedron; see Figures 2.4 and 2.6). For the purpose of this appendix, we use the notation $g_2 = u$, $g_3 = r$ and $g_5 = rur$, in line with [KH89, Fig. 3]. This is suggestive because the index then refers to the order of the rotation axis. The group Y is isomorphic to the alternating group A_5. One explicit isomorphism is given by

$$g_2 = (24)(35) \, , \quad g_3 = (143) \, , \quad g_5 = (12345) \, ,$$

in standard cycle notation. The group Y has order $|Y| = 60$ and contains five conjugacy classes. They derive from the seven classes of the symmetric group S_5, only four of which contain even elements. One of these classes splits into two of equal size. The icosahedral group is *ambivalent* [JK81, Lemma 1.2.12], which means that each group element is conjugate to its inverse. As representatives for the five classes, we choose the elements e (the identity), g_2, g_3, g_5 and g_5^2.

Recall that a *representation* of a group G is a homomorphism ρ from G into $\mathrm{Aut}(V)$, where V is a (finite-dimensional) vector space over a field K. The most important cases in our situation are $K \in \{\mathbb{C}, \mathbb{R}, \mathbb{Q}\}$. By standard results from the representation theory of finite groups over $K = \mathbb{C}$, we know that Y has precisely five *irreducible representations* (irreps), which are the building blocks of all finite-dimensional representations over $\mathbb{C}$.

For a given representation ρ, one needs its *character* $\chi_\rho \colon G \longrightarrow K$, which is defined as $\chi_\rho(g) = \mathrm{tr}(\rho(g))$ for $g \in G$. In particular, $\chi_\rho(e) = \dim(\rho)$ is the *dimension* (or degree) of the representation ρ. Characters are constant on

TABLE A.1. Character table of the proper icosahedral group $Y \simeq A_5$, for representations over $\mathbb{C}$, $\mathbb{R}$ and $\mathbb{Q}$. As before, $\tau = (1 + \sqrt{5})/2$ is the golden ratio, and $\tau' = 1 - \tau$ denotes its algebraic conjugate.

	$\mathcal{C}(e)$	$\mathcal{C}(g_2)$	$\mathcal{C}(g_3)$	$\mathcal{C}(g_5)$	$\mathcal{C}(g_5^2)$	class
	1	15	20	12	12	order
$\chi^{(1)}$	1	1	1	1	1	
$\chi^{(4)}$	4	0	1	-1	-1	
$\chi^{(5)}$	5	1	-1	0	0	
$\chi_+^{(3)}$	3	-1	0	τ	τ'	
$\chi_-^{(3)}$	3	-1	0	τ'	τ	
$\chi_+^{(3)}+\chi_-^{(3)}$	6	-2	0	1	1	
character						

entire conjugacy classes. For a finite group, this leads to its character table, which is shown for $G = Y$ in Table A.1. We use a notation with the dimension as an upper index and, if needed, an extra lower index for uniqueness. All irreps of Y over $\mathbb{C}$ are *real*, in the sense that all representation matrices can be chosen in $\mathrm{GL}(n, \mathbb{R})$. This follows from the fact that the Frobenius–Schur indicator, see [JK81, Eq. 2.6.4], satisfies

$$\frac{1}{|Y|} \sum_{g \in Y} \chi(g^2) = 1$$

for all $\mathbb{C}$-irreducible characters from Table A.1. Consequently, there is a bijection between irreps over $\mathbb{C}$ and irreps over $\mathbb{R}$ for the group Y.

In our context, we are interested in *faithful* representations ρ, meaning that the homomorphism ρ is injective. Among the irreps of Y, all but the trivial representation $\rho^{(1)}$ are faithful, compare [JK81, Thm. 2.1.13]. This also follows from the useful criterion that a finite-dimensional representation ρ of a compact group G is faithful if and only if the equation $\chi_\rho(g) = \dim(\rho)$ holds for $g = e$ only. The previous claim is now obvious from Table A.1.

Moreover, we need to know which representations of Y are compatible with a lattice. This means that we are looking for *integer* representations (where the representation matrices of the generators can be chosen to have integer entries only), for instance by starting from representations over $\mathbb{Q}$. The representations $\rho^{(1)}$, $\rho^{(4)}$ and $\rho^{(5)}$ are rational, which is obvious for the trivial representation and follows from [JK81, p. 369] for the other two.

The two three-dimensional representations $\rho_{\pm}^{(3)}$ cannot be rational, because the golden ratio τ is irrational and thus cannot be the trace of a $GL(3, \mathbb{Q})$ matrix. One explicit choice [KH89] is given by

$$\rho_+^{(3)}(g_2) = \begin{pmatrix} -1 & 0 & 0 \\ 0 & 1 & 0 \\ 0 & 0 & -1 \end{pmatrix}, \qquad \rho_+^{(3)}(g_3) = \frac{1}{2}\begin{pmatrix} -1 & \tau' & \tau \\ -\tau' & \tau & 1 \\ -\tau & 1 & \tau' \end{pmatrix},$$

(A.1)

$$\rho_+^{(3)}(g_5) = \frac{1}{2}\begin{pmatrix} -\tau' & -\tau & 1 \\ \tau & 1 & -\tau' \\ -1 & -\tau' & \tau \end{pmatrix}, \qquad \rho_+^{(3)}(g_5^2) = \frac{1}{2}\begin{pmatrix} -\tau & -1 & -\tau' \\ 1 & \tau' & \tau \\ \tau' & \tau & 1 \end{pmatrix},$$

where we give one example per conjugacy class (apart from $\mathcal{C}(e)$). The corresponding choice for $\rho_-^{(3)}$ is obtained by algebraic conjugation in $\mathbb{Q}(\sqrt{5})$, as defined by $\sqrt{5} \mapsto -\sqrt{5}$, so that $\tau \leftrightarrow \tau'$ in Eq. (A.1).

However, the direct sum $\rho^{(6)} := \rho_+^{(3)} \oplus \rho_-^{(3)}$ is (equivalent to) a rational representation of dimension 6 that is irreducible over $\mathbb{Q}$, and originates from an irreducible representation of S_5. This decomposition is in line with the general relation between the symmetric and the alternating group; see [JK81, Thm. 2.5.7]. So, Table A.1 contains the complete information about irreducible characters over $\mathbb{C}$, $\mathbb{R}$ and $\mathbb{Q}$. This matches the well-known fact that all irreducible representations of the symmetric group are rational [JK81, Thm. 2.1.12]. Each irreducible rational representation of S_5 is equivalent to an integral one, see [JK81, p. 369], wherefore this also holds for the rational representations of $Y \simeq A_5$.

One common choice for $\rho^{(6)}$ is given in [Els86]. Here, to be consistent with our previous setting, we consider the representation matrices

$$\rho^{(6)}(g_2) = \begin{pmatrix} -1 & 0 & 0 & 0 & 0 & 0 \\ 0 & -1 & 0 & 0 & 0 & 0 \\ 0 & 0 & 0 & 1 & 0 & 0 \\ 0 & 0 & 1 & 0 & 0 & 0 \\ 0 & 0 & 0 & 0 & 0 & -1 \\ 0 & 0 & 0 & 0 & -1 & 0 \end{pmatrix}, \quad \rho^{(6)}(g_3) = \begin{pmatrix} 0 & -1 & 0 & 0 & 0 & 0 \\ 0 & 0 & 0 & 0 & 0 & 1 \\ 0 & 0 & 0 & 0 & 1 & 0 \\ 0 & 0 & 1 & 0 & 0 & 0 \\ 0 & 0 & 0 & 1 & 0 & 0 \\ -1 & 0 & 0 & 0 & 0 & 0 \end{pmatrix},$$

$$\rho^{(6)}(g_5) = \begin{pmatrix} 0 & 0 & 0 & 0 & 0 & 1 \\ 0 & 0 & 0 & -1 & 0 & 0 \\ 1 & 0 & 0 & 0 & 0 & 0 \\ 0 & 0 & 1 & 0 & 0 & 0 \\ 0 & 0 & 0 & 0 & 1 & 0 \\ 0 & -1 & 0 & 0 & 0 & 0 \end{pmatrix}, \quad \rho^{(6)}(g_5^2) = \begin{pmatrix} 0 & -1 & 0 & 0 & 0 & 0 \\ 0 & 0 & -1 & 0 & 0 & 0 \\ 0 & 0 & 0 & 0 & 0 & 1 \\ 1 & 0 & 0 & 0 & 0 & 0 \\ 0 & 0 & 0 & 0 & 1 & 0 \\ 0 & 0 & 0 & 1 & 0 & 0 \end{pmatrix},$$

which emerge from [Els86] by a (signed) permutation of the basis vectors. Since these matrices (and hence all matrices $\rho^{(6)}(g)$ for $g \in Y$) are signed permutation matrices, they map the integer lattice $\mathbb{Z}^6$ onto itself. Moreover, they also preserve the root lattice D_6 (up to scale, the face-centred hypercubic lattice) and the weight lattice D_6^* (its body-centred counterpart). These are the only hypercubic lattices in 6-space up to similarity transformations, see Example 3.2. In fact, the above choice gives an explicit embedding of Y, via a six-dimensional representation, into the hypercubic group in 6-space (which is a finite group of order $2^6 \, 6!$; compare [JK81, Baa84]) with the structure of a wreath product.

Viewed over $\mathbb{R}$, the representation $\rho^{(6)}$ is reducible, as is clear from the construction. Consequently, there are two mutually orthogonal three-dimensional invariant subspaces. To be explicit, we set

$$
\pi_1 = \frac{\sqrt{5}}{10}
\begin{pmatrix}
\sqrt{5} & 1 & 1 & -1 & 1 & 1 \\
1 & \sqrt{5} & 1 & -1 & -1 & -1 \\
1 & 1 & \sqrt{5} & 1 & 1 & -1 \\
-1 & -1 & 1 & \sqrt{5} & 1 & -1 \\
1 & -1 & 1 & 1 & \sqrt{5} & 1 \\
1 & -1 & -1 & -1 & 1 & \sqrt{5}
\end{pmatrix}
\quad \text{and} \quad \pi_2 = \left(\pi_1 \right)'.
$$

This defines two projectors which commute with all $\rho^{(6)}(g)$ and satisfy

$$
\pi_1^2 = \pi_1, \quad \pi_2^2 = \pi_2, \quad \pi_1 \pi_2 = \mathbb{O}, \quad \pi_1 + \pi_2 = \mathbb{1}.
$$

The two invariant subspaces are given by $\ker(\pi_1)$ and $\ker(\pi_2)$, and one has $\ker(\pi_1) = \operatorname{im}(\pi_2)$, and vice versa.

To complete this discussion, we also give the explicit basis transformation B for the reduction of $\rho^{(6)}$, such that

$$
B\rho^{(6)}(g)B^{-1} = \left(
\begin{array}{c|c}
\rho_+^{(3)}(g) & \mathbb{O} \\
\hline
\mathbb{O} & \rho_-^{(3)}(g)
\end{array}
\right)
$$

for the above choice of the representation matrices and all $g \in Y$. This gives

$$
(\text{A.2}) \qquad B = \frac{1}{\sqrt{5 + \sqrt{5}}}
\begin{pmatrix}
\tau & \tau & 1 & -1 & 0 & 0 \\
0 & 0 & \tau & \tau & 1 & -1 \\
1 & -1 & 0 & 0 & \tau & \tau \\
-1 & -1 & \tau & -\tau & 0 & 0 \\
0 & 0 & -1 & -1 & \tau & -\tau \\
\tau & -\tau & 0 & 0 & -1 & -1
\end{pmatrix},
$$

which is a matrix in $\mathrm{SO}(6)$, and as such is unique (up to a multiplication by $-\mathbb{1}$). Note that, disregarding the global prefactor, the first three rows columnwise contain the coordinates of the $\mathbb{Z}$-basis of the primitive module $\mathcal{M}_{\mathsf{P}}$ of Example 2.20. The last three rows are obtained from the former ones by algebraic conjugation in $\mathbb{Z}[\tau]$, followed by a multiplication with τ, as mentioned in Example 3.8. The matrix B also satisfies the conjugation relations $B\pi_1 B^{-1} = \operatorname{diag}(1,1,1,0,0,0)$ and $B\pi_2 B^{-1} = \operatorname{diag}(0,0,0,1,1,1)$.

The same matrix B also works for the tetrahedral group T, located as a subgroup of Y as described in Section 2.3.1. However, in this case, it is not unique, and Schur's lemma leaves a degree of freedom to rotate the projections [Kra87b, BJK91]. This is the origin of the possible transition between models with cubic and with icosahedral symmetry; compare Remark 3.5.

The full icosahedral group is $Y_{\mathsf{h}} = Y \times C_2$, where the C_2 corresponds to space inversion. Consequently, the representation theory of Y_{h} follows immediately from this direct product structure.

Let us finally consider the possibility of inflation maps in this setting. If a discrete icosahedral structure in 3-space is constructed via projection from $\mathbb{Z}^6$, D_6 or D_6^*, and displays a (possibly local) scaling relation (or an LIDS in the sense of Definition 5.16), the latter has to lift to a lattice endomorphism I (or automorphism) in 6-space. Moreover, I has to commute with the six-dimensional rotation matrices. Applying Schur's lemma (see [JL01]) and the structure of the basis matrix B, one finds

$$
I \;=\; B^{-1} \left(\begin{array}{c|c} \alpha\,\mathbb{1} & \mathbb{O} \\ \hline \mathbb{O} & \beta\,\mathbb{1} \end{array} \right) B \;=\; \frac{\alpha+\beta}{2}\,\mathbb{1} + \frac{\alpha-\beta}{2}\,(\pi_1 - \pi_2),
$$

where $\sqrt{5}\,(\pi_1 - \pi_2)$ is an integer matrix with zero entries on its diagonal, and ± 1 at all other positions. Since I can be a lattice endomorphism of $\mathbb{Z}^6$ only when it is an integer matrix in this representation, we may conclude that $\alpha+\beta \in 2\mathbb{Z}$ and $\alpha-\beta \in 2\sqrt{5}\,\mathbb{Z}$. This now implies the relation $\alpha, \beta \in \mathbb{Z}[\sqrt{5}\,] = \mathbb{Z}[2\tau]$ together with $\beta = \alpha'$. For I to be a lattice automorphism, α must be a unit in $\mathbb{Z}[2\tau]$. Since $\tau^3 = 2\tau + 1$, this means $\alpha = \pm\tau^m$ for some $m \in 3\mathbb{Z}$, in line with our previous observation for the module $\mathcal{M}_{\mathsf{P}}$. This means that icosahedral models on the basis of $\mathbb{Z}^6$ can never have inflation (or scaling) symmetries with inflation multiplier τ or τ^2.

As one can check explicitly, all choices of $m \in \mathbb{Z}$ turn I into a lattice automorphism of D_6 and D_6^*, which is consistent with our assertion in Example 2.20 that the modules $\mathcal{M}_{\mathsf{B}}$ and $\mathcal{M}_{\mathsf{F}}$ are $\mathbb{Z}[\tau]$-modules.

APPENDIX B

The Dynamical Spectrum

The purpose of this appendix is a brief (and somewhat informal) summary of notions and results around the spectral theory of dynamical systems, where we will also indicate some connections with the theory of diffraction. Dynamical spectra are discussed in many textbooks on ergodic theory; for our context, we refer to [CFS82, Que10], as well as to [Pet83, Wal00, Nad98, Sol98b] for further background and additional examples. For simplicity, we begin with a formulation for the $\mathbb{Z}$-action of the shift in (one-dimensional) symbolic dynamics, and extend to more general translation actions afterwards.

Let $\mathbb{X} \subset \mathcal{A}^{\mathbb{Z}}$ be a (closed) shift space over a finite alphabet $\mathcal{A}$, and hence compact. The action of the shift S on sequences $x \in \mathbb{X}$ is as explained in Chapter 4, so that $(Sw)_i = w_{i+1}$ for $w \in \mathbb{X}$. This generates a continuous $\mathbb{Z}$-action on the compact space $\mathbb{X}$. We assume that a shift-invariant probability measure μ on $\mathbb{X}$ is given (relative to the standard Borel σ-algebra), so that $(\mathbb{X}, \mathbb{Z}, \mu)$ is a measure-theoretic dynamical system. A natural (separable) Hilbert space attached to this system is $\mathcal{H} = L^2(\mathbb{X}, \mu)$, with the standard inner product

$$\langle f \mid g \rangle := \int_{\mathbb{X}} \overline{f(x)}\, g(x)\, \mathrm{d}\mu(x),$$

which is linear in the second argument. The shift S on $\mathbb{X}$ induces the mapping $U_S : \mathcal{H} \longrightarrow \mathcal{H}$ via $f \mapsto U_S f$, where $(U_S f)(x) := f(Sx)$. This defines a unitary operator on $\mathcal{H}$, with the usual spectral theory of unitary operators on Hilbert spaces at one's disposal; see [Koo31, vNe32, HvN44] for the origins and [Que10, Ch. 2] for a systematic exposition.

The spectrum of U_S is known as the *dynamical spectrum* of $(\mathbb{X}, \mathbb{Z}, \mu)$. Let us explain this in more detail. First of all, there is a subspace $\mathcal{H}_{\mathsf{pp}} \subset \mathcal{H}$ that is spanned by the eigenfunctions of U_S, so $\mathcal{H}_{\mathsf{pp}} = \langle f_1, f_2, \ldots \rangle_{\mathbb{C}}$ with $U_S f_i = \lambda_i f_i$ and eigenvalues $\lambda_i \in \mathbb{S}^1$. Note that $f \equiv 1$ is always an eigenfunction (with eigenvalue 1), so that $\mathcal{H}_{\mathsf{pp}} \neq \{0\}$. When an eigenfunction can be chosen to be a continuous function (such as for $\lambda = 1$), one calls the corresponding eigenvalue a *topological* eigenvalue. The topological point spectrum is the set of all topological eigenvalues. In nice cases, such as the hulls of primitive substitutions or the hulls of certain FLC Delone sets, all eigenvalues are topological; see [KS14] for some recent developments. In general, however, there are further eigenvalues, and one has to work with measurable functions.

When $\mathcal{H}_{pp} = \mathcal{H}$, one says that $(\mathbb{X}, \mathbb{Z}, \mu)$ has *pure point dynamical spectrum*; see [CFS82, Ch. 12] for a systematic treatment. This happens, for instance, when $\mathbb{X}$ is the hull of a periodic sequence (which is easy to prove, since $\mathbb{X}$ is then finite by Corollary 4.1 and $L^2(\mathbb{X}, \mu) \simeq \mathbb{C}^{|\mathbb{X}|}$) or when $\mathbb{X}$ is the hull of the silver mean sequence or of other (repetitive) examples that can also be described as model sets. When the dynamical system is ergodic, each eigenvalue has multiplicity 1, and all eigenvalues together form a subgroup of $\mathbb{S}^1$; see [CFS82, Thm. 12.1.1] for more, and [Pet83, Wal00] for general background.

The dynamical spectrum was introduced in [Koo31], while its full potential was recognised by von Neumann [vNe32], wherefore it is sometimes also called the von Neumann spectrum. He observed that two ergodic, measure-theoretic dynamical systems with pure point dynamical spectrum are metrically isomorphic if and only if they have the same spectrum; compare [CFS82, Thm. 12.2.2] or [Que10, Sec. 3.4.1]. The general formulation of this property, together with a representative via group addition on a compact Abelian group, is often referred to as the Halmos–von Neumann theorem [HvN44]. This observation is also a crucial ingredient to the inverse problem of diffraction theory, when dealing with systems with pure point spectrum; see [LM11, TB13, Ter13] for some recent progress.

In the case that $\mathcal{H}_{pp}^{\perp} \neq \{0\}$, there is at least one function g in this complement, and $\mathcal{C}(g) = \overline{\{U_S^n g \mid n \in \mathbb{Z}\}} \subset \mathcal{H}$ is a (closed) cyclic space that is infinite-dimensional. The function defined on $\mathbb{Z}$ by $n \mapsto \langle g \mid U_S^n g \rangle$ is positive definite, and has a unique representation as

$$(B.1) \qquad \langle g \mid U_S^n g \rangle = \int_0^1 e^{2\pi i n u} \, d\nu_g(u) = \overset{\smile}{\nu_g}(n)$$

on $\mathbb{T}$ (viewed as the dual group of $\mathbb{Z}$, as in Section 8.7), with a positive measure ν_g. This is the Herglotz–Bochner theorem for this situation, compare Theorem 8.6, and ν_g is called the *spectral measure* attached to g. In our situation, due to $g \in \mathcal{H}_{pp}^{\perp}$, it is a continuous measure. In general, one can define the (positive) spectral measure ν_g for any $g \in \mathcal{H}$, which is an important ingredient to the spectral representation theorem. Whenever g is an eigenfunction of U_S, with $\langle g \mid g \rangle = 1$ and eigenvalue $\lambda = e^{2\pi i \alpha}$ say, one simply has $\nu_g = \delta_\alpha$, which is a pure point measure.

Let $f, g \in \mathcal{H}$, observe the classic polarisation identity

$$4 \langle f \mid U_S^n g \rangle = \langle f + g \mid U_S^n(f + g) \rangle - \langle f - g \mid U_S^n(f - g) \rangle$$
$$- i \langle f + ig \mid U_S^n(f + ig) \rangle + i \langle f - ig \mid U_S^n(f - ig) \rangle,$$

and define the complex measure

$$\nu_{f,g} = \frac{1}{4}\left(\nu_{f+g} - \nu_{f-g} - \mathrm{i}\nu_{f+\mathrm{i}g} + \mathrm{i}\nu_{f-\mathrm{i}g}\right).$$

This gives the representation $\langle f \,|\, U_S^n g\rangle = \widetilde{\nu_{f,g}}(n)$. The family $\left(\nu_{f,g}\right)_{f,g\in\mathcal{H}}$ is known as the *spectral family* of the unitary operator U_S. It is clear that one has $\nu_{g,g} = \nu_g$ and $\nu_{g,f} = \overline{\nu_{f,g}}$ as well as

$$\nu_{f+g} = \nu_f + \nu_g + \nu_{f,g} + \nu_{g,f},$$

which underlies many calculations with the spectral measures. For any Borel function ψ on $\mathbb{T}$, one now has

$$\langle f \,|\, \psi(U_S)g\rangle = \int_0^1 \psi(u)\,\mathrm{d}\nu_{f,g}(u),$$

which is known as the *spectral representation theorem* for unitary operators; see [Que10, Thm. 2.2] for details.

Let us now recall the *spectral decomposition theorem* for the unitary operator U_S on the (separable) Hilbert space $\mathcal{H}$; compare [Que10, Thm. 2.3]. It states that there is a (possibly finite) sequence $(h_n)_{n\geqslant 1}$ of elements of $\mathcal{H}$ such that $\mathcal{H} = \bigoplus_{n\geqslant 1} \mathcal{C}(h_n)$ with $\mathcal{C}(h_n) \perp \mathcal{C}(h_m)$ for all $n \neq m$, together with the property that ν_{n+1} is absolutely continuous with respect to ν_n for all $n \geqslant 1$ (where ν_n denotes the spectral measure for h_n, and hence for $\mathcal{C}(h_n)$). Here, the spectral measures are essentially determined in the sense that any other sequence $(h'_n)_{n\geqslant 1}$ of functions with the above conditions possesses spectral measures ν'_n each of which is mutually absolutely continuous with its 'partner' measure ν_n. Various simplifications emerge when the spectrum is simple; see [Que10] for details. With respect to the sequence $(\nu_n)_{n\in\mathbb{N}}$ of measures, the spectral type of ν_1 is called the *maximal* spectral type of U_S, which is an important indication of the type of order in the underlying shift space. It is connected to diffraction theory via Fraczek's theorem [Fra97], which states that the maximal spectral measure in our case belongs to a continuous function g. The latter can then be used to define a factor whose diffraction measure essentially is the spectral measure of g.

When one moves from symbolic dynamics to tilings on the line, it is often more natural to consider the (continuous) action of the group $\mathbb{R}$ by translation. This action is defined by the mapping $\alpha\colon (\mathbb{X},\mathbb{R}) \longrightarrow \mathbb{X}$ through $(x,t) \mapsto \alpha_t(x) := t + x$, where $\mathbb{X}$ is now a compact, translation invariant space (such as the continuous hull of a geometric inflation rule, as introduced in Chapter 4). The induced unitary operators on $L^2(\mathbb{X},\mu)$ then act as $\left(U_{\alpha_t}f\right)(x) = f(\alpha_t x) = f(t + x)$. For eigenfunctions, it is now advantageous to employ a generator of the group $\mathbb{R}$, the translation by 1 say, and to write the eigenvalues of U_{α_t} with $t \in \mathbb{R}$ as $\exp(2\pi\mathrm{i}kt)$. The characteristic quantity is then the number k, which is best viewed as an element of the dual group $\widehat{\mathbb{R}}$ (which is isomorphic to $\mathbb{R}$ in this case). In particular, the constant function is then an eigenfunction that corresponds to $k = 0$.

There is no need to stay in one dimension. Indeed, both $\mathbb{Z}^d$ and $\mathbb{R}^d$ act via shift or translation on Euclidean d-space, and this is the usual setting for tilings (or locally equivalent point sets) in Euclidean space; see [Sol97, Sol98b, Rob04, Fra08] and references therein. The systematic connection between discrete and continuous group actions is analysed via sections and suspensions; see [CFS82, Ch. 11.1] for a detailed exposition, where a suspension is an example of a special flow, as well as [EW11]. Another rather useful extension emerges from viewing point sets and tilings as special cases of (translation bounded) measures. This naturally leads to spaces of measures that are compact in the vague topology, and thus to measure dynamical systems; see [BL04] and references therein for details.

Consider a uniquely ergodic dynamical system $(\mathbb{X}, \mathbb{R}^d, \mu)$, where $\mathbb{X}$ is a compact space of translation bounded measures. This system has pure point dynamical spectrum if and only if all elements of $\mathbb{X}$ have pure point diffraction spectrum; see [Schl00, LMS02, BL04, LR07] and references therein for the development of this relation. It can be extended beyond unique ergodicity, for instance by defining the diffraction spectrum for the hull (rather than for its individual elements); compare [Gou03, BL04, BBM10]. Due to this correspondence, the Halmos–von Neumann theorem becomes a powerful tool for the inverse problem of diffraction, as mentioned above.

When the dynamical spectrum comprises continuous components, the simple equivalence statement from the pure point case generally breaks down, as one knows from characteristic examples such as the Thue–Morse chain [vEM92] or the close-packed dimer system [BvE11]; see our treatment of these examples in the main text. The dynamical spectrum is typically richer than the diffraction spectrum. For sufficiently 'nice' systems, such as subshifts over finite alphabets or point set dynamical systems generated from FLC Delone sets, the dynamical spectrum can be recovered from the diffraction spectrum of the system and a suitable (and often small) collection of its factors [BLvE14], as one can explicitly see in several of our key examples.

References

[AW99]   Adkins W.A. and Weintraub S.H. (1999). *Algebra — An Approach via Module Theory*, corr. 2nd printing (Springer, New York).

[Aki12]   Akiyama S. (2012). A note on aperiodic Ammann tiles, *Discr. Comput. Geom.* **48**, 702–710.

[AAP92]   Alimov Sh.A., Ashurov R.R. and Pulatov A.K. (1992). Multiple Fourier series and Fourier integrals. In *Commutative Harmonic Analysis IV*, Khavin V.P. and Nikol'skiĭ N.K. (eds.), pp. 1–95 (Springer, Berlin).

[A-PCG13]   Aliste-Prieto J., Coronel D. and Gambaudo J.-M. (2013). Linearly repetitive Delone sets are rectifiable, *Ann. Inst. H. Poincaré Anal. Non Linéaire* **30**, 275–290. arXiv:1103.5423.

[All97]   Allouche J.-P. (1997). Schrödinger operators with Rudin-Shapiro potentials are not palindromic, *J. Math. Phys.* **38**, 1843–1848.

[ABCD03]   Allouche J.-P., Baake M., Cassaigne J. and Damanik D. (2003). Palindrome complexity, *Theor. Comput. Sci.* **292**, 9–31. arXiv:math.CO/0106121.

[AMF95]   Allouche J.-P. and Mendès France M. (1995). Automatic sequences. In [AG95], pp. 293–367.

[AS94]   Allouche J.-P. and Shallit J. (1994). Complexité des suites de Rudin-Shapiro généralisés, *Bull. Belg. Math. Soc.* **1**, 133–143.

[AS99]   Allouche J.-P. and Shallit J. (1999). The ubiquitous Prouhet-Thue-Morse sequence. In *Sequences and Their Applications: Proceedings of SETA '98*, Ding C., Helleseth T. and Niederreiter H. (eds.), pp. 1–16 (Springer, Berlin).

[AS03]   Allouche J.-P. and Shallit J. (2003). *Automatic Sequences* (Cambridge University Press, Cambridge).

[Alz95]   Alzer H. (1995). Note on an extremal property of the Rudin-Shapiro sequence, *Abh. Math. Sem. Univ. Hamburg* **65**, 243–248.

[AGS92]   Ammann R., Grünbaum B. and Shephard G.C. (1992). Aperiodic tiles, *Discr. Comput. Geom.* **8**, 1–25.

[AGZ10]   Anderson G.W., Guionnet A. and Zeitouni O. (2010). *An Introduction to Random Matrices* (Cambridge University Press, Cambridge).

[AP98]   Anderson J.E. and Putnam I.F. (1998). Topological invariants for substitution tilings and their associated C^*-algebras, *Ergodic Th. & Dynam. Syst.* **18**, 509–537.

[AAR99]   Andrews G.E., Askey R. and Roy R. (1999). *Special Functions* (Cambridge University Press, Cambridge).

[Apo76]   Apostol T.M. (1976). *Introduction to Analytic Number Theory* (Springer, New York).

[AGL74]   Argabright L. and Gil de Lamadrid J. (1974). Fourier analysis of unbounded measures on locally compact Abelian groups, *Memoirs Amer. Math. Soc.*, no. 145 (AMS, Providence, RI).

[Ata75]   Ataman Y. (1975). On positive definite measures, *Monatsh. Math.* **79**, 265–272.

[AP07]   Au-Yang H. and Perk J.H.H. (2007). Q-dependent susceptibility in Z-invariant pentagrid Ising model, *J. Stat. Phys.* **127**, 221–264. arXiv:cond-mat/0409557.

[AB09]   Averkov G. and Bianchi G. (2009). Confirmation of Matheron's conjecture on the covariogram of planar convex bodies, *J. Europ. Math. Soc.* **11**, 1187–1202. arXiv:0711.0572.

[AL12]   Averkov G. and Langfeld B. (2012). On the reconstruction of planar lattice-convex sets from the covariogram, *Discr. Comput. Geom.* **48**, 216–238. arXiv:1011.5530.

[AG95]   Axel F. and Gratias, D. (eds.) (1995). *Beyond Quasicrystals* (Springer, Berlin, and EDP Sciences, Les Ulis).

[Baa84]   Baake M. (1984). Structure and representations of the hyperoctahedral group, *J. Math. Phys.* **25**, 3171–3182.

[Baa97]   Baake M. (1997). Solution of the coincidence problem in dimensions $d \leqslant 4$. In [Moo97], pp. 9–44.

[Baa99]   Baake M. (1999). A note on palindromicity, *Lett. Math. Phys.* **49**, 217–227. arXiv:math-ph/9907011.

[Baa02a]   Baake M. (2002). A guide to mathematical quasicrystals. In [SSH02], pp. 17–48. arXiv:math-ph/9901014.

[Baa02b]   Baake M. (2002). Diffraction of weighted lattice subsets, *Can. Math. Bull.* **45**, 483–498. arXiv:math.MG/0106111.

[BB-A+94]   Baake M., Ben-Abraham S.I., Klitzing R., Kramer P. and Schlottmann M. (1994). Classification of local configurations in quasicrystals, *Acta Cryst. A* **50**, 553–566.

[BBM10]   Baake M., Birkner M. and Moody R.V. (2010). Diffraction of stochastic point sets: Explicitly computable examples, *Commun. Math. Phys.* **293**, 611–660. arXiv:0803.1266.

[BF05]   Baake M. and Frettlöh D. (2005). SCD patterns have singular diffraction, *J. Math. Phys.* **46**, 033510: 1–10. arXiv:math-ph/0411052.

[BFG07]   Baake M., Frettlöh D. and Grimm U. (2007). A radial analogue of Poisson's summation formula with applications to powder diffraction and pinwheel patterns, *J. Geom. Phys.* **57**, 1331–1343. arXiv:math.SP/0610408.

[BG98]   Baake M. and Gähler F. (1998). Symmetry structure of the Elser-Sloane quasicrystal. In *Aperiodic '97*, de Boissieu M., Verger-Gaugry J.-L. and Currat R. (eds.), pp. 63–67 (World Scientific, Singapore). cond-mat/9809100.

[BGG12a]   Baake M., Gähler F. and Grimm U. (2012). Spectral and topological properties of a family of generalised Thue-Morse sequences, *J. Math. Phys.* **53**, 032701: 1–24. arXiv:1201.1423.

[BGG12b]   Baake M., Gähler F. and Grimm U. (2012). Hexagonal inflation tilings and planar monotiles, *Symmetry* **4**, 581–602. arXiv:1210.3967.

[BGG13]   Baake M., Gähler F. and Grimm U. (2013). Examples of substitution systems and their factors, *J. Int. Seq.* **16**, article 13.2.14: 1–18; arXiv:1211.5466.

[BG97]   Baake M. and Grimm U. (1997). Coordination sequences for root lattices and related graphs, *Z. Krist.* **212**, 253–256. arXiv:cond-mat/9706122.

[BG03]   Baake M. and Grimm U. (2003). A note on shelling, *Discr. Comput. Geom.* **30**, 573–589. arXiv:math.MG/0203025.

[BG07]   Baake M. and Grimm U. (2007). Homometric model sets and window covariograms, *Z. Krist.* **222**, 54–58. arXiv:math.MG/0610411.

[BG08]   Baake M. and Grimm U. (2008). The singular continuous diffraction measure of the Thue-Morse chain, *J. Phys. A: Math. Theor.* **41**, 422001: 1–6. arXiv:0809.0580.

[BG09]   Baake M. and Grimm U. (2009). Kinematic diffraction is insufficient to distinguish order from disorder, *Phys. Rev. B* **79**, 020203(R): 1–4 and *Phys. Rev. B* **80**, 029903(E). arXiv:0810.5750.

[BG11]   Baake M. and Grimm U. (2011). Kinematic diffraction from a mathematical viewpoint, *Z. Krist.* **226**, 711–725. arXiv:1105.0095.

[BG12]   Baake M. and Grimm U. (2012). Mathematical diffraction of aperiodic structures, *Chem. Soc. Rev.* **41**, 6821–6843. arXiv:1205.3633.

[BG14]   Baake M. and Grimm U. (2014). Squirals and beyond: Substitution tilings with singular continuous spectrum, *Ergodic Th. & Dynam. Syst.* **34**, 1077–1102. arXiv:1205.1384.

[BGE97]   Baake M., Grimm U. and Elser V. (1997). The entropy of square-free words, *Mathem. Comp. Modelling* **26**, 13–26. arXiv:math-ph/9809010.

[BGHZ08] Baake M., Grimm U., Heuer M. and Zeiner P. (2008). Coincidence rotations for the root lattice A_4, *Europ. J. Combin.* **29**, 1808–1819. arXiv:0709.1341.

[BGJ93]   Baake M., Grimm U. and Joseph D. (1993). Trace maps, invariants, and some of their applications, *Int. J. Mod. Phys. B* **7**, 1527–1550. arXiv:math-ph/9904025.

[BGW94]   Baake M., Grimm U. and Warrington D. (1994). Some remarks on the visible points of a lattice, *J. Phys. A: Math. Gen.* **27**, 2669–2674 and 5041 (erratum). arXiv:math-ph/9903046.

[BHP97]   Baake M., Hermisson J. and Pleasants P.A.B. (1997). The torus parametrization of quasiperiodic LI classes, *J. Phys. A: Math. Gen.* **30**, 3029–3056. mp_arc/02-168.

[BHM08]   Baake M., Heuer M. and Moody R.V. (2008). Similar sublattices of the root lattice A_4, *J. Algebra* **320**, 1391–1408. arXiv:math.MG/0702448.

[BH00]   Baake M. and Höffe M. (2000). Diffraction of random tilings: Some rigorous results, *J. Stat. Phys.* **99**, 219–261. arXiv:math-ph/9901008.

[BJ90]   Baake M. and Joseph D. (1990). Ideal and defective vertex configurations in the planar octagonal quasilattice, *Phys. Rev. B* **42**, 8091–8102.

[BJK91]   Baake M., Joseph D. and Kramer P. (1991). The Schur rotation as a simple approach to the transition between quasiperiodic and periodic phases, *J. Phys. A: Math. Gen.* **24**, L961–L967.

[BJKS90] Baake M., Joseph D., Kramer P. and Schlottmann M. (1990). Root lattices and quasicrystals, *J. Phys. A: Math. Gen.* **23**, L1037–L1041. arXiv:cond-mat/0006062.

[BJS90]   Baake M., Joseph D. and Schlottmann M. (1990). The root lattice D_4 and planar quasilattices with octagonal and dodecagonal symmetry, *Int. J. Mod. Phys. B* **5**, 1927–1953.

[BKS92]   Baake M., Klitzing R. and Schlottmann M. (1992). Fractally shaped acceptance domains of quasiperiodic square-triangle tilings with dodecagonal symmetry, *Physica A* **191**, 554–558.

[BK11]   Baake M. and Kösters H. (2011). Random point sets and their diffraction, *Philos. Mag.* **91**, 2671–2679. arXiv:1007.3084.

[BKM14]   Baake M., Kösters H. and Moody R.V. (2014). Diffraction theory of point processes: Systems with clumping and repulsion, *Preprint* arXiv:1405.4255.

[BKSZ90]   Baake M., Kramer P., Schlottmann M. and Zeidler D. (1990). Planar patterns with fivefold symmetry as sections of periodic structures in 4-space, *Int. J. Mod. Phys. B* **4**, 2217–2268.

[BL04]   Baake M. and Lenz D. (2004). Dynamical systems on translation bounded measures: Pure point dynamical and diffraction spectra, *Ergod. Th. & Dynam. Syst.* **24**, 1867–1893. arXiv:math.DS/0302061.

[BL05]   Baake M. and Lenz D. (2005). Deformation of Delone dynamical systems and pure point diffraction, *J. Fourier Anal. Appl.* **11**, 125–150. arXiv:math.DS/0404155.

[BLM07]   Baake M., Lenz D. and Moody R.V. (2007). Characterisation of model sets by dynamical systems, *Ergod. Th. & Dynam. Syst.* **27**, 341–382. arXiv:math.DS/0511648.

[BLR07]   Baake M., Lenz D. and Richard C. (2007). Pure point diffraction implies zero entropy for Delone sets with uniform cluster frequencies, *Lett. Math. Phys.* **82**, 61–77. arXiv:0706.1677.

[BLvE14]   Baake M., Lenz D. and van Enter A.C.D. (2014). Dynamical versus diffraction spectrum for structures with finite local complexity, *Ergod. Th. & Dynam. Syst.*, in press. arXiv:1307.7518.

[BM13]   Baake M. and Moll M. (2013). Random noble means substitutions. In *Aperiodic Crystals*, Schmid S., Withers R.L. and Lifshitz R. (eds.), pp. 19–27 (Springer, Dordrecht). arXiv:1210.3462.

[BM98]   Baake M. and Moody R.V. (1998). Diffractive point sets with entropy, *J. Phys. A: Math. Gen.* **31**, 9023–9039. arXiv:math-ph/9809002.

[BM99]   Baake M. and Moody R.V. (1999). Similarity submodules and root systems in four dimensions, *Can. J. Math.* **51**, 1258–1276. arXiv:math.MG/9904028.

[BM00a]   Baake M. and Moody R.V. (2000). Self-similar measures for quasicrystals. In [BM00b], pp. 1–42. arXiv:math.MG/0008063.

[BM00b]   Baake M. and Moody R.V. (eds.) (2000). *Directions in Mathematical Quasicrystals*, CRM Monograph Series, vol. 13 (AMS, Providence, RI).

[BM04]   Baake M. and Moody R.V. (2004). Weighted Dirac combs with pure point diffraction, *J. reine angew. Math. (Crelle)* **573**, 61–94. arXiv:math.MG/0203030.

[BMP00]   Baake M., Moody, R.V. and Pleasants P.A.B. (2000). Diffraction for visible lattice points and kth power free integers, *Discr. Math.* **221**, 3–42. arXiv:math.MG/9906132.

[BMRS03] Baake M., Moody, R.V., Richard, C. and Sing, B. (2003). Which distributions of matter diffract? — Some answers, In *Quasicrystals: Structure and Physical Properties*, H.-R. Trebin (ed.), Wiley-VCH, Berlin, pp. 188-207. arXiv:math-ph/0301019.

[BMS98] Baake M., Moody R.V. and Schlottmann M. (1998). Limit-(quasi)periodic point sets as quasicrystals with *p*-adic internal spaces, *J. Phys. A: Math. Gen.* **31**, 5755–5765. arXiv:math-ph/9901008.

[BSJ91] Baake M., Schlottmann M. and Jarvis P.D. (1991). Quasiperiodic tilings with tenfold symmetry and equivalence with respect to local derivability, *J. Phys. A: Math. Gen.* **24**, 4637–4654.

[BS04a] Baake M. and Sing B. (2004). Kolakoski-(3, 1) is a (deformed) model set, *Can. Math. Bull.* **47**, 168–190. arXiv:math.MG/0206098.

[BS04b] Baake M. and Sing B. (2004). Diffraction spectrum of lattice gas models above T_c, *Lett. Math. Phys.* **68**, 165–173. arXiv:math-ph/0405064.

[BvE11] Baake M. and van Enter A.C.D. (2011). Close-packed dimers on the line: Diffraction versus dynamical spectrum, *J. Stat. Phys.* **143**, 88–101. arXiv:1011.1628.

[BW10] Baake M. and Ward T. (2010). Planar dynamical systems with pure Lebesgue diffraction spectrum, *J. Stat. Phys.* **140**, 90–102. arXiv:1003.1536.

[BZ08] Baake M. and Zint N. (2008). Absence of singular continuous diffraction for discrete multi-component particle models, *J. Stat. Phys.* **130**, 727–740. arXiv:0709.2061.

[Bak86] Bak P. (1986). Icosahedral crystals from cuts in six-dimensional space, *Scr. Metall.* **20**, 1199–1204.

[Ban97] Bandt C. (1997). Self-similar tilings and patterns described by mappings. In [Moo97], pp. 45–83.

[BW01] Bandt C. and Wang Y. (2001). Disk-like self-affine tiles in $\mathbb{R}^2$, *Discr. Comput. Geom.* **26**, 591–601.

[BD07] Barge M. and Diamond B. (2007). Proximality in Pisot tiling spaces, *Fund. Math.* **194**, 191–238. arXiv:math/0509051.

[BDH03] Barge M., Diamond B. and Holton C. (2003). Asymptotic orbits of primitive substitutions, *Theor. Comput. Sci.* **301**, 439–450.

[BO14] Barge M. and Olimb C. (2014). Asymptotic structure in substitution tiling spaces, *Erg. Th. & Dynam. Syst.* **34**, 55–94. arXiv:1101.4902.

[BV11] Barnsley M.F. and Vince A. (2011). The chaos game on a general iterated function system, *Erg. Th. & Dynam. Syst.* **31**, 1073–1079. arXiv:1005.0322.

[Bau01] Bauer H. (2001). *Measure and Integration Theory* (de Gruyter, Berlin).

[BF13a] Bédaride N. and Fernique T. (2013). The Ammann-Beenker tilings revisited. In *Aperiodic Crystals*, Schmid S., Withers R.L. and Lifshitz R. (eds.), pp. 59–65 (Springer, Dordrecht). arXiv:1208.3545.

[BF13b] Bédaride N. and Fernique T. (2013). When periodicities enforce aperiodicity, *Preprint* arXiv:1309.3686.

[Bee82] Beenker F.P.M. (1982). Algebraic theory of non-periodic tilings of the plane by two simple building blocks: A square and a rhombus, *TH-Report* 82-WSK-04 (TU Eindhoven).

[Bel95]   Bellissard J. (1995). Gap labelling theorems for Schrödinger's operators. In *From Number Theory to Physics*, 2nd corr. printing, Waldschmidt M., Moussa P., Luck J.-M. and Itzykson C. (eds.), pp. 538–630 (Springer, Berlin).

[BHZ00]   Bellissard J., Herrmann D.J.L. and Zarrouati M. (2000). Hulls of aperiodic solids and gap labeling theorems. In [BM00b], pp. 207–258.

[BIST89]   Bellissard J., Iochum B., Scoppola E. and Testard D. (1989). Spectral properties of one-dimensional quasi-crystals, *Commun. Math. Phys.* **125**, 527–543.

[B-AG99]   Ben-Abraham S.I. and Gähler F. (1999). Covering cluster description of octagonal MnSiAl quasicrystals, *Phys. Rev. B* **60**, 860–864.

[B-AQS13]   Ben-Abraham S.I., Quandt A. and Shapira D. (2013). Multidimensional paperfolding systems, *Acta Cryst. A* **69**, 123–130.

[BBD'E10]   Benassi C., Bianchi G. and D'Ercole G. (2010). Covariogram of non-convex sets, *Mathematika* **56**, 267–284. arXiv:1003.4122.

[Ben85]   Bendersky L. (1985). Quasicrystal with one-dimensional translational symmetry and a tenfold rotation axis, *Phys. Rev. Lett.* **55**, 1461–1463.

[BBK05]   Benedetto J.J., Bernstein E. and Konstantinidis I. (2005). Multiscale Riesz products and their support properties, *Acta Appl. Math.* **88**, 201–227.

[Ber65]   Berberian S.K. (1965). *Measure and Integration* (Macmillan, New York).

[BF75]   Berg C. and Forst G. (1975). *Potential Theory on Locally Compact Abelian Groups* (Springer, Berlin).

[Ber66]   Berger R. (1966). The undecidability of the domino problem, *Memoirs Amer. Math. Soc.*, no. 66, pp. 1–72 (AMS, Providence, RI).

[BD00]   Bernuau G. and Duneau M. (2000). Fourier analysis of deformed model sets. In [BM00b], pp. 33–40.

[BFS12]   Berthé V., Frettlöh D. and Sirvent V. (2012). Selfdual substitutions in dimension one, *Europ. J. Combin.* **33**, 981–1000. arXiv:1108.5053.

[BS07]   Berthé V. and Siegel A. (2007). Purely periodic beta-expansions in the Pisot non-unit case, *J. Number Th.* **153**, 153–172. arXiv:math.DS/0407282.

[BSS+11]   Berthé V., Siegel A., Steiner W., Surer P. and Thuswaldner J.M. (2011). Fractal tiles associated with shift radix systems, *Adv. Math.* **226**, 139–175. arXiv:0907.4872.

[BDG+92]   Bertin M.J., Decomps-Guilloux A., Grandet-Hugot M., Pathiaux-Delefosse M. and Schreiber J.P. (1992). *Pisot and Salem Numbers* (Birkhäuser, Basel).

[Bes54]   Besicovitch A.S. (1954) *Almost Periodic Functions*, reprint (Dover, New York).

[BKMS00]   Bezuglyi S., Kwiatkowski J., Meydynets K. and Solomyak B. (2000). Invariant measures on stationary Bratteli diagrams, *Ergod. Th. & Dynam. Syst.* **30**, 973–1007. arXiv:0812.1088.

[Bia02]   Bianchi G. (2002). Determining convex polygons from their covariograms, *Adv. Appl. Prob.* **34**, 261–266.

[Bia05]   Bianchi G. (2005). Matheron's conjecture for the covariogram problem, *J. London Math. Soc.* (2) **71**, 203–220.

[Bia09]   Bianchi G. (2009). The covariogram determines three-dimensional convex polytopes, *Adv. Math.* **220**, 1771–1808.

[BGK11]   Bianchi G., Gardner R.J. and Kiderlen M. (2011). Phase retrieval for charac-
          teristic functions of convex bodies and reconstruction from covariograms, *J.
          Amer. Math. Soc.* **24**, 293–343. arXiv:1003.4486.

[Bil95]   Billingsley P. (1995). *Probability and Measure*, 3rd ed. (Wiley, New York).

[Boh93]   Bohl P. (1893). *Über die Darstellung von Funktionen einer Variabeln durch
          trigonometrische Reihen mit mehreren einer Variabeln proportionalen Argu-
          menten*, Thesis (E.J. Karow, Dorpat).

[Boh47]   Bohr H. (1947). *Almost Periodic Functions*, reprint (Chelsea, New York).

[BT86]   Bombieri E. and Taylor J.E. (1986). Which distributions of matter diffract?
          An initial investigation, *J. Phys. Colloques* **47**, C3-19–C3-28.

[BS66]   Borevicz Z.I. and Shafarevich I.R. (1966). *Number Theory* (Academic Press,
          New York).

[Bou66]   Bourbaki N. (1966). *General Topology*, parts I and II (Hermann, Paris).

[Brl89]   Brlek S. (1989). Enumeration of factors in the Thue-Morse word, *Discr. Appl.
          Math.* **24**, 83–96.

[Bro65]   Bromwich T.J.I'A. (1965). *An Introduction to the Theory of Infinite Series*,
          2nd rev. ed. (Macmillan, London).

[BBN+78] Brown H., Bülow R., Neubüser J., Wondratschek H. and Zassenhaus H.
          (1978). *Crystallographic Groups of Four-dimensional Space* (Wiley, New
          York).

[BK98]   Burago D. and Kleiner B. (1998). Separated nets in Euclidean space and
          Jacobians of bi-Lipschitz maps, *Geom. Funct. Anal.* **8**, 273–282.
          arXiv:math.DG/9703022.

[CJ94]   Cabo A.J. and Janssen R.H.P. (1994). Cross-covariance functions characterise
          bounded regular closed sets, CWI Report BS-R9426, available at
          http://oai.cwi.nl/oai/asset/5110/5110D.pdf

[CIV03]   Campanino M., Ioffe D. and Velenik Y. (2003). Ornstein-Zernike theory for the
          finite range Ising models above T_c, *Probab. Th. Relat. Fields* **125**, 305–349.
          arXiv:math.PR/0111274.

[CL90]   Carmona R. and Lacroix J. (1990). *Spectral Theory of Random Schrödinger
          Operators* (Birkhäuser, Boston, MA).

[Car00]   Carothers N.L. (2000). *Real Analysis* (Cambridge University Press, Cam-
          bridge).

[Cas71]   Cassels J.W.S. (1971). *An Introduction to the Geometry of Numbers*, 2nd
          corr. printing (Springer, Berlin).

[CS13]   Cellarosi F. and Sinai Ya.G. (2013). Ergodic properties of square-free numbers,
          J. Europ. Math. Soc. **15**, 1343–1374. arXiv:1112.4691.

[CV13]   Cellarosi F. and Vinogradov I. (2013). Ergodic properties of k-free integers in
          number fields, *J. Modern Dynamics* **7**, 461–488. arXiv:1304.0214.

[Cha87]   Champeney D.C. (1987). *A Handbook of Fourier Theorems* (Cambridge Uni-
          versity Press, Cambridge).

[CMP98]   Chen L., Moody R.V. and Patera J. (1998). Non-crystallographic root sys-
          tems. In [Pat98], pp. 135–178.

[CSM88]   Cheng Z., Savit R. and Merlin R. (1988), Structure and electronic properties
          of Thue–Morse lattices, *Phys. Rev. B* **37**, 4375–4382.

[CS03]   Clark A. and Sadun L. (2003). When size matters: Subshifts and
             their related tiling spaces, *Ergod. Th. & Dynam. Syst.* **23**, 1043–1057.
             arXiv:math.DS/0201152.

[CS06]   Clark A. and Sadun L. (2006). When shape matters: Deformation of tiling
             spaces, *Ergod. Th. & Dynam. Syst.* **26**, 69–86. arXiv:math.DS/0306214.

[CE80]   Collet P. and Eckmann J.-P. (1980). *Iterated Maps on the Interval as Dynam-
             ical Systems* (Birkhäuser, Basel).

[CT08]   Connelly B. and Terrell B. (2008). *Highly Symmetric Tensegrity Structures*,
             available at http://www.math.cornell.edu/~tens/

[CKH98]   Conrad M., Krumeich F. and Harbrecht B. (1998). A dodecagonal quasicrys-
             talline chalcogenide, *Angew. Chem. Int. Ed.* **37**, 1383–1386.

[CBG08]   Conway J.H., Burgiel H. and Goodman-Strauss C. (2008). *The Symmetries
             of Things* (A.K. Peters, Wellesley, MA).

[CR98]   Conway J.H. and Radin C. (1998). Quaquaversal tilings and rotations, *Invent.
             Math.* **132**, 179–188.

[CS99]   Conway J.H. and Sloane N.J.A. (1999). *Sphere Packings, Lattices and Groups*,
             3rd ed. (Springer, New York).

[Cor89]   Corduneanu C. (1989). *Almost Periodic Functions*, 2nd English ed. (Chelsea,
             New York).

[CFS82]   Cornfeld I.P., Fomin S.V. and Sinai Ya.G. (1982). *Ergodic Theory* (Springer,
             New York).

[CDP10]   Cortez M.I., Durand F. and Petite S. (2010). Linearly repetitive Delone sys-
             tems have a finite number of non-periodic Delone system factors, *Proc. Amer.
             Math. Soc.* **138**, 1033–1046. arXiv:0807.2907.

[CS11]   Cortez M.I. and Solomyak B. (2011). Invariant measures for non-primitive
             tiling substitutions, *J. d'Analyse Math.* **115**, 293–342. arXiv:1007.1686.

[CH73]   Coven E.M. and Hedlund G.A. (1973). Sequences with minimal block growth,
             Math. Systems Theory **7**, 138–153.

[Cow95]   Cowley J.M. (1995). *Diffraction Physics*, 3rd ed. (North-Holland, Amster-
             dam).

[Cox73]   Coxeter H.S.M. (1973). *Regular Polytopes*, 3rd ed. (Dover, New York).

[CEPT86]   Coxeter H.S.M., Emmer M., Penrose R. and Teuber M.L. (eds.) (1986). *M.C.
             Escher. Art and Science* (North-Holland, Amsterdam).

[CM80]   Coxeter H.S.M. and Moser W.O.J. (1980). *Generators and Relations for Dis-
             crete Groups*, 4th ed. (Springer, Berlin).

[Cul96]   Culik K. (1996). An aperiodic set of 13 Wang tiles, *Discr. Math.* **160**, 245–251.

[Cur84]   Curtis M.L. (1984). *Matrix Groups*, 2nd ed. (Springer, New York).

[DV-J88]   Daley D.J. and Vere-Jones D. (1988). *An Introduction to the Theory of Point
             Processes* (Springer, New York).

[Dam01]   Damanik D. (2001). Schrödinger operators with low-complexity potentials,
             Ferroelectrics **250**, 143–149.

[DL99]   Damanik D. and Lenz D. (1999). Uniform spectral properties of one-
             dimensional quasicrystals, I. Absence of eigenvalues, *Commun. Math. Phys.*
             207, 687–696. arXiv:math-ph/9903011.

[DL06a]   Damanik D. and Lenz D. (2006). Substitution dynamical systems: Characterization of linear repetitivity and applications, *J. Math. Anal. Appl.* **321**, 766–780. arXiv:math.DS/0302231.

[DL06b]   Damanik D. and Lenz D. (2006). Zero-measure Cantor spectrum for Schrödinger operators with low-complexity potentials, *J. Math. Pures Appl.* **85**, 671–686.

[Dan89]   Danzer L. (1989). Three-dimensional analogs of the planar Penrose tilings and quasicrystals, *Discr. Math.* **76**, 1–7.

[Dan95]   Danzer L. (1995). A family of 3D-spacefillers not permitting any periodic or quasiperiodic tiling. In *Aperiodic '94*, Chapuis G. and Paciorek W. (eds.), pp. 11–17 (World Scientific, Singapore).

[DPT93]   Danzer L., Papadopolos Z. and Talis A. (1993). Full equivalence between Socolar's tilings and the (A,B,C,K)-tilings leading to a rather natural decoration, *Int. J. Mod. Phys. B* **7**, 1379–1386.

[DSvOD93] Danzer L., Sonneborn P., van Ophuysen G. and Duitmann S. (1993). *The {A,B,C,K}-Book* (Internal Report, Univ. Dortmund).

[dBr81]   de Bruijn N.G. (1981). Algebraic theory of Penrose's non-periodic tilings of the plane. I & II, *Kon. Nederl. Akad. Wetensch. Proc. Ser. A* **84**, 39–52 and 53–66.

[dGN97a] de Gier J. and Nienhuis B. (1997). The exact solution of an octagonal rectangle triangle random tiling, *J. Stat. Phys.* **87**, 415–437. arXiv:solv-int/9610009.

[dGN97b] de Gier J. and Nienhuis B. (1997). On the integrability of the square-triangle random tiling model, *Phys. Rev. E* **55**, 3926–3933. arXiv:solv-int/9611005.

[dGN98]   de Gier J. and Nienhuis B. (1998). Bethe Ansatz solution of a decagonal rectangle-triangle random tiling, *J. Phys. A: Math. Gen.* **31**, 2141–2154. arXiv:cond-mat/9709338.

[Dek78]   Dekking F.M. (1978). The spectrum of dynamical systems arising from substitutions of constant length, *Z. Wahrscheinlichkeitsth. verw. Geb.* **41**, 221–239.

[Dek97]   Dekking F.M. (1997). What is the long range order in the Kolakoski sequence? In [Moo97], pp. 115–125.

[DM09]   Deng X. and Moody R.V. (2009). How model sets can be determined by their two-point and three-point correlations. *J. Stat. Phys.* **135**, 621–637. arXiv:0901.4381.

[DGS76]   Denker M., Grillenberger C. and Sigmund K. (1976). *Ergodic Theory on Compact Spaces*, LNM 527 (Springer, Berlin).

[dSp99]   de Spinadel V.W. (1999). The family of metallic means, *Visual Math.* **1**, issue 3, paper 4; available online as http://www.mi.sanu.ac.rs/vismath/spinadel/index.html.

[DSS95]   Deuber W.A., Simonovits M. and Sós V.T. (1995). A note on paradoxical metric spaces, *Stud. Sci. Math. Hung.* **30**, 17–23; revised and annotated version available at http://mta.renyi.hu/~miki/walter07.pdf.

[Dev89]   Devaney R.L. (1989). *An Introduction to Chaotic Dynamical Systems*, 2nd ed. (Addison-Wesley, Redwood City).

[dWo74]   de Wolff P.M. (1974). The pseudo-symmetry of modulated crystal structures, *Acta Cryst. A* **30**, 777–785.

[Die70]   Dieudonné J. (1970). *Treatise of Analysis*, vol. II (Academic Press, New York).

498   REFERENCES

[Dod83]   Dodd F.W. (1983). *Number Theory in the Quadratic Field with Golden Section Unit* (Polygonal Publishing House, Passaic, NJ).

[DG72]   Domb C. and Green M.S. (eds.) (1972). *Phase Transitions and Critical Phenomena. Vol. 1. Exact Results* (Academic Press, London).

[DSV00]   Draco B., Sadun L. and Van Wieren D. (2000). Growth rates in the quaquaversal tiling, *Discr. Comput. Geom.* **23**, 419–435. arXiv:math-ph/9812018.

[Dub06]   Dubickas A. (2006). Arithmetical properties of powers of algebraic numbers, *Bull. London Math. Soc.* **38**, 70–80.

[Dur00]   Durand F. (2000). Linearly recurrent subshifts have a finite number of nonperiodic subshift factors, *Ergod. Th. & Dynam. Syst.* **20**, 1061–1078 and **23** (2003), 663–669 (corrigendum and addendum). arXiv:0807.4430.

[Dwo93]   Dworkin S. (1993). Spectral theory and X-ray diffraction, *J. Math. Phys.* **34**, 2965–2967.

[DMcK72] Dym H. and McKean H.P. (1972). *Fourier Series and Integrals* (Academic Press, New York).

[Dys62]   Dyson F. (1962). Statistical theory of the energy levels of complex systems. III, *J. Math. Phys.* **3**, 166–175.

[Ebel02] Ebeling W. (2002). *Lattices and Codes*, 2nd ed. (Vieweg, Braunschweig).

[Ebe49]   Eberlein W.F. (1949). Abstract ergodic theorems and weak almost periodic functions, *Trans. Amer. Math. Soc.* **67**, 217–240.

[Ebe55]   Eberlein W.F. (1955). The point spectrum of weakly almost periodic functions, *Michigan Math. J.* **3**, 137–139.

[ENP07]   Eigen S. Navarro J. and Prasad V. S. (2007). An aperiodic tiling using a dynamical system and Beatty sequences. In *Dynamics, Ergodic Theory, and Geometry*, Hasselblatt B. (ed.), MSRI Publications, vol. 54, pp. 223–241 (Cambridge University Press, Cambridge).

[EW11]   Einsiedler M. and Ward T. (2011). *Ergodic Theory, with a View towards Number Theory* (Springer, London).

[Els85]   Elser V. (1985). Comment on "Quasicrystals: A new class of ordered structures", *Phys. Rev. Lett.* **54**, 1730.

[Els86]   Elser V. (1986). The diffraction pattern of projected structures, *Acta Cryst.* A **42**, 36–43.

[ES87]   Elser V. and Sloane N.J.A. (1987). A highly symmetric four-dimensional quasicrystal, *J. Phys. A: Math. Gen.* **20**, 6161–6168.

[Esc04]   Esclangon E. (1904). *Les fonctions quasi-périodiques*, Thesis (Gauthier-Villars, Paris).

[Ete81]   Etemadi N. (1981). An elementary proof of the strong law of large numbers, *Z. Wahrscheinlichkeitsth. verw. Geb.* **55**, 119–122.

[Fek23]   Fekete M. (1923). Über die Verteilung der Wurzeln bei gewissen algebraischen Gleichungen mit ganzzahligen Koeffizienten, *Math. Z.* **17**, 228–249.

[Fel72]   Feller W. (1972). *An Introduction to Probability Theory and Its Applications*, vol. II, 2nd ed. (Wiley, New York).

[FeS12]   Fernique T. and Sablik M. (2012). Local rules for computable planar tilings, *EPTCS* **90**, 133–141. arXiv:1208.2759.

[FR03]   Fisher B.N. and Rabson D.A. (2003). Applications of group cohomology to the classification of quasicrystal symmetries, *J. Phys. A: Math. Gen.* **36**, 10195–10214. arXiv:math-ph/0105010.

[For10]   Forrester P.J. (2010). *Log-Gases and Random Matrices* (Princeton University Press, Princeton).

[Fra97]   Fraczek K.M. (1997). On a function that realizes the maximal spectral type, *Studia Math.* **124**, 1–7.

[Fra03]   Frank N.P. (2003). Substitution sequences in $\mathbb{Z}^d$ with a non-simple Lebesgue component in the spectrum, *Ergod. Th. & Dynam. Syst.* **23**, 519–532.

[Fra05]   Frank N.P. (2005). Multi-dimensional constant-length substitution sequences, *Topol. Appl.* **152**, 44–69.

[Fra08]   Frank N.P. (2008). A primer of substitution tilings of the Euclidean plane, *Expo. Math.* **26**, 295–326. arXiv:0705.1142.

[Fra09]   Frank N.P. (2009). Spectral theory of bijective substitution sequences, *MFO Reports* **6**, 752–756.

[FR08]   Frank N.P. and Robinson E.A. (2008). Generalized β-expansions, substitution tilings, and local finiteness, *Trans. Amer. Math. Soc.* **360**, 1163–1177. arXiv:math.DS/0506098.

[FS09]   Frank N.P. and Sadun L. (2009). Topology of some tiling spaces without finite local complexity, *Discr. Cont. Dyn. Syst. A* **23**, 847–865. arXiv:math.DS/0701424.

[FS14]   Frank N.P. and Sadun L. (2014). Fusion tilings with infinite local complexity, *Top. Proc.* **43**, 235–276. arXiv:1201.3911.

[FSo01]   Frank N.P. and Solomyak B. (2001). A characterization of planar pseudo-self-similar tilings, *Discr. Comput. Geom.* **26**, 289–306.

[FW11]   Frank N.P. and Whittaker M.F. (2011). A fractal version of the pinwheel tiling, *Math. Intelligencer* **33** (2), 7–17. arXiv:1001.2203.

[Fre02]   Frettlöh D. (2002). *Nichtperiodische Pflasterungen mit ganzzahligem Inflationsfaktor*, PhD thesis (Univ. Dortmund).

[Fre08]   Frettlöh D. (2008). Substitution tilings with statistical circular symmetry, *Europ. J. Combin.* **29**, 1881–1893. arXiv:0704.2521.

[Fre11]   Frettlöh D. (2011). A fractal fundamental domain with 12-fold symmetry, *Symmetry: Culture and Science* **22**, 237–246.

[Fre13]   Frettlöh D. (2013). Highly symmetric fundamental cells for lattices in $\mathbb{R}^2$ and $\mathbb{R}^3$, *Preprint* arXiv:1305.1798.

[FG13]   Frettlöh D. and Garber A. (2013). Bi-Lipschitz equivalence and wobbling equivalence of Delone sets, in preparation.

[FH13]   Frettlöh D. and Harriss E. (2013). Parallelogram tilings, worms, and finite orientations, *Discr. Comput. Geom.* **49**, 531–539. arXiv:1202.4686.

[FR14]   Frettlöh D. and Richard C. (2014). Dynamical properties of almost repetitive Delone sets, *Discr. Cont. Dynam. Syst. A* **34**, 531–556. arXiv:1210.2955.

[FSi07]   Frettlöh D. and Sing B. (2007). Computing modular coincidences for substitution tilings and point sets, *Discr. Comput. Geom.* **37**, 381–407. arXiv:math.MG/0601067.

[Gäh88]   Gähler F. (1988). *Quasicrystal Structures from the Crystallograhic Viewpoint*, PhD thesis no. 8414 (ETH Zürich).

[Gäh93]   Gähler F. (1993). Matching rules for quasicrystals: The composition-decomposition method, *J. Non-Cryst. Solids* **153**&**154**, 160–164.

[Gäh10]   Gähler F. (2010). MLD relations of Pisot substitution tilings, *J. Phys.: Conf. Ser.* **226**, 012020: 1–6. arXiv:1001.2744.

[Gäh13]   Gähler F. (2013). Substitution rules and topological properties of the Robinson tilings. In *Aperiodic Crystals*, Schmid S., Withers R.L. and Lifshitz R. (eds.), pp. 67–73 (Springer, Dordrecht). arXiv:1210.6468.

[GGB-A03] Gähler F., Gummelt P. and Ben-Abraham S.I. (2003). Generation of quasiperiodic order by maximal cluster covering. In [KP03], pp. 63–95.

[GJS12]   Gähler F., Julien A. and Savinien J. (2012). Combinatorics and topology of the Robinson tiling, *C. R. Math. Acad. Sci. Paris* **350**, 627–631. arXiv:1203.1387.

[GK97]   Gähler F. and Klitzing R. (1997). The diffraction pattern of self-similar tilings. In [Moo97], pp. 141–174.

[GM13]   Gähler F. and Maloney G. (2013). Cohomology of one-dimensional mixed substitution tiling spaces, *Topol. Appl.* **160**, 703–719. arXiv:1112.1475.

[GR86]   Gähler F. and Rhyner J. (1986). Equivalence of the generalised grid and projection methods for the construction of quasiperiodic tilings, *J. Phys. A: Math. Gen.* **19**, 267–277.

[GS93]   Gähler F. and Stampfli P. (1993). The dualisation method revisited: Dualisation of product Laguerre complexes as a unifying framework, *Int. J. Mod. Phys. B* **6–7**, 1333–1349.

[Gan59]   Gantmacher F.R. (1959). *The Theory of Matrices*, reprint (AMS Chelsea, New York).

[Gar77]   Gardner M. (1977). Extraordinary nonperiodic tiling that enriches the theory of tiles, *Sci. Amer.* **236**, 110–121.

[GGZ05]   Gardner R.J., Gronchi P. and Zong C. (2005). Sums, projections, and sections of lattice sets, and the discrete covariogram, *Discr. Comput. Geom.* **34**, 391–409.

[Geo11]   Georgii H.-O. (2011). *Gibbs Measures and Phase Transitions*, 2nd ed. (de Gruyter, Berlin).

[Geo12]   Georgii H.-O. (2012). *Stochastics: An Introduction to Probability and Statistics*, 2nd ed. (de Gruyter, Berlin).

[GLA90]   Gil de Lamadrid J. and Argabright L.N. (1990). Almost periodic measures, *Memoirs Amer. Math. Soc.* **85**, no. 428 (AMS, Providence, RI).

[Gin65]   Ginibre J. (1965). Statistical ensembles of complex, quaternion, and real matrices, *J. Math. Phys.* **6**, 440–449.

[Gli10]   Glied S. (2010). *Coincidence and Similarity Isometries of Modules in Euclidean Space*, PhD thesis (Univ. Bielefeld).

[Gne98]   Gnedenko B.V. (1998). *Theory of Probability*, 6th ed. (CRC press, Amsterdam).

[God89]   Godrèche C. (1989). The sphinx: A limit-periodic tiling of the plane, *J. Phys. A: Math. Gen.* **22**, L1163–L1166.

[GL89]   Godrèche C. and Luck J.M. (1989). Quasiperiodicity and randomness in tilings of the plane, *J. Stat. Phys.* **55**, 1–28.

[Gol49]   Golay M.J.E. (1949). Multislit spectroscopy, *J. Opt. Soc. Amer.* **39**, 437–444.

[Goo98]   Goodman-Strauss C. (1998). Matching rules and substitution tilings, *Ann. Math.* **147**, 181–223.

[Goo99a]   Goodman-Strauss C. (1999). A small aperiodic set of planar tiles, *Europ. J. Combin.* **20**, 375–384.

[Goo99b]   Goodman-Strauss C. (1999). An aperiodic pair of tiles in $\mathbb{E}^n$ for all $n \geqslant 3$, *Europ. J. Combin.* **20**, 385–395.

[Goo03]   Goodman-Strauss C. (2003). Matching rules for the sphinx substitution tiling, unpublished notes, available at http://comp.uark.edu/~strauss/.

[Goo11]   Goodman-Strauss C. (2011). Tazzellazioni. In *La Matematica vol III: Suoni, Forme, Parole*, Bartocci C. and Odifreddi P. (eds.), pp. 249–285 (Einaudi, Torino). English version: Tessellations, available at http://mathfactor.uark.edu/downloads/tessellations.pdf.

[Gou03]   Gouéré J.-B. (2003). Diffraction and Palm measure of point processes, *C. R. Acad. Sci. Paris* **342**, 141–146. arXiv:math.PR/0208064.

[Gou05]   Gouéré J.-B. (2005). Quasicrystals and almost periodicity, *Commun. Math. Phys.* **255**, 655–681. arXiv:math-ph/0212012.

[Gou97]   Gouvêa F.Q. (1997). *p-adic Numbers — An Introduction*, 2nd ed. (Springer, Berlin).

[GKQ95]   Gratias D., Katz A. and Quiquandon M. (1995). Geometry of approximant structures in quasicrystals, *J. Phys.: Condens. Matter* **7**, 9101–9125.

[Gri99]   Grimmett G. (1999). *Percolation*, 2nd ed. (Springer, Berlin).

[GS87]   Grünbaum B. and Shephard G.C. (1987). *Tilings and Patterns* (Freeman, New York).

[GM95]   Grünbaum F.A. and Moore C.C. (1995). The use of higher-order invariants in the determination of generalized Patterson cyclotomic sets, *Acta Cryst. A* **51**, 310–323.

[GS85]   Guiaşu S. and Shenitzer A. (1985). The principle of maximum entropy, *Math. Intelligencer* **7**, 42–48.

[Gum96]   Gummelt P. (1996). Penrose tilings as coverings of congruent decagons, *Geom. Dedicata* **62**, 1–17.

[Gum99]   Gummelt P. (1999). *Aperiodische Überdeckungen mit einem Clustertyp*, PhD thesis, Univ. Greifswald (Shaker, Aachen).

[H-AE+09]   Haji-Akbari A., Engel M., Keys A.S., Zheng X., Petschek R.G., Palffy-Muhoray P. and Glotzer S.C. (2009). Disordered, quasicrystalline and crystalline phases of densely packed tetrahedra, *Nature* **462**, 773–777. arXiv:1012.5138.

[Hal44]   Halmos P.R. (1944). In general a measure preserving transformation is mixing, *Ann. Math.* **45**, 786–792.

[HvN44]   Halmos P.R. and von Neumann J. (1944), Operator methods in classical mechanics. II. *Ann. Math.* **43**, 332–350.

[HW08]   Hardy G.H. and Wright E.M. (2008). *An Introduction to the Theory of Numbers*, 6th ed., revised by Heath-Brown D.R. and Silverman J.H. (Oxford University Press, Oxford).

[HFonl]   Harriss E. and Frettlöh D. (eds.). *Tilings Encyclopedia*, available at http://tilings.math.uni-bielefeld.de/.

[H–Y11]   Hayashi H., Kawachi Y., Komatsu K., Konda A., Kurozoe M., Nakano F., Odawara N., Onda R., Sugio A. and Yamauchi M. (2011). Notes on vertex atlas of Danzer tiling, *Nihonkai Math. J.* **22**, 49–58.

[HZ74]   Hecht E. and Zajac A. (1974). *Optics* (Addison-Wesley, Reading, MA).

[Hen86]   Henley C.L. (1986). Sphere packings and local environments in Penrose tilings, *Phys. Rev. B* **34**, 797–816.

[Hen98]   Henley C.L. (1998). Cluster maximization, non-locality, and random tilings. In *Proccedings of the 6th International Conference on Quasicrystals*, Takeuchi S. and Fujiwara T. (eds.), pp. 27–30 (World Scientific, Singapore).

[Hen99]   Henley C.L. (1999). Random tiling models. In *Quasicrystals: The State of the Art*, 2nd ed., DiVincenzo D.P. and Steinhardt P.J. (eds.), pp. 459–560 (World Scientific, Singapore).

[Her49]   Hermann C. (1949). Kristallographie in Räumen beliebiger Dimensionszahl. I. Die Symmetrieoperationen, *Acta Cryst.* **2**, 139–145.

[HRB97]   Hermisson J., Richard C. and Baake M. (1997). A guide to the symmetry structure of quasiperiodic tiling classes, *J. Phys. I France* **7**, 1003–1018. mp_arc/02-180.

[Her13]   Herning J.L. (2013). *Spectrum and Factors of Substitution Dynamical Systems*, PhD thesis (George Washington University, Washington, DC).

[HR97]   Hewitt E. and Ross K.A. (1997). *Abstract Harmonic Analysis I*, 2nd ed., corr. 3rd printing (Springer, New York).

[Hil85]   Hiller H. (1985). The crystallographic restriction in higher dimensions, *Acta Cryst. A* **41**, 541–544.

[Hof95a] Hof A. (1995). On diffraction by aperiodic structures, *Commun. Math. Phys.* **169**, 25–43.

[Hof95b] Hof A. (1995). Diffraction by aperiodic structures at high temperatures. *J. Phys. A: Math. Gen.* **28**, 57–62.

[Hof97a] Hof A. (1997). Diffraction by aperiodic structures. In [Moo97], pp. 239–268.

[Hof97b] Hof A. (1997). On scaling in relation to singular spectra, *Commun. Math. Phys.* **184**, 567–577.

[HKS95]   Hof A., Knill O. and Simon B. (1995). Singular continuous spectrum for palindromic Schrödinger operators, *Commun. Math. Phys.* **174**, 149–159.

[Höf00]   Höffe M. (2000). Diffraction of the dart-rhombus random tiling, *Math. Sci. Eng. A* **294–296**, 373–376. arXiv:math-ph/9911014.

[Höf01]   Höffe M. (2001). *Diffraktionstheorie stochastischer Parkettierungen*, PhD thesis, Univ. Tübingen (Shaker, Aachen).

[HB00]   Höffe M. and Baake M. (2000). Surprises in diffuse scattering, *Z. Krist.* **215**, 441–444. arXiv:math-ph/0004022.

[HKPM97] Hohneker C., Kramer P., Papadopolos Z. and Moody R.V. (1997). Canonical icosahedral quasilattices for the F-phase generated by coherent phases in physical space, *J. Phys. A: Math. Gen.* **30**, 6493–6507.

[HS03] Hollander M. and Solomyak B. (2003). Two-symbol Pisot substitutions have pure discrete spectrum, *Ergod. Th. & Dynam. Syst.* **23**, 533–540.

[HRS05] Holton C., Radin C. and Sadun L. (2005). Conjugacies for tiling dynamical systems, *Commun. Math. Phys.* **254**, 343–359.

[HKPV09] Hough J.B., Krishnapur M., Peres Y. and Virág B. (2009). *Zeros of Gaussian Analytic Functions and Determinantal Point Processes* (AMS, Providence, RI).

[Huc09] Huck C. (2009). Discrete tomography of icosahedral model sets, *Acta Cryst. A* **65**, 240–248. arXiv:0705.3005.

[HuB14] Huck C. and Baake M. (2014). Dynamical properties of k-free lattice points *Acta Phys. Pol. A*, in press. arXiv:1402.2202.

[Hum92] Humphreys J.E. (1992). *Reflection Groups and Coxeter Groups*, 2nd corr. printing (Cambridge University Press, Cambridge).

[Hur19] Hurwitz A. (1919). *Vorlesungen über die Zahlentheorie der Quaternionen* (Springer, Berlin).

[Hut81] Hutchinson J.E. (1981). Fractals and self-similarity, *Indiana Univ. Math. J.* **30**, 713–743.

[Ing99] Ingersent K. (1999). Matching rules for quasicrystalline tilings. In *Quasicrystals: The State of the Art*, 2nd ed., DiVincenzo D.P. and Steinhardt P.J. (eds.), pp. 197–224 (World Scientific, Singapore).

[INF85] Ishimasa T., Nissen H.-U. and Fukano Y. (1985). New ordered state between crystalline and amorphous in Ni-Cr particles, *Phys. Rev. Lett.* **55**, 511–513.

[IK04] Iwaniec H. and Kowalski E. (2004). *Analytic Number Theory* (AMS, Providence, RI).

[JK69] Jacobs K. and Keane M. (1969). 0-1-sequences of Toeplitz type, *Z. Wahrscheinlichkeitsth. verw. Geb.* **13**, 123–131.

[JK81] James G. and Kerber A. (1981). *The Representation Theory of the Symmetric Group* (Addison-Wesley, Reading, MA).

[JL01] James G. and Liebeck M. (2001). *Representations and Characters of Groups*, 2nd ed. (Cambridge University Press, Cambridge).

[JJ77] Janner A. and Janssen T. (1977). Symmetry of periodically distorted crystals, *Phys. Rev. B* **15**, 643–658.

[Jan94] Janot C. (1994). *Quasicrystals: A Primer*, 2nd ed. (Clarendon Press, Oxford).

[JCdB07] Janssen T., Chapuis G. and de Boissieu M. (2007). *Aperiodic Crystals — From Modulated Phases to Quasicrystals* (Clarendon Press, Oxford).

[Jar88] Jarić M.V. (ed.) (1988). *Introduction to Quasicrystals*, Aperiodicity & Order, vol. 1 (Academic Press, San Diego, CA).

[Jar89] Jarić M.V. (ed.) (1989). *Introduction to the Mathematics of Quasicrystals*, Aperiodicity & Order, vol. 2 (Academic Press, San Diego, CA).

[JG89] Jarić M.V. and Gratias D. (eds.) (1989). *Extended Icosahedral Structures*, Aperiodicity & Order, vol. 3 (Academic Press, San Diego, CA).

[Jay82]   Jaynes E.T. (1982). On the rationale of maximum entropy methods, *Proc. IEEE* **70**, 939–952.

[JS94]   Jeong H.-C. and Steinhardt P.J. (1994). Cluster approach for quasicrystals, *Phys. Rev. Lett.* **73**, 1943–1946.

[JM97]   Johnson A. and Madden K. (1997). Putting the pieces together: Understanding Robinson's nonperiodic tilings, *College Math. J.* **28**, 172–181.

[Jul10]   Julien A. (2010). Complexity and cohomology for cut-and-projection tilings, *Ergodic Th. & Dynam. Syst.* **30**, 489–523. arXiv:0804.0145.

[Kak72]   Kakutani S. (1972). Strictly ergodic symbolic dynamical systems. In *Proceedings of the Sixth Berkeley Symposium on Mathematical Statistics and Probability*, Le Cam L.M., Neyman J. and Scott E.L. (eds.), pp. 319–326 (University of California Press, Berkeley).

[Kal94]   Kalugin P.A. (1994). The square-triangle random-tiling model in the thermodynamic limit, *J. Phys. A: Math. Gen.* **27**, 3599–3614.

[Kal05]   Kalugin P. (2005). Cohomology of quasiperiodic patterns and matching rules, *J. Phys. A: Math. Gen.* **38**, 3115–3132.

[Kar96]   Kari J. (1996). A small aperiodic set of Wang tiles, *Discr. Math.* **160**, 259–264.

[KT75]   Karlin S. and Taylor H.M. (1975). *A First Course in Stochastic Processes*, 2nd ed. (Academic Press, San Diego, CA).

[Kar91]   Karr A.F. (1991). *Point Processes and their Statistical Inference*, 2nd ed. (Dekker, New York).

[Kas63]   Kasteleyn P.W. (1963). Dimer statistics and phase transitions, *J. Math. Phys.* **4**, 287–293.

[KH95]   Katok A. and Hasselblatt B. (1995). *Introduction to the Modern Theory of Dynamical Systems* (Cambridge University Press, Cambridge).

[Kat88]   Katz A. (1988). Theory of matching rules for the 3-dimensional Penrose tilings, *Commun. Math. Phys.* **118**, 263–288.

[Kat95]   Katz A. (1995). Matching rules and quasiperiodicity: The octagonal tilings. In [AG95], pp. 141–189.

[KG94]   Katz A. and Gratias D. (1994). Tilings and quasicrystals. In *Lectures on Quasicrystals*, Hippert F. and Gratias D. (eds.), pp. 187–264 (Les Editions des Physique, Les Ulis).

[Kat04]   Katznelson Y. (2004). *An Introduction to Harmonic Analysis*, 3rd ed. (Cambridge University Press, Cambridge).

[Kea68]   Keane M. (1968). Generalized Morse sequences, *Z. Wahrscheinlichkeitsth. verw. Geb.* **10**, 335–353.

[KP06]   Keane M. and Petersen K. (2006). Easy and nearly simultaneous proofs of the ergodic theorem and maximal ergodic theorem. In *Dynamics & Stochastics: Festschrift in honor of M.S. Keane*, Denteneer D., den Hollander F. and Verbitskiy E. (eds.), pp. 248–251 (Institute of Mathematical Statistics, Beachwood, OH).

[Kel95]   Kellendonk J. (1995). Noncommutative geometry of tilings and gap labelling, *Rev. Math. Phys.* **7**, 1133–1180. arXiv:cond-mat/9403065.

[KS14]   Kellendonk J. and Sadun L. (2014). Meyer sets, topological eigenvalues, and Cantor fiber bundles, *J. London Math. Soc.* **89**, 114–130. arXiv:1211.2250.

[Kel98]   Keller G. (1998). *Equilibrium States in Ergodic Theory* (Cambridge University Press, Cambridge).

[Ken97]   Kenyon R. (1997). Local statistics of lattice dimers, *Ann. Inst. H. Poincaré B* **33**, 591–618. arXiv:math.CO/0105054.

[Ken00]   Kenyon R. (2000). The planar dimer model with boundary: A survey. In [BM00b], pp. 307–328.

[KS10]   Kenyon R. and Solomyak B. (2010). On the characterization of expansion maps for self-affine tilings, *Discr. Comput. Geom.* **43**, 577–593. arXiv:0801.1993.

[Kep19]   Kepler J. (1619). *Harmonices Mundi Libri V* (J. Planck for G. Tampach, Linz).

[Khi57]   Khinchin A.I. (1957). *Mathematical Foundations of Information Theory* (Dover, New York).

[KS09]   Killip R. and M. Stoiciu M. (2009). Eigenvalue statistics for CMV matrices: From Poisson to clock via random matrix ensembles, *Duke Math. J.* **146**, 361–399. arXiv:math-ph/0608002.

[KET12]   Kiselev A., Engel M. and Trebin H.-R. (2012). Confirmation of the random tiling hypothesis for a decagonal quasicrystal, *Phys. Rev. Lett.* **109**, 225502: 1–4. arXiv:1210.4227.

[Kit98]   Kitchens B.P. (1998). *Symbolic Dynamics* (Springer, Berlin).

[Kle84]   Klein F. (1884). *Vorlesungen über das Ikosaeder und die Auflösung der Gleichungen vom fünften Grade* (Teubner, Leipzig).

[KSB93]   Klitzing R., Schlottman M. and Baake M. (1993). Perfect matching rules for undecorated triangular tilings with 10-, 12-, and 8-fold symmetry, *Int. J. Mod. Phys. B* **7**, 1455–1473.

[Kni98]   Knill O. (1998). Singular continuous spectrum and quantitative rates of weak mixing, *Discr. Cont. Dynam. Syst.* **4**, 33–42.

[Kob96]   Koblitz N. (1996). *p-adic Numbers, p-adic Analysis, and Zeta-Functions*, corr. 2nd ed. (Springer, New York).

[KR91]   Koecher M. and Remmert R. (1991). Hamilton's quaternions. In *Numbers*, Ebbinghaus H.-D. *et al.* (eds.), pp. 189–220 (Springer, New York).

[Kol65]   Kolakoski W. (1965). Self-generating runs; problem 5304, *Amer. Math. Monthly* **72**, 674.

[Koo31]   Koopman B.O. (1931). Hamiltonian systems and transformations in Hilbert space, *Proc. Nat. Acad. Sci. USA* **17**, 315–318.

[Kow38]   Kowalewski G. (1938). *Der Keplersche Körper und andere Bauspiele* (Koehlers Antiquarium, Leipzig).

[Kra82]   Kramer P. (1982). Non-periodic central space filling with icosahedral symmetry using copies of seven elementary cells, *Acta Cryst. A* **38**, 257–264.

[Kra87a]   Kramer P. (1987). Atomic order in quasicrystals is supported by several unit cells, *Mod. Phys. Lett. B* **1**, 7–18.

[Kra87b]   Kramer P. (1987). Continuous rotation from cubic to icosahedral order, *Acta Cryst. A* **43**, 486–489.

[Kra88]   Kramer P. (1988). Space-group theory for a nonperiodic icosahedral quasilattice, *J. Math. Phys.* **29**, 516–524.

[KH89]   Kramer P. and Haase R.W. (1989). Group theory of icosahedral quasicrystals. In [Jar89], pp. 81–146.

[KN84]   Kramer P. and Neri R. (1984) On periodic and non-periodic space fillings of $\mathbb{E}^m$ obtained by projection, *Acta Cryst. A* **40**, 580–587 and *Acta Cryst. A* **41** (1985), 619 (erratum).

[KP03]   Kramer P. and Papadopolos Z. (eds.) (2003). *Coverings of Discrete Quasiperiodic Sets* (Springer, Berlin).

[KPSZ94] Kramer P., Papadopolos Z., Schlottmann M. and Zeidler D. (1994). Projection of the Danzer tiling, *J. Phys. A: Math. Gen.* **27**, 4505–4517.

[KS89]   Kramer P. and Schlottmann M. (1989). Dualisation of Voronoi domains and Klotz construction: A general method for the generation of proper space fillings, *J. Phys. A: Math. Gen.* **22**, L1097–L1102.

[KN74]   Kuipers L. and Niederreiter H. (1974). *Uniform Distribution of Sequences* (Wiley, New York). Reprint (Dover, New York, 2006).

[Kül03a] Külske C. (2003). Universal bounds on the selfaveraging of random diffraction measures, *Probab. Th. Relat. Fields* **126**, 29–50. arXiv:math-ph/0109005.

[Kül03b] Külske C. (2003). Concentration inequalities for functions of Gibbs fields with application to diffraction and random Gibbs measures, *Commun. Math. Phys.* **239**, 29–51.

[Kwa11]   Kwapisz J. (2011). Rigidity and mapping class group for abstract tiling spaces, *Ergodic Th. & Dynam. Syst.* **31**, 1745–1783.

[Lag96]   Lagarias J.C. (1996). Meyer's concept of quasicrystal and quasiregular sets, *Commun. Math. Phys.* **179**, 365–376.

[Lag99]   Lagarias J.C. (1999). Geometric models for quasicrystals I. Delone sets of finite type, *Discr. Comput. Geom.* **21**, 161–191.

[LP02]   Lagarias J.C. and Pleasants P.A.B. (2002). Local complexity of Delone sets and crystallinity, *Can. Math. Bull.* **48**, 634–652. arXiv:math.MG/0105088.

[LP03]   Lagarias J.C. and Pleasants P.A.B. (2003). Repetitive Delone sets and quasicrystals, *Ergod. Th. & Dynam. Syst.* **23**, 831–867. arXiv:math.DS/9909033.

[LW96]   Lagarias J.C. and Wang Y. (1996). Self-affine tiles in $\mathbb{R}^n$, *Adv. Math.* **121**, 21–49.

[LW03]   Lagarias J.C. and Wang Y. (2003). Substitution Delone sets, *Discr. Comput. Geom.* **29**, 175–209. arXiv:math.MG/0110222.

[LZ12]   Lagarias J.C. and Zong C. (2012). Mysteries in packing regular tetrahedra, *Notices Amer. Math. Soc.* **59**, 1540–1549.

[LB88]   Lançon F. and Billard L. (1988). Two-dimensional system with a quasicrystalline ground state, *J. Physique (France)* **49**, 249–256.

[Lan93]   Lang S. (1993). *Real and Functional Analysis*, 3rd ed. (Springer, New York).

[Lan02]   Lang S. (2002). *Algebra*, rev. 3rd ed. (Springer, New York).

[Le95]   Le T.T.Q. (1995). Local rules for pentagonal quasi-crystals, *Discr. Comput. Geom.* **14**, 31–70.

[Le97]   Le T.T.Q. (1997). Local rules for quasiperiodic tilings. In [Moo97], pp. 331–366.

[Led78]   Ledrappier F. (1978). Un champ markovien peut être d'entropie nulle et mélangeant, *C. R. Acad. Sci. Paris Sér. A-B* **287**, A561–A563.

[LM01]   Lee J.-Y. and Moody R.V. (2001). Lattice substitution systems and model
          sets, *Discr. Comput. Geom.* **25**, 173–201. arXiv:math.MG/0002019.

[LM08]   Lee J.-Y. and Moody R.V. (2008). Deforming Meyer sets, *Europ. J. Combin.*
          29, 1919–1924. arXiv:0910.4446.

[LM13]   Lee J.-Y. and Moody R.V. (2013). Taylor-Socolar hexagonal tilings as model
          sets, *Symmetry* **5**, 1–46. arXiv:1207.6237.

[LMS02]   Lee J. Y., Moody R.V. and Solomyak B. (2002). Pure point dynamical and
          diffraction spectra, *Ann. H. Poincaré* **3**, 1003–1018. arXiv:0910.4809.

[Len02]   Lenz D. (2002). Uniform ergodic theorems on subshifts over a finite alphabet,
          Ergod. Th. & Dynam. Syst. **22**, 245–255. arXiv:math.DS/0005067.

[Len09]   Lenz D. (2009). Continuity of eigenfunctions of uniquely ergodic dynamical
          systems and intensity of Bragg peaks, *Commun. Math. Phys.* **287**, 225–258.
          arXiv:math-ph/0608026.

[LM09]   Lenz D. and Moody R.V. (2009). Extinctions and correlations for uniformly
          discrete point processes with pure point dynamical spectra, *Commun. Math.
          Phys.* **289**, 907–923. arXiv:0902.0567.

[LM11]   Lenz D. and Moody R.V. (2011). Stationary processes with pure point diffrac-
          tion, *Preprint* arXiv:1111.3617.

[LR07]   Lenz D. and Richard C. (2007). Pure point diffraction and cut and project
          schemes for measures: The smooth case, *Math. Z.* **256**, 347–378.
          arXiv:math.DS/0603453.

[LS84]   Levine D. and Steinhardt P.J. (1984). Quasicrystals: A new class of ordered
          structures, *Phys. Rev. Lett.* **53**, 2477–2480.

[Lev88]   Levitov L.S. (1988). Local rules for quasicrystals, *Commun. Math. Phys.* **119**,
          627–666.

[Lif96]   Lifshitz R. (1996). The symmetry of quasiperiodic crystals, *Physica A* **232**,
          633–647.

[Lif97]   Lifshitz R. (1997). Theory of color symmetry for periodic and quasiperiodic
          crystals, *Rev. Mod. Phys.* **69**, 1181–1218.

[LD07]   Lifshitz R. and Diamant H. (2007). Soft quasicrystals — why are they stable?
          Philos. Mag. **87**, 3021–3030. arXiv:cond-mat/0611115.

[Lin04]   Lind D. (2004). Multi-dimensional symbolic dynamics, *Proc. Sympos. Appl.
          Math.* **60**, 61–79.

[LM95]   Lind D.A. and Marcus B. (1995). *An Introduction to Symbolic Dynamics and
          Coding* (Cambridge University Press, Cambridge).

[LSW90]   Lind D.A., Schmidt K. and Ward T. (1990). Mahler measure and entropy for
          commuting automorphisms of compact groups, *Inv. Math.* **101**, 593–629.

[Loo11]   Loomis L.H. (2011). *Introduction to Abstract Harmonic Analysis*, reprint
          (Dover, New York).

[LRK00]   Lord E.A., Ranganathan S. and Kulkarni U.D. (2000). Tilings, coverings,
          clusters and quasicrystals, *Current Sci.* **78**, 64–72.

[Lot02]   Lothaire M. (2002). *Algebraic Combinatorics on Words* (Cambridge Univer-
          sity Press, Cambridge).

[LGJJ93]   Luck J.M., Godrèche C., Janner A. and Janssen T. (1993). The nature of the atomic surfaces of quasiperiodic self-similar structures, *J. Phys. A: Math. Gen.* **26**, 1951–1999.

[Lüc88]   Lück R. (1988). Description of imperfections in the icosahedral Penrose tiling. In *Quasicrystalline Materials*, Janot Ch. and Dubois J.M. (eds.), pp. 308–317 (World Scientific, Singapore).

[Lüc93]   Lück R. (1993). Basic ideas of Ammann bar grids, *Int. J. Mod. Phys. B* **7**, 1437–1453.

[Lüc00]   Lück R. (2000). Dürer–Kepler–Penrose, the development of pentagonal tilings, *Mat. Sci. Eng. A* **294–296**, 263–267.

[LP92]   Lunnon W.F. and Pleasants P.A.B. (1992). Characterization of two-distance sequences, *J. Austral. Math. Soc. (Series A)* **53**, 198–218.

[LAT04]   Luo J., Akiyama S. and Thuswaldner J.M. (2004). On the boundary connectedness of connected tiles, *Math. Proc. Cambridge Philos. Soc.* **137**, 397–410.

[Mac81]   Mackay A.L. (1981). De nive quinquangula: On the pentagonal snowflake, *Sov. Phys. Cryst.* **26**, 517–522.

[Mac82]   Mackay A.L. (1982). Crystallography and the Penrose pattern, *Physica A* **114**, 609–613.

[MKS76]   Magnus W., Karrass A. and Solitar D. (1976). *Combinatorial Group Theory*, rev. ed. (Dover, New York).

[Mah27]   Mahler K. (1927). The spectrum of an array and its application to the study of the translation properties of a simple class of arithmetical functions. Part II: On the translation properties of a simple class of arithmetical functions, *J. Math. Massachusetts* **6**, 158–163.

[Mak92]   Makovicky E. (1992). 800-year-old pentagonal tiling from Marāgha, Iran, and the new varieties of new aperiodic tiling it inspired. In *Fivefold Symmetry*, Hargittai I. (ed.), pp. 67–86 (World Scientific, Singapore).

[Mat75]   Matheron G. (1975). *Random Sets and Integral Geometry* (Wiley, New York).

[Mat86]   Matheron G. (1986). Le covariogramme géometrique des compacts convexes de $\mathbb{R}^2$, *Technical Report* 2/86 (Centre de Géostatistique, École des Mines de Paris).

[MN83]   Mathew J. and Nadkarni M.G. (1983). A measure preserving transformation whose spectrum has Lebesgue component of multiplicity two, *Bull. London Math. Soc.* **16**, 402–406.

[Mat10]   Matzutt K. (2010). *Diffraction of Point Sets with Structural Disorder*, PhD thesis (Univ. Bielefeld).

[McM98]   McMullen C.T. (1998). Lipschitz maps and nets in Euclidean space, *Geom. Funct. Anal.* **8**, 304–314.

[MS02]   McMullen P. and Schulte E. (2002). *Abstract Regular Polytopes* (Cambridge University Press, Cambridge).

[Meh04]   Mehta M.L. (2004). *Random Matrices*, 3rd ed. (Elsevier, Amsterdam).

[MRW87]   Mermin N.D., Rokhsar D.S. and Wright D.C. (1987). Beware of 46-fold symmetry: The classification of two-dimensional quasicrystallographic lattices, *Phys. Rev. Lett.* **58**, 2099–2101.

[Mey70]   Meyer Y. (1970). *Nombres de Pisot, Nombres de Salem et Analyse Harmonique*, LNM 117 (Springer, Berlin).

[Mey72]   Meyer Y. (1972). *Algebraic Numbers and Harmonic Analysis* (North Holland, Amsterdam).

[Mey12]   Meyer Y. (2012). Quasicrystals, almost periodic patterns, mean-periodic functions and irregular sampling, *African Diaspora J. Math.* **13**, 1–45.

[Mię93]   Miękisz J. (1993). *Quasicrystals: Microscopic Models on Nonperiodic Structures* (Leuven University Press, Leuven).

[MvE90]   Miękisz J. and van Enter A.C.D. (1990). Breaking of periodicity at positive temperatures, *Commun. Math. Phys.* **134**, 647–651.

[MSR+10]   Mikhael J., Schmiedeberg M., Rausch S., Roth J., Stark H. and Bechinger C. (2010), Proliferation of anomalous symmetries in colloidal monolayers subjected to quasiperiodic light fields, *PNAS* **107**, 7214–7218. `arXiv:1005.2120`.

[Mol14]   Moll M. (2014). Diffraction of random noble means words, *J. Stat. Phys.* **156**, in press. `arXiv:1404.7411`.

[Moo97]   Moody R.V. (ed.) (1997). *The Mathematics of Long-Range Aperiodic Order*, NATO ASI Series C 489 (Kluwer, Dordrecht).

[Moo97a]   Moody R.V. (1997). Meyer sets and their duals. In [Moo97], pp. 403–441.

[Moo00]   Moody R.V. (2000). Model sets: A Survey. In *From Quasicrystals to More Complex Systems*, Axel F., Dénoyer F. and Gazeau J.P. (eds.), pp. 145–166 (EDP Sciences, Les Ulis, and Springer, Berlin). `arXiv:math.MG/0002020`.

[Moo02]   Moody R.V. (2002). Uniform distribution in model sets, *Can. Math. Bull.* **45**, 123–130.

[MP92]   Moody R.V. and Patera J. (1992). Voronoi and Delone cells of root lattices: Classification of their faces and facets by Coxeter–Dynkin diagram, *J. Phys. A: Math. Gen.* **25**, 5089–5134.

[MP93]   Moody R.V. and Patera J. (1993). Quasicrystals and icosians, *J. Phys. A: Math. Gen.* **26**, 2829–2853.

[MP96]   Moody R.V. and Patera J. (1996). Dynamical generation of quasicrystals, *Lett. Math. Phys.* **36**, 291–300.

[MW94]   Moody R.V. and Weiss A. (1994). On shelling E_8 quasicrystals, *J. Number Th.* **47**, 405–412.

[MH38]   Morse M. and Hedlund G.A. (1938). Symbolic dynamics, *Amer. J. Math.* **60**, 815–866.

[MH40]   Morse M. and Hedlund G.A. (1940). Symbolic dynamics II, *Amer. J. Math.* **62**, 1–42.

[Mou10]   Moustafa H. (2010). PV cohomology of the pinwheel tilings, their integer group of coinvariants and gap-labeling, *Commun. Math. Phys.* **298**, 369–405. `arXiv:0906.2107`.

[Moz89]   Mozes S. (1989). Tilings, substitution systems and dynamical systems generated by them, *J. d'Analyse Math.* **53**, 139–186.

[MR13]   Müller P. and Richard C. (2013). Ergodic properties of randomly coloured point sets, *Can. J. Math.* **65**, 349–402. `arXiv:1005.4884`.

[Nad95]   Nadkarni M.G. (1995). *Basic Ergodic Theory*, 2nd ed. (Birkhäuser, Basel).

[Nad98]   Nadkarni M.G. (1998). *Spectral Theory of Dynamical Systems* (Birkhäuser, Basel).

[NM87]   Niizeki N. and Mitani H. (1987). Two-dimensional dodecagonal quasilattices, *J. Phys. A: Math. Gen.* **20**, L405–L410.

[NHL13]   Nikola N., Hexner D. and Levine D. (2013). Entropic commensurate-incommensurate transition, *Phys. Rev. Lett.* **110**, 125701: 1–4. arXiv:1211.6972.

[Nil12a]   Nilsson, J. (2012). On the entropy of a family of random substitution systems, *Monatsh. Math.* **166**, 1–15. arXiv:1001.3513 and arXiv:1103.4777.

[Nil12b]   Nilsson, J. (2012). A space efficient algorithm for the calculation of the digit distribution in the Kolakoski sequence, *J. Integer Sequences* **15**, 12.6.7: 1–13. arXiv:1110.4228.

[ND96]   Nischke K.-P. and Danzer L. (1996). A construction of inflation rules based on n-fold symmetry, *Discr. Comput. Geom.* **15**, 221–236.

[Off08]   Offner C.D. (2008). Repetitions of words and the Thue-Morse sequence, unpublished notes; available from http://www.cs.umb.edu/~offner/.

[Old39]   Oldenburger R. (1939). Exponent trajectories in symbolic dynamics, *Trans. Amer. Math. Soc.* **46**, 453–466.

[O'Mea73] O'Meara O.T. (1973). *Introduction to Quadratic Forms*, 3rd. corr. printing (Springer, Berlin); reprint (Springer, Berlin, 2000).

[OH93]   Oxborrow M. and Henley C.L. (1993). Random square-triangle tilings: A model for twelvefold-symmetric quasicrystals, *Phys. Rev. B* **48**, 6966–6998.

[Oxt52]   Oxtoby J.C. (1952). Ergodic sets, *Bull. Amer. Math. Soc.* **58**, 116–136.

[PHK00]   Papadopolos Z., Hohneker C. and Kramer P. (2000). Tiles-inflation rules for the class of canonical tilings $\mathcal{T}^{*(2F)}$ derived by the projection method. *Discr. Math.* **221**, 101–112.

[PKK97]   Papadopolos Z., Klitzing R. and Kramer P. (1997). Quasiperiodic icosahedral tilings from the six-dimensional bcc lattice, *J. Phys. A: Math. Gen.* **30**, L143–L147.

[Par67]   Parthasarathy K.R. (1967). *Probability Measures on Metric Spaces* (Academic Press, New York).

[Pat97]   Patera J. (1997). Noncrystallographic root systems and quasicrystals. In [Moo97], pp. 443–465.

[Pat98]   Patera J. (ed.) (1998). *Quasicrystals and Discrete Geometry*, Fields Institute Monographs, vol. 10 (AMS, Providence, RI).

[Patt44]   Patterson A.L. (1944). Ambiguities in the X-ray analysis of crystal structures, *Phys. Rev.* **65**, 195–201.

[PK87]   Pavlovitch A. and Kléman M. (1987). Generalised 2D Penrose tilings: Structural properties, *J. Phys. A: Math. Gen.* **20**, 687–702.

[Ped95]   Pedersen G.K. (1995). *Analysis Now*, rev. printing (Springer, New York).

[PP58a]   Penrose L.S. and Penrose R. (1958). Impossible objects: A special type of visual illusion, *Brit. J. Psychol.* **49**, 31–33; reprinted in *Roger Penrose Collected Works*, vol. 1, pp. 355–358 (2011, Oxford University Press, Oxford).

[PP58b]   Penrose L.S. and Penrose R. (1958). Puzzles for Christmas, *New Scientist* (Dec. 25) 1580–1581 and 1597–1598; reprinted in *Roger Penrose Collected Works*, vol. 1, pp. 359–363 (2011, Oxford University Press, Oxford).

[Pen74]   Penrose R. (1974). The rôle of aesthetics in pure and applied mathematical research, *Bull. Inst. Math. Appl.* **10**, 266–271.

[Pen78]   Penrose R. (1978). Pentaplexity, *Eureka* **39**, 16–22; reprinted in *Math. Intelligencer* **2** (1979), 32–38.

[Pen97]   Penrose R. (1997). Remarks on tiling: Details of a $(1 + \varepsilon + \varepsilon^2)$-aperiodic set. In [Moo97], pp. 467–497; a supplement appeared in the *Twistor Newsletter* (TN **41** (1996) 37, TN **42** (1997) 25–26 and TN **43** (1998) 34); all reprinted in *Roger Penrose Collected Works*, vol. 6 (2011, Oxford University Press, Oxford).

[Per54]   Perron O. (1954). *Die Lehre von den Kettenbrüchen*, 3rd ed. (Teubner, Stuttgart).

[Pet83]   Petersen K. (1983). *Ergodic Theory* (Cambridge University Press, Cambridge).

[Pin02]   Pinsky M.A. (2002). *Introduction to Fourier Analysis and Wavelets* (Brooks/Cole, Pacific Grove, CA).

[Ple00]   Pleasants P.A.B. (2000). Designer quasicrystals: Cut-and-project sets with pre-assigned properties. In [BM00b], pp. 95–141.

[PH13]   Pleasants P.A.B. and Huck C. (2013). Entropy and diffraction of the k-free points in n-dimensional lattices, *Discr. Comput. Geom.* **50**, 39–68. arXiv:1112.1629.

[Pow14]   Power S.C. (2014). Crystal frameworks, matrix-valued functions and rigidity operators. In *Concrete Operators, Spectral Theory, Operators in Harmonic Analysis and Approximation*, Cepedello Boiso M., Hedenmalm H., Kaashoek M., Montes Rodríguez A. and Treil S. (eds.), pp. 405–420 (Birkhäuser, Basel). arXiv:1111.2943.

[Pra94]   Prasolov V.V. (1994). *Problems and Theorems in Linear Algebra* (AMS, Providence, RI).

[PF02]   Pytheas Fogg N. (2002). *Substitutions in Dynamics, Arithmetics and Combinatorics*, LNM 1794 (Springer, Berlin).

[Que95]   Queffélec M. (1995). Spectral study of automatic and substitutive sequences. In [AG95], pp. 369–414.

[Que10]   Queffélec M. (2010). *Substitution Dynamical Systems — Spectral Analysis*, LNM 1294, 2nd ed. (Springer, Berlin).

[Rad87]   Radin C. (1987). Low temperature and the origin of crystalline symmetry, *Int. J. Mod. Phys. B* **1**, 1157–1191.

[Rad99]   Radin C. (1999). *Miles of Tiles* (AMS, Providence, RI).

[RS98]   Radin C. and Sadun L. (1998). Subgroups of SO(3) associated with tilings, *J. Algebra* **202**, 611–633. arXiv:math.GR/9601202.

[RS99]   Radin C. and Sadun L. (1999). On 2-generator subgroups of SO(3), *Trans. Amer. Math. Soc.* **351**, 4469–4480.

[RW92]   Radin C. and Wolff M. (1992). Space tilings and local isomorphism, *Geom. Dedicata* **42**, 355–360.

[RS80]   Reed M. and Simon B. (1980). *Methods of Modern Mathematical Physics I: Functional Analysis*, 2nd ed. (Academic Press, San Diego, CA).

[RG03]   Reichert M. and Gähler F. (2003). Cluster model of decagonal tilings, *Phys. Rev. B* **68**, 214202: 1–10.

[Rei03]   Reiner I. (2003). *Maximal Orders*, 2nd ed. (Clarendon Press, Oxford).

[ReSt00]   Reiter H. and Stegeman J.D. (2000). *Classical Harmonic Analysis and Locally Compact Groups* (Clarendon Press, Oxford).

[Rho05]   Rhoads G.C. (2005). Planar tilings by polyominoes, polyhexes, and polyiamonds, *J. Comput. Appl. Math.* **174**, 329–353.

[Ric99a]   Richard C. (1999). An alternative view on random tilings, *J. Phys. A: Math. Gen.* **32**, 8823–8829. arXiv:cond-mat/9907262.

[Ric99b]   Richard C. (1999). *Statistische Physik stochastischer Parkettierungen*, PhD thesis, Univ. Tübingen (UFO Verlag, Allensbach).

[Ric03]   Richard C. (2003). Dense Dirac combs in Euclidean space with pure point diffraction, *J. Math. Phys.* **44**, 4436–4449. arXiv:math-ph/0302049.

[RHHB98]   Richard C., Höffe M., Hermisson J. and Baake M. (1998). Random tilings — concepts and examples, *J. Phys. A: Math. Gen.* **31**, 6385–6408. arXiv:cond-mat/9712267.

[Rob96]   Robinson E.A. (1996). The dynamical properties of Penrose tilings, *Trans. Amer. Math. Soc.* **348**, 4447–4464.

[Rob99]   Robinson E.A. (1999). On the table and the chair, *Indag. Math.* **10**, 581–599.

[Rob04]   Robinson E.A. (2004). Symbolic dynamics and tilings of $\mathbb{R}^d$, *Proc. Sympos. Appl. Math.* **60**, 81–119.

[Rob71]   Robinson R.M. (1971). Undecidability and nonperiodicity for tilings of the plane, *Invent. Math.* **12**, 177–209.

[RMW87]   Rokhsar D.S., Mermin N.D. and Wright D.C. (1987). Rudimentary quasicrystallography: The icosahedral and decagonal reciprocal lattices, *Phys. Rev. B* **35**, 5487–5495.

[Rot93]   Roth J. (1993). The equivalence of two face-centered icosahedral tilings with respect to local derivability, *J. Phys. A: Math. Gen.* **26**, 1455–1461.

[Rud59]   Rudin W. (1959). Some theorems on Fourier coefficients, *Proc. Amer. Math. Soc.* **10**, 855–859.

[Rud62]   Rudin W. (1962). *Fourier Analysis on Groups* (Wiley, New York).

[Rud87]   Rudin W. (1987). *Real and Complex Analysis*, 3rd ed. (McGraw Hill, New York).

[Rud90]   Rudolph D.J. (1990). ×2 and ×3 invariant measures and entropy, *Ergod. Th. & Dynam. Syst.* **10**, 395–406.

[Sad98]   Sadun L. (1998). Some generalizations of the pinwheel tiling, *Discr. Comput. Geom.* **20**, 79–110.

[Sad08]   Sadun L. (2008). *Topology of Tiling Spaces* (AMS, Providence, RI).

[SW03]   Sadun L. and Williams R.F. (2003). Tiling spaces are Cantor set fibre bundles, *Ergod. Th. & Dynam. Syst.* **23**, 307–316.

[Sal63]   Salem R. (1963). *Algebraic Numbers and Fourier Analysis* (D. C. Heath and Company, Boston, MA).

[Sar11]   Sarnak P. (2011). Three lectures on the Möbius function randomness and dynamics (Lecture 1), unpublished notes, available at http://publications.ias.edu/sarnak/paper/512.

[Sas86]   Sasisekharan V. (1986). A new method for generation of quasi-periodic struc-
            tures with n fold axes: Application to five and seven folds, *Pramana J. Phys.*
            26, L283–L293.

[SW99]   Schaefer H.H. and Wolff M.P. (1999). *Topological Vector Spaces*, 2nd ed.
            (Springer, New York).

[Schl93a]   Schlottmann M. (1993). Periodic and quasi-periodic Laguerre tilings, *Int. J.
            Mod. Phys. B* **6–7**, 1351–1363.

[Schl93b]   Schlottmann M. (1993). *Geometrische Eigenschaften quasiperiodischer Struk-
            turen*, PhD thesis (Univ. Tübingen).

[Schl98]   Schlottmann M. (1998). Cut-and-project sets in locally compact Abelian
            groups. In [Pat98], pp. 247–264.

[Schl00]   Schlottmann M. (2000). Generalised model sets and dynamical systems.
            In [BM00b], pp. 143–159.

[Schm95]   Schmidt K. (1995). *Dynamical Systems of Algebraic Origin* (Birkhäuser,
            Basel).

[SV09]   Schmidt K. and Verbitskiy E. (2009). Abelian sandpiles and the harmonic
            model, *Commun. Math. Phys.* **292**, 721–759. arXiv:0901.3124.

[Schw98]   Schwartz L. (1998). *Théorie des Distributions*, 3rd ed. (Hermann, Paris).

[Schw80]   Schwarzenberger R.L.E. (1980). *N-Dimensional Crystallography* (Pitman,
            London).

[Sen95]   Senechal M. (1995). *Quasicrystals and Geometry* (Cambridge University
            Press, Cambridge).

[Sen04]   Senechal M. (2004). The mysterious Mr. Ammann, *Math. Intelligencer* **26**,
            10–21.

[Sen06]   Seneta E. (2006). *Non-negative Matrices and Markov Chains*, 2nd rev. ed.
            (Springer, New York).

[SAH06]   Shannon A.G., Anderson P.G. and Horadam A.F. (2006). Properties of Cor-
            donnier, Perrin and van der Laan numbers, *Int. J. Math. Educ. Sci. Techn.*
            37, 825–831.

[Sha51]   Shapiro H.S. (1951). *Extremal Problems for Polynomials and Power Series*,
            Masters thesis (MIT, Boston).

[SBGC84]   Shechtman D., Blech I., Gratias D. and Cahn J.W. (1984). Metallic phase with
            long-range orientational order and no translational symmetry, *Phys. Rev. Lett.*
            53, 1951–1953.

[Sie03]   Siegel A. (2003). Représentation des systèmes dynamiques substitutifs non
            unimodulaires, *Ergodic Th. & Dynam. Syst.* **23**, 1247–1273.

[ST09]   Siegel A. and Thuswaldner J.M. (2009). *Topological Properties of Rauzy Frac-
            tals*, Mém. Soc. Math. France **118** (Société Mathématiques de France, Paris).

[Sie44]   Siegel, C.L. (1944). Algebraic numbers whose conjugates lie in the unit circle,
            Duke Math. J. **11**, 597–602.

[Sim95]   Simon B. (1995). Operators with singular continuous spectrum. I. General
            operators, *Ann. Math.* **141**, 131–145.

[Sin02]   Sing B. (2002). *Spektrale Eigenschaften der Kolakoski-Sequenzen*, Diploma
            thesis (Univ. Tübingen).

[Sin03]   Sing B. (2003). Kolakoski-$(2m, 2n)$ are limit-periodic model sets, *J. Math. Phys.* **44**, 899–912. arXiv:math-ph/0207037.

[Sin06]   Sing B. (2006). *Pisot Substitutions and Beyond*, PhD thesis (Univ. Bielefeld).

[Sin11]   Sing B. (2011). More Kolakoski sequences, *Integers* **11B**, A14: 1–17. arXiv:1009.4061.

[SW06]   Sing B. and Welberry T.R. (2006). Deformed model sets and distorted Penrose tilings, *Z. Krist.* **221**, 621–634. mp_arc/06-199.

[Sin85]   Singh P. (1985). The so-called Fibonacci numbers in ancient and medieval India, *Hist. Math.* **12**, 229–244.

[SS02]   Sirvent V.F. and Solomyak B. (2002). Pure discrete spectrum for one-dimensional substitution systems of Pisot type, *Can. Math. Bull.* **45**, 697–710.

[SW02]   Sirvent V.F. and Wang Y. (2002). Self-affine tiling via substitution dynamical systems and Rauzy fractals, *Pacific J. Math.* **206**, 465–485.

[Sla81]   Slawny J. (1981). Ergodic properties of equilibrium states, *Commun. Math. Phys.* **80**, 477–483.

[Sloane]   Sloane N.J.A. (ed.). *The On-Line Encyclopedia of Integer Sequences*, available at http://oeis.org/.

[Soc89]   Socolar J.E.S. (1989). Simple octagonal and dodecagonal quasicrystals, *Phys. Rev. B* **39**, 10519–10551.

[Soc90]   Socolar J.E.S. (1990). Weak matching rules for quasicrystals, *Commun. Math. Phys.* **129**, 599–619.

[SS86]   Socolar J.E.S. and Steinhardt P.J. (1986). Quasicrystals. II. Unit-cell configurations. *Phys. Rev. B* **34**, 617–647.

[ST11]   Socolar J.E.S. and Taylor J.M. (2011). An aperiodic hexagonal tile, *J. Comb. Theory A* **118**, 2207–2231. arXiv:1003.4279.

[Sol11]   Solomon Y. (2011). Substitution tilings and separated nets with similarities to the integer lattice, *Israel J. Math.* **181**, 445–460. arXiv:0810.5225.

[Sol97]   Solomyak B. (1997). Dynamics of self-similar tilings, *Ergod. Th. & Dynam. Syst.* **17**, 695–738 and *Ergod. Th. & Dynam. Syst.* **19** (1999), 1685 (erratum).

[Sol98a]   Solomyak B. (1998). Nonperiodicity implies unique composition for self-similar translationally finite tilings, *Discr. Comput. Geom.* **20**, 265–279.

[Sol98b]   Solomyak B. (1998). Spectrum of dynamical systems arising from Delone sets. In [Pat98], pp. 265–275; for an erratum, see the author's homepage.

[Sol07]   Solomyak B. (2007). Pseudo-self-similar tilings in $\mathbb{R}^d$, *J. Math. Sci. (N.Y.)* **140**, 452–460. arXiv:math/0510014.

[Sos00]   Soshnikov A. (2000). Determinantal random point fields, *Russian Math. Surveys* **55**, 923–975. arXiv:math.PR/0002099.

[SO87]   Steinhardt P.J. and Ostlund S. (eds.) (1987). *The Physics of Quasicrystals* (World Scientific, Singapore).

[Ste17]   Steinhaus H. (1917). Nowa własność mnogości Cantora, *Wektor* **7**, 1–3. English translation: A new property of the Cantor set. In *Hugo Steinhaus: Selected Papers*, Hartman S. and Urbanik K. (eds.), pp. 205–207 (Polish Academic Publishers, Warsaw, 1985).

[Stc04] Steurer W. (2004). Twenty years of structure research on quasicrystals. Part I. Pentagonal, octagonal, decagonal and dodecagonal quasicrystals, *Z. Krist.* **219**, 391–446.

[SD09] Steurer W. and Deloudi S. (2009). *Crystallograhy of Quasicrystals: Concepts, Methods and Structures* (Springer, Berlin).

[SSW07] Steurer W. and Sutter-Widmer D. (2007). Photonic and phononic quasicrystals, *J. Phys. D: Appl. Phys.* **40**, R229–R247.

[Stew89] Stewart I. (1989). *Galois Theory*, 2nd ed. (Chapman and Hall, London).

[ST79] Stewart I.N. and Tall D.O. (1979). *Algebraic Number Theory* (Chapman and Hall, London).

[STJ89] Strandburg K.J., Tang L.-H. and Jarić M.V. (1989). Phason elasticity in entropic quasicrystals, *Phys. Rev. Lett.* **63**, 314–317.

[Str05a] Strungaru N. (2005). *Diffraction Symmetries for Point Sets*, PhD thesis (Univ. of Alberta, Edmonton).

[Str05b] Strungaru N. (2005). Almost periodic measures and long-range order in Meyer sets, *Discr. Comput. Geom.* **33**, 483–505.

[Str11] Strungaru N. (2011). Positive definite measures with discrete Fourier transform and pure point diffraction, *Can. Math. Bull.* **54** (2011), 544–555.

[Str13a] Strungaru N. (2013). Almost periodic measure and Bragg diffraction, *J. Phys. A: Math. Theor.* **46**, 125205: 1–11. arXiv:1209.2168.

[Str13b] Strungaru N. (2013). On the Bragg diffraction spectra of a Meyer set, *Can. J. Math.* **65**, 675–701. arXiv:1003.3019.

[SSH02] Suck J.-B., Schreiber M. and Häussler P. (eds.) (2002). *Quasicrystals — An Introduction to Structure, Physical Properties and Applications*, Springer Series in Materials Science, vol. 55 (Springer, Berlin).

[Tan07] Tan B. (2007). Mirror substitutions and palindromic sequences, *Theor. Comput. Sci.* **389**, 118–124.

[Tan90] Tang L.-H. (1990). Random-tiling quasicrystal in three dimensions, *Phys. Rev. Lett.* **64**, 2390–2393.

[Tay10] Taylor J.M. (2010). Aperiodicity of a functional monotile, available at http://www.math.uni-bielefeld.de/sfb701/preprints/view/420.

[Tem92] Tempelman A. (1992). *Ergodic Theorems for Group Actions* (Kluwer, Dordrecht).

[Ter13] Terauds V. (2013). The inverse problem of pure point diffraction — examples and open questions, *J. Stat. Phys.* **152**, 954–968. arXiv:1303.3260.

[TB13] Terauds V. and Baake M. (2013). Some comments on the inverse problem of pure point diffraction. In *Aperiodic Crystals*, Schmid S., Withers R.L. and Lifshitz R. (eds.), pp. 35–41 (Springer, Dordrecht). arXiv:1210.3460.

[Tre03] Trebin H.-R. (ed.) (2003). *Quasicrystals: Structure and Physical Properties* (Wiley-VCH, Weinheim).

[Ush99] Ushakov N.G. (1999). *Selected Topics in Characteristic Functions* (Brill Academic Publishers, Utrecht).

[Vai28] Vaidyanathaswamy R. (1928). Integer roots of the unit matrix, *J. London Math. Soc.* **3**, 121–124; On the possible periods of integer matrices, *J. London Math. Soc.* **3**, 268–272.

[vDan30]   van Dantzig D. (1930). Ueber topologisch homogene Kontinua, *Fund. Math.*
            15, 102–125.

[vEdG11]   van Enter A.C.D. and de Groote E. (2011). An ultrametric state space with
            a dense discrete overlap distribution: Paperfolding sequences, *J. Stat. Phys.*
            142, 223–228. arXiv:1010.2338.

[vEHM92]   van Enter A.C.D., Hof A. and Miękisz J. (1992). Overlap distributions for
            deterministic systems with many pure states, *J. Phys. A: Math. Gen.* **25**,
            L1133–L1137.

[vEM92]   van Enter A.C.D. and Miękisz J. (1992). How should one define a (weak)
            crystal? *J. Stat. Phys.* **66**, 1147–1153.

[vEMZ98]   van Enter A.C.D., Miękisz J. and Zahradník M. (1998). Non-periodic long-
            range order for fast decaying interactions at positive temperatures, *J. Stat.
            Phys.* **90**, 1441–1447. arXiv:cond-mat/9711206.

[vEZ97]   van Enter A.C.D. and Zegarlinski B. (1997). Non-periodic long-range order
            for one-dimensional pair interactions, *J. Phys. A: Math. Gen.* **30**, 501–505.
            mp_arc/96-509.

[vSm07]   van Smaalen, S. (2007). *Incommensurate Crystallography* (Oxford University
            Press, Oxford).

[vNe32]   von Neumann, J. (1933). Zur Operatorenmethode in der klassischen
            Mechanik, *Ann. Math. (2)* **33**, 587–642.

[Vig80]   Vignéras M.-F. (1980). *Arithmétique des Algèbres de Quaternions*, LNM 800
            (Springer, Berlin).

[Vin00]   Vince A. (2000). Digit tiling of Euclidean space. In [BM00b], pp. 329–370.

[vQu79]   von Querenburg B. (1979) *Mengentheoretische Topologie*, 2nd ed. (Springer,
            Berlin)

[Wal00]   Walters P. (2000). *An Introduction to Ergodic Theory*, reprint (Springer, New
            York).

[WCK87]   Wang M., Chen H. and Kuo K.H. (1987). Two-dimensional quasicrystal with
            eightfold rotational symmetry, *Phys. Rev. Lett.* **59**, 1010–1013.

[Wan50]   Wannier G.H. (1950). Antiferromagnetism. The triangular Ising net, *Phys.
            Rev.* **79**, 357–364 and *Phys. Rev. B* **7** (1973), 5017 (erratum).

[Was97]   Washington L.C. (1997). *Introduction to Cyclotomic Fields*, 2nd ed. (Springer,
            New York).

[Wel04]   Welberry T.R. (2004). *Diffuse X-Ray Scattering and Models of Disorder*
            (Clarendon Press, Oxford).

[Wey16]   Weyl H. (1916). Über die Gleichverteilung von Zahlen mod. Eins, *Math. An-
            nalen* **77**, 313–352.

[Wic91]   Wicks K.R. (1991). *Fractals and Hyperspaces*, LNM 1492 (Springer, Berlin).

[Wid93]   Widom M. (1993). Bethe ansatz solution of the square-triangle random tiling
            model, *Phys. Rev. Lett.* **70**, 2094–2097.

[Wie50]   Wielandt H. (1950). Unzerlegbare, nicht negative Matrizen, *Math. Z.* **52**, 642–
            648.

[Wie27]   Wiener N. (1927). The spectrum of an array and its application to the study
            of the translation properties of a simple class of arithmetical functions. Part
            I: The spectrum of an array, *J. Math. Massachusetts* **6**, 145–157.

[Wil74]   Williams R.F. (1974). Expanding attractors, *Publ. Math. IHES* **43**, 169–203.

[Wit05]   Withers R.L. (2005). Disorder, structured diffuse scattering and the transmission electron microscope, *Z. Krist.* **220**, 1027–1034.

[Wu09]   Wu F.Y. (2009). *Exactly Solved Models: A Journey in Statistical Mechanics* (World Scientific, Singapore).

[WMTB76]   Wu T.T., McCoy B.M., Tracy C.A. and Barouch E. (1976). Spin-spin correlation functions for the two-dimensional Ising model: Exact theory in the scaling region, *Phys. Rev. B* **13**, 316–374.

[Zag84]   Zagier D.B. (1984). *Zetafunktionen und quadratische Körper* (Springer, Berlin).

[Zak02]   Zaks M.A. (2002). On the dimensions of the spectral measure of symmetric binary substitutions, *J. Phys. A: Math. Gen.* **35**, 5833–5841.

[Zin09]   Zint N. (2009). *Influence of Randomness in mathematical models for diffraction and T-cell recognition*, PhD thesis (Univ. Bielefeld).

[Zob93a]   Zobetz E. (1993). Homometric cubic point configurations, *Z. Krist.* **205**, 177–199.

[Zob93b]   Zobetz E. (1993). One-dimensional quasilattices: Fractally shaped atomic surfaces and homometry, *Acta Cryst. A* **49**, 667–676.

List of Definitions

List of Examples

List of Remarks

Index

Printed in the United States
By Bookmasters